RETURNED
29. APR 94
18 MAY 1999
7/9/07
22 APR 1994
RETURNED
RETURNED
RETURNED
27 OCT 2003
11 NOV 1997
RETURNED
15 APR 2005
13 FEB 1999
RETURNED

The Chemistry of β-Lactams

The Chemistry of β-Lactams

Edited by

MICHAEL I. PAGE
Head of Department of Chemical Sciences
The University of Huddersfield

BLACKIE ACADEMIC & PROFESSIONAL
An Imprint of Chapman & Hall
London · Glasgow · New York · Tokyo · Melbourne · Madras

Published by
Blackie Academic & Professional
An Imprint of Chapman & Hall
Wester Cleddens Road, Bishopbriggs, Glasgow G64 2NZ, UK

Chapman & Hall, 2–6 Boundary Row, London SE1 8HN, UK

Blackie Academic & Professional, Wester Cleddens Road, Bishopbriggs, Glasgow G64 2NZ, UK

Chapman & Hall, 29 West 35th Street, New York NY 10001, USA

Chapman & Hall Japan, Thomson Publishing Japan, Hirakawacho Nemoto Building, 6F, 1-7-11 Hirakawa-cho, Chiyoda-ku, Tokyo 102, Japan

DA Book (Aust.) Pty Ltd, 648 Whitehorse Road, Mitcham 3132, Victoria, Australia

Chapman & Hall India, R. Seshadri, 32 Second Main Road, CIT East, Madras 600 035, India

---

First edition 1992

Typeset in 10/12pt Times by EJS Chemical Composition, Bath

Printed in Great Britain by the University Press, Cambridge

ISBN 0 7514 0061 0
0 412 03331 3 (USA)

A catalogue record for this book is available from the British Library.

Library of Congress Cataloging-in-Publication Data available.

Contributors

Professor J.E. Baldwin	The Dyson Perrins Laboratory and the Oxford Centre for Molecular Sciences, South Parks Road, Oxford OX1 3QY, UK
Dr E.W. Colvin	Chemistry Department, University of Glasgow, Glasgow G12 8QQ, UK
Dr R.D.G. Cooper	Lilly Research Laboratories, Eli Lilly and Company, Indianapolis, IN 46285, USA
Dr J. Coyette	Centre d'Ingénierie des Protéines et Laboratoire d'Enzymologie, Institut de Chimie B6, Université de Liège au Sart Tilman, B-4000 Liège 1, Belgium
Professor J.-M. Frère	Centre d'Ingénierie des Protéines et Laboratoire d'Enzymologie, Institut de Chimie B6, Université de Liège au Sart Tilman, B-4000 Liège 1, Belgium
Dr B. Joris	Centre d'Ingénierie des Protéines et Laboratoire d'Enzymologie, Institut de Chimie B6, Université de Liège au Sart Tilman, B-4000 Liège 1, Belgium
Dr L.N. Jungheim	Lilly Research Laboratories, Eli Lilly and Company, Indianapolis, IN 46285, USA
Dr M. Nguyen-Distèche	Centre d'Ingénierie des Protéines et Laboratoire d'Enzymologie, Institut de Chimie B6, Université de Liège au Sart Tilman, B-4000 Liège 1, Belgium
Professor H.C. Neu	College of Physicians and Surgeons, Columbia University, 630 West 168th Street, New York, New York 10032, USA
Professor M.I. Page	Department of Chemical Sciences, The University of Huddersfield, Huddersfield HD1 3DH, UK

Professor R.F. Pratt Department of Chemistry, Wesleyan University, Middletown, CT 06459, USA

Dr C. Schofield The Dyson Perrins Laboratory and the Oxford Centre for Molecular Sciences, South Parks Road, Oxford OX1 3QY, UK

Dr R.J. Ternansky Lilly Research Laboratories, Eli Lilly and Company, Indianapolis, IN 46285, USA

Dr S.G. Waley University of Oxford, Laboratory of Molecular Biophysics, Rex Richards Building, South Parks Road, Oxford OX1 3QU and Oxford Centre for Molecular Sciences, Oxford, UK

Contents

Editorial Introduction

It is over sixty years since Alexander Fleming observed antibiosis between a *Penicillium* mould and bacterial cultures and gave the name penicillin to the active principle. Although it was proposed in 1943 that penicillin (**1**) contained a β-lactam ring, this was not generally accepted until an X-ray crystallographic determination of the structure had been completed.

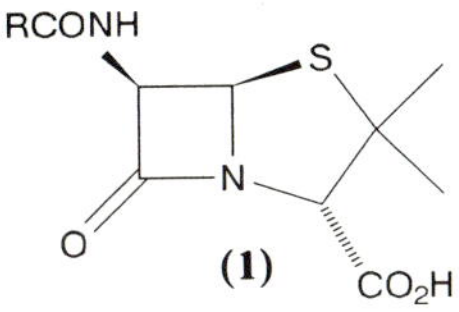

RCONH S N O CO_2H

(**1**)

Penicillin was the first naturally occurring antibiotic to be characterised and used in clinical medicine. It is now seen as the progenitor of the β-lactam family of antibiotics, which are characterised by the possession of the four-membered β-lactam ring. Chapters in this book will describe how current research has demonstrated that other lactam structures may also show antibacterial activity. Penicillins and cephalosporins (**2**), the second member of the β-lactam antibiotic family, were both originally discovered in fungi but later detected in streptomycetes.

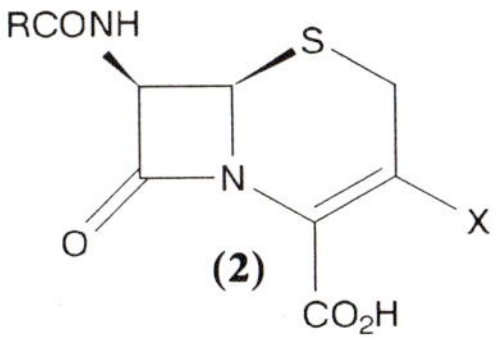

RCONH S N O X CO_2H

(**2**)

Until 1970 penicillins and cephalosporins were the only examples of naturally occurring β-lactam antibiotics. The discovery of 7-α-methoxy-cephalosporins (**3**) from *Streptomyces* in 1971 stimulated the search for novel β-lactam antibiotics from microbes, both by using sensitive new screening procedures, and by laboratory synthesis. At present, β-lactam antibiotics can be classified into several groups according to their structure:

- Penicillins (penams) (**1**)
- Cephalosporins (cephems) (**2**)
- Cephamycins (**3**)
- Oxacephems (**4**)

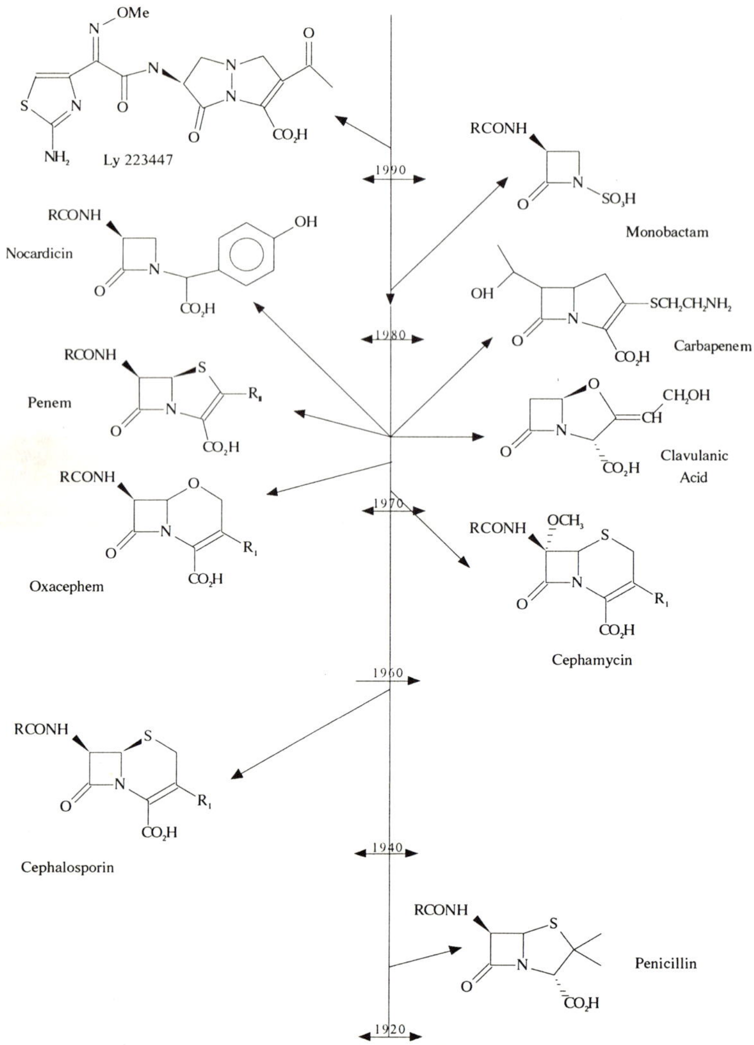

Figure 1

- Penems (**5**)
- Oxapenams such as clavulanic acid (**6**)
- Carbapenems such as thienamycin (**7**)
- Nocardicins (**8**)
- Monobactams (**9**)

The chronology of these antibiotics is illustrated in Figure 1. Some trivial names used to describe the ring structures are shown in Tables 1 and 2. β-Lactams have now been found in eukaryotic fungi, actinomycetes and

Table 1 Names and structures of common penicillins.

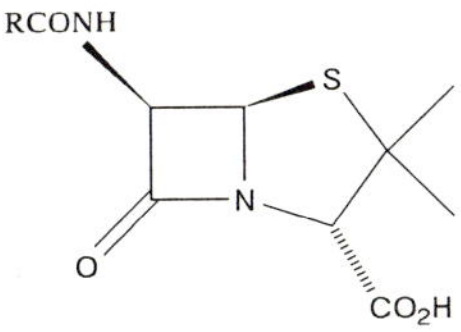

RCONH, S, N, O, CO_2H

Penicillin	R	Penicillin	R
Benzyl penicillin (Pen G)	$PhCH_2$–	Carbenicillin	$PhCH_2$– CO_2H
Pen F	$C_2H_5CH{=}CHCH_2$–	Oxacillin	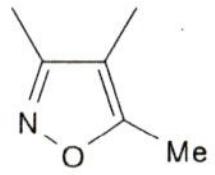 N O Me
Pen X	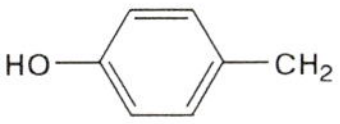 HO CH_2	Pen V	$PhOCH_2$–
Pen K	$CH_3(CH_2)_6$–	Cloxacillin	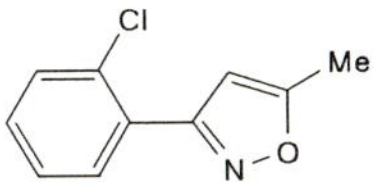 Cl Me N O
Pen N	H_3N CH–$(CH_2)_3$– CO_2	Dicloxacillin	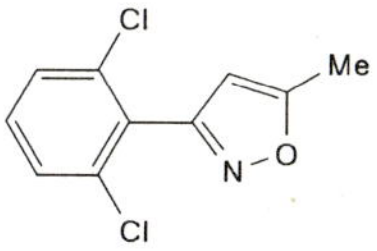 Cl Me N O Cl
Propicillin	PhOCH(Et)–	Phenbenicillin	PhOCH(Ph)–
Diphenicillin	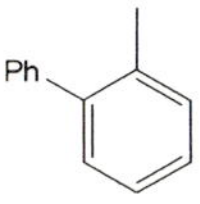 Ph	Nufcillin	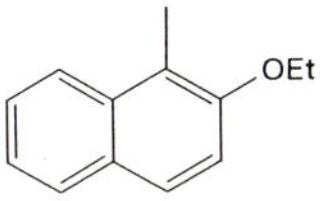 OEt
Methicillin	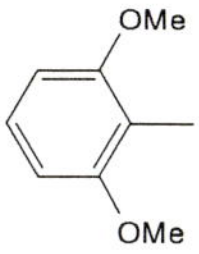 OMe OMe	Quinacillin	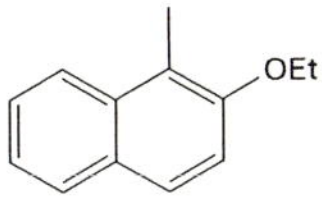 OEt
Ampicillin	PHCH– NH_2	Ticarcillin	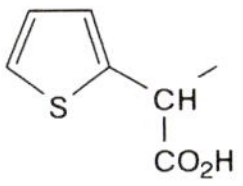 S CH CO_2H

even in bacteria. γ-Lactams such as lactivicin (**10**), and β-lactones, which show antibacterial activity, have also been isolated from various micro-organisms.

A large number of nuclear analogues of the β-lactam antibiotics have been

Table 2 Names and structures of common cephalosporins.

RCONH S N O CH_2L CO_2H

Cephalosporin	R	L
Oral cephalosporins		
cephaloglycin	$PhCH(NH_2)$–	–OAC
cephalexin	$PhCH(NH_2)$–	–H
cefaclor	$PhCH(NH_2)$–	–Cl
cefatrizine	$HOC_6H_4CH(NH_2)$–	–S– N N N H
First-generation cephalosporins		
cephalothin	S CH_2	–OAc
cephaloridine	S CH_2	+ N
cefazolin	N=N N N CH_2	N–N S S CH_3
Second-generation cephalosporins		
cefamandole	PhCH(OH)–	N–N S N CH_3
cefuroxime	O C N OMe	–O–$CONH_2$
cephamycin	HO_2C $CH(CH_2)_3$– NH_2	–O–$CONH_2$
Third-generation cephalosporins		
cefotaxime	OMe N N H_2N S	–OAc
cefoperazone	N–N N S N CH_3	CH NHCO HO O N O N C_2H_5

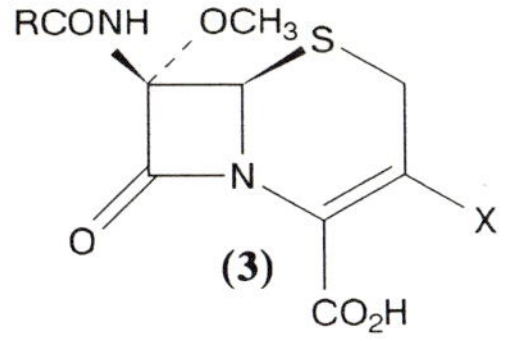

prepared by complete chemical synthesis, or by partial synthesis starting from a naturally occurring β-lactam. Most semisynthetic penicillins are made by the acylation of 6-β-aminopenicillanic acid (**11**), whilst cephalosporins can be prepared to give a variety of side chain substituents at C-7 and C-3 (**12**). Classically, β-lactam antibiotics have been named using the ring sulphur as position-1. Hence, for penicillins and cephalosporins the numbering is as illustrated in (**13**) and (**12**), respectively. Similarly, substituents in these bicyclic systems are usually described by the prefix α- or β- rather than the equivalent *exo* and *endo* description. The normal configuration at the three asymmetric centres in penicillin is therefore 3*S*, 5*R* and 6*R*.

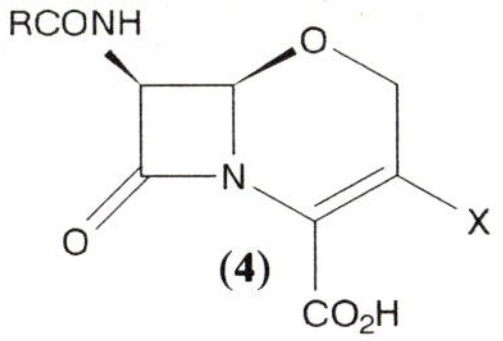

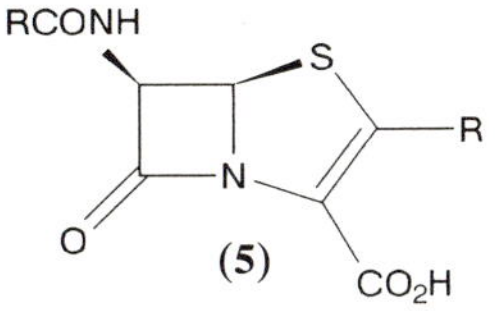

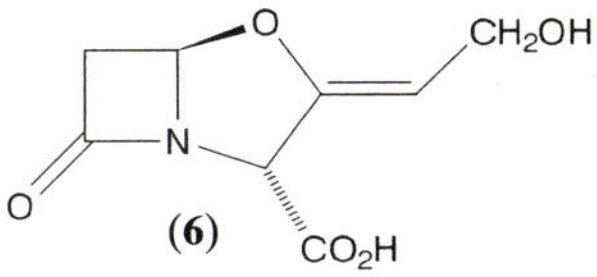

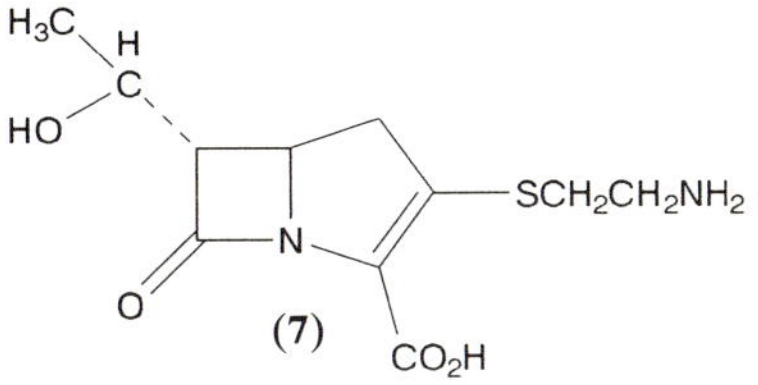

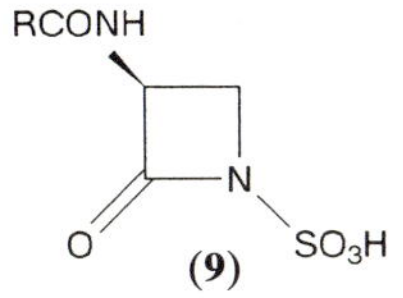

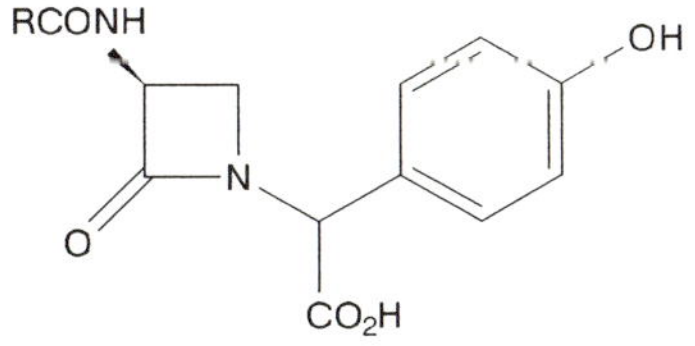

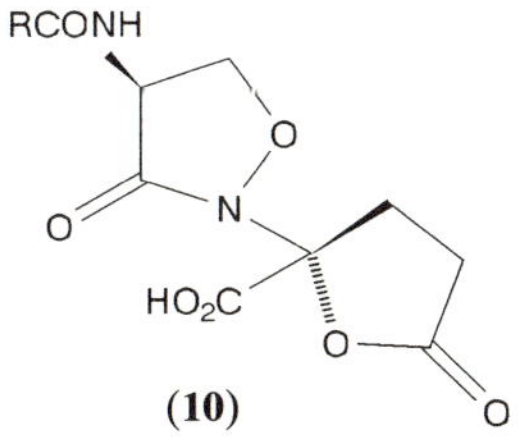

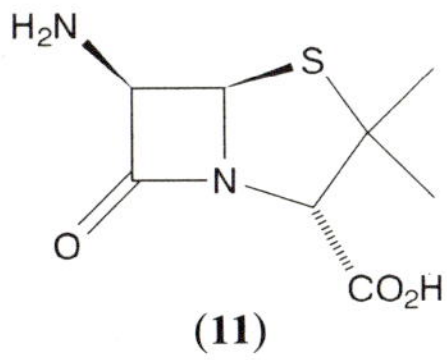

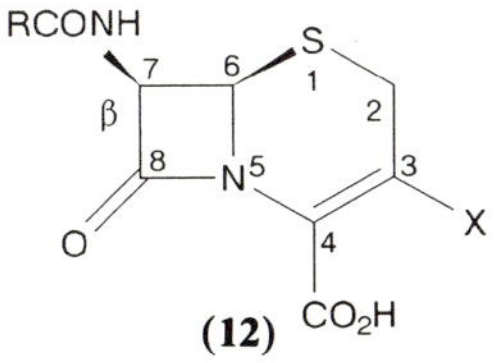

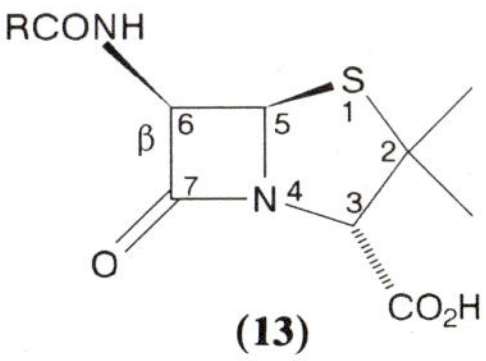

Some examples of clinically used penicillins and cephalosporins are given in Tables 1 and 2, respectively. The so-called first- and second-generation injectable cephalosporins are 3,7-modified derivatives (Table 2). Although 7-β-side chains may improve potency and breadth of antibacterial activity, their potential is usually only expressed when the molecule also contains an appropriate 3-side chain. Third-generation cephalosporins differ in their increased spectrum of activity, potency and high cost.

It is generally accepted that the biochemical targets for the β-lactams are some of the enzymes concerned with cell-wall synthesis. However, the complexity of drug action increases as one moves from target enzyme to target cell, and from target cell in a culture tube to target pathogen in the complex environment of an infected host. An understanding of the mode of action of any drug required a knowledge of:

(i) the pathway by which the drug reaches its biochemical target;
(ii) the identity and structure of the biochemical target, and the chemistry of how the target is inhibited;
(iii) the cell's physiological response to the inhibition of the drug-sensitive reaction; and
(iv) how and why inhibition of this reaction leads to interference of the life cycle of the cell.

β-Lactams exert their lethal action only on growing bacterial cells. Within a short time of adding penicillin to a bacterial culture, small bulges form in the bacteria — often near mid-cell where cell division is expected to occur. With time, the bulge enlarges and ultimately the cell membrane ruptures, resulting in death of the cell. These morphological and lytic effects of penicillin are accomplished by the covalent binding of the antibiotic to proteins, called penicillin binding proteins (PBPs). The PBPs are located on the inner membrane and, to reach them, the antibiotic has to traverse a variety of chemical and physical barriers. In gram-negative bacteria the diffusion of the β-lactam through the outer (plasma) membrane and across the periplasm may be a rate-limiting factor in determining antibacterial effectiveness.

There are at least three families of enzymes that specifically recognise the β-lactam antibiotics such as the penicillins (**1**) and cephalosporins (**2**). One class is the *transpeptidase enzymes*, which are the inhibitory targets for antibiotics when they kill bacteria. The intermediates formed between transpeptidases and the β-lactam antibiotics are relatively stable, and are responsible for inhibiting the enzyme and preventing cell-wall synthesis. The second group are *β-lactamases*, which catalyse the hydrolysis of the β-lactam — yielding biologically inactive products. Plasmid-mediated formation of β-lactamases by bacteria is largely responsible for the resistance of many bacteria to the normally lethal action of β-lactam antibiotics. Gram-positive bacteria excrete a β-lactamase in the culture medium during growth but they also have a distinct enzyme bound to the membrane. In gram-negative bacteria, β-lactamases are often located in the periplasmic space and appear indistinguishable from the membrane-bound enzyme. There has been much discussion on the evolutionary relationship between β-lactamases and the transpeptidases. The biosynthesis of penicillin involves the intermediate formation of a linear tripeptide, which then undergoes an enzyme catalysed desaturative bicyclisation, catalysed by a penicillin *synthase*. Although the β-lactam is the product of this reaction, there must again be some form of recognition between the protein and the antibiotic.

All three classes of enzymes that interact with the β-lactam antibiotics are described in this book. In addition to molecular recognition of the β-lactam, these enzymes are also involved in bond-making and -breaking processes. The mechanisms by which reactions of β-lactams occur are thus of obvious interest. However, understanding the reactivity of β-lactams is also of importance in the design of new antibiotics. This understanding has been seminal in the development of non-β-lactam structures, which may also be effective as killing agents of bacteria. These aspects are discussed, together with the synthetic methodology used to make this exciting class of compounds.

1 The biosynthesis of β-lactams

J.E. BALDWIN and C. SCHOFIELD

1.1 Introduction

Until relatively recently, progress in our understanding of the machinery responsible for the biosynthesis of β-lactams has been slow. However, in the last decade or so, many interesting new developments have taken place within the field of β-lactam biosynthesis. In this chapter, it is hoped to convey some of the excitement and enjoyment that the authors have felt during the course of their studies.

1.2 Penicillin and cephalosporin biosynthesis

Our current understanding of the biosynthetic pathways leading to the penicillins, cephalosporins, and cephamycins is summarised in Scheme 1.1. The names of the various genes encoding the biosynthetic enzymes that have been identified are indicated in parentheses, and follow the nomenclature suggested by Ingolia and Queener.[1,2]

1.2.1 Early studies

Almost 40 years ago it was proposed that the nucleus of the penicillins was derived from a peptide containing L-cysteine and valine.[3,4] The discovery of isopenicillin N (**1**) in *Penicillium chrysogenum*,[5,6] led to the hypothesis that δ-(α-aminoadipoyl)-cysteinyl-valine, which had also been isolated in minute amounts,[7] was the precursor of the penicillins. This tripeptide was also isolated from *Cephalosporium acremonium* and the stereochemistry was determined to be: δ-(L-α-aminoadipoyl)–L-cysteinyl-D-valine [ACV (**2**)].[8] In addition to ACV (**2**), two other tetrapeptides were also isolated, both containing an N-terminal glycine residue, and one of which contained a 3-hydroxyvaline residue in place of valine.[8] The tripeptide from *P. chrysogenum* was subsequently rigorously shown to be identical to ACV (**2**).[9,10] ACV (**2**) was also detected together with δ-(L-α-aminoadipoyl)–L-cysteine

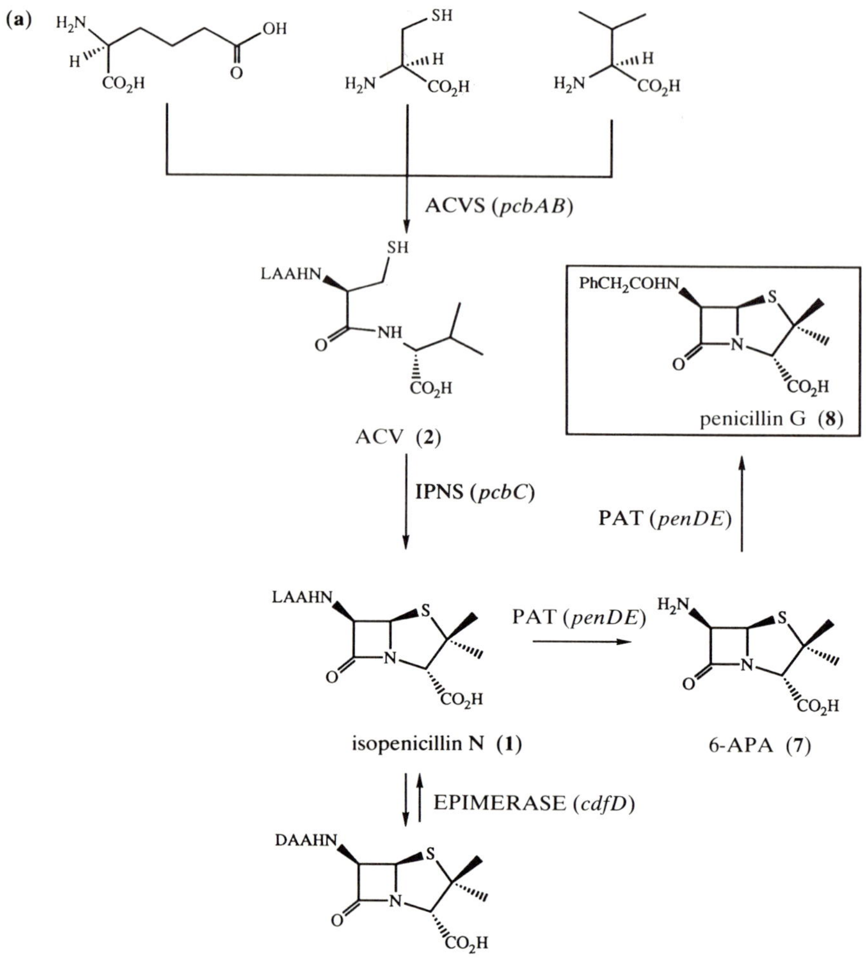

(a)

ACVS (*pcbAB*)

ACV (**2**)

IPNS (*pcbC*)

isopenicillin N (**1**)

PAT (*penDE*)

6-APA (**7**)

PAT (*penDE*)

penicillin G (**8**)

EPIMERASE (*cdfD*)

penicillin N (**3**)

Scheme 1.1 Biosynthetic pathways leading to (a) the penicillins; and (b) the cephalosporins and cephamycins. ACV = δ-(L-α-aminoadipoyl)–L-cysteinyl–D-valine (**2**); ACVS = ACV synthetase; LAAHN = δ-(L-α-aminoadipoyl); DAAHN = δ-(D-α-aminoadipoyl); IPNS = isopenicillin N synthase; DAOC = deacetoxycephalosporin C (**4**); DAC = deacetylcephalosporin C (**5**); DAOCS = DAOC synthase; DACS = DAC synthase; 6-APA = 6-aminopenicillanic acid (**7**); PAT = isopenicillin N aminohydrolase/phenylacetyl-coenzyme A:6-APA acyltransferase.

(**b**)

DAAHN

penicillin N (**3**)

C. acremonium: DAOCS/DACS (*cefEF*)
Streptomyces spp. DAOCS (*cefE*)

DAAHN

DAOC (**4**)

C. acremonium: DAOCS/DACS (*cefEF*)
Streptomyces spp. DACS (*cefF*)

DAAHN

DAC (**5**)

DAAHN

o-Carbamoyl DAC (**9**)

DAAHN

Cephalosporin C (**6**)

DAAHN OMe

Cephamycin C (**10**)

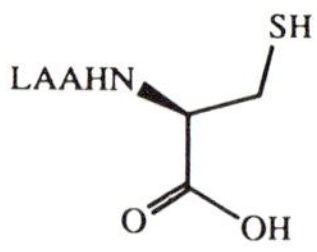

SH
LAAHN
O OH

AC (**11**)

[AC (**11**)] in a species of *Streptomyces* that also produced both penicillins and cephamycins.[11] That the α-aminoadipoyl side chain was crucial for the biosynthesis of the penicillins was further substantiated by the observation that a mutant of *P. chrysogenum*, blocked in the biosynthesis of α-aminoadipic acid, only produced penicillin G (**8**) when α-aminoadipic acid was added to the fermentation medium.[12] Since ACV (**2**) itself was found not to be transported into the mycelial cells, in order to unequivocally establish whether it was the precursor of the penicillins it was necessary to develop a cell-free system capable of converting ACV (**2**) into a penicillin. In the key experiment, Abraham and coworkers demonstrated that a crude cell lysate was able to transform labelled ACV (**2**) into either penicillin N (**3**) or isopenicillin N (**1**).[13] Subsequent studies established that the latter was the major product and that the carbon skeleton of ACV (**2**) remained intact during the conversion to isopenicillin N (**1**).[14,15] Thus ACV (**2**) was established as the first key intermediate in the biosynthesis of the penicillins.

1.2.2 ACV biosynthesis

Some of the factors affecting the biosynthetic production of ACV (**2**) *in vivo* have been examined.[16–21] It has been proposed that ACV (**2**) biosynthesis is the rate-limiting step in the biosynthesis of penicillins, and studies on the effect of ammonium ions,[17,18] phosphate ions,[19] and glucose[16,21] on ACV (**2**) and antibiotic production have been made. In addition, the biosynthesis of ACV (**2**) was found to be stimulated by the addition of cycloheximide or anisomicin, consistent with the suggestion that it was biosynthesised by a non-ribosomal process.[20]

In an early study Bauer[22] reported the biosynthesis of a tripeptide using a cell-free preparation made from *P. chrysogenum*; however, the stereochemistry of the product was not determined. Definitive descriptions of cell-free extracts capable of catalysing the biosynthesis of ACV (**2**) were made by Abraham and Loder.[23,24] They reported that, in the presence of ATP and Mg^{2+}, cell-free extracts from *C. acremonium* were capable of biosynthesising ACV (**2**) from AC (**11**) and L-valine. No synthesis of the tripeptide (**2**) from L-α-aminoadipic acid plus L-cysteinyl–L-valine, D-α-aminoadipic acid plus L-cysteinyl-L-valine, or AC (**11**) plus D-valine was found to occur. Subsequently, using a β-lactam defective mutant[25] of *C. acremonium*, a soluble activity that biosynthesised ACV (**2**) from the L-enantiomers of the three constituent amino acids was prepared.[26] In these experiments, small amounts of AC (**11**) were also detected; in contrast there was no evidence for the formation of cysteinyl–valine. Partial purifications of putative AC (**11**) synthetases from *P. chrysogenum*[27] and *C. acremonium* C-10[28] were also reported and, in the latter case, Banko *et al.* made the observation that the putative AC (**11**) synthetase activity was stabilised by the addition of glycerol. The same authors subsequently found that the conversion of the individual amino acids, L-α-aminoadipic acid, L-cysteine and L-valine, to ACV (**2**) was much more rapid than the conversion of AC (**11**) plus L-valine to ACV (**2**).[29] The identity of the product tripeptide as ACV (**2**) was established by HPLC comparison with authentic standards, and by isolation of the peptide and conversion of it to isopenicillin N (**1**) using isopenicillin N synthase (IPNS). Incubation of L,L,L-ACV (**12**) with the purified extracts, followed by incubation with IPNS did not result in the production of isopenicillin N (**1**). Similar results were obtained with the ACV synthetase isolated from *S. clavuligerus*.[30] The bacterial enzyme was also immobilised on to an anion exchange resin and shown to catalyse the production of ACV (**2**).[31]

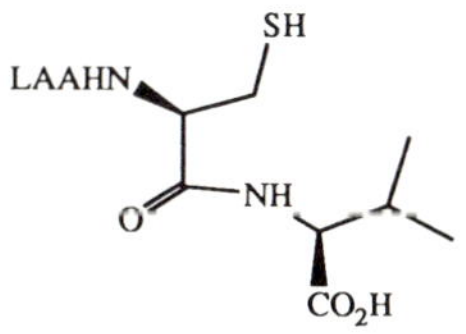

SH
LAAHN
O
NH
CO_2H

(**12**)

These studies led Banko *et al.*[29] to propose that ACV (**2**) biosynthesis in *C. acremonium* is mediated by a single, multifunctional enzyme (multienzyme), similar to those responsible for the biosyntheses of other peptide antibiotics.[32,33] Firm evidence for such a proposal came with the purification of ACV synthetase (ACVS) from *Aspergillus nidulans* by von Dohren and coworkers.[34,35] By using buffer containing a high glycerol content, together with a grinding method for the lysis of the mycelium, they were able to extract and purify an active form of ACVS. The highly purified protein was found to catalyse the formation of ACV (**2**) from the L-enantiomers of the constituent amino acids in the presence of ATP, Mg^{2+} and dithioerythritol. Furthermore, it catalysed ATP-pyrophosphate exchange in the presence of each of the three amino acids independently, consistent with a mechanism involving formation of the amino acid acyl adenylates, and with a single multienzyme being responsible for the formation of the tripeptide. In the presence of [^{14}C]-valine, without cysteine and α-aminoadipic acid, and the appropriate cofactors, protein-bound radioactivity was detected; if ATP was omitted from the reaction mixture no radioactivity was bound to the enzyme. The enzyme-bound valine could be released from the enzyme by the addition of performic acid, but the radioactivity remained bound if formic acid was substituted for performic acid. These data were interpreted as being in support of a mechanism by which the valine is bound to a sulphydryl group of the enzyme in a thioester linkage,[35] in similar manner to that proposed for other multienzyme complexes.[33]

Subsequent to the purification by the Berlin group, protocols for the effective purification of ACVS from *C. acremonium*[36,37] and *S. clavuligerus*[36,38,39] have been reported. It is of interest to compare the sizes for the ACVSs based on electrophoretic mobilities. The Berlin group estimated that the size of the synthetase from *A. nidulans* was 220 kDa, both from gel filtration and polyacrylamide gel electrophoresis. The enzymes from *C. acremonium* and *S. clavuligerus* were estimated to be significantly larger (283–360 kDa) under denaturing conditions, whilst analysis under non-denaturing indicated that the ACVSs, from *C. acremonium*[36] and *S. clavuligerus*[36,39] are probably dimers in the native state. Other multienzyme complexes catalysing peptide biosyntheses have also been shown to have very large subunits[32,33] and it is possible that the reported discrepancies in size of the ACVSs are due either to methods of analysis, or to partial processing of a single multienzyme (*vide infra*). Intriguingly, in addition to a large protein (reported M_r under denaturing conditions = 283 kDa) Jensen *et al.*[39] also reported the co-purification of a much smaller protein (M_r = 32 000), and concluded that the ACVS from *S. clavuligerus* is probably a multimer composed of non-identical subunits. During the purification of the enzyme from *C. acremonium*, one of the present authors has also noted that a relatively small protein co-purifies with ACVS.[40] Jensen has suggested that the high molecular weight component may represent the peptide-forming component, while the smaller protein is a thioesterase required to release the completed tripeptide from the complex. In this regard it is of interest that the operon which encodes the biosynthetic genes for gramicidin S synthetase also contains a gene encoding a 29 kDa protein that shows homology to fatty acid thioesterases.[41]

The apparent similarity between ACVS and related multienzymes involved in peptide biosynthesis prompted investigations to determine if phosphopantothenic acid was present in the purified multienzyme. Two different methods were utilised. In the first, the protein was treated with alkali and subsequently digested with alkaline phosphatase to release pantothenic acid. The amount of pantothenate released was then assayed using the pantothenate-requiring organism, *Lactobacillus plantarum*. The results indicated that approximately one mole of pantothenate was associated with ACVS from both *C. acremonium* and *S. clavuligerus*. In the second method, *C. acremonium* was grown in the presence of radioactive [^{14}C]-pantothenate, and radioactivity was found to be associated with the electrophoresis band corresponding to ACVS.[36]

A mechanistic scheme for the biosynthesis of ACV (**2**) has been proposed, in which each of the amino acids is activated as the corresponding acyl

E'SH + ATPMg^{2+} + LAAOH ⇌ E'SH(LAA–AMP) + FPi + Mg^{2+} → E'S–AAL + AMP
E^{2}SH + ATPMg^{2+} + CYS ⇌ E^{2}SH(CYS–AMP) + PPi + Mg^{2+} → E^{2}S–CYS + AMP
E^{3}SH + ATPMg^{2+} + L-VAL → E^{3}SH(L-VAL–AMP) + PPi + Mg^{2+} → E^{3}S–L-VAL + AMP
or
E^{3}S–D-VAL + AMP
E^{3}S–D-VAL–CYS–AAL
Thioesterase
ACV (2)

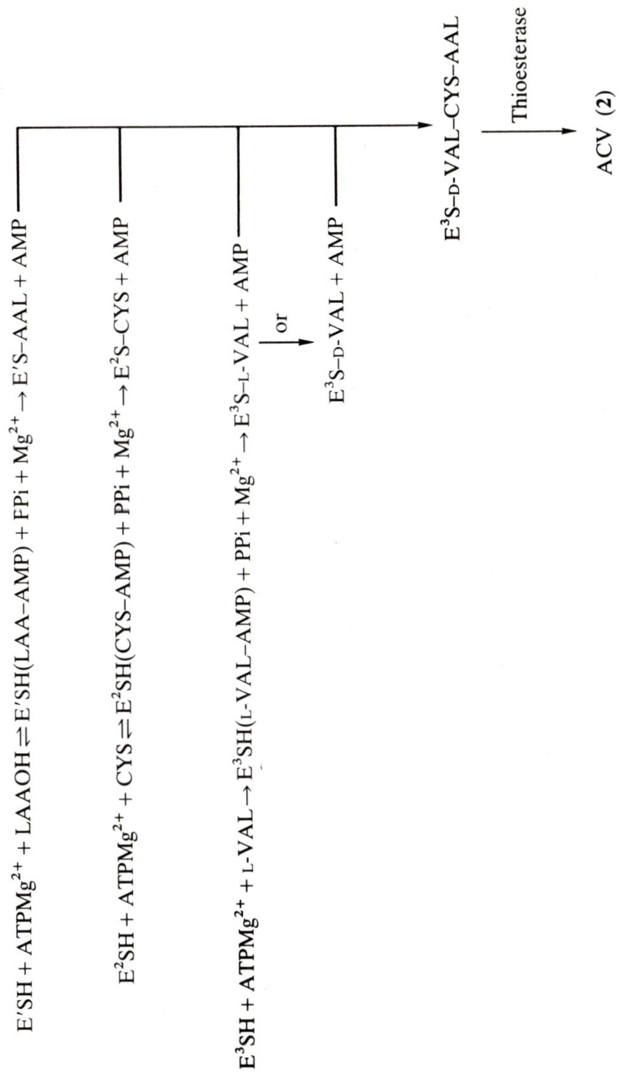

Scheme 1.2

adenylate before binding to the multienzyme by a thioester linkage.[33] It is envisaged that the valinyl residue would be epimerised at the thioester stage (either before or after formation of the tripeptide). Intermediate peptidyl thioesters are transported to the next acyl residue by the 4′-phosphopantothenate arm, where the peptide bond is formed by a transpeptidation reaction. Finally, the completed tripeptide is cleaved from the enzyme by a thioesterase (Scheme 1.2).

Exogenous feeding experiments, using a mutant of *C. acremonium*,[16] of L-[$^{18}O_2$]-valine demonstrated that ACV (**2**) formation occurs with an intracellular exchange of one and both valine oxygens, consistent with a thioester intermediate.[42,43] It remains to be seen whether exchange of only one of the oxygens of the carboxyl group of valine will be observed using ACVS *in vitro*. Experiments to clarify this point are in progress.[40]

Some preliminary substrate analogue studies have been carried out on ACVS. Thus, tripeptide analogues were formed when L-carboxymethylcysteine[29] was added in place of L-α-aminoadipic acid, or when L-*allo*-isoleucine or L-α-aminobutyrate were added in place of valine, utilising enzymes from both *C. acremonium*[29,36] and *S. clavuligerus*.[30,36] The observation that L-α-aminobutyrate is a substrate for ACVS is of interest, since the β-methyl penam (**13**) and an associated cepham (**14**) have been isolated from *Streptomyces* ACC 13285.[44] Presumably they are derived from the reaction of δ-(L-α-aminoadipoyl)–L-cysteinyl–D-aminobutyrate (**15**) with IPNS (*vide infra*), followed by epimerisation of the side chain. The isolation of δ-(L-α-aminoadipoyl)–L-serinyl–D-valine (**16**) and (α-aminoadipoyl)-serinyl-isodehyrovaline (**17**) from *P. chrysogenum*[45] may initially suggest that the ACVS will accept serine in place of cysteine; however, it would also seem possible that these peptides are by-products resulting from the action of the oxygenases subsequently operative in the pathway.

Recently there have been several reports describing the identification of the ACVS gene from both fungal and bacterial sources. Smith *et al.*[46] reported that a cosmid clone containing the penicillin biosynthetic gene cluster from *P. chrysogenum* was used to transform *Neurospora crassa* and *Aspergillus niger*, species of filamentous fungi which do not produce β-lactams. The transformed hosts were shown to produce penicillin V (**18**) and also to contain ACVS, IPNS and 6-aminopenicillanic acid acyltransferase activities. Subsequently, the nucleotide sequence of the *P. chrysogenum* ACVS gene (*pcbAB*) was determined and found to correspond to an open reading frame of 11 238 base pairs (bp) encoding a protein of 3746 amino acids with a predicted molecular weight of 421 073 Da.[47] Diez *et al.*[48] also reported the cloning of the ACVS gene from *P. chrysogenum* by a strategy that involved complementation of DNA fragments into mutants blocked in the ACV biosynthesis step. They also sequenced the region containing the ACVS gene and reported that it included an open reading frame of 11 376 bp, corresponding to a protein of 425 971 Da. A protein of

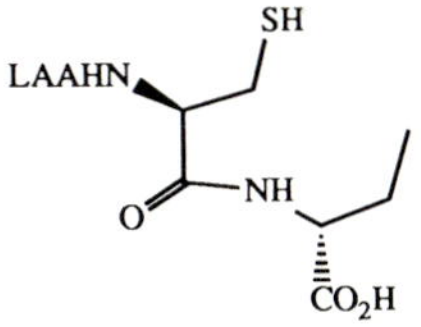

(15) 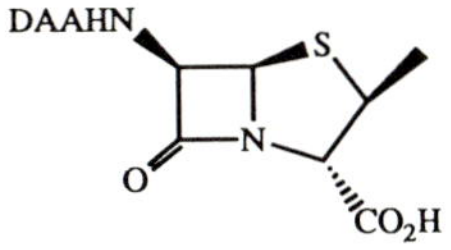 (13) 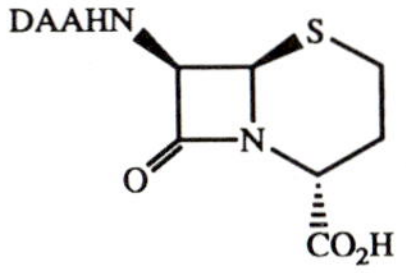 (14)

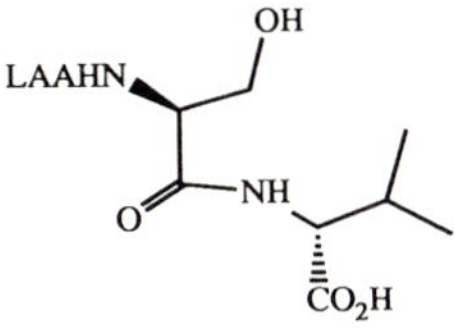

(16) 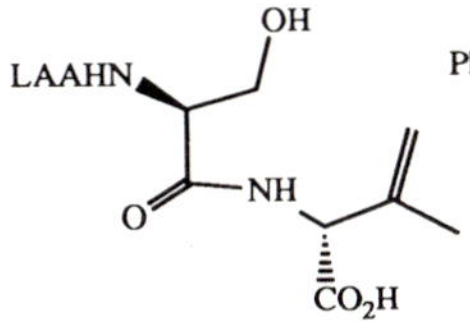 (17) 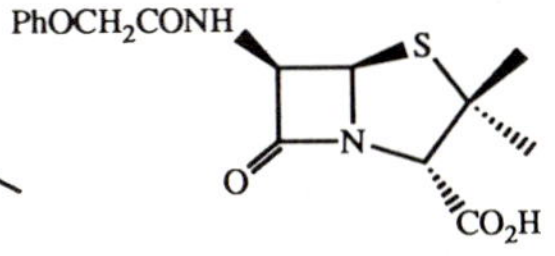 Penicillin V (18)

about 250 kDa that was absent in the untransformed mutants was observed in transformants believed to contain the ACVS gene. It was proposed that the discrepancy between the observed molecular weight and that calculated from the gene sequence may be due to processing of the intact enzyme into two subunits.[48] Similarly, the report by van Liempt *et al.* of a molecular weight of 220 kDa for the ACVS from *A. nidulans*[35] may be explained by cleavage of an intact multienzyme complex. Both Smith *et al.*[47] and Diez *et al.*[48] identified three domains within the protein that have potentially significant homology with each other (*ca.* 40%) and also with other peptide synthetases (gramicidin S synthetase 1 and tyrocidine synthetase 1) from *Bacillus brevis*. It was speculated that each of the three domains may bind and activate one of the constituent amino acids,[47] and the expression of the individual domains should address this question. It is of interest that, as yet, no one has identified a putative thioesterase region within the ACVS requence, and, given Jensen's observation of the purification of a small protein with IPNS,[39] it is possible that this is encoded for separately.

Based on the knowledge that, whereas the IPNS genes from *S. clavuligerus* and *P. chrysogenum* hybridise only poorly to each other, and that the IPNS gene from *Flavobacterium* hybridises to IPNS from both species, Smith *et al.*[49] proposed that the ACVS gene in *Flavobacterium* might be identified by cross-hybridisation experiments against *P. chrysogenum* DNA. The location of the ACVS gene in *Flavobacterium* was determined, allowing the isolation of the corresponding genes from *S. clavuligerus, P. chrysogenum*, and *A. nidulans*. In each case the ACVS gene was found to be closely linked to the IPNS gene. Using oligonucleotide probes based on amino acid sequence data derived from purified ACVS, the identification of the ACVS gene within the penicillin biosynthesis gene cluster of *A. nidulans* has also been reported by MacCabe *et al.*[50,51] The ACVS gene has also been identified in *C. acremonium.*[52] In the latter case the genomic region immediately upstream of the IPNS gene (*pcbC*) was targeted for gene disruption, and transformants that lacked the ability to produce β-lactams were obtained. It was subsequently shown that these transformants lacked ACVS activity. The location of the ACVS gene was confirmed by comparison of amino acid sequences obtained from purified ACVS from *C. acremonium,*[36] with that predicted from the DNA sequence. Like the *P. chrysogenum* enzyme, the *C. acremonium* ACVS also showed significant homology to other peptide synthetase enzymes from *B. brevis*.

Previous mechanistic investigations into the peptide synthetases and the related machinery of fatty acid biosynthesis have been hampered by the size and lability of the multienzyme complexes. It has been pointed out that ACVS is probably the most simple peptide synthetase yet discovered,[35] and thus is possibly the best choice for further investigations into the general mechanisms of activation, epimerisation and coupling of the amino acids by peptide synthetases.

1.2.3 Isopenicillin N synthase

Following the demonstration that cell-free extracts of *C. acremonium* possessed ring expansion activity (*vide infra*), and were capable of converting penicillin N (**3**) into deacetoxycephalosporin C [DAOC (**4**)], efforts were made to detect the activity responsible for converting ACV (**2**) into a penicillin. Again using *C. acremonium*, it was possible to demonstrate cell-free synthesis of isopenicillin N (**1**).[53,54,55] The enzyme was purified[56,57] and identified as a 38 kDa protein (IPNS) whose catalytic activity was strongly enhanced by ferrous iron and ascorbic acid, and required dioxygen as a cosubstrate. As studies proceeded it became clear that, like the ring expansion activity, ferrous iron and dioxygen played a central role in the biosynthesis of the penicillin nucleus.

The gene for this enzyme was cloned[58] and successfully over-expressed[59,60] at high levels in *Escherichia coli*. The gene sequences of seven different IPNS isozymes have subsequently been determined from fungal and bacterial species[1,2] Homology between the eukaryotic and the prokaryotic sequences is about 75% within each group, but there is only *ca.* 55% homology between IPNS isozymes from different groups. This fact has led to the suggestion that, following a gene transfer event from a *Streptomyces* precursor to a fungal precursor, the sequences evolved separately to their present position. The acquisition by fungi of the biosynthetic machinery for penicillin biosynthesis would thereby provide them with an evolutionary advantage over bacteria in the soil environment, since their cell walls are relatively inert to the effects of these antibiotics.

As soon as crude cell-free preparations of IPNS were available, mechanistic studies on the conversion of ACV (**2**) into isopenicillin N (**1**) were begun by the present authors.[61] It was shown that ferrous iron was essential for activity, and the stoichiometry of iron binding was later confirmed as 1 : 1.[62] The stoichiometry of dioxygen consumption in the reaction was also found to be 1 : 1 with the production of isopenicillin N (**1**).[63] A series of studies using isotopically labelled ACV (**2**), following earlier work in intact cells with free amino acids, revealed that four hydrogen atoms were stereospecifically removed and that the two new ring structures were formed with retention of configuration.[64,65,66] Concomitantly the dioxygen molecule is reduced to water, but none of the oxygen atoms of ACV (**2**) was found to undergo exchange (Scheme 1.3).[67,68] Thus, penicillin biosynthesis utilises the most exothermic non-photochemical reaction available to biological systems, i.e. the complete reduction of dioxygen to water. This large driving force is necessary to provide the means whereby the strained ring system of penicillin is formed.

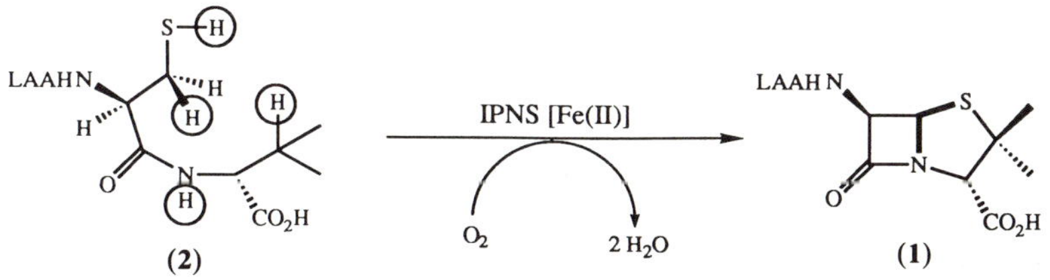

Scheme 1.3

Initial mechanistic studies focused on the possibility of observing intermediates in the reaction by ^{1}H nmr spectroscopy. However, these experiments failed to reveal the presence of an enzyme-free intermediate.[69] Since at least two chemically reasonable intermediates had been previously discussed in the literature it was decided to prepare them by synthesis and test them with IPNS. Thus, the β-hydroxyvaline peptide (**19**), and the thiazepinone peptide (**20**) and its sulphoxide (**21**) were tested. None of them showed any activity as penicillin precursors with IPNS.[70] The remaining possibility, the azetidinone thiol (**22**), presented a more substantial synthetic challenge, but was finally obtained.[71,72] It was discovered that at the incubation pH of 7.5, the thiol (**22**) underwent a rapid, non-enzymatic, ring opening isomerisation to the enethiol (**23**), although it (**22**) could be stabilised at low pH. A search for the enethiol (**23**) and so-derived products during the enzymatic conversion of ACV (**2**) into isopencillin N (**1**) failed to reveal their presence. Thus, it was concluded that if the thiol was indeed an intermediate it was not 'free' during the reaction cycle.

The next approach involved the determination of kinetic isotope effects (KIEs) in the removal of hydrogens from the two non-exchangeable sites of

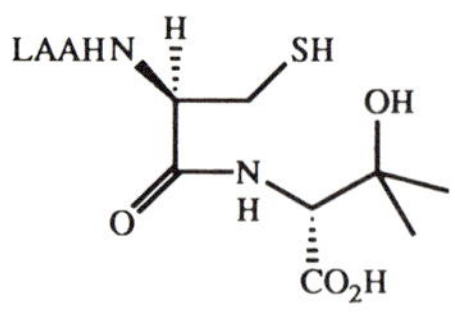

(19) 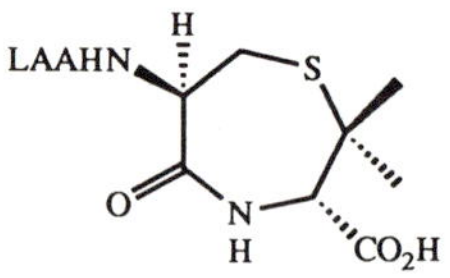 **(20)** 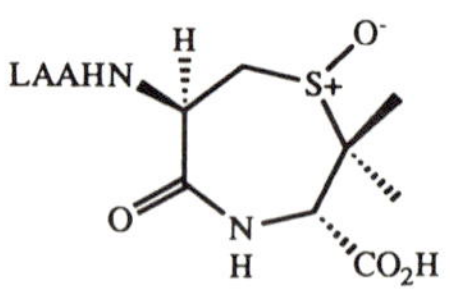 **(21)**

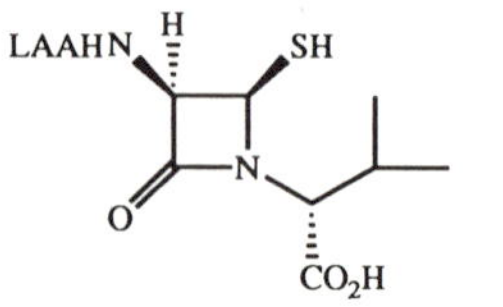

(22) 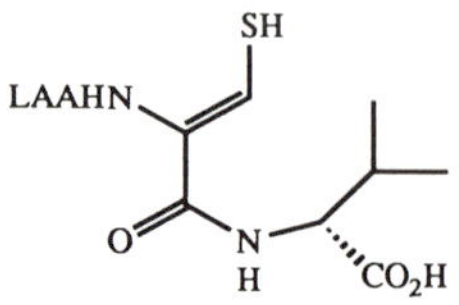 **(23)**

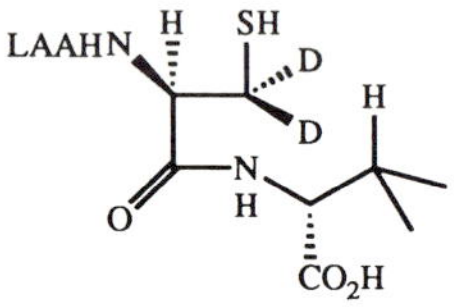

(24) 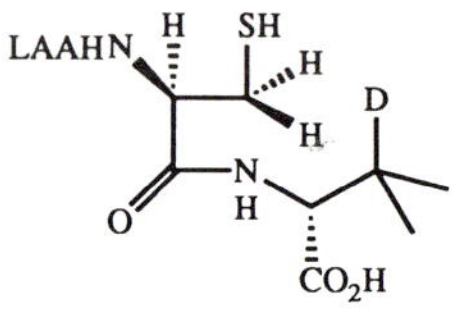 **(25)** 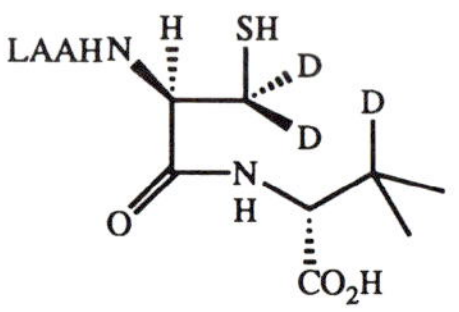 **(26)**

(**2**), i.e. the *pro*-3-*S*-hydrogen in the cysteinyl residue and the β-hydrogen in the valine unit. Three suitably deuterated peptides were prepared, (**24**), (**25**), and (**26**), and the KIEs for the IPNS reaction determined for the two parameters V_{max}/K_M and V_{max}.[73] The results are shown in Scheme 1.4.

These results are consistent with a two-step mechanism, the first step of which involves the removal of the *pro*-3-*S*-hydrogen of the cysteine. This is followed by the secondary removal of the β-valinyl atom. The putative intermediate (**27**) between these events, which remains in the valley of the energy profile (Scheme 1.5), must also remain enzyme-bound as is shown by

$\boldsymbol{V_{max}/K}$ $\qquad \boldsymbol{D_{V/K} = (V/K)_H/(V/K)_D}$

(24)	1.8 ± 0.2
(25)	1.0 ± 0.1

$\boldsymbol{V_{max}}$ $\qquad \boldsymbol{D_V = k_H/k_D}$

	$D_{V(1)}$	$D_{V(2)}$	$D_{V(1,2)}$
(24)	5.6 ± 0.3		
(25)		13 ± 2	
(26)			18 ± 2

Value of $\boldsymbol{D_{V(1,2)} = D_{V(1)} + D_{V(2)} - 1 = 17.6 \pm 2.3}$

Scheme 1.4

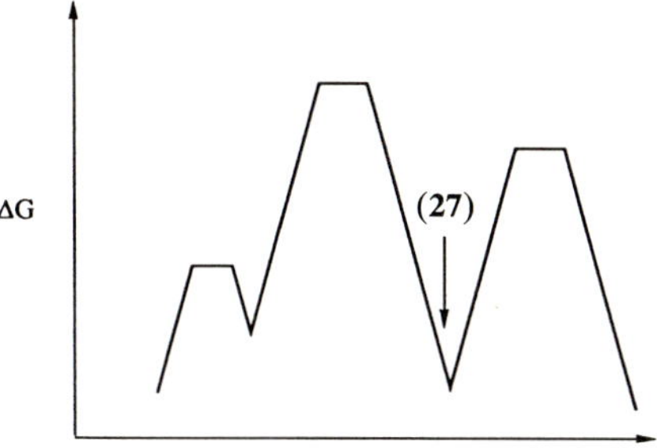

Scheme 1.5

the coupling of the V_{max} isotope effects in the trideuterated substrate (**26**) which, under steady-state conditions, is given by:

$$D_{V(1,2)} = D_{V(1)} + D_{V(2)} - 1.$$

The working hypothesis for this two-step mechanism is shown in Scheme 1.6, wherein a covalent linkage occurs between the thiol group of ACV to the ferrous iron/dioxygen complex in the active site. In this way the sulphur atom acts as a conductor of electrons from the peptide to the iron dioxygen redox centre.[74] Subsequently, absorption spectroscopy and EXAFS have confirmed this proposal of a iron–sulphur link between the substrate and IPNS.[75,76,77]

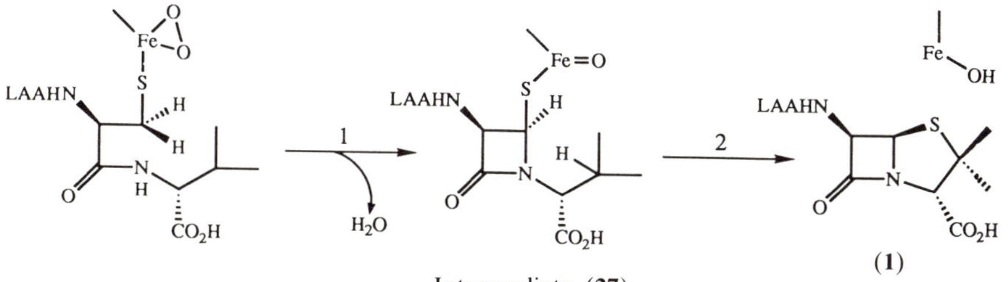

(**1**)

Intermediate (**27**)

Scheme 1.6

In order to probe the nature of the intermediate it was decided to use the large V_{max} isotope effect on step 2 (Scheme 1.4) to encourage release of the intermediate from the enzyme and permit its interception in solution (Scheme 1.7). This would be a kinetic, isotope-induced branching of the reaction pathway. It is of interest to note that the proposed intermediate (**27**) is in fact an iron(IV)-oxo-ester of the previously synthesised azetidine thiol (**22**). In the event a substance was released from IPNS during catalytic

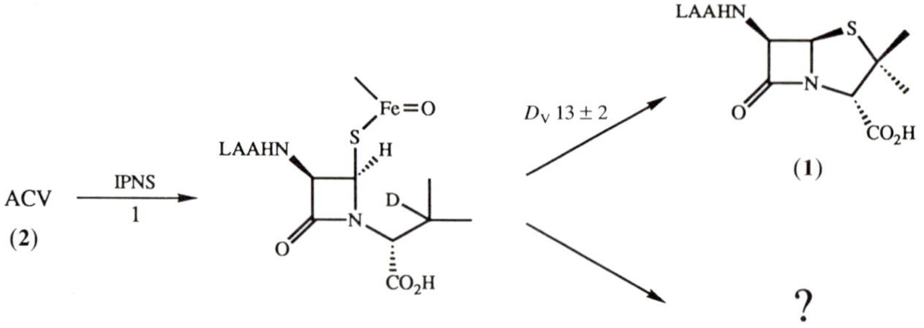

ACV (2) → IPNS 1 → ... → D_V 13 ± 2 → (1); → ?

Scheme 1.7

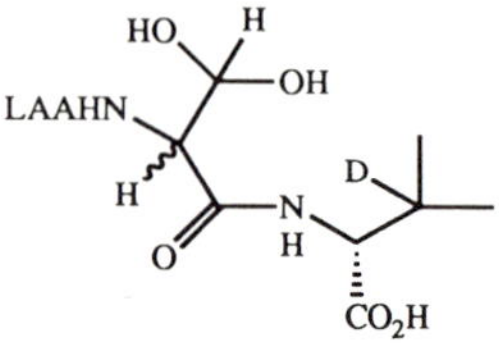

(28)

turnover, subject to a deuterium isotope effect on step 2, which was identified as the diol (**28**).[78] The loss of sulphur is believed to arise from the alternative iron(IV)-oxo- driven fragmentation (Scheme 1.8) which, consequent to the isotope effect, now begins to compete with the ring closure. It should be noted that it is proposed that the iron centre returns to iron(II) concomitant with the production of atomic sulphur, and so the catalytic cycle continues, i.e. the diol is a product of the catalytic turnover of IPNS.

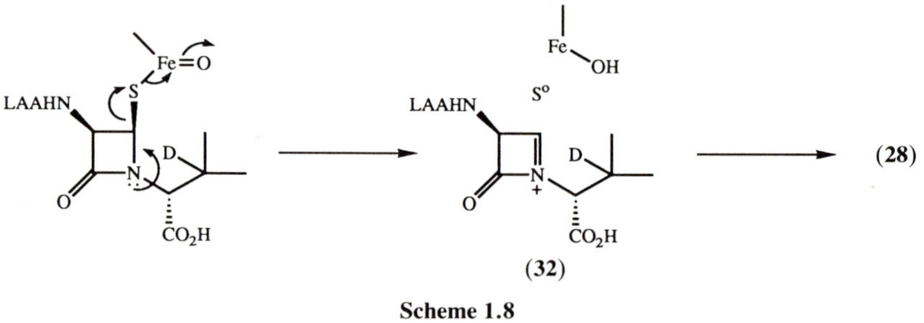

(32) → (28)

Scheme 1.8

Support for this interpretation was accrued by utilising a homocysteine-containing peptide (**29**) as a substrate for IPNS, whereupon the epimeric monocyclic lactams were formed (**31**) (Scheme 1.9).[79] In this case the lower ring strain encountered by the five-membered iminium cation (**30**) versus the four-membered case (**32**) permits monocyclisation to be the sole path-

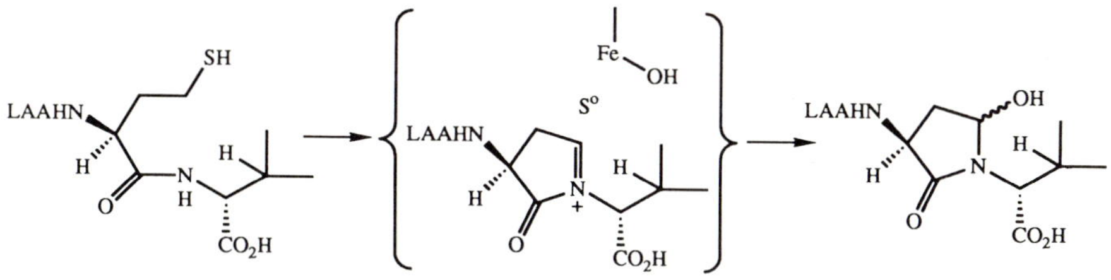

(29) (30) (31)

Scheme 1.9

way. Utilisation of $^{18}O_2$ as a co-substrate gave the partially ^{18}O-hydroxy-labelled γ-lactams (**31**), indicating that the quenching hydroxyl was attached to the iron centre, derived from the reduction of dioxygen. This fact supports the suggestion that the cysteine sulphur atom is directly bound to the iron throughout the reaction cycle. The loss of hydrogen from the γ-carbon of the homocysteinyl moiety of (**29**) was, as in the natural substrate ACV (**2**), stereospecific.[79]

Further evidence for the proposed acyl imminium cation in these so-called 'shunt' pathways was obtained by placing two strongly electron-withdrawing substituents adjacent to the sulphur-bearing carbon on the homocycsteinyl moiety. If indeed a cation were involved in the sulphur fragmentation such substituents should hinder the process. Thus, the difluoro homocysteine peptide (**33**) was prepared and challenged with IPNS. The product of the reaction was the thiocarboxylic acid (**34**) (Scheme 1.10).[80] This is entirely in accord with the authors' mechanism since, in the absence of fragmentation,

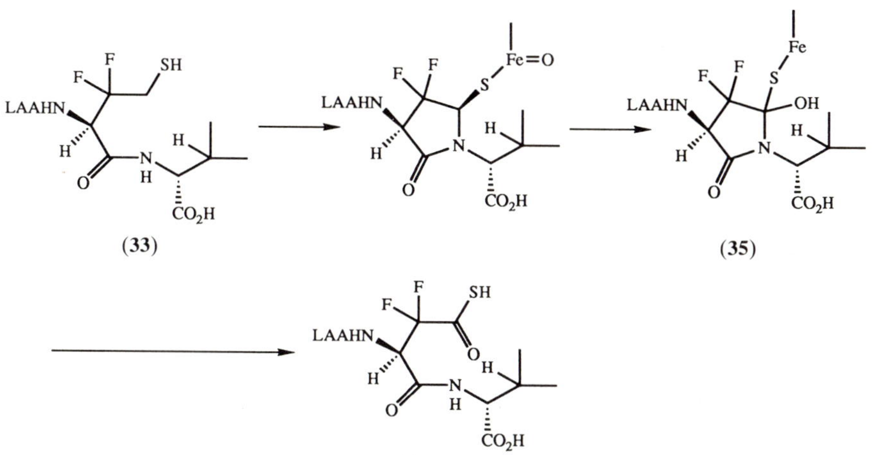

(33) (35)

(34)

Scheme 1.10

inhibited by the fluorine inductive effect, the iron-oxo species now hydroxylates the remaining hydrogen and gives, after ring-opening of (**35**), the thiocarboxylic acid (**34**). The use of $^{18}O_2$ gas gave the thiocarboxylic acid (**34**), with >95% incorporation of a single ^{18}O label, confirming the hydroxylation hypothesis.[80]

In the proposed intermediates the cysteinyl or homocysteinyl sulphur is attached to a strongly electron-withdrawing iron(IV)-oxo species and, as such, should be sensitive to disulphide formation with other thiols. This propensity was revealed in the dithiopeptide (**36**), which was cyclised to the bicyclic disulphide (**37**) (Scheme 1.11).[81]

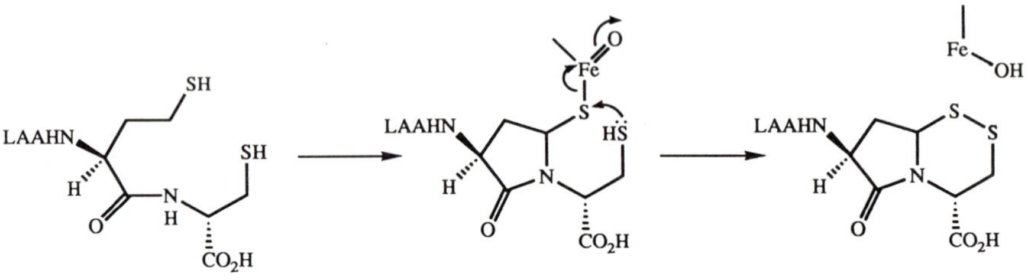

Scheme 1.11

All of these experiments support the notion of direct attachment of the thiol function to the iron-dioxygen centre, and provide insights into the way in which such systems can evolve chemically, dependent on the structural environment and the various isotope effects.

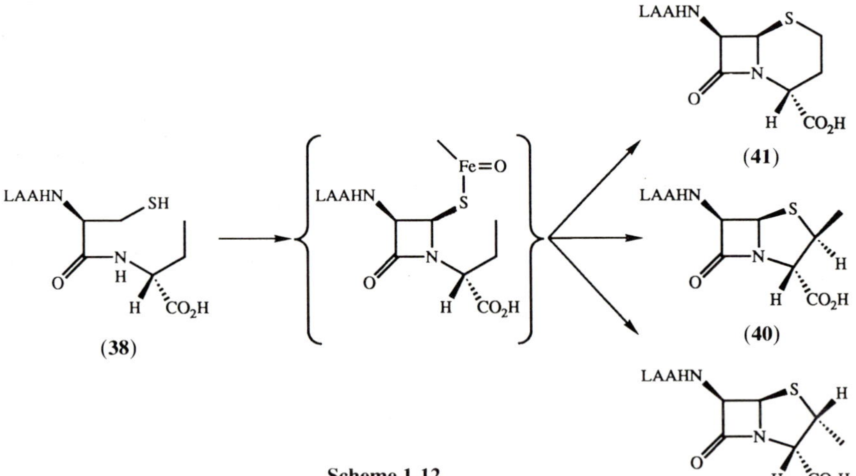

Scheme 1.12

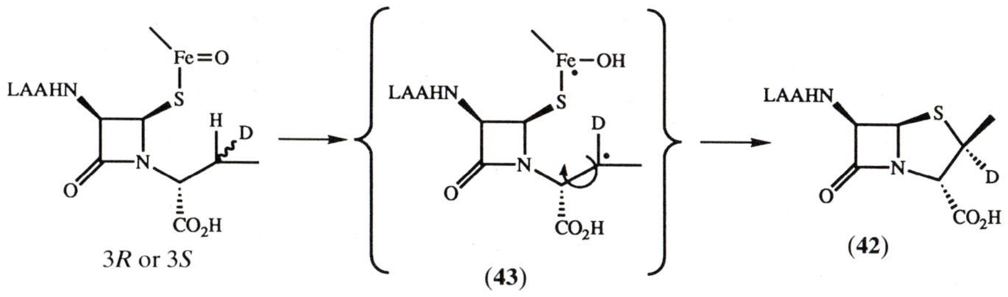

LAAHN Fe=O S H D N O CO_2H 3*R* or 3*S*
LAAHN Fe–OH S D N O CO_2H (43)
LAAHN S D N O CO_2H (42)

Scheme 1.13

Given the proposed monocyclic ferryl intermediate (**27**), the strategy for investigation of the second ring closure was to vary the substitution of the valinyl moiety, in order to develop a profile of reactivity versus structure. The very first example studied, the aminobutyrate peptide (**38**), simultaneously gave three bicylic products (Scheme 1.12).[82] Interestingly, two of these, (**40**) and (**41**), were later discovered as their δ-(D-α-aminoadipoyl) derivatives, (**13**) and (**14**), in a *Streptomyces* species (*vide supra*).[44] The isolation of the desmethyl penicillins, (**39**) and (**40**), permitted investigations into the stereospecificity of the second ring closure in this case. Using 3*R*- and 3*S*-monodeuterated aminobutyrate peptides, it was found that the 3*R*-isomer gave the deuterated penam (**42**) with retention whereas the 3*S*-isomer gave the same penam with inversion![83] In order to explain how two monodeuterated peptides gave the same product, but via opposite stereochemical courses, the intermediacy of a free radical (**43**), formed by *hydrogen* abstraction (isotope effect), and which could rotate faster than ring closure, was suggested (Scheme 1.13).

As a probe for carbon free radicals the cyclopropylcarbinyl test[84] was selected (Scheme 1.14). The cyclopropyl unit was incorporated into the valine moiety of two peptides, both in fact giving ring opened products.[85] The conversion of one of these peptides (**44**) (Scheme 1.15) permitted a very precise analysis of the stereochemistry of the ring opening process, with some highly unexpected results. When (**44**) was specifically labelled in either of the prochiral methyl groups it underwent an entirely regiospecific ring opening (Scheme 1.16). However, when the peptide labelled stereospecifically with a single deuterium on the migrating methyl group (**45**) was synthesised and incubated with IPNS, it was shown to lose its stereochemistry during migration (Scheme 1.17).[86]

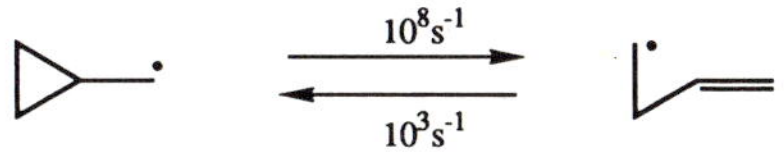

$10^8 s^{-1}$ $10^3 s^{-1}$

Scheme 1.14

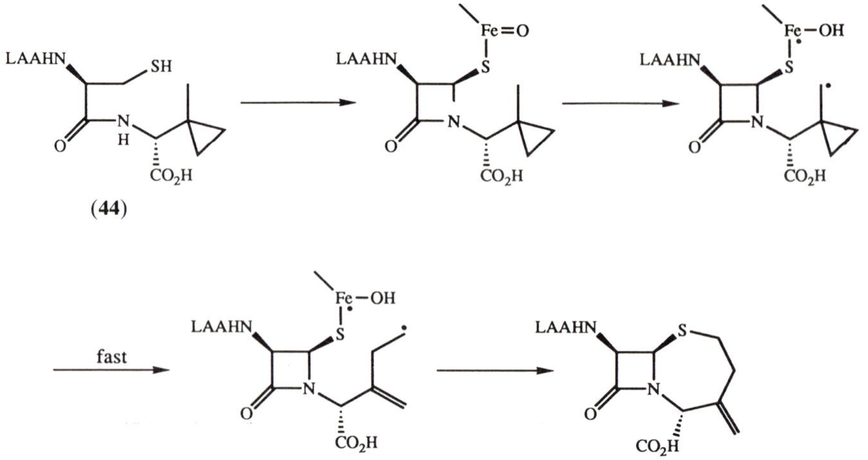

LAAHN SH Fe=O Fe–OH S N H O CO_2H (**44**)

fast

Scheme 1.15

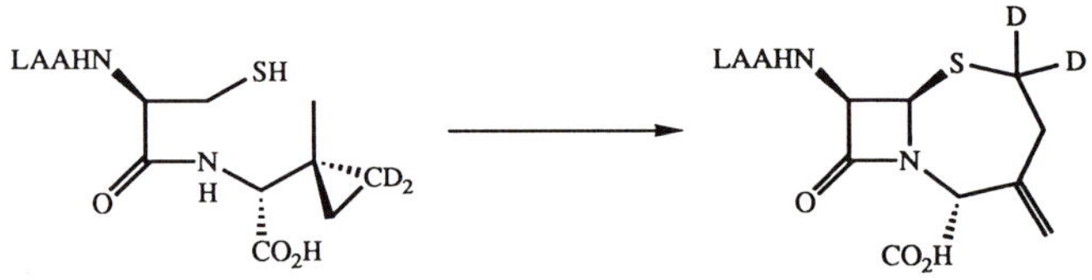

CD_2 D

Scheme 1.16

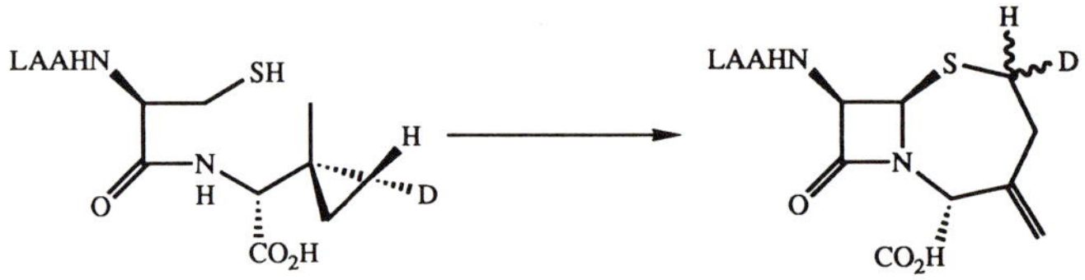

H D (**45**)

Scheme 1.17

In the authors' view, these results, along with those in which the remaining methylene group was chirally labelled and showed integrity of stereochemistry in the bicyclic product, are not consistent with an initial homolytic C–H bond cleavage, since such a cyclopropyl carbinyl radical, given its reactivity, would be expected to break *both* C–C bonds with equal facility, in contrast to what was experimentally observed. Rather, it is suggested that initial C–H bond cleavage involves an *insertion* process to form an iron–carbon bond (**46**), which is followed by a stereospecific sigmatropic rearrangement process (path a, Scheme 1.18) and, finally, by

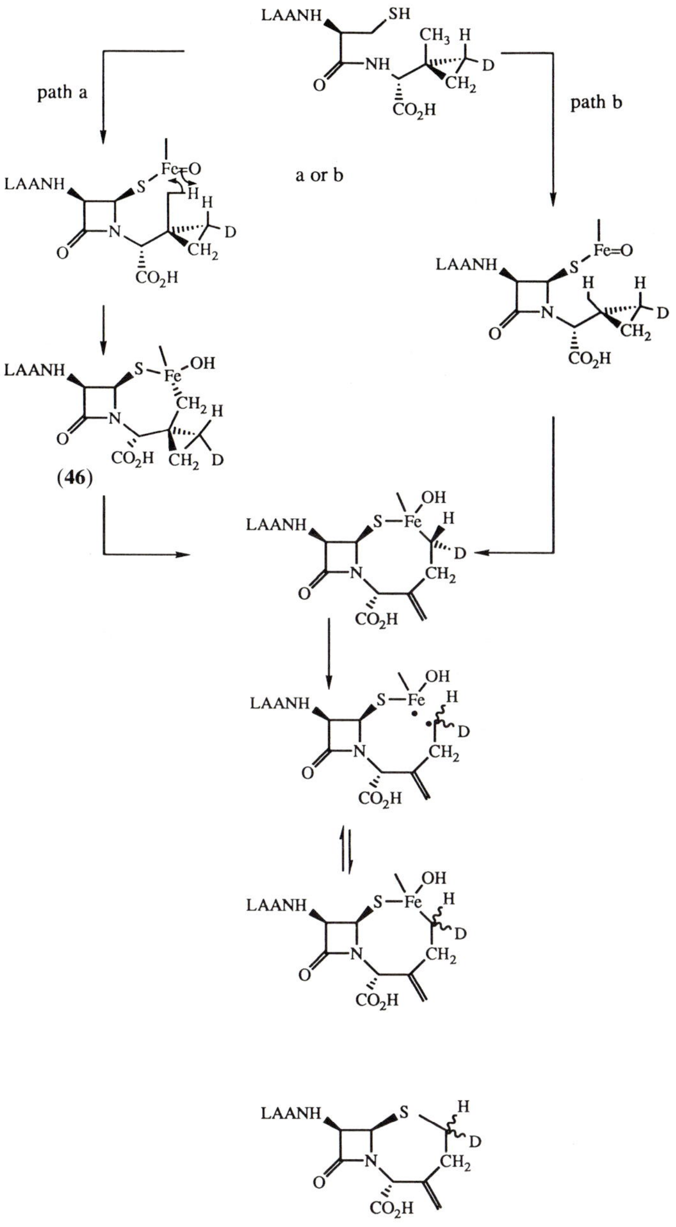

path a

path b

a or b

(46)

Scheme 1.18

homolysis and C–S bond formation (alternatively, a rearrangement may occur directly — path b, Scheme 1.18) . The proposal of an analogous species containing an iron–carbon bond during the conversion of ACV (**2**) to isopenicillin N (**1**) was first made by the authors in 1982.[87]

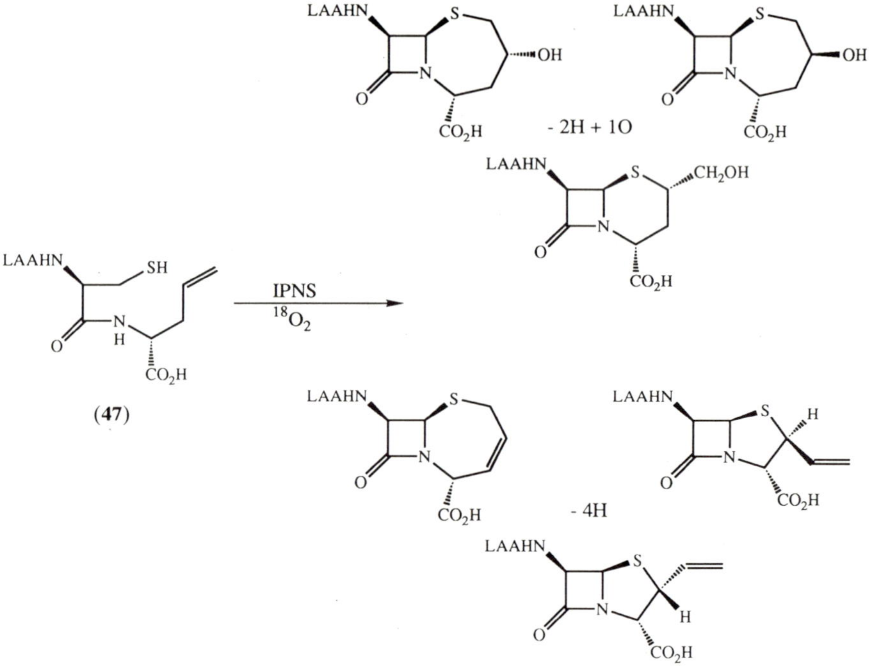

Scheme 1.19

Probing the mechanism of IPNS with tripeptides containing unsaturated amino acids in place of valine has also led to interesting and surprising results. The allyl glycine tripeptide (**47**) was transformed by IPNS into no fewer than six new bicyclic products (Scheme 1.19).[88,89] These products fall into two groups: (i) those involving overall loss of four hydrogens (desaturation), which also occurs with the natural substrate ACV (**2**); and (ii) those that involve the loss of three hydrogen atoms and the gain of one oxygen, the latter from the co-substrate dioxygen.[90] This latter and new pathway involves both desaturation (during formation of the β-lactam ring) and hydroxylation (second ring closure). Stereochemical determinations using stereospecifically-deuterated allylglycine-containing tripeptides have been done for all of these different types of product. These results, which may be found in the original article,[91] suggest that the proposed iron(IV)-oxo species can undergo four distinct types of reaction on closing the second ring. The selection between these is determined by a balance of reactivity of the substrate versus the geometrical relationship between the side chain with respect to the putative iron-oxo species (Scheme 1.20). It follows from Scheme 1.20 that the stereochemistry of carbon–sulphur bond formation will be determined by the competition between the rate of rotation of the radical versus ring closure. Thus, in the case of isopenicillin N (**1**) formation

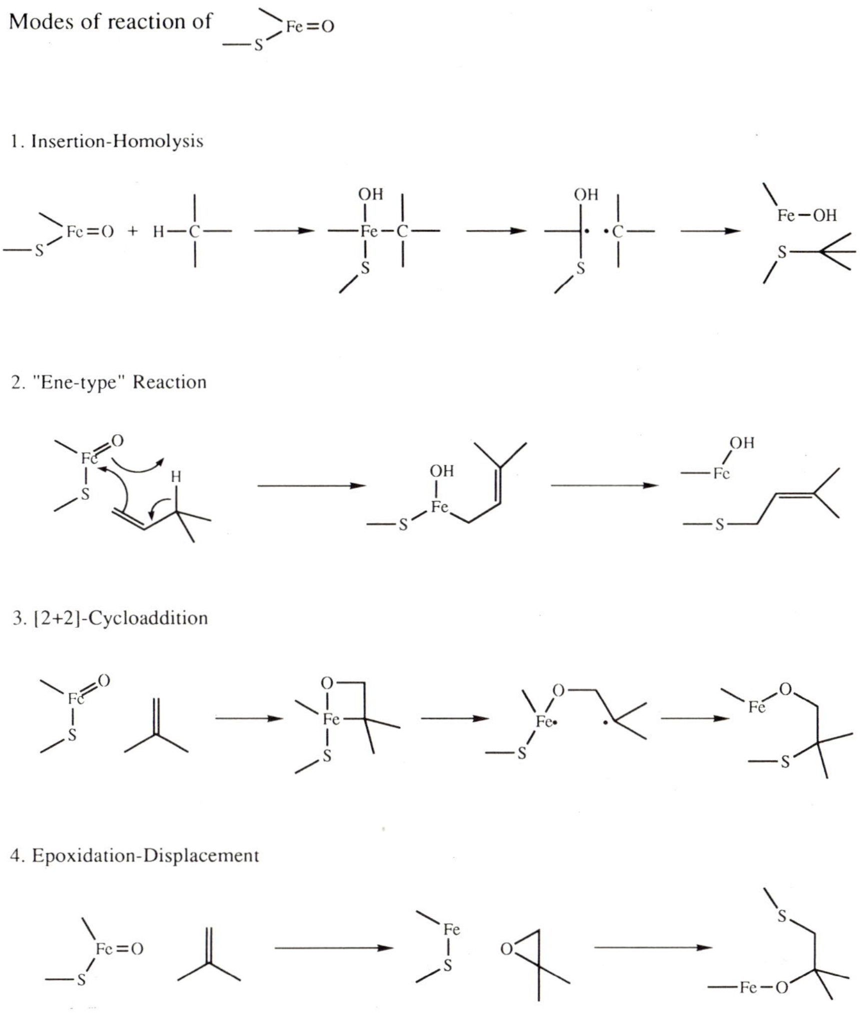

Modes of reaction of —S—Fe=O

1. Insertion-Homolysis

2. "Ene-type" Reaction

3. [2+2]-Cycloaddition

4. Epoxidation-Displacement

Scheme 1.20

the reaction is clearly stereospecific with retention, as a consequence of a slowly rotating tertiary radical,[66] whereas the α-aminobutyrate tripeptide (**38**) (both inversion and retention) showed the consequence of a rapidly rotating radical.[83] On the other hand, the ene-reaction pathway gave clean carbon–sulphur stereochemistry, probably due to resistance to rotation in allyl radicals. In contrast to these, the formation of the carbon–oxygen bonds, formed via the hydroxylative pathway from unsaturated substrates, is apparently *always* stereospecific. A very good example of this as well as

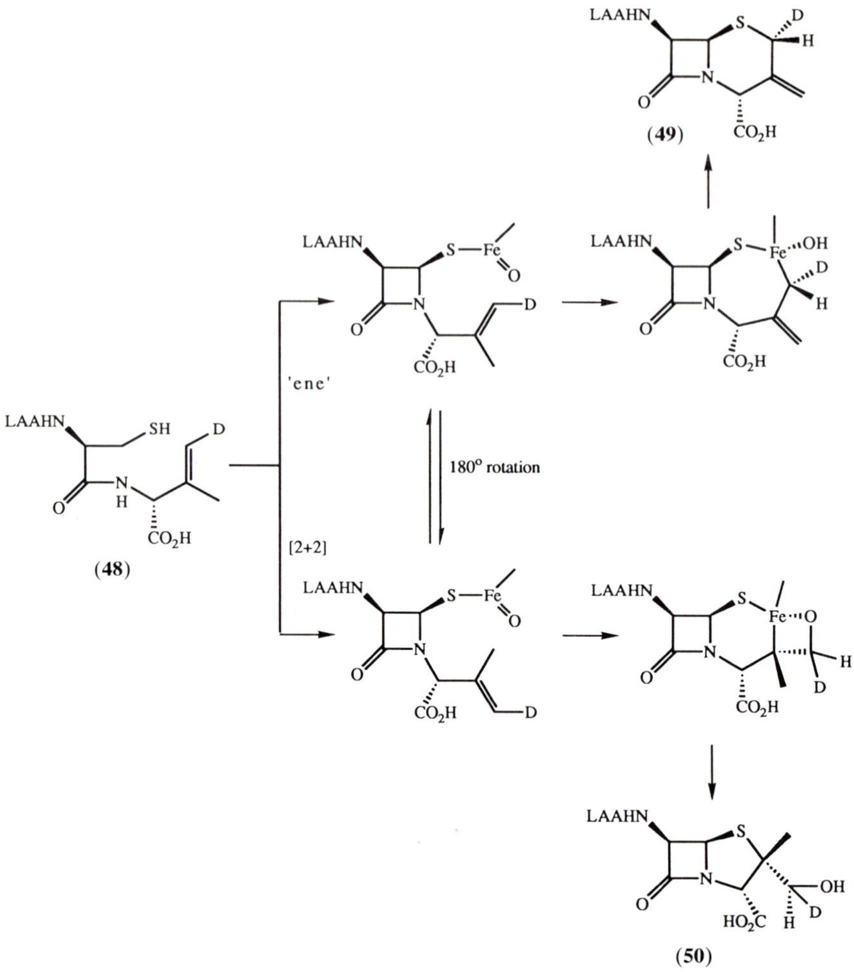

Scheme 1.21

the geometric factor in the balance of reaction modes is revealed by the unsaturated peptide (**48**) (Scheme 1.21). In this case two products, an exomethylene cephalosporin (**49**) and an α-hydroxymethylpenam (**50**), are clearly derived from the two 180° rotamers of the isopropenyl group in the intermediate monocyclic lactam. The rotamers present either of the two faces of the double bond to the iron-oxo species, thereby providing the correct geometry for either [2 + 2] cycloaddition or an ene-type reaction. The precise stereochemistry predicted by this analysis was confirmed by deuterium labelling experiments as shown in Scheme 1.21.[92]

The enzyme IPNS will tolerate structural variations in the key cysteine unit. Thus, methylation at the 2-position (**51**) or the 3*R*-position (**52**) produced tripeptide substrates for IPNS (to give (**53**) and (**54**) respectively),[93] as did methoxylation at the 2-position (**55**) (to give (**56**)) (Scheme 1.22).[94] The use of homocysteine variants has already been discussed. A sample of the seleno variant of ACV (**2**) was synthesised and shown to readily and oxidatively form a dehydroalanine peptide, although, in the presence of IPNS, a small amount of a bioactive penicillinase-sensitive material was produced, presumably seleno-isopenicillin N, though this could not be isolated.[95]

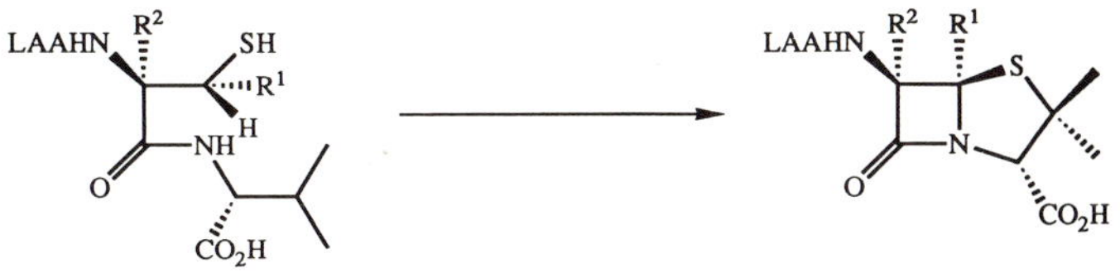

(**52**) R^1 = Me, R^2 = H
(**51**) R^1 = H, R^2 = Me
(**55**) R^1 = H, R^2 = OMe

(**54**) R^1 = Me, R^2 = H
(**53**) R^1 = H, R^2 = Me
(**56**) R^1 = H, R^2 = OMe

Scheme 1.22

Changes in the α-aminoadipoyl moiety have also been extensively investigated. The results indicate that the α-aminoadipoyl residue may be replaced with a six-carbon side chain, or equivalent, terminating with a carboxyl function to give an efficient substrate.[96,97,98] For example, a very good substrate is the naphthyl derivative (**57**), which supports the view that the side chain resides in an extended, staggered conformation. Although the phenoxyacetyl and phenylacetyl side-chained penicillins can be synthesised from the corresponding tripeptides with IPNS, the conversions are poor.[99] The steady-state parameters suggest that these substrates bind to the active site, via the sulphur atom, but that the catalytic events are retarded; this possibly indicates that, after binding, conformational changes occur within the protein, which are facilitated by the formation of a salt bridge at the carboxyl terminus of the aminoadipoyl moiety. Use of diazirine-substituted α-aminoadipoyl variants permitted photolabelling of IPNS.[100–102] These studies will become of considerable interest when a three-dimensional structure of IPNS is determined.

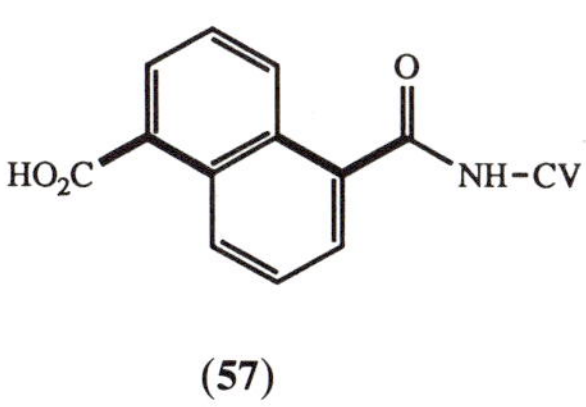

(**57**)

Based on the above experimental work, the detailed reaction cycle shown in Scheme 1.23 is proposed as a working hypothesis. In this scheme are indicated the two isotopically-sensitive steps (steps 1 and 2) and the isotopically-sensitive branching point, giving rise to the 'shunt' product. The immediate environment of the iron(II) species at the active site has been characterised by physical studies, particularly of the nitric oxide complex of the ACV–IPNS species, as the octahedral ligand arrangement (**58**), in which three histidines participate in the primary attachment of the iron(II)

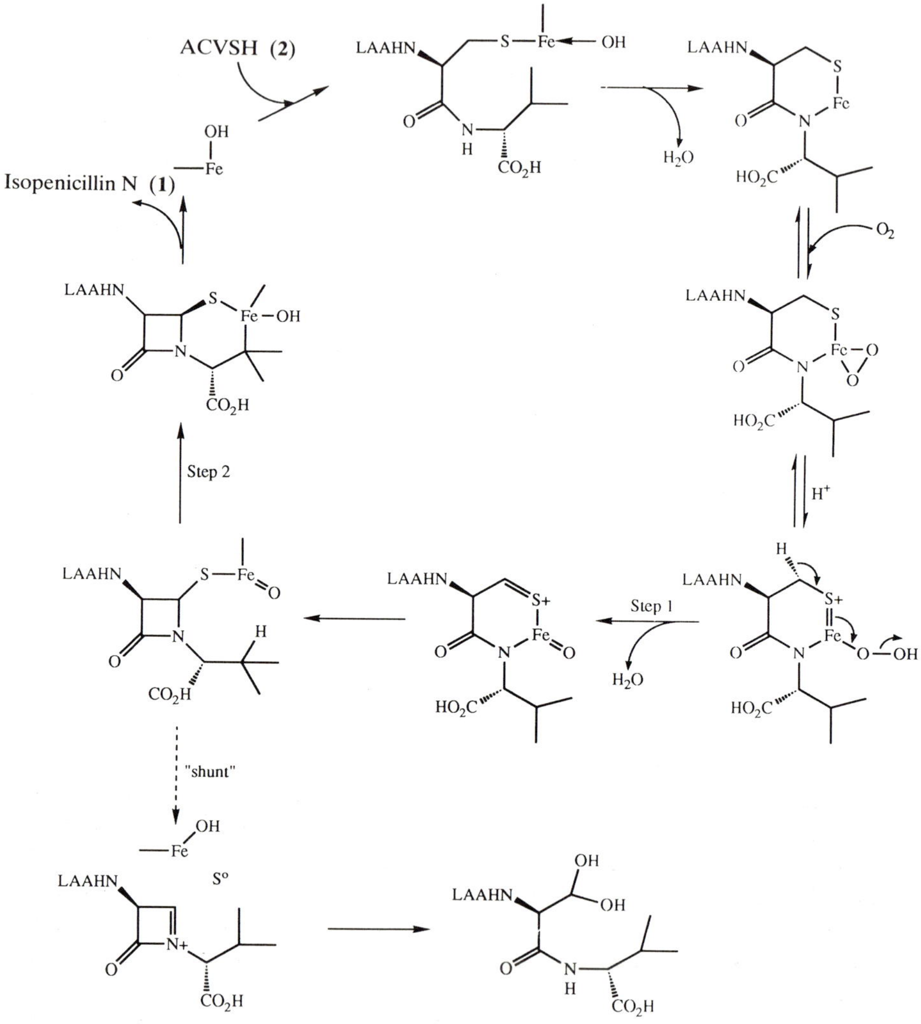

Scheme 1.23

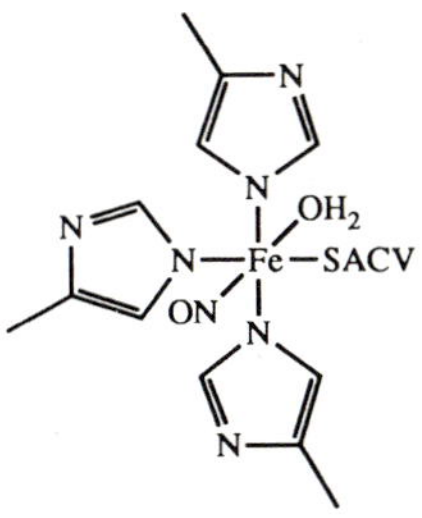

(58)

to the protein.[75,76] However, it is not yet known which of the histidines of the *C. acremonium* IPNS are implicated in this structure.

The question must arise as to whether this intriguing reaction could occur in the absence of the protein. To date, all experiments with ACV (**2**) and iron(II)/oxygen or iron(II)/peroxide systems have not produced penicillins. However, the second step in the proposed mechanism, i.e. the second ring closure, has been successfully mimicked in a non-enzymatic system. Thus, treatment of peptide (**59**) with iron(II)/oxygen/ascorbic acid or, preferably, iron(II)/hydrogen peroxide/ascorbic acid gave modest yields of a penicillin (**60**) along with a cepham (**61**) (Scheme 1.24).[103] Stereochemical studies showed that in this case, carbon–sulphur bond formation was stereorandom, as may have been anticipated in the absence of the chiral environment provided by the active site of IPNS.

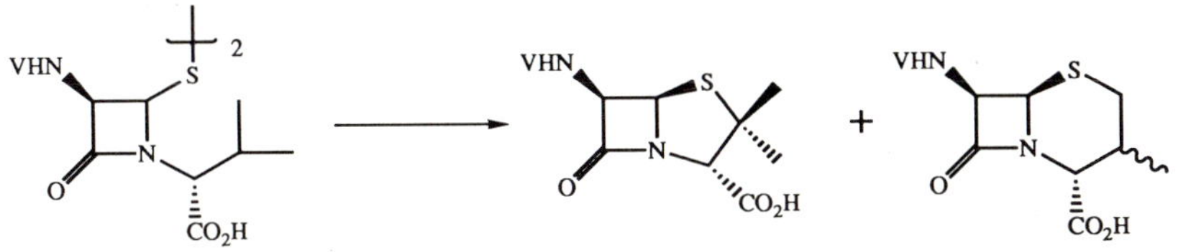

(59) (60) (61)

Scheme 1.24

Another fundamental question that arises concerns the evolution of such a reaction sequence. A number of constraints, imposed by the chemistry, are significant in these considerations. First, at present levels of atmospheric oxygen, ferrous complexes rapidly and irreversibly react via μ-peroxidases[104] (Scheme 1.25). The products are μ-oxo-bridged ferric complexes. This reaction can be slowed down by sterically inhibiting ligands around the iron,[105] such that formation of the first iron(II)-dioxygen complex is reversible.

A second consideration is the oxygen-driven oxidative coupling of thiols to disulphides, catalysed by iron salts. Presumably this occurs by a

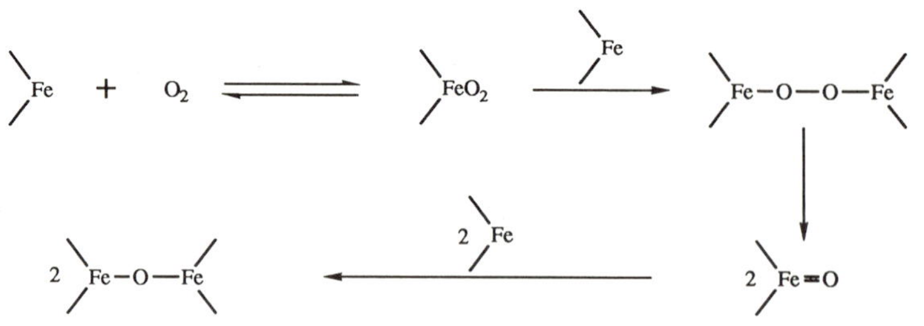

Scheme 1.25

mechanism such as shown in Scheme 1.26. In this case a bimolecular event again results in an irreversible step — formation of a disulphide bond. Taking all of these factors into consideration it would seem not unreasonable that an iron(II) complex of an ACV(**2**)-like tripeptide could have significant steric hindrance at the iron centre, thereby avoiding bimolecular reactions. If such a complex were to encounter dioxygen, *at low concentrations*, it is not unreasonable to expect that the types of reactions described above for IPNS could take place in the absence of the protein. As it is generally believed that the present atmospheric level of oxygen arose from the photosynthetic splitting of water by living organisms, beginning about three billion years ago, it may be that the origins of penicillin are ancient indeed.

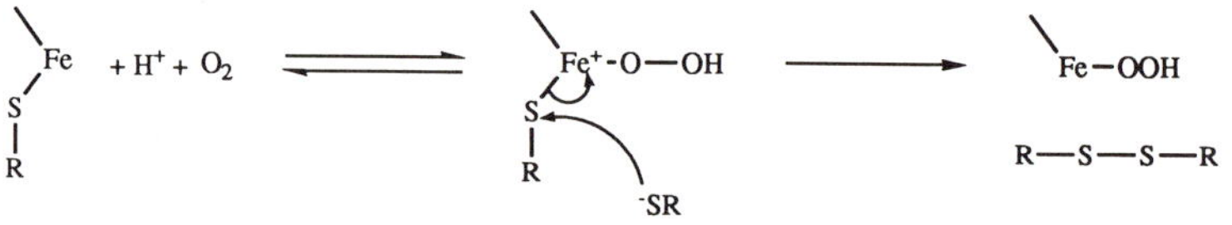

Scheme 1.26

1.2.4 Isopenicillin N/penicillin N epimerase

The formation of penicillin N (**3**) represents a branching point between those organisms that produce cephalosporins and those that produce penicillins only (Scheme 1.1). It has not been found in any of the latter. It was first suggested by Abraham that penicillin N (**3**) is produced by the epimerisation of isopenicillin N (**1**). Circumstantial evidence was provided for such a process when it was discovered that trace amounts of cephalosporins were produced on incubation of ACV (**2**) with crude cell lysates of *C. acremonium*.[106] Subsequently it was discovered that isopenicillin N (**1**) was converted into cephalosporins using lysates derived from protoplasts of *C. acremonium*, but that a component of the system was unstable at −80°C. Since the component of the crude mixture that converted penicillin N (**3**) to cephalosporins was shown to be stable at this temperature, it was concluded that isopenicillin N (**1**) was converted into penicillin N (**3**), by an unstable epimerase.[107] Direct evidence for the epimerisation was provided when it was shown that isopenicillin N (**1**) was converted into penicillin N (**3**) by a highly labile epimerase.[108]

The lability of the *C. acremonium* epimerase has apparently precluded any further studies, and efforts have been concentrated on the enzyme from *S. clavuligerus*. The bacterial epimerase was found to be significantly more stable[109] than the fungal enzyme, and hence has been amenable to further investigation. Jensen *et al.*[109,110] reported the purification and partial characterisation of an epimerase of molecular weight 60 kDa (estimated by gel chromatography, although no protein of that molecular weight could be detected by SDS-PAGE). Experiments in which the L-α-aminoadipoyl side chain of isopenicillin N (**1**) was replaced with L-carboxymethylcysteine to give the analogue (**62**) (in order to facilitate HPLC analysis) also revealed that the epimerase catalyses a reversible reaction (Scheme 1.27). Equilibrium was reached when a 1 : 1 mixture of the isopenicillin N analogue (**62**) and the penicillin N analogue (**63**) was present.[111] The conversion of analogues of ACV (**2**) to cephalosporins using semi-purified extracts of *S. clavuligerus* and the appropriate cofactors also provided evidence that the epimerase will tolerate certain modifications in the penam nucleus.[112]

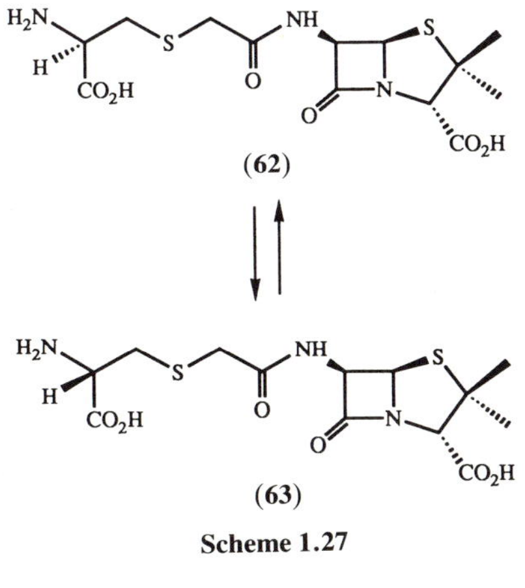

(**62**)

(**63**)

Scheme 1.27

The purification to homogeneity of the *S. clavuligerus* epimerase has been reported by Usui and Yu.[113] The purified enzyme was found to be monomeric with a molecular weight of 47–50 kDa, and to catalyse complete epimerisation of the side chains of both isopenicillin N (**1**) and penicillin N (**3**), such that an equimolar mixture of the two penicillins was produced. The K_m for isopenicillin N (**1**) (0.30 mM) was found to be lower than that for penicillin N (**3**) (0.78 mM). Therefore, *in vivo*, one would anticipate that the bulk of the penicillin N (**3**) produced by the epimerase would be efficiently processed through subsequent steps in the pathway before re-epimerisation to isopenicillin N (**1**). Highly purified epimerase was found to contain absorption maxima at 280 and 420 nm and to contain one molecule of pyridoxal-5′-phosphate per molecule of enzyme. The cofactor could be removed by the treatment with hydroxylamine, followed by gel filtration to give an apoprotein, with no absorption maxima at 420 nm. Activity could be restored to the apoprotein by the addition of pyridoxal-5′-phosphate. Treatment of the holo enzyme with sodium borohydride destroyed all activity, consistent with linkage of the cofactor to the enzyme via an imine linkage. Thus, isopenicillin N/penicillin N epimerase appears to be similar to some other bacterial amino acid racemases, for example alanyl racemase from *Salmonella typhimurium*.[114] The epimerase activity was found to be partially stimulated by the addition of thiol reducing agents and was strongly inhibited by sulphydryl alkylating agents. The authors speculated that a thiol group may play a key role in the catalysis of the epimerisation.

Using oligonucleotide probes based on the N-terminal sequence of the purified epimerase, cosmid clones that were known to contain β-lactam biosynthesis genes from *S. clavuligerus* were screened for hybridisation by Southern blotting.[115] The epimerase gene (*cefD*) was located immediately upstream of the expandase gene (*cefE*). It was sequenced and shown to encode a protein of predicted molecular weight 44 kDa. The *cefD* gene was expressed in *E. coli* and shown to comigrate with the wild type material from *S. clavuligerus*. Northern blot analysis indicated that the bacterial epimerase (*cefD*) and expandase (*cefE*) genes constituted part of an operon, and when *cefD* and *cefE* were placed in an expression vector, concomitant production of both epimerase and expandase was observed.

1.2.5 The ring expansion and hydroxylation reactions

The sequence of biosynthetic intermediates in the latter stages of the biosynthesis of cephalosporin C (**6**), as depicted in Scheme 1.1, was established in the 1970s. Thus, Koshaka and Demain[116] described the preparation of a cell-free system, from *C. acremonium*, which catalysed the conversion of penicillin N (**3**) into a penicillinase-resistant material. The cephalosporin product was subsequently identified as DAOC (**4**).[117–119] That DAOC (**4**) is the precursor of DAC (**5**) was unequivocally demonstrated by the conversion of [3-$^{3}H_3C$]-DAOC (**4**) to [3-$HO^{3}H_2C$]-DAC (**5**) with cell-free extracts of *C. acremonium*[120] and *S. clavuligerus*.[121] By this time cephalosporin C (**6**) mutants that accumulated DAC (**5**) had been isolated,[122,123] and Fujisawa *et al*. demonstrated that buffered extracts of the parent organism of revertants of the mutants could convert DAC (**5**) into cephalosporin C (**6**) in the presence of actyl coenzyme A and Mg^{2+}.[124]

The early studies identified that the cell-free conversion of penicillin N (**3**) into cephalosporins was stimulated by the addition of ferrous ions, ascorbate, and α-ketoglutarate.[125,126] That dioxygen is directly involved in cephalosporin biosynthesis was first demonstrated by Stevens *et al.* who showed that when *C. acremonium* was grown under an atmosphere of $^{18}O_2$, there was partial incorporation of the label into cephalosporin C (**6**).[127] ATP was also reported to stimulate the ring expansion process; however, it has little or no stimulatory effect on highly purified material from *C. acremonium,*[128,129] and high concentrations have been reported to inhibit the activity from *S. clavuligerus.*[130] Similar cofactor requirements were identified for the conversion of DAOC (**4**) to DAC (**5**) in *C. acremonium* and *S. clavuligerus.*[120,121]

Scheidegger *et al.* reported that the DAOCS and DACS activities from *C. acremonium* could not be separated from each other by ion exchange chromatography. Since the two enzyme activities had the same cofactor requirement and catalysed sequential reactions in the biosynthetic pathway, it was proposed that the two reactions were mediated by the same enzyme.[128] Further studies by the authors[130] and Dotzlaf and Yeh,[131] reported the purification to electrophoretic homogeneity of the DAOCS and DACS activities from *C. acremonium*. It was found that the activities could not be separated by ion exchange, dye ligand, gel filtration, or hydrophobic chromatography. Dotzlaf and Yeh reported that throughout the purification the two activities remained associated and that the ratio of DAOCS activity to DACS activity was constant (at 7 : 1).[132] It was concluded that a single monomeric protein (DAOCS/DACS) was responsible for the biosynthesis of DAOC (**4**) and DAC (**5**). Based on SDS-PAGE analysis, molecular weight estimates of 40–41 kDa were reported for the purified protein.[130,131] A more recent analysis by electrospray mass spectrometry has shown that the true molecular weight of the wild-type DAOCS/DACS from *C. acremonium* is *ca.* 36.5 kDa.[133]

In contrast, using cell-free extracts from *S. clavuligerus*, Jensen *et al.* have shown that the DAOCS and DACS activities could be readily separated by anion exchange chromatography.[134,135] The DAOCS and DACS activities have also been subsequently separated in extracts from *S. lactamdurans.*[136] The disovery of the separation of the activities in *Streptomyces* was important, since it was the first significant divergence between the eukaryotic and prokaryotic biosynthetic pathways to the cephalosporins and cephamycins to be identified.

Unambiguous evidence that DAOC and DAC biosynthesis is mediated by a single bifunctional enzyme in *C. acremonium* came with the cloning and expression of the requisite gene (*cefEF*) in *E. coli.*[137,138] Although the *C. acremonium* enzyme was found to be refactory to N-terminal sequencing, probably as a result of a modification to the amino terminus,[132,133] tryptic digestion was used to expose a sequencable fragment. On the basis of the amino acid sequence, probes were constructed and used to identify the *cefEF* gene. A single open reading frame was expressed in *E. coli*, and the resultant protein was shown to catalyse both DAOCS and DACS activities, in the presence of the appropriate cofactors.[137]

High-level expression of the *cefEF* gene in *E. coli* unfortunately resulted in the formation of inclusion bodies, containing insoluble DAOCS/DACS. However, treatment of these bodies with high concentrations of urea dissociated them, with the production of active enzyme after dialysis. This refolded enzyme showed the same substrate specificity as the wild-type enzyme.[137,139] Despite the success of this procedure, the yields of active DAOCS/DACS obtained from the inclusion bodies were relatively low compared to those obtained for recombinant IPNS, which has been expressed in a soluble form.[59,60] Consequently it is desirable that a soluble expression system for DAOCS/DACS is developed and efforts are under-way to do so.[140] The molecular weight of the recombinant DAOCS/DACS has been analysed by electrospray mass spectrometry and found to be approximately 60 Da lower than that of the wild-type material from *C. acremonium*, consistent with an N-terminal blockage in the latter.[133]

The genes for DAOCS (*cefE*) and DACS (*cefF*) from *S. clavuligerus* have also been cloned and separately expressed in *E. coli*.[138,141] A small amount of DACS activity was found with expression of DAOCS, and vice versa, indicative of a close evolutionary relationship.[141] Extensive homology was noted between the *cefE*, *cefF* and *cefEF* genes. In particular, sequence homology (67% at the DNA level and 57% at the protein level) was observed between the *C. acremonium* DAOCS/DACS and the *S. clavuligerus* DAOCS, consistent with a proposal that there was a horizontal transfer of the entire pathway from bacteria to fungi.[138] Homology between the sequences obtained for IPNS isozymes and those of DAOCS/DACS, DAOCS and DACS has also been observed.[137,142]

In view of the similarity in the chemistry carried out by IPNS and the enzymes of cephalosporin biosynthesis, it is tempting to speculate on the nature of a common precursor. However, the enzymes catalysing the ring expansion and hydroxylation processes are also part of a larger class of oxygenases that utilise α-ketoglutarate, of which prolyl hydroxylase and γ-butyrobetaine hydroxylase[143] are the most well studied. Several other members of this family, for example thymine-7-hydroxylase, also share—in common with DAOCS/DACS from *C. acremonium*—the ability to catalyse sequential oxidations.[144] Despite many years of investigation, the mechanism of these oxygenases remains obscure, and no structural studies have been reported. The cloning and high-level expression of the cephalosporin biosynthesis enzymes thus represent an exciting opportunity for the study of this family of oxygenases. In addition, the sequence homology observed between the IPNS isozymes and the α-ketoglutarate-dependent oxygenases associated with cephalosporin biosynthesis strongly suggest that the latter are part of a larger class of non-haem ferrous-dependent oxygenases. Such a proposal gains further support from the recent observation that the ethylene-forming enzyme, which does not utilise α-ketoglutarate, shows homology to flavanone-3-hydroxylase, which does.[145]

The stereochemistry of the ring expansion process has been examined by the feeding of valines, stereospecifically labelled with ^{13}C in the isopropyl group.[146,147] When (3*R*)-[2-^{13}C]valine was fed to *C. acremonium*, it was found that [2-^{13}C]cephalosporin C (**6**) was produced,[147] whereas the addition of (3*S*)-[2-^{13}C]valine gave [3-$HO^{13}CH_2$]cephalosporin C (**6**) and [2-α-$^{13}CH_3$]penicillin N (**3**). Subsequent *in vitro* studies, using a stereospecifically labelled penicillin N (**3**), have confirmed the deduction that the β-methyl group of penicillin N (**3**) is incorporated into the endocyclic methylene group of DAOC (**4**).[148] (Scheme 1.28)

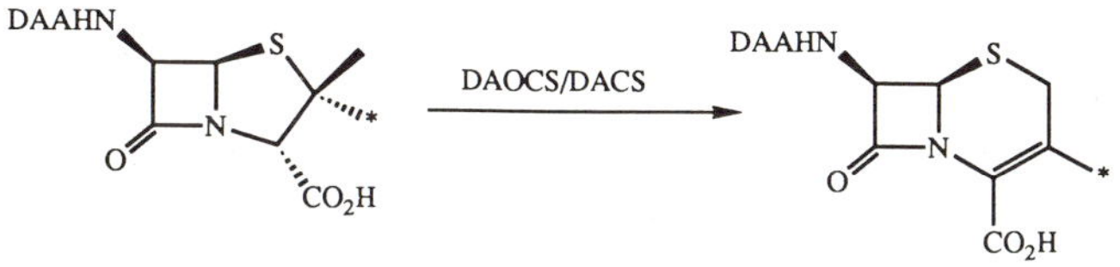

DAOCS/DACS

(**3**) (**4**)

Scheme 1.28

These findings were of interest to chemists, since a chemical process for the ring expansion of penicillins to cephalosporins had been previously invented by Morin *et al.*[149] In this process a penicillin β-sulphoxide is converted into a cephalosporin via a Pummerer-type process, possibly involving an intermediate episulphonium ion. Inspired by this chemical analogy, a number of possible intermediates were envisaged for the biological ring expansion process. However, neither the α- (**65**) nor the β- (**64**) sulphoxides of penicillin N were found to be substrates for DAOCS/DACS.[130] The β-methylenehydroxy penam (**66**), which had also been proposed as an intermediate in the biotransformation[150] (a conversion for which there is also synthetic precedent[151]) was also synthesised.[131,152] No conversion to penicillinase-resistant material was observed; however, under the incubation conditions, the β-methylenehydroxy penam (**66**) was found

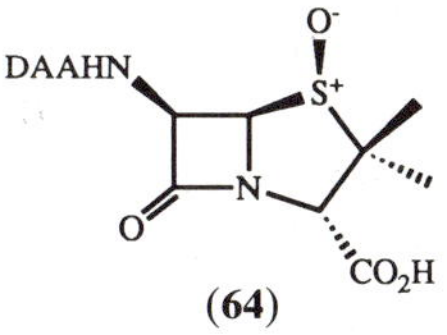

(**64**) 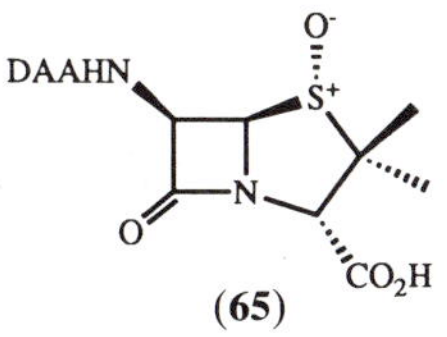 (**65**) 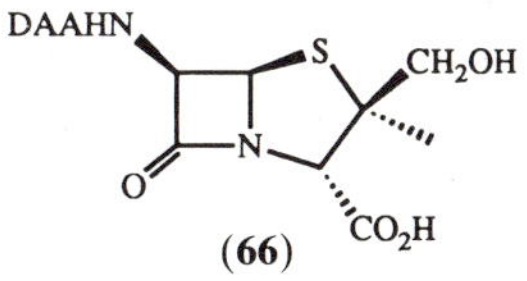 (**66**)

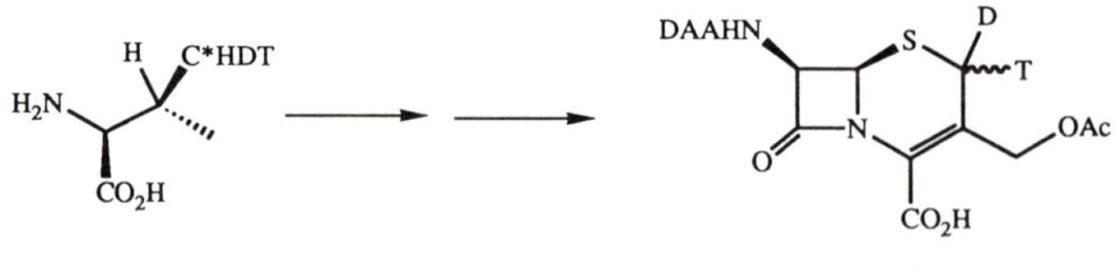

Scheme 1.29

to be highly unstable, probably ruling it out as an enzyme-free intermediate.[131]

Townsend and coworkers,[153,154] and Abraham and coworkers[155] have investigated the stereochemical outcome of the ring expansion process, by feeding valine bearing a chiral methyl group in the (3-*pro-R*) position. The independent conclusion of both groups was that the incorporation of the (3-*pro-R*) methyl group into the endocyclic methylene group of cephalosporin C (**6**) occurs with complete randomisation of stereochemistry (Scheme 1.29). In contrast, the conversion of the (3-*pro-S*)methyl group of valine into the exocyclic methylene group of cephalosporin C (**6**) was shown to occur with retention of stereochemistry. (Scheme 1.30).[154,156]

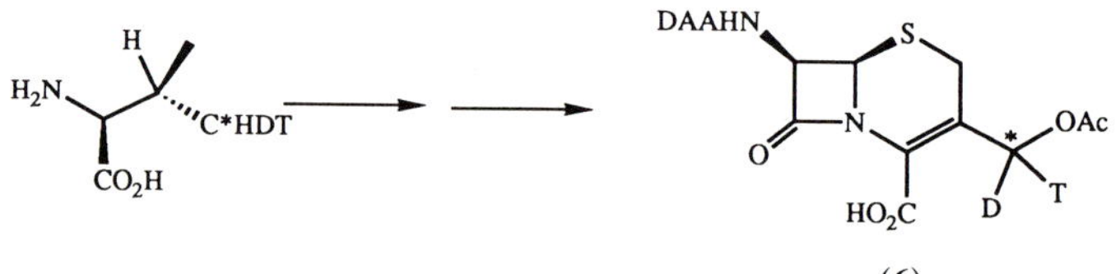

Scheme 1.30

The observed loss of stereochemistry in the ring expansion step prompted the suggestion that the 3-β-methyl radical (**67**), which can undergo rotation before ring expansion, is an intermediate. The chemical feasibility of a ring expansion mechanism via a radical process has been tested.[157,158] Thus, reductive debromination of the β-bromomethylene penam (**68**) under radical chain conditions (Ph_3SnH/azobisisobutyronitrile) gave a mixture of cephams (**69**) (Scheme 1.31, path a). The same ratio of cephams was also produced by reaction of the β-bromo cepham (**70**) under radical chain conditions (Scheme 1.31, path b). When Ph_3SnH was replaced with an allyl stannane the S-alkylated material (**71**) was obtained (Scheme 1.31, path c). These results were rationalised by invoking an equilibrating mixture of isomeric radicals, and the radical ring expansion was thus proposed as a biomimetic synthesis.[157,158]

The conversion of penicillin N (**3**) to DAOC (**4**) involves the loss of one hydrogen from the β-methyl group and one from the C-3 position of penicillin N (**3**). In order to determine the order of the hydrogen losses, V_{max}/K_m

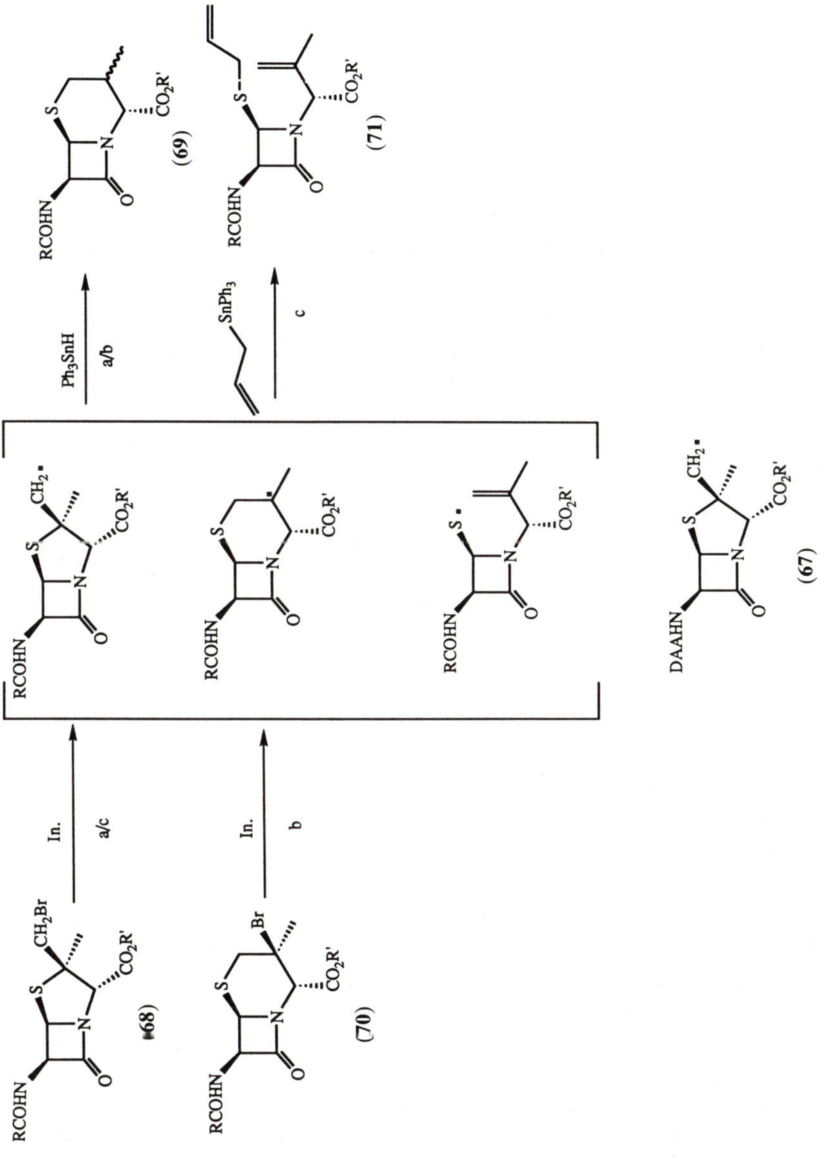

Scheme 1.31

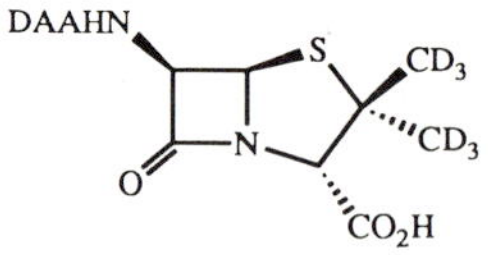

DAAHN S CD_3 N CD_3 O CO_2H

(72)

DAAHN S N O D CO_2H

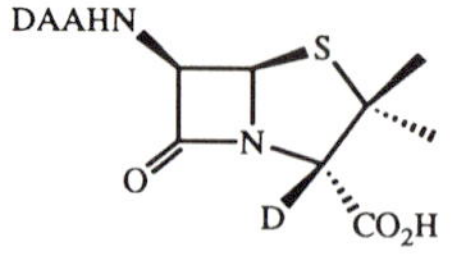

(73)

isotope effects have been conducted. The requisite, specifically deuterated penicillin N derivatives, (**72**) and (**73**), were synthesised *in vitro*, from the corresponding D,L,D-tripeptides using IPNS. The competitive (V_{max}/K_m) isotope effects for 1 : 1 mixtures of (**72**) : (**3**) and (**73**) : (**3**) were then determined. The ratios of deuterated to protiated penicillin N in the incubation were determined at various time points. A competitive isotope effect was observed only on the methyl-labelled species.[159]

During these investigations the formation of a minor product was observed in crude ^1H nmr spectra, acquired during the study of the ring expansion of penicillin N (**3**) to DAOC (**4**). On incubation of the C-3 deuterated penicillin N (**73**), this minor component became a major product, which was readily isolated and shown to be the 3-β-hydroxy cepham (**74**)[160] (Scheme 1.32). This compound had been previously isolated from a filtered broth of *C. acremonium*, prompting speculation as to its intermediacy in the ring expansion process.[161] Clearly, the bias in the ratio of products produced was due to the operation of a primary isotope effect.[160] The origin of the hydroxyl group was shown to be partly derived from molecular oxygen by the appropriate labelling experiment. Further studies demonstrated that the stereochemistry of the ring expansion of the deuterated penicillin N (**73**) to produce the hydroxy cepham (**74**) was analogous to that observed in the conversion of penicillin N (**3**) to DAOC (**4**) (Scheme 1.32).[148]

The V_{max}/K_m studies, coupled with the observation of the hydroxy cepham

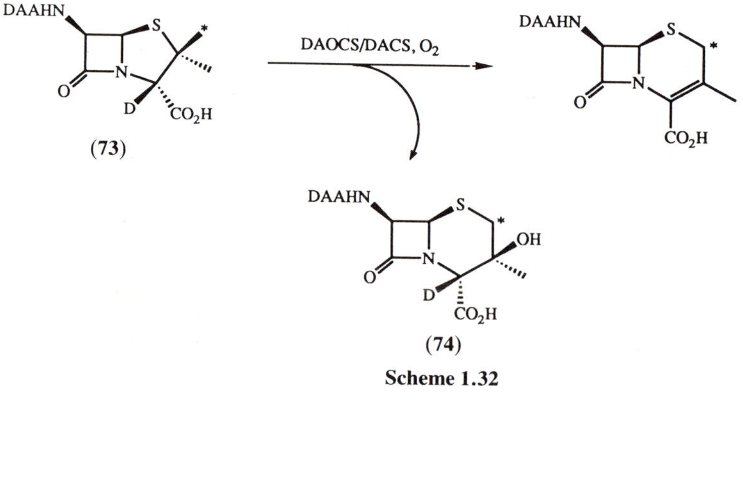

DAAHN S * N O D CO_2H

(73)

DAOCS/DACS, O_2

DAAHN S * N O CO_2H

DAAHN S * OH N O D CO_2H

(74)

Scheme 1.32

(**74**), can be best rationalised by a sequence of events in which there is an initial irreversible loss of the 2β-methyl hydrogen, prior to that at the C-3 position of penicillin N (**3**). Furthermore, the bias in the pathway towards production of the hydroxy cepham (**74**), in the case of the deuterated material (**73**), indicates a common intermediate prior to the branch point. Possible structures for this intermediate are the bridged cation or radical (**75**) (Scheme 1.33).

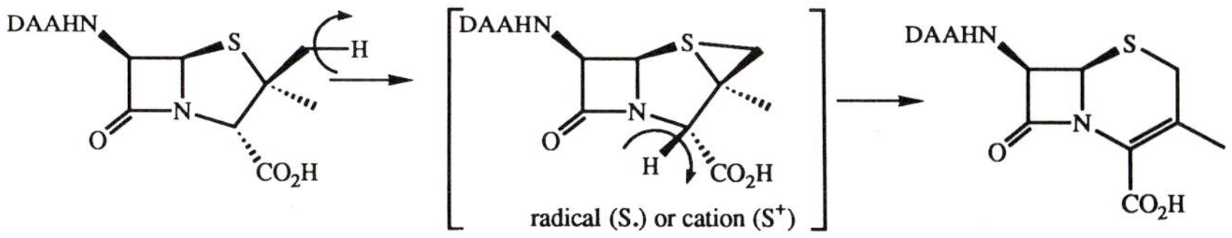

radical (S.) or cation (S^+)
(**75**)

Scheme 1.33

The stoichiometry of the consumption of α-ketoglutarate and dioxygen has been examined, for both the ring expansion and hydroxylation steps for DAOCS/DACS from *C. acremonium*. In the presence of saturating β-lactam substrate (either (**3**) or (**4**)) each step was found to require one equivalent of α-ketoglutarate and dioxygen.[130,162] Using specifically labelled [2-^{13}C]-α-ketoglutarate (**76**),[163] it was shown that succinate is produced stoichiometrically with the consumption of α-ketoglutarate.[162] Furthermore, incubations of the [2-^{13}C]-α-ketoglutarate under an atmosphere of $^{18}O_2$ demonstrated that the incorporation into succinate was regiospecific and greater than 95%. In the same incubation the level of incorporation into DAC (**5**) was found to be only *ca.* 50%, indicating an exchange of ^{18}O for ^{16}O during the catalytic process (Scheme 1.34).

In contrast to a prior report,[126] it has been found that in the absence of β-lactam substrate there is still some 'uncoupled' turnover of α-ketoglutarate to succinate and CO_2. This uncoupled turnover was found to be stoichiometric with the consumption of ascorbate in the incubation mixture, and to occur at *ca.* 5% of the rate of the 'coupled' turnover.[164] A similar uncoupled turnover of α-ketoglutarate has been observed in the case of other α-ketoglutarate-dependent oxygenases.[165]

The success of synthetic substrate analogues as mechanistic and active site probes with IPNS has prompted a similar approach with DAOCS/DACS. The side chain specificity of the purified enzyme has been studied;[166,167] penicillins in which there is an adipoyl (**77**) or *m*-carboxyphenylacetyl (**78**) side chain underwent ring expansion to the corresponding cephems, (**79**) and (**80**) respectively, at about half the rate of penicillin N (**3**). A lower rate was observed when the side chain was D-glutamyl and a much lower rate when it was D-glutaryl. No conversion into cephalosporin was

Scheme 1.34

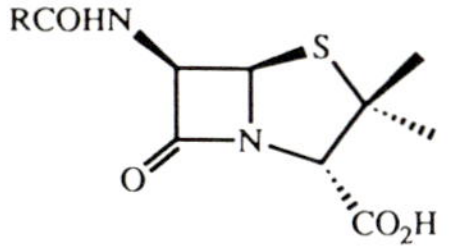

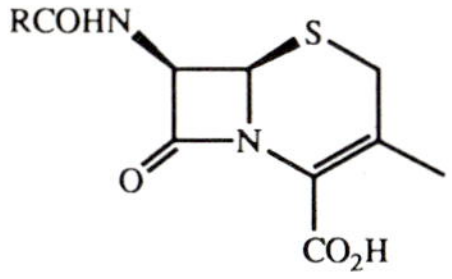

(77) R = HO_2C

(78) R =

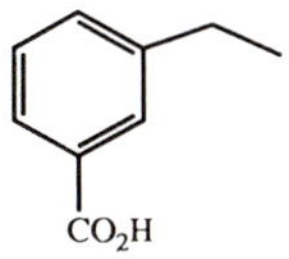

(79) R = HO_2C

(80) R =

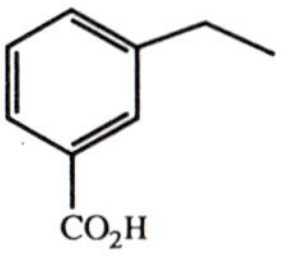

observed when the penicillin side chain was δ-(L-α-aminoadipoyl) (**1**), 5-aminopentanoyl, phenylacetyl (**8**), or phenoxyacetyl; neither was 6-APA (**7**) converted into 7-aminodeacetoxycephalosporanic acid.[166,167] The D-(5-carboxymethylcysteine) side-chained penicillin (**63**) has also been reported as an efficient substrate for DAOCS/DACS.[111] These results indicated a similar requirement for DAOCS/DACS to that previously observed for IPNS. However, IPNS utilises tripeptides with both L- and D-N-terminal δ-α-aminoadipoyl side chains, whereas DAOCS/DACS accepts only the latter.

Several modified penicillins have been reported as substrates for both the fungal DAOCS/DACS and the bacterial DAOCS acivities. Using extracts from *S. clavuligerus* it was reported that tripeptide analogues were converted to cephalosporins.[112] Thus the tripeptides containing D-amino-butyrate and D-*allo*-isoleucine in place of valine were shown to give penicillinase resistant products.[112,168] Using recombinant DAOCS/DACS, the pure β-methyl penam (**13**) was shown to give the cephem (**81**)[169] (Scheme 1.35).

The incubation of modified cephams and cephems has been more interesting. The ethyl cephem (**82**) was converted through to a single hydroxylated product (**83**), as yet of undefined stereochemistry (Scheme 1.36).[170] Several unsaturated analogues have also been tested — the Δ^2-isomer (**84**) of DAOC (**4**) was hydroxylated, to give the hydroxylated Δ^2-

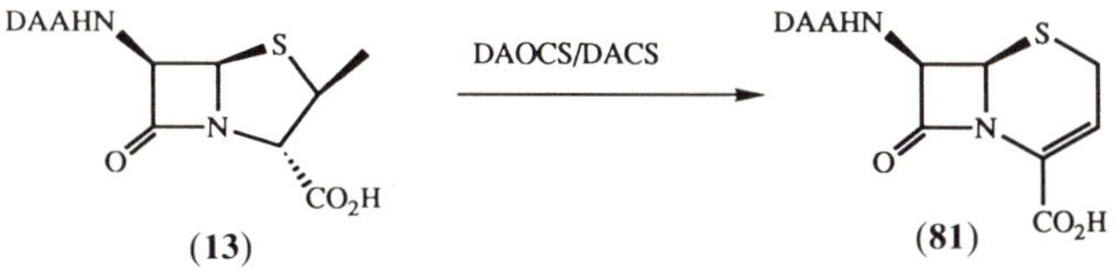

DAAHN — DAOCS/DACS → DAAHN

(13) (81)

Scheme 1.35

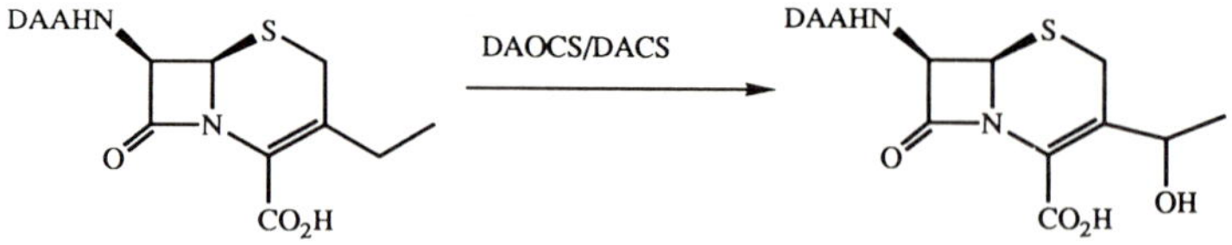

DAOCS/DACS

(82) (83)

Scheme 1.36

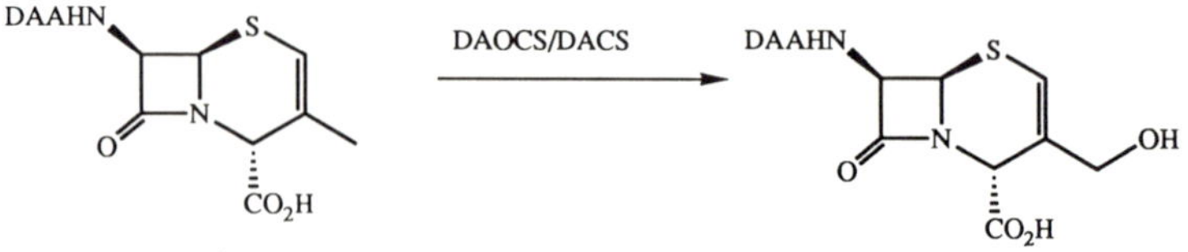

DAOCS/DACS

(84) (85)

Scheme 1.37

cephem (**85**) (Scheme 1.37).[139] Incubations of the exomethylene cepham (**86**) with DAOCS/DACS produced DAC (**5**) (Scheme 1.38). DAOC (**4**) was not found to be an intermediate and the hydroxyl group of the product DAC (**5**), derived from (**86**), was shown to be partially derived from dioxygen.[171,172] The possibility of an epoxide or related intermediate is currently under investigation.[173] The cyclopropyl-containing substrates (**88**), (**89**) and (**90**) have been recently synthesised and tested as mechanistic probes. In the cases of (**89**) and (**90**), no new β-lactam products were detected on incubation with DAOCS/DACS. However, in the case of (**88**), conversion to the alcohol (**91**) was observed (Scheme 1.39). Again, incubation under an atmosphere of $^{18}O_2$ showed that the oxygen of the hydroxyl group was partially derived from dioxygen.[174]

It will be apparent from the mechanistic studies, thus far carried out, that

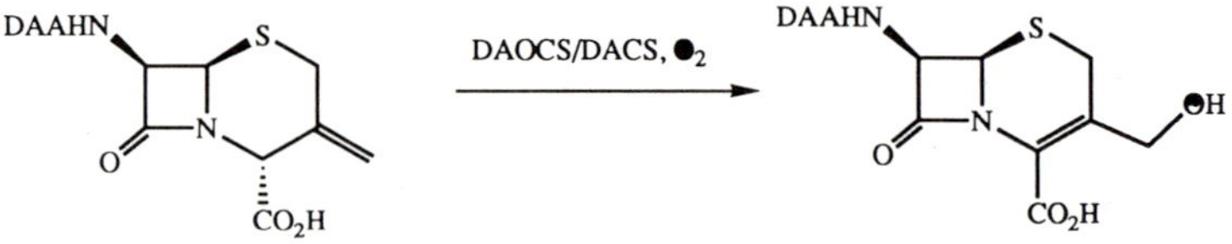

DAOCS/DACS, ●$_2$

(86) (87)

Scheme 1.38

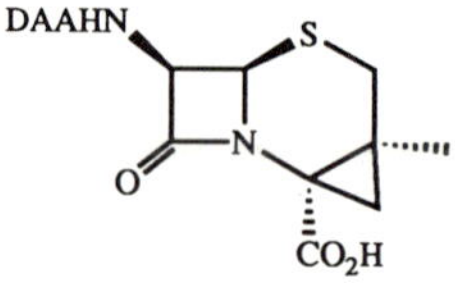

(89) 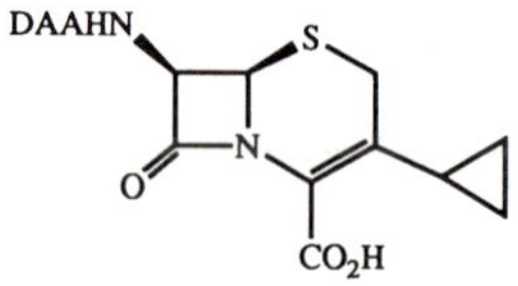 (90)

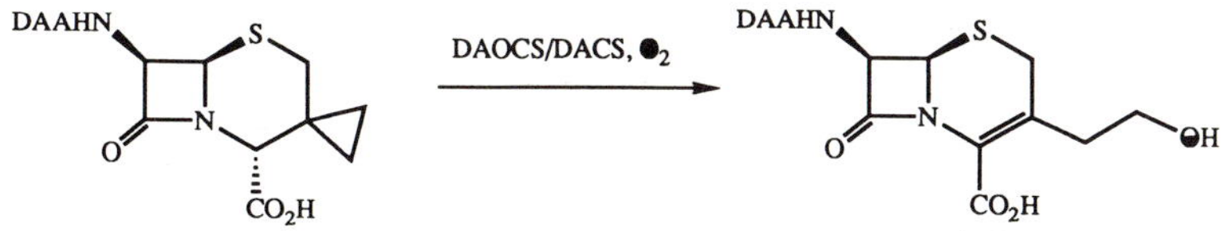

DAOCS/DACS, $\bullet_2$

(88) (91)

Scheme 1.39

the chemistry of DAOCS/DACS bears a strong similarity to that of IPNS. A major difference is in the stoichiometry of dioxygen utilisation by the two enzymes. For IPNS, an overall four electron oxidation of ACV (**2**) is achieved, with conversion of one molecule of dioxygen into water. In the case of DAOCS/DACS, for each consecutive step, only a two electron oxidation of the β-lactam substrate is achieved from each molecule of dioxygen used up, since, in each step, one molecule of α-ketoglutorate is transformed into carbon dioxide and succinate. DAOCS/DACS, like other α-ketoglutarate-dependent oxygenases, utilises only the ferryl state in its attack on the substrate, having produced this by initial splitting of α-keto-glutarate and dioxygen (Scheme 1.40).[175]

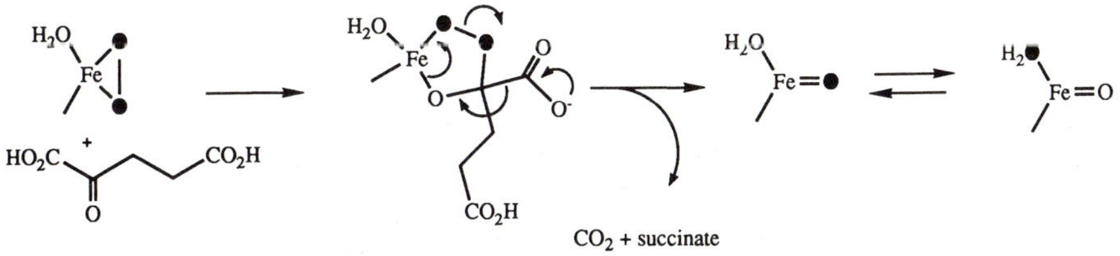

CO_2 + succinate

Scheme 1.40

It is envisaged that for penicillin N (**3**), the reactive ferryl species inserts into a C–H bond of the 2-β-methyl group, to form a weak C–Fe bond (**92**), which dissociates to form the radical (**93**); this can then undergo rearrangement followed by β-elimination to form DAOC(**4**) (Scheme 1.41). The hydroxy cepham (**74**) results from a reductive elimination process occurring after the rearrangement. The functionalisation of the methyl group of DAOC (**4**) to give DAC (**5**) then results from a second attack of a regenerated ferryl species on this centre, followed by reductive elimination, before a dissociation radical can occur (Scheme 1.42).

In the case of the cyclopropyl analogue (**88**), it is envisaged that the insertion of the ferryl species into the C–H bond at C-4 may result in the formation of a radical (**94**), which can undergo cyclopropyl ring opening, followed by recombination and reductive elimination to give the alcohol (**91**) (Scheme 1.43). Alternatively, an ene-type reaction may give intermediate (**95**) directly.

Scheme 1.41

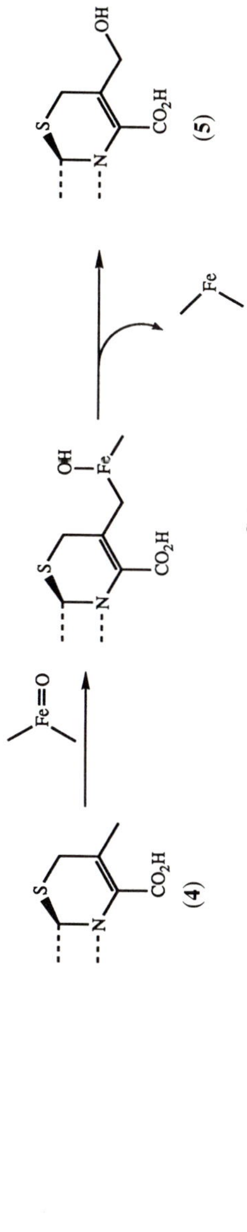

Scheme 1.42

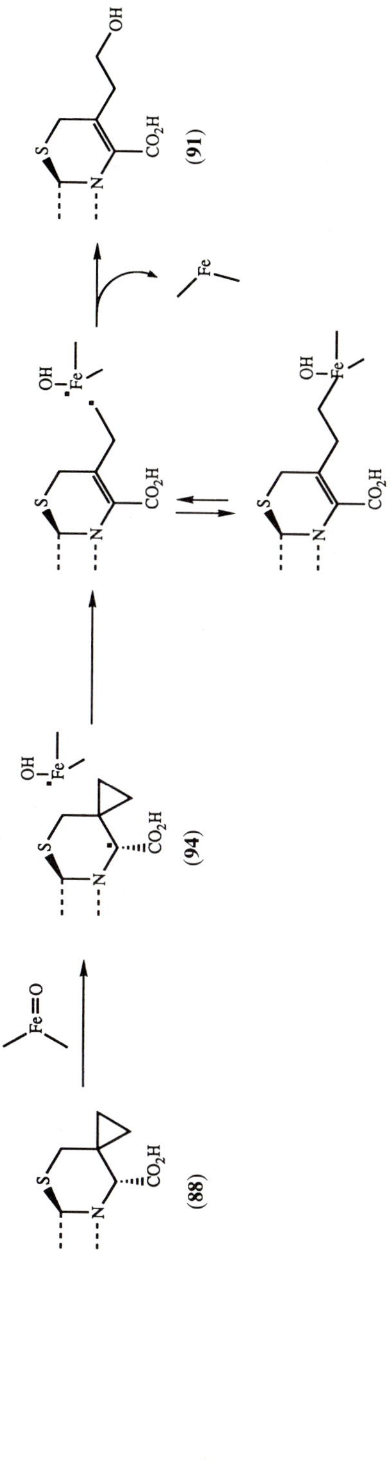

Scheme 1.43

1.2.6 Acetylation and carbamoylation of deacetylcephalosporin C

Many naturally occurring cephems result from apparent functionalisation of the hydroxyl group of DAC (**5**). It is beyond the scope of this review to speculate on the biological processes leading to all of them, but the two specific examples for which *in vitro* studies have been carried out will be briefly discussed.

The final step in cephalosporin C (**6**) biosynthesis is the acetylation of DAC (**5**). Cell-free extracts containing acetyl coenzyme A: DAC *o*-acetyl-transferase have been prepared (Scheme 1.44, path a). For optimal conversion, a pH of 6.5 and Mg^{2+} ions were required.[176] The crude activity was found to have specificity for a 3-hydroxymethyl cephem, whilst the presence or absence of an α-aminoadipoyl side chain apparently had little effect on the rate of reaction. The activity has been further purified by Liersch *et al.*,[177] however, no sequence information has yet been reported.

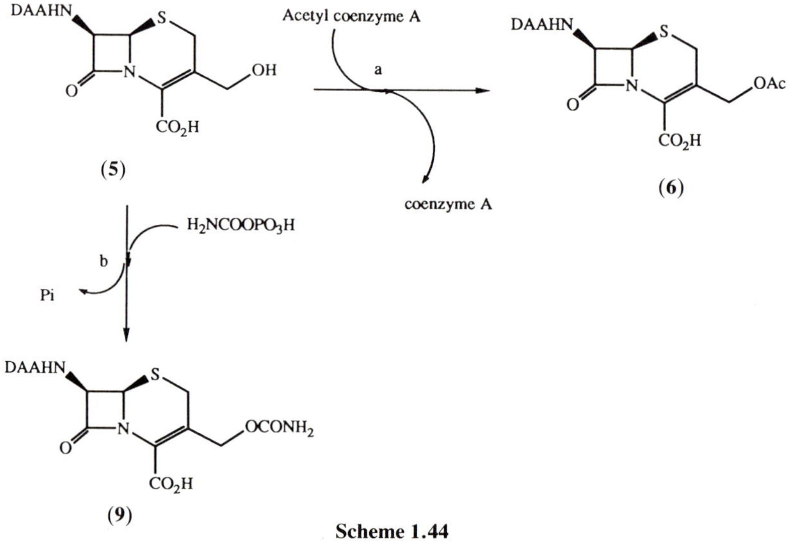

DAAHN S N O OH CO_2H (**5**)
Acetyl coenzyme A a coenzyme A
DAAHN S N O OAc CO_2H (**6**)
$H_2NCOOPO_3H$ b Pi
DAAHN S N O $OCONH_2$ CO_2H (**9**)

Scheme 1.44

The hydrolysis of cephalosporin C (**6**) to DAC (**5**) by an extracellular exterase has been reported in both eukaryotic[178,179] and prokaryotic organisms.[180] The enzyme from *C. acremonium* is reportedly produced in the stationary phase of growth and is repressible by glucose and other carbohydrates.[180]

It is worthwhile noting in this section that the aldehyde (**96**), together with DAC (**5**), was found to accumulate in the broth of a *C. acremonium* mutant blocked in the acetyl transfer step.[176] It was suggested that the aldehyde (**96**) may result from the 'over oxidation' of DAC (**5**) by the action of DAOCS/DACS followed by hydrolysis of the β-lactam ring.

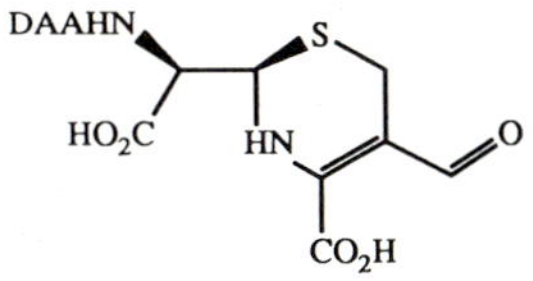

(96)

O-Carbamoyldeacetylcephalosporin C (**9**) is produced by some strains of *S. clavuligerus*.[181] Whereas whole cells were ineffective in accomplishing the conversion of (**5**) to (**9**),[182] a partial purification of an intracellular *o*-carbamoyltransferase has been reported by Brewer *et al.*[183] (Scheme 1.44, path b). The purified enzyme required carbamoyl phosphate as a co-substrate and ATP, Mn^{2+} and Mg^{2+} for optimal activity. Substrate analogue studies demonstrated that the carbamoyl transferase will catalyse the transfer of a carbamoyl group to a range of 3-hydroxymethyl cephems, including 7-α-methoxycephems. Cephalosporins with neutral side chains at the 7-position were reportedly the best substrates. Whether or not ATP actually activates the substrate is at present not clear. However, the semi-purified extracts did catalyse the conversion of ATP to ADP and AMP.[183]

1.2.7 7-α-Functionalised cephalosporins

In addition to the cephamycins,[181,184,185] which contain a 7-α-methoxy group, several cephalosporins functionalised with a 7-α-formamide group (**97**) have also been reported.[186] Although the biological origins of the latter remain obscure, O'Sullivan, Abraham and coworkers have investigated the biogenesis of the 7-α-methoxyl substituted cephems, exemplified by cephamycin C (**10**).

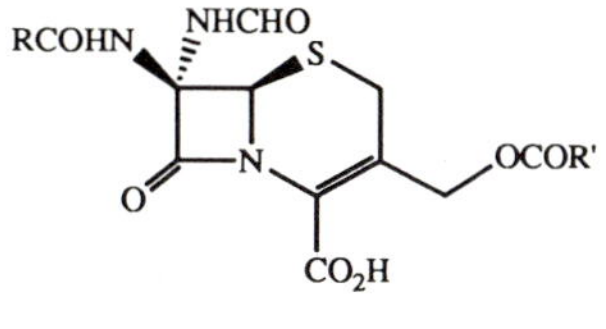

(97)

By feeding L-[methyl-$^3H_3$]methionine it was demonstrated that the methyl group of (**10**) is derived from methionine.[187] Growth of whole cells of *S. clavuligerus* under an atmosphere of $^{18}O_2$ showed that the oxygen of the methoxyl group of (**10**) is derived, at least in part, from dioxygen.[187] These workers went on to describe cell-free extracts from *S. clavuligerus* that catalysed the efficient conversion of *o*-carbamoyl DAC (**9**) to cephamycin

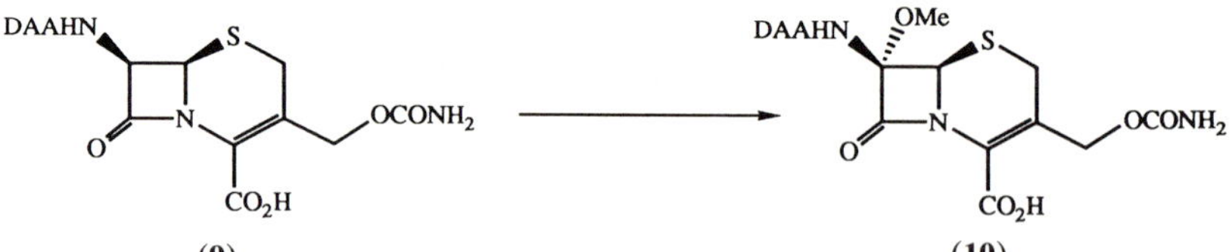

(9) (10)

Scheme 1.45

(**10**) in the presence of Fe^{2+}, α-ketoglutarate and S-adenosylmethionine, (Scheme 1.45).[187] DAOC (**4**) was found to be a relatively poor substrate for the cell-free extracts and no methoxylation of DAC (**5**) was observed. These observations would imply that the major *in vivo* route to cephamycin C (**10**) is DAOC (**4**) $\rightarrow$ DAC (**5**) $\rightarrow$ *o*-carbamoyl DAC (**9**) $\rightarrow$ 7-α-hydroxy-*o*-carbamoyl DAC (**98**) $\rightarrow$ cephamycin C (**10**) (Scheme 1.1), although other possibilities cannot be ruled out. It is obviously of great interest to obtain sequence data of the 7-α-hydroxylase for comparison with the other oxygenase enzymes of the pathway.

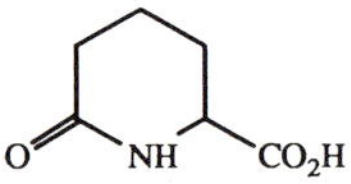

(99) 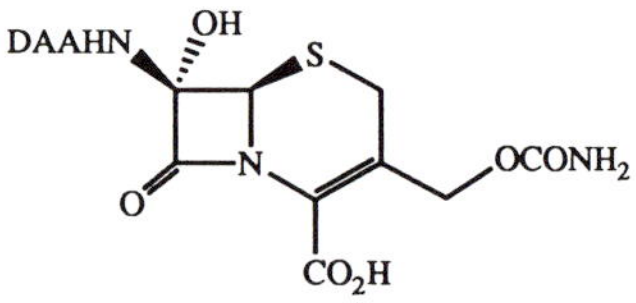 (98)

1.2.8 The biosynthesis of penicillins with hydrophobic side chains

The multiplicity of penicillins produced with different side chains, by *P. chrysogenum*, was one of the first indications that an acylation step may be involved in their biosynthesis.[188] All the hydrophobic penicillins produced have side chains that correspond to monosubstituted acetic acids that exist at low intracellular concentrations in *P. chrysogenum*. In fact, the range of organisms known to produce 'penicillium-type' penicillins is not limited to this species; *Aspergillus* and *Trichophyton* spp. are also known to produce significant quantities of hydrophobic penicillins.[189]

With the acceptance of Arnstein's theory that the tripeptide ACV (**2**) was an obligate intermediate in the biosynthesis of all penicillins and cephalosporins, Demain[190] suggested that the hydrophobic penicillins arose from acylation of 6-APA (**7**), produced by side chain hydrolysis of isopenicillin N (**1**). In 1967, Pruess and Johnson[191] demonstrated that a cell-free extract of *P. chrysogenum* could catalyse transfer of the acyl group of various penicillins, including penicillin G (**8**), to [^{35}S]-6-APA (**7**). Spencer confirmed these results and suggested that the acyl group was introduced via the intermediacy of an acyl-coenzyme A thioester.[192]

Brunner *et al.*[193] were the first to demonstrate the direct transfer of the acyl group from phenylacetyl-coenzyme A and phenoxyacetyl-coenzyme A to 6-APA (**7**). Moreover, the crude extracts used in these experiments also contained a coenzyme A ligase activity, which could catalyse the formation of the coenzyme A thioesters of phenyl and phenoxyacetic acids from the appropriate acid, ATP and coenzyme A. It was also shown that partially purified enzyme would accept alternative side chains, such as *p*-methoxyphenylacetyl and octanyl, via their coenzyme A thioesters.[194] Further to these studies, Spencer and Maung[195] communicated descriptions of a purified preparation from *P. chrysogenum* which contained four different catalytic activities: (i) acyl-coenzyme A : 6-APA acyltransferase; (ii) hydrophobic penicillin amidohydrolase; (iii) exchange of [^{35}S] label between hydrophobic penicillins and 6-APA (**7**); and (iv) phenylacetyl-coenzyme A hydrolase.

The first substantive evidence for the hitherto elusive activity capable of converting isopenicillin N (**1**) to penicillin G (**8**) was described by Loder in 1972.[196] A crude extract of *P. chrysogenum* was shown to catalyse the incorporation of ^{14}C from [^{14}C]-phenylacetyl-coenzyme A into penicillin G (**8**), in the presence of isopenicillin N (**1**). In confirmation of these results, Abraham and coworkers[197,198] showed that when isopenicillin N (**1**), labelled with ^{3}H in the 3-β-methyl group, was incubated with extracts of *P. chrysogenum* in the presence of phenylacetyl-coenzyme A, [^{3}H]-penicillin G (**8**) was produced. These workers went on to show that, in the absence of phenylacetyl-coenzyme A, when cell-free extracts from *P. chrysogenum* were incubated with isopenicillin N (**1**), 6-APA (**7**) was produced.[199,200] The soluble fractions that catalysed the acyl-coenzyme A : isopenicillin N (**1**) acyl-transfer and isopenicillin N (**1**) amidohydrolase activities were also shown to catalyse acyl-coenzyme A : 6-APA (**7**) acyl transfer with approximately equal efficiency. About this time, the observation that 6-oxopiperidine-2-carboxylic acid (**99**) was produced by *P. chrysogenum* concomitantly with penicillin V (**18**) was taken as further evidence that an intermediate with an α-aminoadipoyl side chain was crucial to the biosynthesis of hydrophobic penicillins.[201]

The first researchers to report purification of the acyl-coenzyme A : 6-APA acyl transferase to a high degree of purity were Kogekar and Deshpande.[202] However, no molecular weight was attributed to the enzyme. These workers also described the isolation of an ATP-dependent monosubstituted acetic acid coenzyme A ligase from *P. chrysogenum*.[203] Subsequent to this Luengo, Revilla and coworkers[204–206] achieved a partial purificiation of the acyl-coenzyme A : 6-APA acyl transferase and described optimal fermentation conditions for its production. The enzyme was reported to be inhibited by Hg^{2+}, Zn^{2+} and N-ethylmaleimide, and to require dithioerythreitol for activity. On the basis of these observations, it was speculated that thiol groups are present at the active site.[207] The minimal and maximal chain lengths of the monosubstituted straight chain acyl-coenzyme A thioesters that were accepted by the enzyme were determined; hexanoyl-coenzyme A and decanoyl-coenzyme A were found to be the extremes allowed, with octanoyl-coenzyme A being the most efficient substrate.[208] The *in vitro* constraints are consistent with the range of penicillins isolated from *Penicilium* fermentations. These studies have been recently extended, and have resulted in the semisynthesis of a number of penicillins from 6-APA (**7**) and the appropriate acyl-coenzyme A.[209] In each case, co-substrates corresponding to a known precursable penicillin were found to be substrates for the acyl-coenzyme A : 6-APA acyl transferase enzyme; conversely, those coenzyme A thioesters corresponding to non-precursable penicillins were not substrates. No mention was given as to whether or not isopenicillin N (**1**) was a substrate for the enzyme. In a different report, Alvarez *et al.*[210] stated that the acyltransferase is a single polypeptide migrating with a relative molecular mass of 29 kDa on SDS-PAGE. In addition, isopenicillin N (**1**) was reported to be a substrate for this single component preparation, whilst hydrolysis of the phenylacetyl side chain of penicillin G (**8**) was not catalysed by the enzyme. This report constituted the first substantive indication that the acyl-coenzyme A : isopenicillin and 6-PA acyltransferases are one and the same protein.

Whiteman *et al.*[211] also reported the purification of acyltransferases from both *P. chrysogenum* and *A. nidulans* to apparent homogeneity. HPLC gel permeation chromatography in the presence of 30% ethylene glycol gave a molecular mass of 40 kDa. However, on SDS-PAGE, two bands (with molecular masses of 30 and 10 kDa) were observed. With the corresponding enzyme from *Aspergillus*, two similar bands (28 and 10.5 kDa) were also observed. In the earlier report of Alvarez *et al.*[210] it is possible that the lower molecular mass band was either not observed or had been purified away. In addition to the transfer of the phenylacetyl group from phenylacetyl-coenzyme A to 6-APA (**7**) the enzyme was also found to catalyse the phenylacetyl coenzyme A : isopenicillin (**8**) acyl transfer, and the hydrolysis of isopenicillin N (**1**) to 6-APA (**7**). In the presence of SDS the hetero-dimeric complex was found to dissociate, with loss of activity, but the phenylacetyl-coenzyme A : 6-APA acyltransferase activity was restored on recombination of the fragments in the absence of SDS.[211] The purification of a penicillin V (**18**) amidohydrolase was also reported by Whiteman *et al.*[211]

Based on N-terminal amino acid sequence information,[210,211] the acyl-transferases genes (*penDE*) from *P. chrysogenum*[212,213] and *A. nidulans*[213] have been cloned. Barredo *et al.*[212] were the first to report the complementation of mutants of *P. chrysogenum*, deficient in acyltransferase activity, with the *penDE* gene. Tobin *et al.*[213] have expressed the *penDE* genes from both *Aspergillus* and *Penicillium* in *E. coli*. The expressed proteins in *E. coli* were shown to have both phenylacetyl-coenzyme A : 6-APA and phenylacetyl-coenzyme A : isopenicillin acyltransferase activities. Barredo *et al.* reported that three proteins of 40, 11 and 29 kDa, were identified in the complemented *P. chrysogenum* mutants.[212] For both *Penicillium* and *Aspergillus* the *penDE* gene was located downstream from the IPNS gene (*pbcC*) and consisted of four exons encoding a protein of 357 amino acids[212,213] (*ca.* 40 kDa) and three introns. The DNA and amino acid sequences of the two genes were found to be well conserved (*ca.* 75% similarity). Thus, the *penDE* gene is the first β-lactam biosynthesis gene found to contain introns. N-terminal amino acid sequences of the reported 30 kDa and 11.5 kDa proteins from *P. chrysogenum* and *A. nidulans*[211,213] were correlated with sequences encoded within the *penDE* genes. The cleavage point of the smaller and larger proteins was assigned as theonine–glycine or glycine–cysteine by Barredo *et al.*[212] and Tobin *et al.*[213] respectively. The mechanism of processing (protease or autocatalytic) of the 40 kDa protein into the heterodimer, and indeed of the acyltransferase and amidohydrolase, remains to be investigated.

Studies into the transport system of phenylacetic acid in *P. chrysogenum,*[214–216] have been initiated and have shown that it is an inducible system, regulated by the carbon and nitrogen sources used for growth.

1.3 Clavulanic acid biosynthesis

Clavulanic acid (**100**) is a naturally occurring, bicyclic β-lactam which, although it only has weak antibacterial activity, is a potent inhibitor of β-lactamases from both gram-positive and gram-negative bacteria.[217] There is an obvious structural analogy between clavulanic acid (**100**) and the penams, in that they each contain a strained and highly reactive, fused β-lactam ring system. Despite this similarity, it has now been shown that they are not derived from common precursors, although it is known that the biogenesis of the highly functionalised ring system of clavulanic acid (**100**) is mediated, at least in part, by an α-ketoglutarate-dependent, non-haem oxygenase. This indicates at least some mechanistic similarity in the use of oxygenase enzymes for the elaboration of commonly available natural precursors for the *in vivo* production of both of these secondary metabolites.

Early whole-cell feeding experiments to *S. clavuligerus* by the Beechams group demonstrated that the 3-carbon skeleton (C-5, C-6, C-7) of clavulanic acid (**100**) is derived from the C_3 pool via glycerol (**101**).[218,219] Subsequently, Townsend and Ho[220] showed that D-glycerate (**102**) was incorporated into (**100**), consistent with its intermediacy between glycerol (**101**) and clavulanic acid (**100**). Gutman *et al.*[221] have reported the specific incorporation of β-hydroxypropionate (**103**) into clavulanic acid (**100**). Townsend *et al.*[222] however, were unable to repeat this work, finding 'only low, non-specific

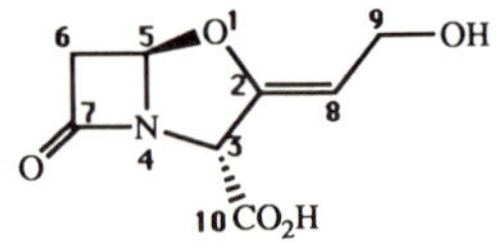

(**100**) Clavulanic acid 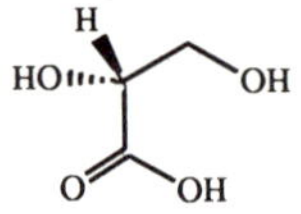 (**102**) 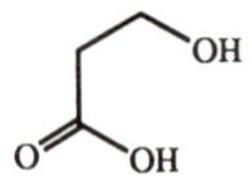 (**103**)

incorporation of radiolabelled β-hydroxypropionate (**103**) into clavulanic acid (**100**)'. A possible explanation for the discrepancy in these results is that the growth medium used by Gutman *et al.* was different from the more usual triglyceride-based media used by Townsend *et al.* and the Beechams group.

Using specifically labelled glycerols it has been shown that the hydrogen at C-5 of (**100**) is derived from a hydrogen attached to a terminal carbon of glycerol.[220,222,223] The stereochemical course of glycerol incorporation was further investigated by the feeding of glycerols, stereospecifically labelled with tritium in the *pro*-1*R* (i.e. 1[^{3}H]*R*, 2*R* glycerol (**101a**)) and *pro*-1*S* (**101b**) positions, which led to the production of (**100**) labelled with tritium, only in the case of the former (Scheme 1.46).[224]

; (**101a**) → (**100**)

(**101**) R = R′ = H
(**101a**) R = H, R′ = ^{3}H
(**101b**) R = ^{3}H, R′ = H

Scheme 1.46

Classical [^{13}C]-acetate labelling experiments indicated that the C-5 skeleton (C-10, C-3, C-2, C-8, C-9) of clavulanic acid (**100**) was derived from the tricarboxylic acid cycle via α-ketoglutarate.[218] These observations led to the search for a more advanced C-5 precursor, and it was subsequently discovered that feeding of racemic [3,4-$^{13}C_2$]-glutamate (**104**) led to the isolation of (**100**) labelled at the C-2 and C-8 positions.[225] However, other carbon centres of (**100**) were also labelled by this precursor, indicating non-specific or indirect incorporation of glutamate (**104**). Incorporation of labelled α-amino-δ-hydroxynorvalines (**105**) into (**100**) has been reported, but the observed levels of incorporations were very low.[225,226] In contrast, Townsend and Ho[226] have reported the high-level incorporation of ornithine (**106**) into (**100**) and suggest that it acts as a direct precursor of the C-5 unit of (**100**). Elegant experiments using ornithines chirally tritiated at C-5 in both *pro*-(*R*) (**106a**) and *pro*-(*S*) (**106b**) forms, demonstrated that the label from the former, but not the latter, was incorporated into (**100**). Further degradation studies showed that the incorporation of the *pro*-*R* form (**106a**)

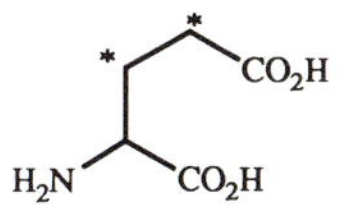

(104) $* = {}^{13}C$

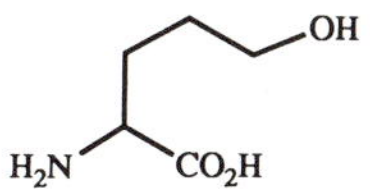

(105)

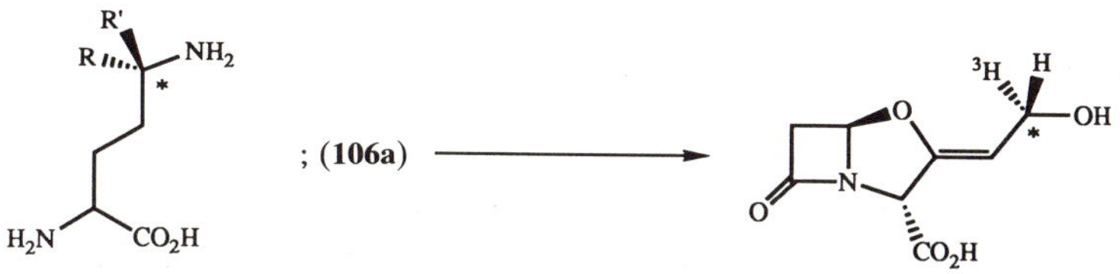

; **(106a)** →

(100)

(106a) $R = H, R' = {}^{3}H, * = {}^{14}C$
(106b) $R = {}^{3}H, R' = H, * = {}^{14}C$

Scheme 1.47

occurred with inversion of stereochemistry (Scheme 1.47). These results suggest that an aldehyde may be an intermediate which, on reduction, could lead to the hydroxymethylene group of **(100)**.[227] In an experiment using racemic [4-H_2; 5-^{13}C]-ornithine **(106c)** it was found that the carbon label was well incorporated into **(100)**, but that there was no detectable incorporation of deuterium at C-8 (Scheme 1.48), also consistent with the intermediacy of an aldehyde, or its equivalent.[228] When cultures of *Streptomyces clavuligerus* were grown under an atmosphere of $^{18}O_2$ gas, ^{18}O was found to be efficiently incorporated into the oxygen of the hydroxyl group, consistent with the involvement of an oxidative deamination process to form an aldehyde, followed by subsequent reduction during the *in vivo* functional group interconversion at the C-5 of ornithine. In the same experiment the oxygen of the oxazolidine ring was also shown to be derived from molecular oxygen.[229]

The next significant advance in elucidating the biological pathway to clavulanic acid **(100)** came when it was discovered that crude, cell-free extracts of *S. clavuligerus* were capable of producing, when aerobically incubated in the presence of α-ketoglutarate and ferrous iron, a new bicyclic β-lactam, clavaminic acid **(107)**.[230] Remarkably, this acid was shown to have

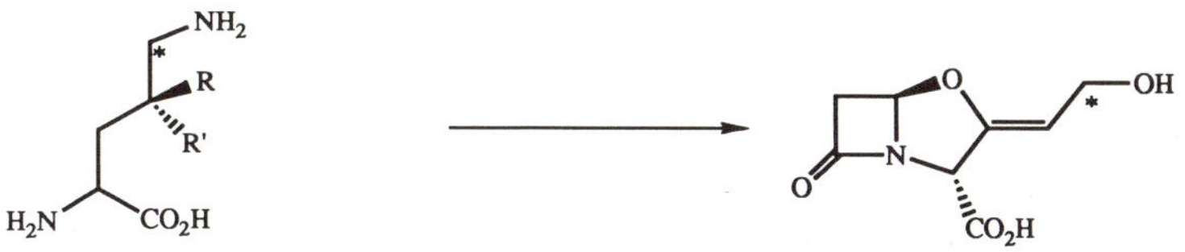

(106c) $R = R' = {}^{2}H, * = {}^{13}C$

(100)

Scheme 1.48

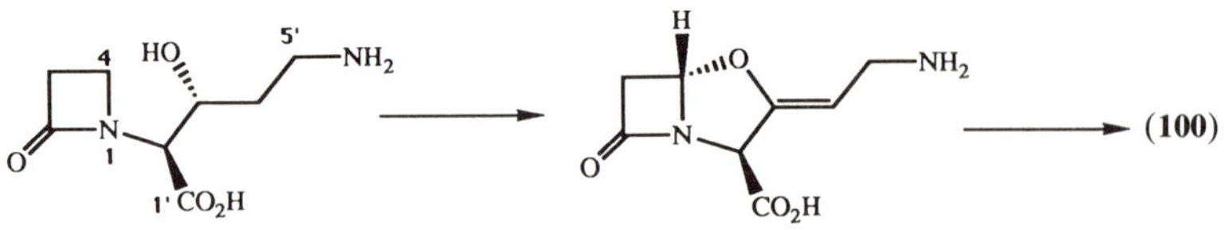

(108) → (107) → (100)

Scheme 1.49

the antipodal stereochemistry (i.e. 3*S*, 5*S*) to that of clavulanic acid (**100**) itself (3*R*, 5*R*). In addition to clavaminic acid (**107**), a monocyclic β-lactam was also discovered, which was later shown to be a precursor of clavaminic acid (**107**), named proclavaminic acid (**108**) (Scheme 1.49).[231] The absolute stereochemistry of the latter has been confirmed by total synthesis[232] and, using synthetically labelled materials, both (**107**) and (**108**) have been shown to be converted through to clavulanic acid (**100**) (Scheme 1.49).[233] Furthermore, a mutant was isolated which was blocked for the production of clavulanic acid, but instead produced acylated derivatives (**109**) of clavaminic acid, presumably indicating that the pathway is blocked between clavaminic (**107**) and clavulanic acids (**100**).[234] When racemic [4-$^2H_2$; 5-^{13}C]-ornithine (**106c**) was fed to the mutant culture and the resultant (**109**) was examined by nmr, it was found that both the carbon label and one deuterium label were incorporated (Scheme 1.50). Thus, the remaining deuterium must be lost in the conversion of clavaminic acid (**107**) to clavulanic acid (**100**).[228]

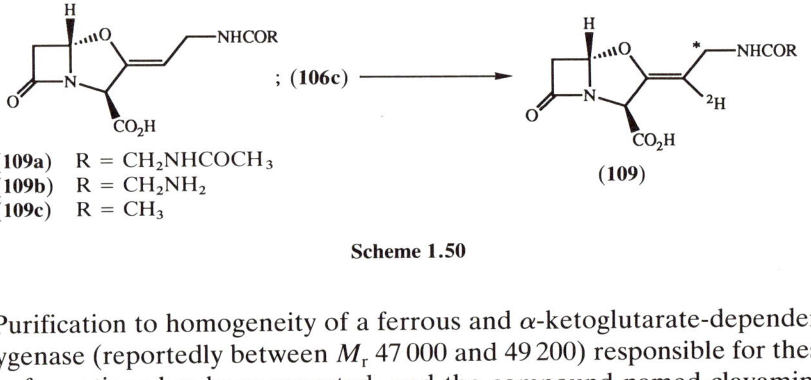

(**109a**) R = $CH_2NHCOCH_3$
(**109b**) R = CH_2NH_2
(**109c**) R = CH_3

; (**106c**) → (**109**)

Scheme 1.50

Purification to homogeneity of a ferrous and α-ketoglutarate-dependent oxygenase (reportedly between M_r 47 000 and 49 200) responsible for these transformations has been reported, and the compound named clavaminic acid synthase (CAS).[230,235,236] An [^{18}O]-labelling experiment carried out by Townsend and coworkers,[237] has demonstrated that the oxygen of the oxazolidine ring derives from the hydroxyl group of proclavaminic acid (**108**) and not from the molecular dioxygen utilised by CAS — note the similarity between this ring closure and that of the thiazolidine ring of the penicillins by IPNS. The stereochemistry of the closure of the oxazolidine ring has also been determined and shown to go with net retention of configuration — as might have been expected for an oxidative cyclisation of this type.[222] The cyclisation has also been shown to proceed without exchange of any label from C-3 (ruling out a Michael-type reaction) or C-2′ of the proclavaminic acid (**108**) substrate.[238]

In the conversion of proclavaminic acid (**108**) to clavaminic acid (**107**) it was hypothesised that two chemically distinct desaturations occur, that is closure of the oxazolidine ring and formation of the exocyclic double bond (Scheme 1.51). Evidence in support of such a theory is provided by the fact that two equivalents of α-ketoglutarate are required for the stoichiometric conversion of proclavaminic acid (**108**) to clavaminic acid (**107**).[230] The bicyclic clavam (**110**) and the 3′-ketone (**111**) were considered as potential intermediates.

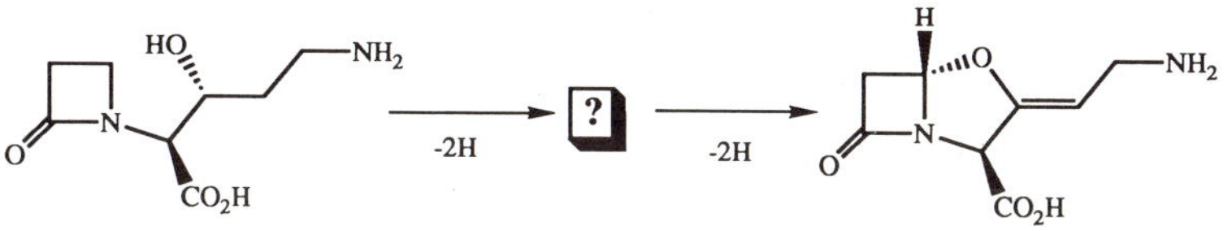

HO NH$_2$ N O CO$_2$H (**108**) —-2H→ ? —-2H→ H O NH$_2$ N O CO$_2$H (**107**)

Scheme 1.51

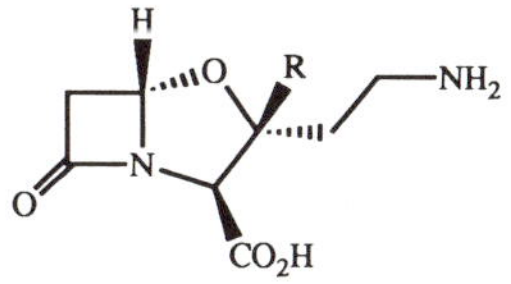

H O R NH$_2$ N O CO$_2$H

(**110**) R = H

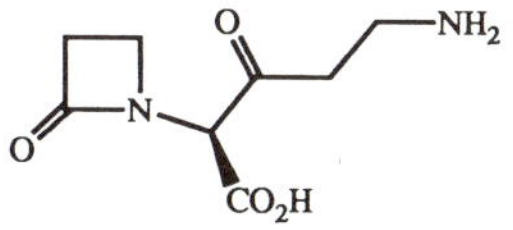

O NH$_2$ N O CO$_2$H

(**111**)

A ^{1}H nmr study of the incubation of proclavaminic acid (**108**) with purified extracts containing CAS found that, in addition to the signals corresponding to clavaminic acid (**107**), a minor resonance at *ca.* 5.4 ppm, of approximately 5–10% of the intensity of the clavaminic acid (**107**) resonances was observed. Attempts were made to isolate this material by HPLC (at pH 7–8); however, on the small scale on which the work was carried out, it proved impossible to obtain a sample of sufficient purity for unequivocal characterisation. The signal at 5.4 ppm occurred as a doublet (J 2.5 Hz) and was thus reminiscent of the C-5 proton of the 1′,2-dihydroclavulanates (**112**).[239] It was reasoned that if dihydroclavaminic acid (**110**) was indeed an intermediate between proclavaminic (**108**) and clavaminic acids (**107**), then incubation of proclavaminic acid deuterated at C-3′ may result in the operation of a primary isotope effect; this would slow down the conversion of dihydroclavaminic acid (**110**) to clavaminic acid (**107**), thus biasing the ratio in the crude incubation product to favour the former and thereby facilitating its isolation (Scheme 1.52).

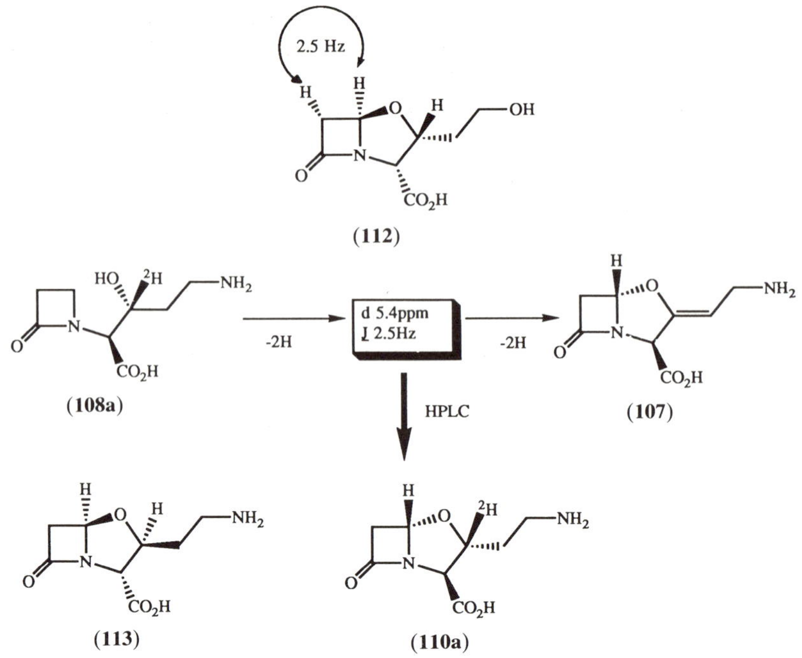

Scheme 1.52

Racemic deuterated proclavaminic acid (**108a**) was then incubated with CAS; the resulting nmr analysis showed a significant enhancement of the resonance at 5.4 ppm. Purification of the crude mixture was initially carried out by HPLC, at pH 7.5, to yield clavaminic acid (**107**) and a material consistent with the deuterated dihydroclavaminic acid (**110a**). A pure sample of the latter was obtained when HPLC purification was carried out under mildly acidic conditions, whence the clavam was found to be more stable. The molecular weight of the new metabolite was determined by mass spectral analysis and found to be consistent with (**110a**), and the relative stereochemistry as drawn was inferred from nuclear Overhauser experiments. The stereochemistry of (**110**) was further confirmed by the total synthesis of its enantiomer (**113**) from clavulanic acid. A pure sample of (**108a**) was re-fed to a highly purified sample of CAS and shown to be efficiently converted to clavaminic acid (**107**), confirming it to be a true intermediate in the conversion of (**108**) to (**107**) rather than a shunt metabolite.[240]

Townsend and coworkers[236] have published details of a partial kinetic characterisation of CAS and propose that two stepwise oxidations of proclavaminic acid occur. This is consistent with the authors' results. Townend and coworkers also conclude that there is not a substantial release of an intermediate from CAS. This statement is inconsistent with the authors' observation of the production of dihydroclavaminic acid (**110**) from incubations with fully protiated proclavaminic acid (**108**). The authors have now isolated and characterised (**110**) itself. It is possible that the discrepancy may be explained by the existence of more than one form of CAS, in a similar fashion to the discovery that there are two forms of DAOC/DAC synthase in *Streptomyces* spp. (see section 1.2.5). Further investigations are in progress. A recent mechanistic hypothesis for the operation of CAS, involving sequential oxidations carried out by a ferryl iron species, generated by reaction of CAS with α-ketoglutarate and dioxygen, is shown in Scheme 1.53.

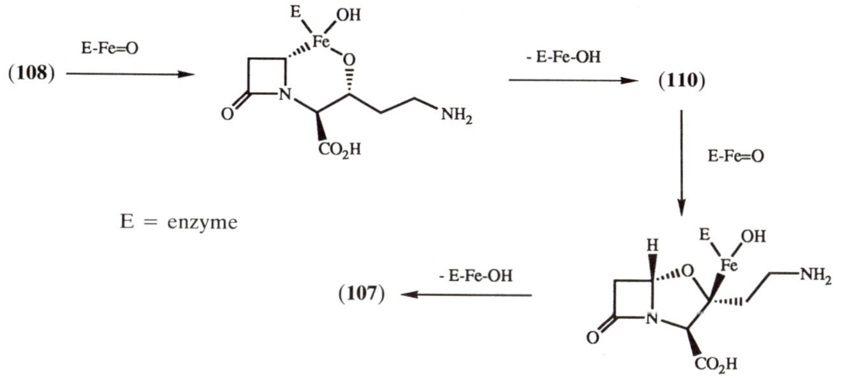

E = enzyme

Scheme 1.53

Cloning of the CAS gene[241] and a genetic locus involved in the biosynthesis of clavulanic acid (**100**)[242] have now been reported. It is hoped that this may facilitate investigation into the mechanisms by which the proposed progenitors glycerol (**101**) and ornithine (**106**) combine to give proclavaminic acid (**108**), as well as the remarkable *in vivo* double epimerisation involved in the conversion of clavaminic acid (**107**) into clavulanic acid (**100**).

1.4 Carbapenem biosynthesis

The thienamycins (**114**) were the first of a group (now over 50) of naturally occurring compounds, containing the carbapenem ring systems, to be isolated from *Streptomyces* spp.[243] Many of the carbapenems that have been subsequently discovered are either derivatives or epimers of thienamycin, for example the olivanic acids (**115**).[244] Most recently, the parent nucleus carbapenem-3-carboxylate (**116**) has been isolated from the species *Erwina* and *Serratia*, as its *p*-nitrobenzyl ester.[246] In addition there have been several

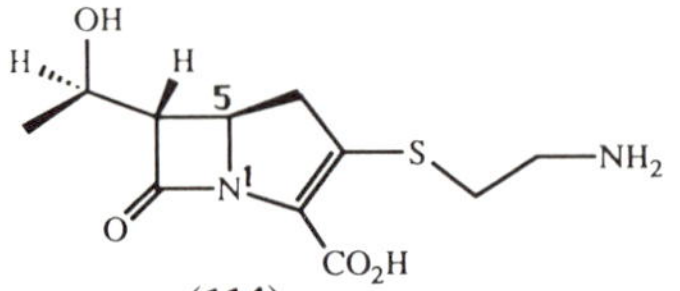

(114) 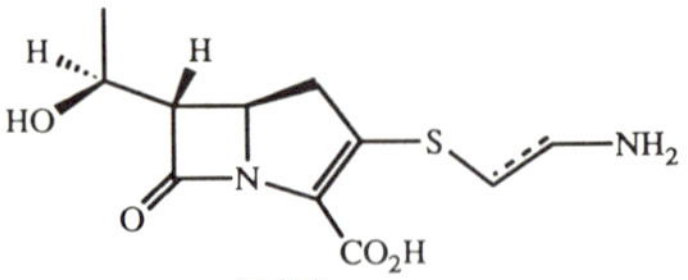 (115)

reports of the production of carbapenams by *Streptomyces* spp, which also produce carbapenems (*vide infra*).

The elucidation of the biosynthetic pathways leading to the carbapenems will be an especially challenging problem not only because of their complex and diverse structures, but also due to the low biological titres in which they are produced; it is remarkable that to date commerically utilised carbapenems are produced most efficiently by multi-step total synthesis rather than by fermentation or by semisynthesis.

1.4.1 Biosynthesis of the β-lactam nucleus

Early studies by the Merck group, using both stable and radio-isotopes, established that the pyrrolidine ring of thienamycin (**114**) is derived from glutamic acid.[244] In subsequent experiments using [^{13}C]-labelled acetates they demonstrated that the C-7 and C-6 carbons of the β-lactam ring are derived from acetate.[245] The isolation of the parent carbapenem (**116**) has prompted Bycroft *et al.*[246] to unequivocally demonstrate that glutamate and acetate combine to form the carbon skeleton of the nucleus. Furthermore, they also initially reported the isolation of two new isomeric carbapenams (**117**) and (**118**).[247] The (3*S*, 5*R*) configuration of the minor *cis*-isomer (**117**) was established by correlation with (**116**), which was of known chirality.[248] Subsequent total synthesis of the (3*R*, 5*R*) *trans*-isomer (**118**) demonstrated it to have a c.d. spectrum with a differential absorbance equal and opposite to that of the natural product. Thus, the naturally occurring compound must be the enantiomer of (**118**), i.e. has the (3*S*, 5*S*) stereochemistry (**119**).[249] Both (**117**) and (**119**), which were found in strains of *Serrantia* and *Erwina* spp., and which also produce carbapenems, including (**116**), were also shown to be derived from acetate (Scheme 1.54) and glutamate.[246]

It would seem probable that the biogenesis of the functionalised carbapenems is intimately linked to that of the nuclei (**116**) and (**119**). Williamson[250] has proposed that acetyl coenzyme A condenses with γ-glutamyl phosphate or glutamine to form a monocyclic intermediate (**120**), which is then tautomerised to (**121**), this can cyclise to form the β-lactam ring in a 'reverse' β-lactamase-type reaction — the thermodynamic driving force being the hydrolysis of the coenzyme A thioester bond (Scheme 1.55, path a). However, it would seem highly unlikely that the carbapenams (**119**)

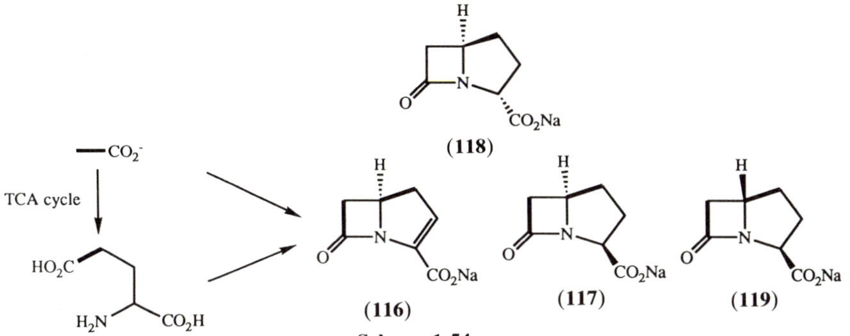

Scheme 1.54

and (**117**) are sequential intermediates on a biosynthetic route (although by analogy with clavulanic acid biosynthesis (see section 1.3) such a possibility cannot be *entirely* ruled out). Whilst it is possible to envisage (**117**) being derived from (**116**) by reduction, it would seem more probable that a common intermediate (such as (**122**) produced as a mixture of epimers at C-5) —which may be derived either by reduction of (**120**) or by condensation of acetyl-S-CoA and glutamate semi-aldehyde (paths b and b′, Scheme 1.55) — gives rise to both (**117**) and (**119**). Carbapenam (**116**) may then be preferentially oxidised, over (**119**), to give carbapenem (**116**) (Scheme 1.55, path c), allowing carbapenam (**119**) to accumulate as a 'shunt metabolite'— thereby explaining why it was isolated in higher yield than carbapenam (**117**) ((6) : (4), 9 : 1).[249]

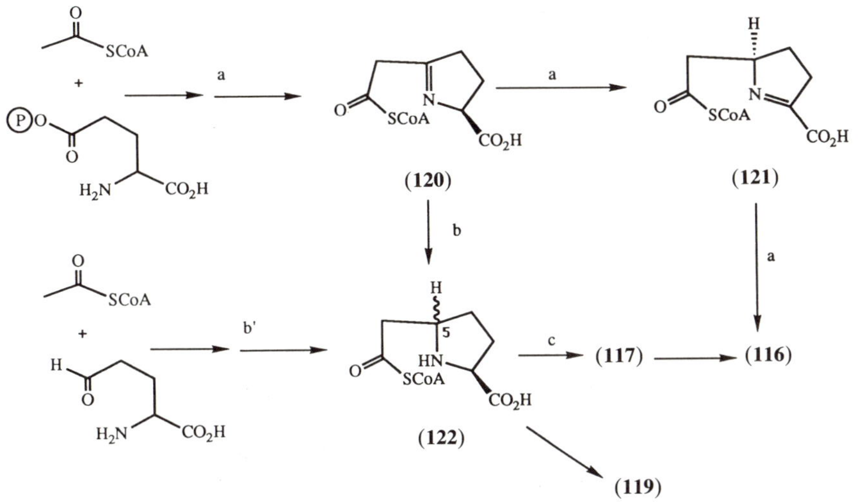

Scheme 1.55

1.4.2 Biosynthesis of the C-6 side chain

The isolation of northienamycin (**123**),[251,252] which contains a single carbon at C-6, prompted the Merck group to examine the possibility that the hydroxyethyl group of (**114**) was formed by two single carbon transfers. Using labelled methionines they demonstrated that both of the carbons of the C-6 side chain of thienamycin are derived from L-methionine. In a double labelling experiment it was also demonstrated that [methyl-^{14}C,^{3}H]-methionine was incorporated into (**114**) with 58% tritium retention, relative to ^{14}C, corresponding to 87% of the maximum value for retention of four of the six hydrogens of the two methyl groups.

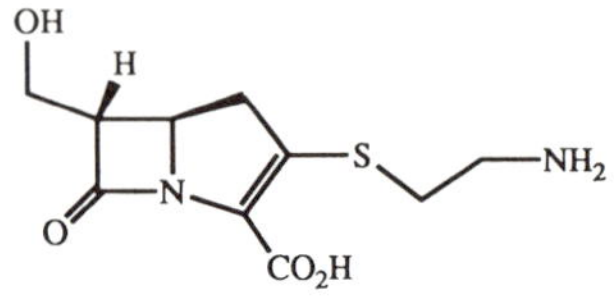

(**123**) northienamycin

Floss and coworkers[253] have extended these studies to examine the stereochemical course of the transfer from methionine of the C-9 methyl group of (**114**). A preliminary feeding experiment using [methyl-^{2}H$_3$C]-methionine clearly demonstrated that ^{13}C is incorporated into C-9 with retention of all three deuterium atoms and into C-8 with retention of one deuterium atom. When a methione bearing either an *R* or *S* chiral methyl group was subsequently fed to *S. cattleya*, it was found that thienamycin (**114**) in which the methyl group had the same configuration as that contained in the starting methionine was produced (Scheme 1.56).

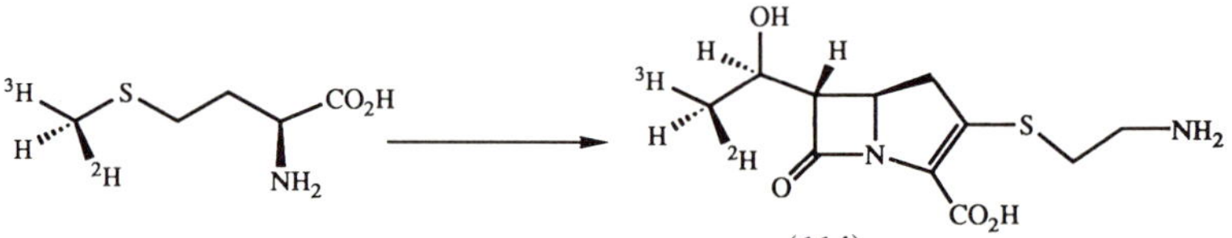

(**114**)

Scheme 1.56

Thus, the transfer of the C-9 methyl group from methionine or its activated form, S-adenosylmethionine occurs with apparent net overall retention of configuration. This result was somewhat surprising, since all other S-adenosylmethionine transferases thus far studied appear to catalyse methyl transfer with inversion of configuration, which has been interpreted as indicative of direct methyl transfer via an S_N2-type mechanism.[254] Whilst it is possible to imagine other scenarios, it would seem most likely that C-9 of

(114) is introduced by a process involving two sequential methyl transfers. Floss and coworkers[258] have pointed out that these processes would have an analogy in the biosynthesis of methionine from 5-methyltetrahydrofolate, by a B_{12}-dependent synthase, which also proceeds with net retention of configuration. In this case the postulated mechanism invokes initial transfer of the methyl group from 5-methylcobalmin, followed by a second transfer to produce methionine. In this regard, it is of interest that carbapenem biosynthesis has been reported to be enhanced by the addition of cobalt or B_{12}

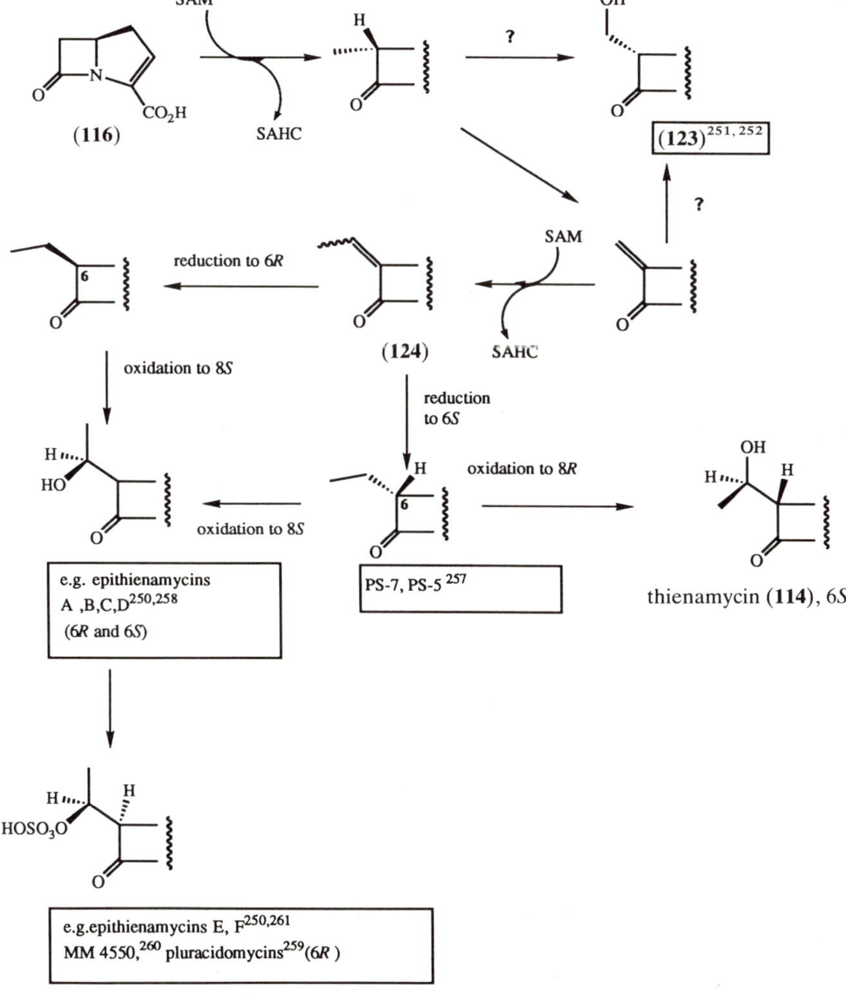

Scheme 1.57 Hypothetical biosynthetic pathway for the introduction of a 2-C side chain to C-6 of carbapenems. The structure of the 2-C side chain is not shown. Isolated compounds are 'boxed'. SAM = S-adenosyl methionine; SAHC = S-adenosyl homocysteine.

to the growth media. Furthermore, B_{12} has been reported to be excreted from a blocked mutant that does not produce thienamycin (**114**), and also to restore activity to other blocked mutants that produce thienamycin (**114**).

Based largely on the structures of known isolated carbapenems, the Merck group have postulated a hypothetical biosynthetic pathway for the introduction of a 2-C side chain at C-6, and its subsequent elaborations (Scheme 1.57). Intermediate (**124**) was given a pivotal role, since it was envisaged as the branching point for the production of the isopropyl carbapenems (Scheme 1.58). The proposal that the isopropyl carbapenems are derived from the ethyl carbapenems has some support in studies by Nozaki *et al.*[256] using blocked mutants.

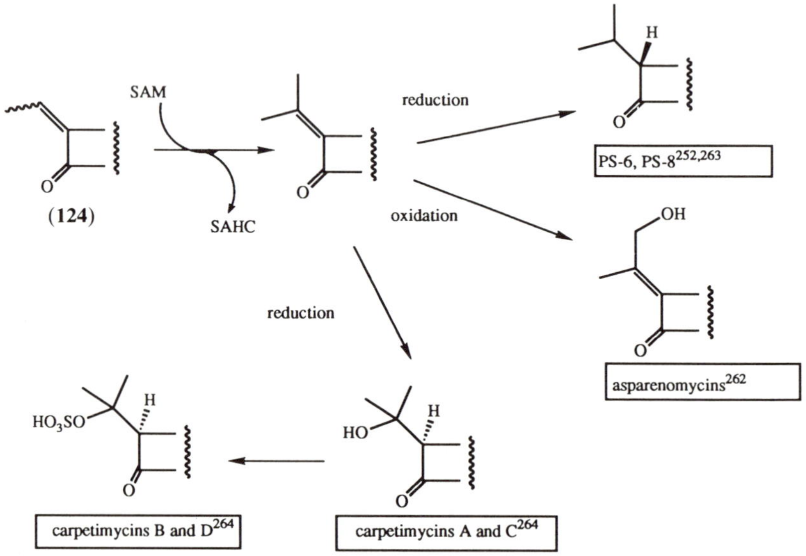

(**124**) SAM SAHC reduction oxidation reduction

PS-6, PS-8[252,263]

asparenomycins[262]

carpetimycins A and C[264]

carpetimycins B and D[264]

Scheme 1.58

Early studies by the Beecham group appear to demonstrate that the sulphated carbapenems are formed from the corresponding alcohols. Thus, the addition of sodium sulphate to the production medium increased the yields of sulphate carbapenems at the expense of the hydroxylated forms.[265] Furthermore, under conditions of reduced aeration, a mutant strain of *S. olivaceus* — blocked in the sulphation step — produced a carbapenem lacking a hydroxyl group, possibly indicating that the hydroxyl group is derived from dioxygen.[266] A similar dependence on the extent of hydroxylation on dissolved oxygen concentration has been observed for *S. fulvoviridis*.[267]

More recently, using a blocked mutant of *S. fulvoviridis* A933 that produces the OA-6129 class of carbapenems (A, B1, B2 and C (**125**)–(**128**),

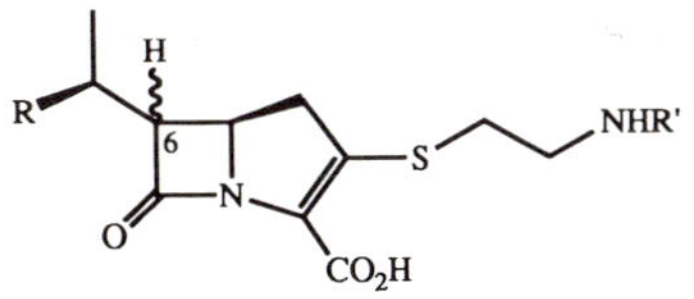

Compound	R	C-6 configuration	
OA-6129 A (**125**)	H	*R*	R′ =
OA-6129 B1 (**126**)	OH	*R*	
OA-6129 B2 (**127**)	OH	*S*	
OA-6129 C (**128**)	OSO_3H	*R*	

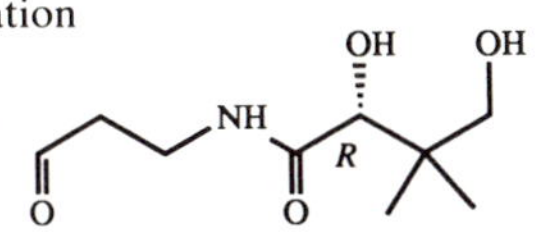

Compound	R	C-6 configuration	
(**132**)	H	*R*	R′ =
(**133**)	OH	*R*	
(**134**)	OH	*S*	
(**135**)	OSO_3H	*R*	

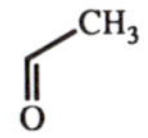

Scheme 1.59

Scheme 1.59), Fukagawa and coworkers[267] have described experiments that suggest a similar sequence of events to those proposed by the Merck group.[245] They were able to produce three further types of mutant capable of producing OA-6129 class carbapenems. Type I produced OA-6129 A (**125**), B1 (**126**), and B2 (**127**); Type II produced OA-6129 (**125**) and B1 (**126**), whilst Type III produced OA-6129 (**125**) only. These results are consistent with and supportive of a sequence of events in which hydroxylation is followed by epimerisation and then by sulphation, as depicted in Scheme 1.60. However, it is clear that further experimentation is required in order to establish the sequence of events more clearly in the early stages of the biosynthesis of the C-6 side chain.

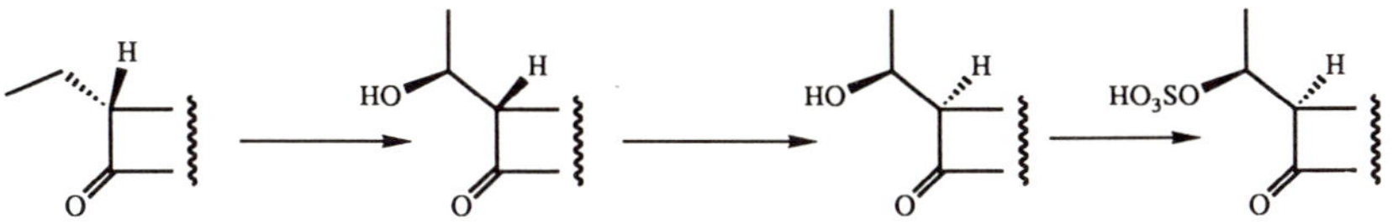

(**125**) (**126**) (**127**) (**128**)

Scheme 1.60

1.4.3 Biosynthesis of the C-2 side chain

Studies by the Merck group[245] in which radiolabelled cystine was fed to resting cells of *S. cattleya* showed a high level of specific incorporation (>70%) of the radioactivity into the cysteaminyl side chain of (**114**). [^{35}S]-cysteamine itself was found, however, to be incorporated at only approximately one-tenth the level of [^{35}S]cystine, indicating that decarboxylation probably takes place after attachment to the ring system.[245] A number of carbapenems with an unsaturated C-3 side chain are known,[268] and the possibility that these compounds might be precursors to thienamycin has been investigated.[245] When a mixture of [3,3′-$^{3}H_2$]-cystine and [^{35}S]-cystine was fed to *S. cattleya*, the thienamycin (**114**) produced was found to have the same ^{3}H/^{35}S ratio as the starting mixture, consistent with the proposal that a dehydro compound is not an intermediate. Firm supportive evidence for such a proposal was provided by Box *et al.*[265] who fed MM22380 (**129**) to a blocked mutant of *S. olivaceus* and demonstrated conversion to the sulphated olefins MM13902 (**130**) and MM4550 (**131**), consistent with the desaturation step occurring after formation of the bicyclic nucleus.

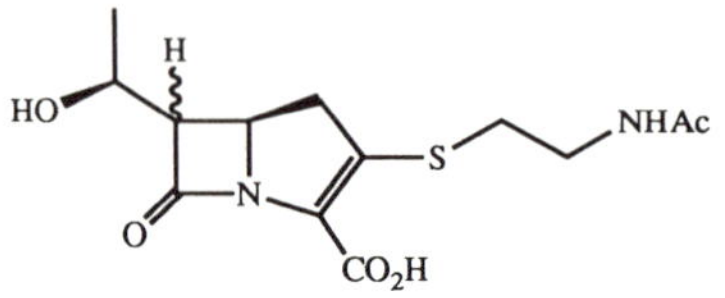

(**129**) 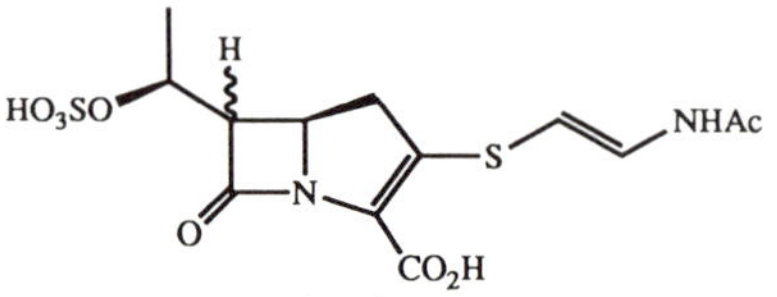 (**130**)

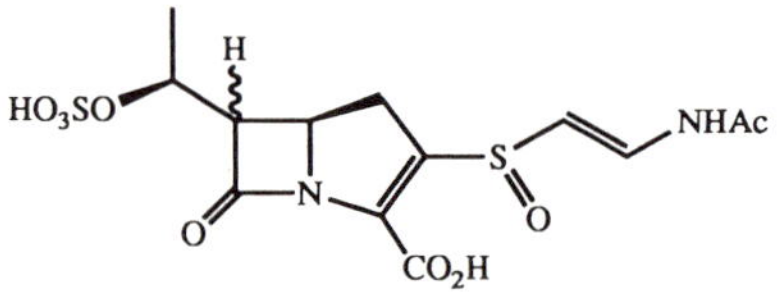

(**131**)

Early studies reported that when [^{35}S]-pantetheine was tested as a precursor of (**114**) from *S. cattleya*, only a low level of radioactivity was incorporated into (**114**). Neither did the addition of pantetheine to the growth medium reduce the level of incorporation of [^{35}S]-cystine, implying that in *S. cattleya*, pantetheine itself is not a direct precursor of the cysteaminyl side chain of (**114**). However, several carbapenems that contain the pantetheinyl side chain at C-3 have been isolated (e.g. (**125**)–(**128**), Scheme 1.59). Fukagawa *et al.*[269] hypothesised that these might be precursors of the analogous carbapenems with cysteaminyl side chains. They mutated a strain of *S. fulvoviridis* which normally produces PS-5 (**132**), epithienamycins A (**133**) and C (**134**), and MM17880 (**135**), amongst other carbapenems. A blocked mutant was found that produced (**125**)–(**128**) instead of (**132**)–(**135**), respectively (Scheme 1.59). These workers also demonstrated that the mutant was blocked in the acylase activity, which removes the pantotheinyl side chain and converts OA-6129 A (**125**) to NS-5 (**136**) (Scheme 1.61, path a).

The acylase from *S. fulvoviridis* A933 has been isolated, and partially purified and characterised. Similar enzyme activities were also detected in *S. cattleya*, *S. cremeus* and *S. argenteolis* — bacterial strains which also

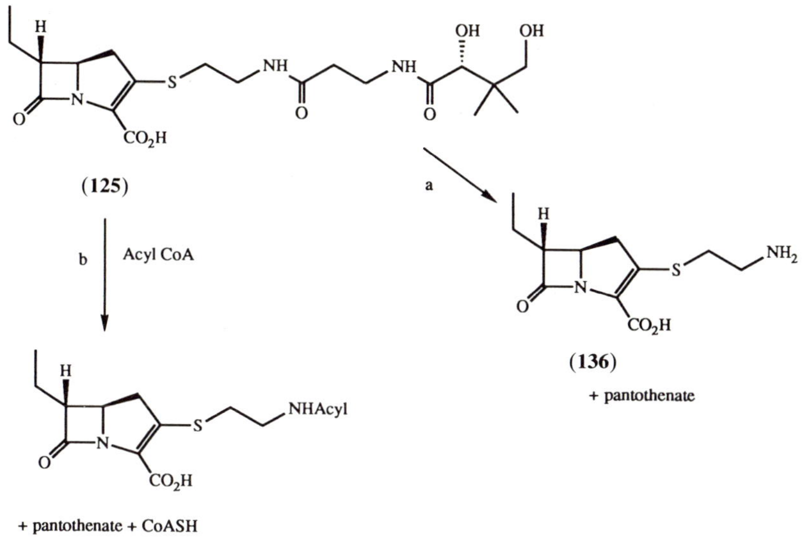

(125)

a

(136)

+ pantothenate

b Acyl CoA

+ pantothenate + CoASH

Scheme 1.61

produce carbapenems.[270] Unfortunately, the enzyme from *S. cattleya* proved to be labile and was not amenable to purification; however, it appeared to have a similar substrate specificity to the A933 acylase. The purified A933 acylase was found to catalyse the depantothenylation of the OA-6129 carbapenems (**125**)–(**128**) and, in addition, catalysed the exchange of the pantothenyl group with acyl coenzyme A's (Scheme 1.61, paths a and b, respectively). As may be anticipated by analogy with the penicillin transacylase activity, the A933 acylase catalysed the acylation of deacylated carbapenems with acyl coenzyme A's (Scheme 1.62). The enzyme was also found to catalyse the deacylation of *N*-acetyl L-amino acids, but not *N*-acetyl D-amino acids. Neither was the deacetylation of PS-5 (**132**) to NS-5 (**136**) observed. The parallel between this acylase and the isopenicillin N (**1**)/ penicillin G (**8**) transacylase (see section 1.2.8) is obvious, and it was even shown that the A933 acylase was capable of acylating 6-aminopenicillinate with acyl coenzyme A's.

Physical characterisation and mechanisitic investigations into the A933 acylase are at an early stage.[271] The enzyme is large (M_r *ca.* 100 kDa), and

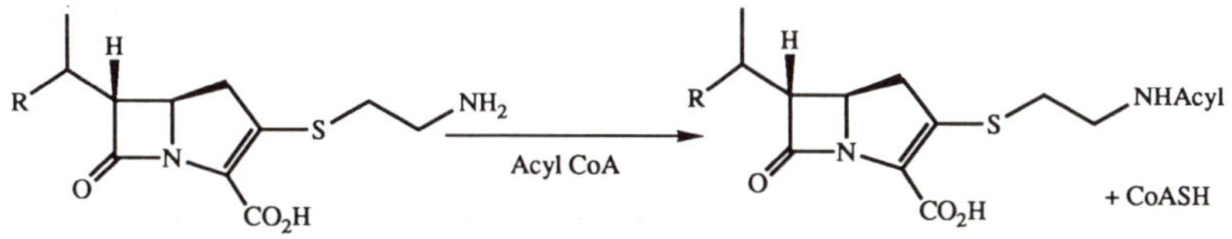

Acyl CoA

+ CoASH

Scheme 1.62

has a pI of 5.2 and pH optimum of *ca* 7–7.5. A number of other L-amino acid acylases were also studied for comparison to the A933 enzyme. The A933 acylase was the only one to carry out the depantothenylation of OA-6129 A (**125**) to give NS-5 (**136**). Also, unlike other acylases studied, it was severely inhibited by cobalt ions and *p*-chloromercuribenzoate, implying that it has an active thiol group.[271,272]

Radiolabelled β-alanine has been shown to be specifically incorporated into the C-3 side pantetheinyl side chain of the OA-6129 carbapenems, whilst labelled pantothenate was not taken up by the mycelia.[273] It was also shown that the extracellular concentrations of β-alanine and pantothenate increased with increased production of carbapenems. Thus, whilst the possibility that the C-3-aminoethylthio side chain is directly acylated with pantothenate cannot be eliminated, it would seem more likely that β-alanine is incorporated into the mycelia and converted first into pantothenate; this in turn condenses with cysteine and is subsequently decarboxylated to give pantetheine before incorporation into the C-2 side chain. Carbapenams containing a pantetheinyl side chain have been isolated and it would thus seem reasonable to suppose that desaturation of the carbapenams to the carbapenems in the pantetheinyl series can occur while the pantetheinyl side chain is still attached. Subsequently, the acylation may be manipulated by the A933 acylase or a related acylase activity. A recent biomimetic approach to the carbapenems has followed just such a strategy.[274]

It is apparent that, compared to the biosynthesis of the penicillins and cephalosporins, investigations into carbapenem biosynthesis are at an early stage. An obvious speculation — given the intimate role that pantetheine plays in the biosynthesis of at least some carbapenem species — is that the thiol-template mechanism involved in the biosynthesis of non-ribosomal peptides (e.g. ACV (**2**)) and fatty acids might also be crucially involved in carbapenem biosynthesis; the outcome of such speculation relies on further experimentation.

1.5 Monocyclic β-lactam biosynthesis

1.5.1 The monobactams

The monobactams (**137**) are a structurally distinct class of monocyclic β-lactams produced by bacteria, which were independently discovered by scientists working in the Takeda[275] and Squibb[276] laboratories. They all contain an N-sulphamate group, but differ in the nature of the acyl side chain and also in the presence or absence of a methoxyl group at the 3-position.

Sulfazecin (SQ 26,445) (**138**), the most common member of the class, SQ 26,180 (**139**) and SQ 26,812 (**140**) have been the subject of biosynthetic

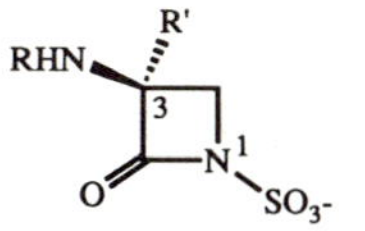

RHN R' 3 N 1 O SO_3^-

(**137**) Monobactam
(**138**) R = D-γ-glutamyl-D-alanyl, R′ = OMe
(**139**) CH_3CO
(**140**) R =

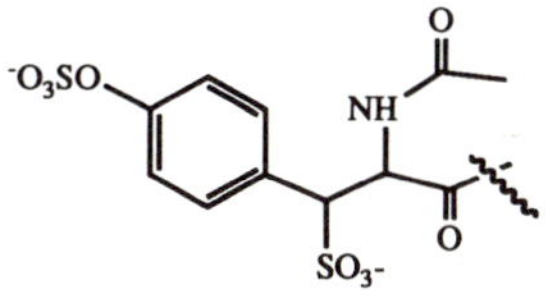

^-O_3SO NH O O SO_3^-

(**141**) R = D-γ-glutamyl–D-alanyl, R′ = H
(**142**) R = D-γ-glutamyl–L-alanyl, R′ = OMe

investigations by the Squibb group. Using doubly radiolabelled amino acids they have demonstrated, for each of (**138**), (**139**) and (**140**), that the carbon atoms of the β-lactam ring are specifically derived from serine.[277] Furthermore, when [3-$^3H_2$]-serine was fed, the monobactams were produced without significant loss of tritium — indicating the oxidation level at the serine 3-positions — is unaltered during the conversion to the β-lactam, consistent with a mechanism in which the β-lactam ring is formed by an S_N2 displacement of an activated serine hydroxyl.

Experiments to determine the origin of the sulphamate sulphur of the monobactams have been inconclusive:[278] inorganic sulphur was found to be the only sulphur source that was utilised by all three organisms studied. Cell-free extracts from *Agrobacterium radiobacter* were shown to produce adenosine phosphosulphate by incubation of sulphate and ATP with the appropriate cofactors, consistent with sulphate activation preceding formation of the sulphamic acid group.

The discovery[276] of monobactams (e.g. (**141**)) without a methoxy group at the C-3 position would imply that this group is introduced only after formation of the β-lactam ring. Indeed, the methyl group of the methoxyl group has been shown to derive specifically from methione,[277] which is analogous with earlier studies on the origin of the analogous methyl group of the cephamycins (see section 1.2.7). It would thus seem a reasonable hypothesis that the oxygen of the methoxyl moiety is introduced by an α-ketoglutarate oxygenase, although this remains to be proven.

Since the monobactams are produced with a variety of side chains, it may seem plausible to suppose that they derive from a common biosynthetic precursor. However, *in vitro* studies[279] have thus far failed to reveal the existence of an acylase/transacylase, as exists for penicillin biosynthesis, in *P. chrysogenum* and *A. nidulans* (see section 1.2.8). Squibb workers have reported that bacterial monobactam producers always produced the same spectrum of compounds, and that some of the producers gave rise to more than one class of side chain; they have suggested that these results imply that the substrates for the ring closure may be different in each organism.

In addition to sulfazecin (**138**), which has a D-γ-(glutamyl)-D-alanyl side chain, the Takeda scientists also isolated isosulfazecin (**142**), with the D-γ-(glutamyl)-L-alanyl side chain, consistent with the prior formation of the latter followed by epimerisation to (**138**).

Clearly, further work needs to be carried out in order to elucidate the sequence of events in monobactam biosynthesis. An intriguing preliminary report mentioning the isolation from the sulfazecin producer *Acetobacter* of a sulphated peptide, which contained N-terminal glutamate, alanine and serine, may be the spur to encourage such studies.[279]

1.5.2 The norcardicins

In 1976, scientists at Fujisawa reported the isolation of the nocardicins (A to G, (**143**)–(**149**)), which were the first naturally occurring monocyclic β-lactams to be discovered.[280] Like the monobactams, the norcardicins contain a 3-amido monocyclic β-lactam ring; they differ, however, in the

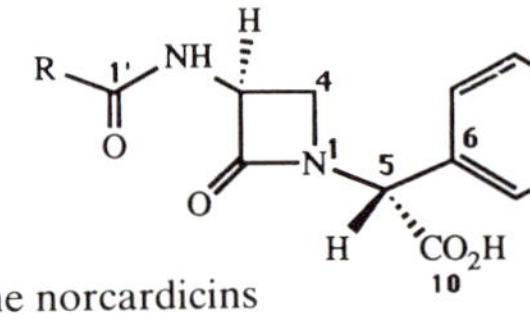

The norcardicins

(**143**) A: R =

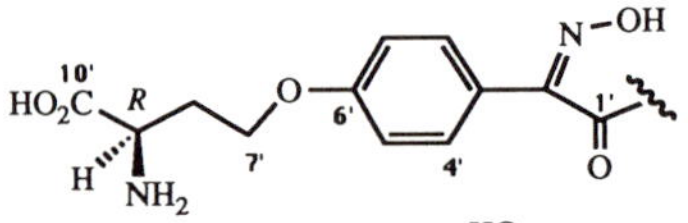

(**144**) B: R = "

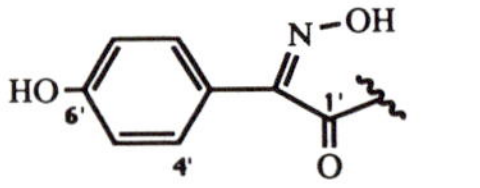

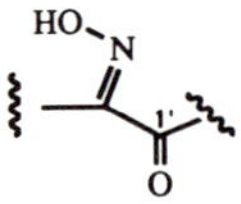

(**145**) C:R = "

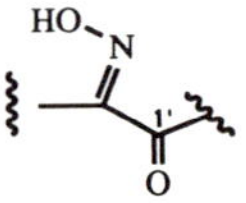

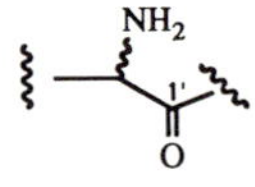

(**146**) D: R = "

(**147**) E: R =

(**148**) F: R = "

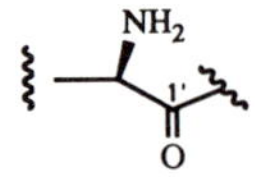

(**149**) G: R = "

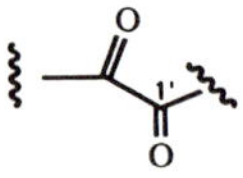

(**150**) R = "

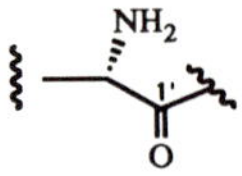

(**152**) R =

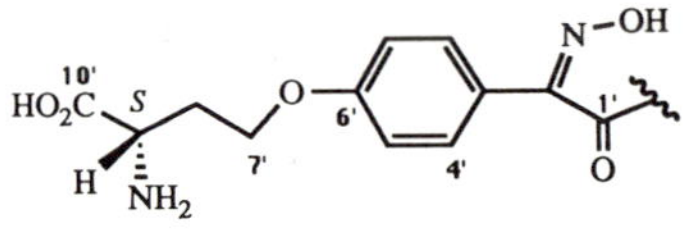

absence of the sulphamic acid group, and in the presence of the unusual ether-linked homoserinyl and *p*-hydroxyphenyl units.

Initial experiments by Hosada *et al.*[280] indicated that, like the penicillins, the C-framework of the norcardicins is derived from amino acids. By feeding radiolabelled precursors they demonstrated that tyrosine was incorporated into the two aromatic residues and that serine was incorporated into the β-lactam ring of norcardicin A. More recently, norcardicin biosynthesis has been the subject of extensive investigations by Townsend and co-workers. By feeding both radiolabelled and stable isotopes of potential precursors to growing cells of *Norcardia uniformis*, they demonstrated that the homoserinyl, β-lactam and aromatic functionalities could be specifically precursed by L-methionine, L-serine and L-(*p*-hydroxyphenyl)glycine respectively (Scheme 1.63).[281] Furthermore, using [$^3H/^{14}C$] double label experiments it was also shown that the β-lactam ring formation takes place without loss of tritium from the β-carbon of serine.

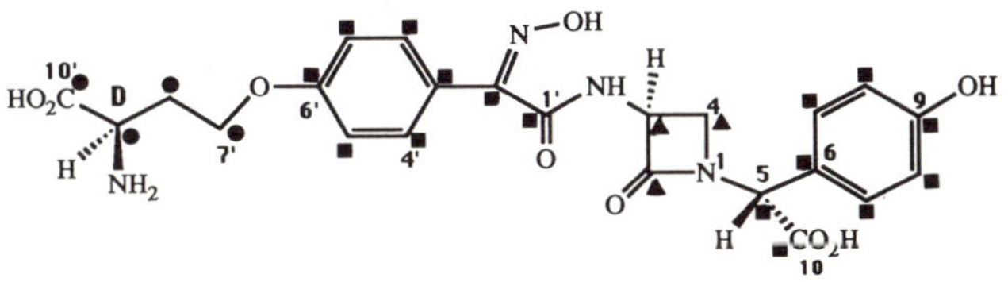

Scheme 1.63 Biosynthesis of norcardicins. ● = L-methionine derived; ■ = L-*p*-(hydroxyphenyl)glycine derived; ▲ = L-serine derived.

Subsequent investigations using L- and D-serines, chirally labelled with tritium at the β-carbon, indicated that L-serine was much better incorporated than D-serine and, significantly, that the β-lactam ring closure proceeds with inversion of stereochemistry at the serine β-carbon (Scheme 1.64).[282] This observation can be most easily rationalised by invoking an S_N2 displacement of the activated seryl hydroxyl group by the amide nitrogen during formation of the β-lactam ring. Support for such a mechanism has also come from a series of biomimetic transformations of serine-containing peptides, using Mitsonobu-type conditions, to produce the norcardicins and related structures (Scheme 1.65).[283,284]

The homoserinyl portion of (**2**) was shown to be most efficiently precursed

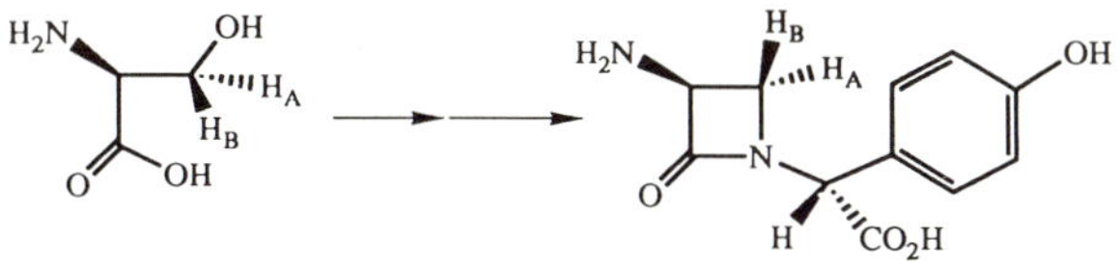

(**143**)

Scheme 1.64

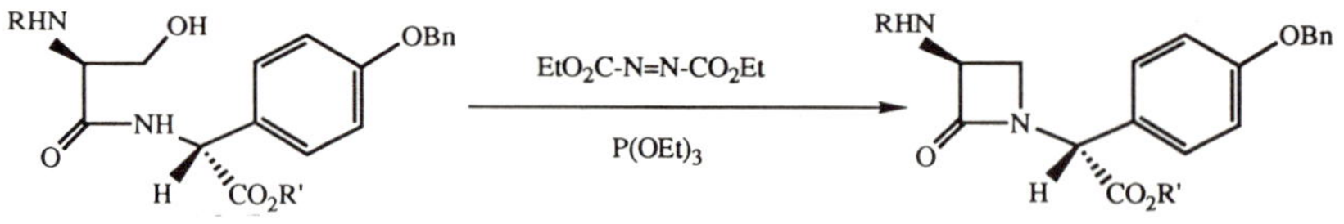

EtO_2C-N=N-CO_2Et
$P(OEt)_3$

Scheme 1.65

by L-methionine, homoserine itself only being incorporated at a much lower level.[281] Using methionines chirally deuterated at the 4-position, Townsend and coworkers have investigated the stereochemistry of the formation of the *o*-(C-7′) oxygen bond of norcardidcin A (**143**).[285] The reaction was shown to proceed with inversion of configuration, probably via direct nucleophilic displacement of S-adenosylmethione. Such a transfer with inversion has precedent in polyamine biosynthesis,[286] and in the transfer of a methyl group from S-adenosyl methionine.[287]

Radiolabelling experiments further established that the most efficient precurser of the aromatic units of the norcardicins is L-(*p*-hydroxyphenyl)-glycine, which, in turn, is derived from L-tyrosine (D-tyrosine being incorporated at a much lower level.)[281] By feeding racemic [2-^{13}C, ^{15}N]-(*p*-hydroxyphenyl)glycine it was also established that both the nitrogen of the β-lactam ring, and the nitrogen of the oxime of (**143**) were derived from this amino acid.[288] Furthermore, using L-[2-^{3}H, 1-^{14}C]-*p*-(hydroxyphenyl)-glycine it was found that there was essentially no incorporation of tritium into the 5-position of (**143**), consistent with epimerisation of *p*-(hydroxyphenyl)glycine after formation of the peptide, in an analogous fashion to that of the valinyl residue by ACV synthetase during the formation of the tripeptide ACV (**2**) in penicillin biosynthesis.

Townsend and coworkers have also demonstrated, in whole cell experiments, the conversion of labelled norcardicin G (**149**), the simplest norcardicin, to norcardicin A (**143**). Under the same conditions there was no evidence for the conversion of the 2′-epimer (**150**) to (**143**), except via degradation to *p*-(hydroxyphenyl)glycine.[289] These experiments are consistent with the simplest norcardicin, G (**149**), being the first formed β-lactam in the pathway. By analogy to penicillin biosynthesis, a logical immediate precursor to (**149**) would be the tripeptide D,L,D-*p*-(hydroxyphenyl)glycine (**151**). Possibly this tripeptide is assembled (and cyclised?) by a multienzyme complex similar to ACV synthetase.

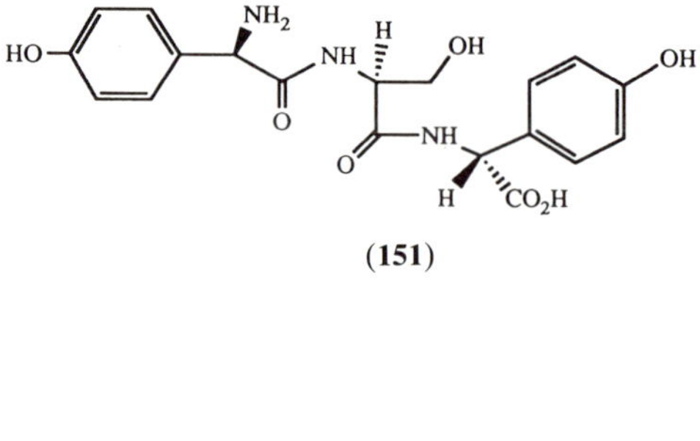

(**151**)

Recently, cell-free extracts that convert norcardicin E (**147**) to isonorcardicin A (**152**) in the presence of S-adenosylmethionine have been reported. Furthermore, the extracts also catalysed the epimerisation of isonorcardicin A (**152**) to norcardicin A (**143**) and vice versa. Based on these findings Townsend and coworkers proposed an overall biosynthetic pathway to norcardicin A, which has some considerable parallels with penicillin and cephalosporin biosynthesis (Scheme 1.66).

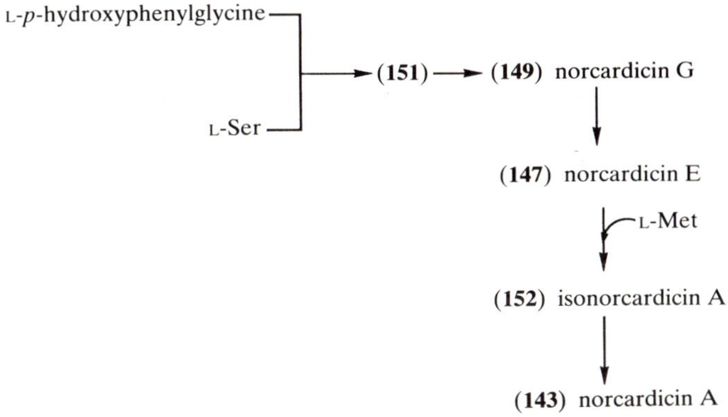

L-*p*-hydroxyphenylglycine
L-Ser
→ (**151**) → (**149**) norcardicin G
↓
(**147**) norcardicin E
↓ L-Met
(**152**) isonorcardicin A
↓
(**143**) norcardicin A

Scheme 1.66

1.5.3 Tabtoxin

The monocyclic β-lactam, tabtoxin (**153**) is an interesting dipeptide exotoxin produced by *Psuedomonas tabacai*, the organism responsible for the wildfire disease of the tobacco plant.[291] *In vivo* peptidase activity in the plant releases the tabtoxin-β-lactam (**154**), which is a potent inhibitor of glutamine synthetase.[292]

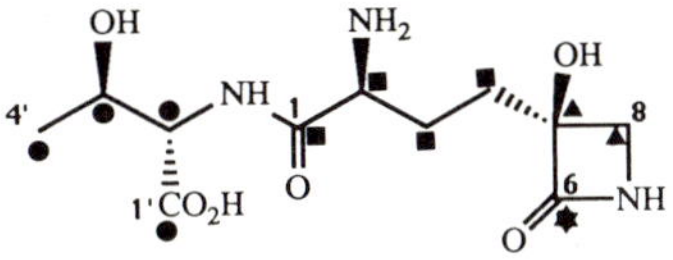

Scheme 1.67 Biosynthesis of tabtoxin (**153**). ● = L-threonine derived; ■ = L-aspartate derived; ▲ = pyruvate derived; ★ = derived from the Me group of methionine.

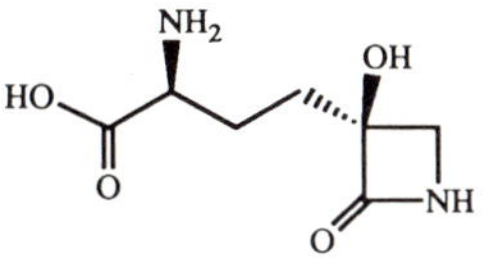

(**154**)

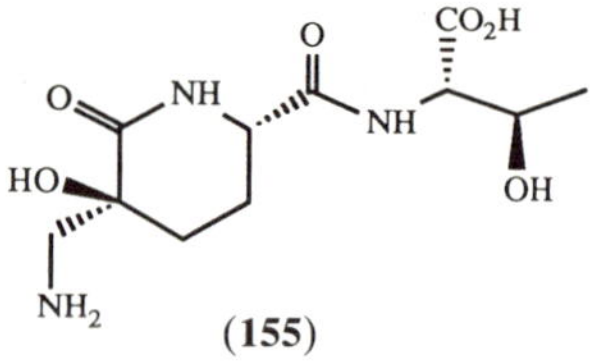

(155)

Tamm and coworkers have studied the biosynthesis of (**153**) by isotopically feeding labelled precursors to *P. tabacai*. In these experiments, (**153**) itself was not isolated, since it was found to be unstable, but was converted during work-up to the stable product of intramolecular transacylation — isotabtoxin (**155**). During initial experiments[293] it was demonstrated, using ^{13}C-labelled materials, that three different amino acids were involved in the formation of (**153**). The side chain (C-1′ to C-4′) was shown to be derived from L-threonine and L-aspartate (**156**) (C-1 to C-4). Intriguingly, the carbonyl of the β-lactam ring itself was shown to be derived from the methyl group of L-methione (cf. thienamycin biosynthesis, section 1.4.1), whilst the remaining C_2 unit (C-5, C-8), was not derived from acetate, but from the C_3 pool. Subsequent experiments using [2,3-^{13}C] pyruvate demonstrated that C-2 and C-3 of the pyruvate were incorporated as an intact unit into (**154**).[294] Previously undertaken complementary experiments using ^{13}C-labelled glucose, concurred with there results.[295] These experiments are summarised in Scheme 1.67.

The manner in which both pyruvate and aspartate are incorporated into (**153**) has led to the suggestion that the biogenesis of (**153**) is related to that of lysine (**157**) (Scheme 1.68). Lysine (**157**) itself, however, is not incorporated, hence Tamm and coworkers[294] have proposed that tabtoxin biosynthesis deviates from that of lysine at an unknown common intermediate stage.

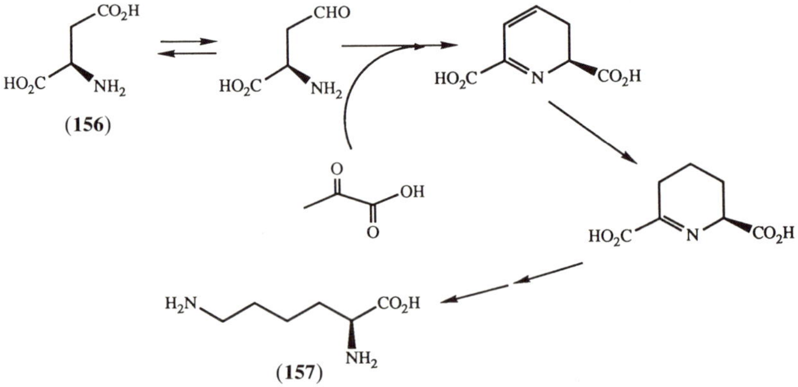

(156)

(157)

Scheme 1.68

The mechanism of β-lactam ring closure in the case of (**153**) is a problem of some interest, since it is the only naturally occurring β-lactam yet discovered in which the lactam carbonyl group is apparently derived from the C_1 pool. Eschenmoser[293] has proposed that an N-formylated intermediate such as (**158**), (R = $(CH_2)_2CH(NH_2)CO_2H$) might be a possible intermediate, since analogous compounds have been shown to undergo ring closure to form 3-hydroxy β-lactams (**159**) in a potentially biomimetic manner (Scheme 1.69).[296] An alternative speculation is that the hydroxyl group of (**153**) is derived from dioxygen — in a reaction catalysed by an α-ketoglutarate oxygenase — in a similar manner to which the methoxyl oxygen group is introduced into the 7-α-methoxyl cephalosporins.

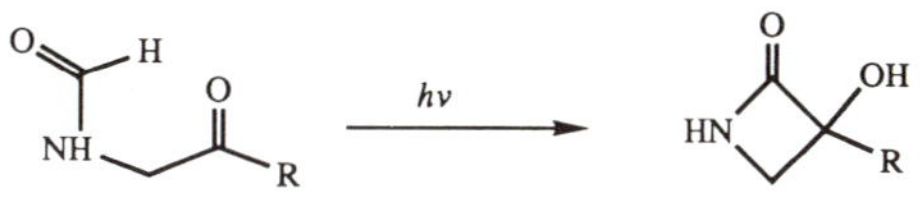

(**158**) —hν→ (**159**)

Scheme 1.69

Acknowledgements

We thank our many coworkers for their efforts and dedication to our studies in the β-lactam biosynthesis project. Their names appear in the references. Dr M.E. Wood and Mrs P.P.L. Schofield are thanked for their assistance in the preparation of this manuscript.

References

1. T.D. Ingolia and S.W. Queener, *Med. Res. Rev.* (1989) **9** 245.
2. S.W. Queener, *Antimicrob. Agents Chemotherap.* (1990) **34** 949.
3. H.R.V. Arnstein and P.T. Grant, *Biochem. J.* (1954) **57** 360.
4. H.R.V. Arnstein and M.E. Clubb, *Biochem. J.* (1957) **65** 618.
5. E.H. Flynn, M.H. McCormick, M.C. Stamper, H. De Valeria and C.W. Godzeski, *J. Am. Chem. Soc.* (1962) **84** 4594.
6. M. Cole and F.R. Batchelor, *Nature (London)* (1963) **198** 383.
7. H.R.V. Arnstein and D. Morris, *Biochem. J.* (1960) **76** 357.
8. P.B. Loder and E.P. Abraham, *Biochem. J.* (1971) **123** 471.
9. P. Adriaens, B. Meesschaart, W. Wuyts, H. Vanderhaeghe, and E. Eyssen, *Antimicrob. Agents Chemotherap.* (1975) **8** 638; P.A. Fawcett, J.J. Usher, J.A. Huddleston, R.C. Bleaney, J.J. Nisbet and E.P. Abraham, *Biochem. J.* (1976) **157** 651.
10. J.A. Chan, F.-C. Huang, and C.J. Sih, *Biochemistry.* (1976) **15** 177.
11. P.A. Fawcett, L. Kilbride, J. O'Sullivan and E.P. Abraham, unpublished results (1979).
12. S.A. Goulden and F.W. Chattaway, *Biochem. J. Proc.* (1968) **110** 55.
13. P.A. Fawcett, P.B. Loder, M.J. Duncan, T.J. Beesley and E.P. Abraham, *J. Gen. Microbiol.* (1973) **79** 293.
14. J. O'Sullivan, R.C. Bleaney, J.A. Huddleston and E.P. Abraham, *Biochem. J.* (1979) **184** 421.

15. T. Konomi, S. Herchen, J.E. Baldwin, M. Yoshida, N.A. Hunt and A.L. Demain, *Biochem. J.* (1979) **184** 427.
16. G. Revilla, F.R. Ramos, M.J. Lopez-Nieto, E. Alvarez and J.F. Martin, *J. Bacteriol.* (1986) **168** 947.
17. J. Zhang, S. Wolfe and A.L. Demain, *J. Antibiot.* (1987) **40** 1746.
18. *Idem, Can. J. Microbiol.* (1989) **35** 399.
19. *Idem, FEMS Microbiol. Lett.* (1989) **57** 145.
20. M.J. Lopez-Nieto, F.R. Ramos, J.M. Luengo and J.F. Martin, *Appl. Microbiol. Biotechnol.* (1985) **22** 343.
21. J. Cortes, P. Liras, J.M. Castro and J.F. Martin, *J. Gen. Microbiol.* (1986) **132** 1805.
22. K.Z. Bauer, *Naturforsch. B.* (1970) **25B** 1125.
23. P.B. Loder and E.P. Abraham, *Biochem. J.* (1971) **123** 477.
24. P.A. Fawcett and E.P. Abraham, *Methods Enzymol.* (1975) **43** 471.
25. H. Shirafuji, Y. Fujisawa, M. Kide, T. Kanzaki and M. Yoneda, *Agric. Biol. Chem.* (1979) **43** 155.
26. R.M. Adlington, J.E. Baldwin, M. Lopez-Nieto, J.A. Murphy and N. Patel, *Biochem. J.* (1983) **213** 573.
27. F. Lara, Rdc. Mateos, G. Vazquez, and S. Sanchez, *Biochem. Biophys. Res. Commun.* (1982) **105** 172.
28. G. Banko, S. Wolfe, and A.L. Demain, *ibid.* (1986) **137** 528.
29. *Idem, J. Am. Chem. Soc.* (1987) **109** 2858.
30. S.E. Jensen, D.W.S. Westlake and S. Wolfe, *FEMS Microbiol. Lett.* (1988) **49** 213.
31. S.E. Jensen, S. Wolfe, and D.W.S. Westlake, *Appl. Microbiol. Biotechnol.* (1989) **30** 111.
32. For a recent review see *Biochemistry of Peptide Antibiotics* (Eds H. Kleinkauf and H. von Döhren), W. de Gruyter, Berlin (1990).
33. H. Kleinkauf and H. von Döhren, *Eur. J. Biochem.* (1990) **192** 1.
34. H. Kleinkauf, H. Palissa and H. von Döhren, poster presented at the *Molecular Biology and Genetics of Industrial Microorganisms Meeting*, Bloomington, Indiana, USA (1988).
35. H. van Liempt, H. von Döhren and H. Kleinkauf, *J. Biol. Chem.* (1989) **264** 3680.
36. J.E. Baldwin, J.W. Bird, R.A. Field, N.M. O'Callaghan and C.J. Schofield, *J. Antibiot.* (1990) **43** 1055; *idem, ibid.* (In Press).
37. J. Zhang and A.L. Demain, *Biochem. Biophys. Res. Commun.* (1990) **169** 1145.
38. *Idem, Biotechnol. Letts.* (1990) **12** 649.
39. S.E. Jensen, A. Wong, M.J. Rollins and D.W.S. Westlake, *J. Bacteriol.* (1990) **172** 7269.
40. J.E. Baldwin and R.A. Field, unpublished results.
41. J. Kratzschmar, M. Krause and M.A. Marahiel, *J. Bacteriol.* (1989) **171** 5422.
42. J.S. Delderfield, E. Mtetwa, R. Thomas and T.E. Tyobeka, *J. Chem. Soc., Chem. Commun.* (1981) **650**.
43. J.E. Baldwin, R.M. Adlington, J.W. Bird and C.J. Schofield, *J. Chem. Soc., Chem. Commun.* (1989) 1616.
44. D.C. Aldridge, D.M. Carr, D.H. Davies, A.J. Hudson, R.D. Nolan, J.P. Poysner and C.J. Strawson, *J. Chem. Soc., Chem. Commun.* (1985) 1513.
45. N. Neuss, R.D. Miller, C.A. Affolder, W. Nakatsukasu, J. Mabe, L.L. Huckstep, N. De La Higuera, A.H. Hunt, J.L. Occlowitz and J.H. Gilliam, *Helv. Chim. Acta* (1980) **63** 1119.
46. D.J. Smith, M.K.R. Burnham, J. Edwards, A.J. Earl and G. Turner, *Bio/Technol.* (1990) **8** 39.
47. D.J. Smith, A.J. Earl and G. Turner, *EMBO J.* (1990) **9** 2743.
48. B. Diez, S. Guttierez, J.L. Barredo, P. van Solingen, L.H.M. van der Voort and J.F. Martin, *J. Biol. Chem.* (1990) **265** 16358.
49. D.J. Smith, M.K.R. Burnham, J.H. Bull, J.E. Hodgson, J.M. Ward, P. Browne, J. Brown, B. Barton, A.J. Earl and G. Turner, *EMBO J.* (1990) **9** 741.
50. A.P. MacCabe, M.B.R. Riach, S.E. Unkles and J.R. Kinghorn, *EMBO J.* (1990) **9** 279.
51. A.P. MacCabe, M.B.R. Riach and J.R. Kinghorn, *J. Biotechnol.* (1991) **17** 91.
52. J. Hoskins, N. O'Callaghan, S.W. Queener, C.A. Cantwell, J.S. Wood, V.J. Chen and P.L. Skatrud, *Curr. Genet.* (1990) **18** 523.
53. J.E. Baldwin, S. Herchen, T. Konomi, M. Yoshida, N.A. Hunt and A.L. Demain, *Biochem. J.* (1979) **184** 427.

54. J.E. Baldwin, Y. Sawada, P.D. Singh, N.A. Solomon and A.L. Demain, *Antimicrob. Agents and Chemother.* (1980) **18** 465.
55. J.E. Baldwin, B.L. Johnson, J.J. Usher, E.P. Abraham, J.A. Huddleston and R.L. White, *J. Chem. Soc., Chem. Commun.* (1980) 1271.
56. J.E. Baldwin, E.P. Abraham, R.M. Adlington, B. Chakravarti, G.S. Jayatilake, C.-P. Pang, H.-H. Ting and R.L. White, *Biochem. J.* (1984) **222** 789.
57. I.J. Hollander, Y.-Q. Shen, J. Heim, A.L. Demain and S. Wolfe, *Science* (1984) **224** 610.
58. S.M. Samson, R. Belagaje, D.T. Blankenship, J.L. Chapman, D. Perry, P.L. Skatrud, R.M. VanFrank, E.P. Abraham, J.E. Baldwin, S.W. Queener and T.D. Ingolia, *Nature* (1985) **318** 191.
59. J.E. Baldwin, S.J. Killin, A.J. Pratt, J.D. Sutherland, N.J. Turner, M. J.C. Crabbe, E.P. Abraham and A.C. Willis, *J. Antibiotics* (1987) **40** 652.
60. J.E. Baldwin, J.M. Blackburn, C.J. Schofield and J.D. Sutherland, *FEMS Microbiol. Lett.* (1990) **68** 45.
61. This chapter covers only key and recent substrate analogue studies with IPNS. For recent more extensive reviews of investigations in this area, see J.E. Baldwin and M. Bradley, *Chem. Rev.* (1990) **90** 1079; J.E. Baldwin and E.P. Abraham, *Nat. Prod. Rep.* (1989) **5** 129; and J.E. Baldwin, *J. Heterocycl. Chem.* (1990) **27** 91.
62. J.E. Baldwin, R.T. Aplin, Y. Fujishima, C.J. Schofield, B.N. Green and S.A. Jarvis, *FEBS Lett.* (1990) **264** 215.
63. J.E. Baldwin, R.L. White, E.-M.M. John and E.P. Abraham, *Biochem. J.* (1982) **203** 791.
64. J.E. Baldwin, R.M. Adlington, H.-H. Ting, D. Arigoni, P. Graf and D. Martinoni, *Tetrahedron* (1985) **41** 3339.
65. J.E. Baldwin, E.P. Abraham, G.L. Burge and H.-H. Ting, *J. Chem. Soc., Chem. Commun.* (1985) 1808.
66. J.E. Baldwin, R.M. Adlington, B.P. Domayne-Hayman, H.-H. Ting and N.J. Turner, *J. Chem. Soc., Chem. Commun.* (1986) 110.
67. J.E. Baldwin, R.M. Adlington, R.T. Aplin, L.D. Field, E.-M. John, E.P. Abraham and R.L. White, *J. Chem. Soc., Chem. Commun.* (1982) 137.
68. J.E. Baldwin, R.M. Adlington, R.T. Aplin, B. Chakravarti, L.D. Field, E.-M.M. John, Sir E.P. Abraham and R.L. White, *Tetrahedron* (1983) **39** 1061.
69. J.E. Baldwin, G. Bahadur, L.D. Field, E.-M.M. Lehtonen, J.J. Usher, C.A. Vallejo, E.P. Abraham and R.L. White, *J. Chem. Soc., Chem. Commun.* (1981) 917.
70. J.E. Baldwin, G. Bahadur, T. Wan, M. Jung, Sir E.P. Abraham, J.A. Huddleston and R.L. White, *J. Chem. Soc., Chem. Commun.* (1981) 1146.
71. J.E. Baldwin, Sir E.P. Abraham, R.M. Adlington, M.J. Crimmin, L.D. Field, G.S. Jayatilake and R.L. White, *J. Chem. Soc., Chem. Commun.* (1982) 1130.
72. J.E. Baldwin, Sir E.P. Abraham, R.M. Adlington, M.J. Crimmin, L.D. Field, G.S. Jayatilake, R.L. White and J.J. Usher, *Tetrahedron* (1984) **40** 1907.
73. J.E. Baldwin, R.M. Adlington, S.E. Moroney, L.D. Field and H.-H. Ting, *J. Chem. Soc., Chem. Commun.* (1984) 984.
74. A comprehensive proposal relating to the concept of sulphur as a biological conductor is presented elsewhere: J.E. Baldwin, G.M. Morris and W.G. Richards, *Proc. Roy. Soc. (B)* (1991) (In Press).
75. L.-J. Ming, L. Que, A. Kriauciunas, C.A. Frolik and V.J. Chen, *Inorg. Chem.* (1990) **29** 1111.
76. V.J. Chen, A.M. Orville, M.R. Harpel, K.K. Surerus, E. Munck and J.D. Lipscomb, *J. Biol. Chem.* (1989) **264** 21677.
77. J.E. Baldwin, J. Charnock, D. Garner, Y. Fujishima and C.J. Schofield, unpublished results.
78. J.E. Baldwin, R.M. Adlington, M. Bradley, W.J. Norris, N.J. Turner and A. Yoshida, *J. Chem. Soc., Chem. Commun.* (1988) 1125.
79. J.E. Baldwin, W.J. Norris, R.T. Freeman, M. Bradley, R.M. Adlington, S. Long-Fox and C.J. Schofield, *J. Chem. Soc., Chem. Commun.* (1988) 1128.
80. J.E. Baldwin, G. Lynch and C.J. Schofield, *J. Chem. Soc., Chem. Commun.* (1991) 736.
81. J.E. Baldwin, J.M. Blackburn, M. Sako and C.J. Schofield, *J. Chem. Soc., Chem. Commun.* (1989) 970.
82. J.E. Baldwin, E.P. Abraham, R.M. Adlington, B. Chakravarti, A.E. Derome, J.A.

Murphy and (in part) L.D. Field, N.B. Green, H.-H. Ting and J.J. Usher, *J. Chem. Soc., Chem. Commun.* (1983) 1317.
83. J.E. Baldwin, E.P. Abraham, R.M. Adlington, J.A. Murphy and (in part) N.B. Green, H.-H. Ting and J.J. Usher, *J. Chem. Soc., Chem. Commun.* (1983) 1319.
84. B. Maillard, D. Forrest and K.U. Ingold, *J. Am. Chem. Soc.* (1976) **98** 7024.
85. J.E. Baldwin, R.M. Adlington, B.P. Domayne-Hayman, G. Knight and H.-H. Ting, *J. Chem. Soc., Chem. Commun.* (1987) 1661.
86. J.E. Baldwin, R.M. Adlington, A.R. Pitt, A. Russell and D. Marquess, unpublished results.
87. J.E. Baldwin, A.P. Davis and L.D. Field, *Tetrahedron* (1982) **38** 2777.
88. J.E. Baldwin, R.M. Adlington, A.E. Derome, H.-H. Ting and N.J. Turner, *J. Chem. Soc., Chem. Commun.* (1984) 1211.
89. J.E. Baldwin, R.M. Adlington, M. Bradley, N.J. Turner and A.R. Pitt, *J. Chem. Soc., Chem. Commun.* (1989) 978.
90. J.E. Baldwin. *Proceedings of the 4th International Symposium on Recent Advances in the Chemistry of β-Lactam Antibiotics*, (eds P.H. Bentley and R. Southgate), The Royal Society of Chemistry, London (1988) p. 1.
91. J.E. Baldwin, M. Bradley, N.J. Turner, R.M. Adlington, A.R. Pitt and H. Sheridan, *Tetrahedron* (1991) (In Press).
92. J.E. Baldwin, R.M. Adlington, L.G. King, M.F. Parisi, W.J. Sobey, J.D. Sutherland and H.-H. Ting, *J. Chem. Soc., Chem. Commun.* (1988) 1635.
93. J.E. Baldwin, R.M. Adlington, N. Moss and N.G. Robinson, *J. Chem. Soc., Chem. Commun.* (1987) 1664.
94. J.E. Baldwin, R.M. Adlington and N. Moss, *Tetrahedron* (1989) **45** 2841.
95. J.E. Baldwin, S.L. Long-Fox and C.J. Schofield, unpublished observations.
96. J.E. Baldwin, E.P. Abraham, R.M. Adlington, G.A. Bahadur, B. Chakravarti, B.P. Domayne-Hayman, L.D. Field, S.L. Flitsch, G.S. Jayatilake, A. Spakovskis, H.-H. Ting, N.J. Turner, R.L. White and J.J. Usher, *J. Chem. Soc., Chem. Commun.* (1984) 1225.
97. J.E. Baldwin, R.M. Adlington, T. Nomoto and C.J. Schofield, *J. Chem. Soc., Chem. Commun* (1987) 806.
98. J.E. Baldwin, C.J. Schofield and B.D. Smith, *Tetrahedron* (1990) **46** 3019.
99. J.E. Baldwin, E.P. Abraham, G.L. Burge and H.-H. Ting, *J. Chem. Soc., Chem. Commun.* (1985) 1808.
100. J.E. Baldwin, A.J. Pratt and M.G. Moloney, *Tetrahedron* (1987) **43** 2565.
101. J.E. Baldwin, J.B. Coates, J.B. Halpern, M.G. Moloney and A.J. Pratt, *Biochem. J.* (1989) **261** 197.
102. J.E. Baldwin, J.B. Coates, M.G. Moloney, A.J. Pratt and A.C. Willis, *Biochem. J.* (1990) **266** 561.
103. J.E. Baldwin, R.M. Adlington and R. Bohlmann, *J. Chem. Soc., Chem. Commun.* (1985) 357.
104. J.E. Baldwin, J. Almog, R.L. Dyer, J. Huff and C.J. Wilkerson, *J. Am. Chem. Soc.* (1974) **96** 5600.
105. J.E. Baldwin and J. Huff, *J. Am. Chem. Soc.* (1973) **95** 5757.
106. T. Konomi, S. Herchen, J.E. Baldwin, M. Yoshida, N.A. Hunt and A.L. Demain, *Biochem. J.* (1979) **184** 427.
107. J.E. Baldwin, J.W. Keeping, P.D. Singh and C.A. Vallejo, *Biochem. J.* (1981) **194** 649.
108. G.S. Jayatilake, J.A. Huddlestone and E.P. Abraham, *Biochem. J.* (1981) **194** 645.
109. S.E. Jensen, D.W.S. Westlake and S. Wolfe, *J. Antibiot.* (1982) **35** 483.
110. *Idem, Can. J. Microbiol.* (1983) **29** 1526.
111. R.J. Bowers, S.E. Jensen, L. Lyubechansky, D.W.S. Westlake and S. Wolfe, *Biochem. Biophys. Res. Commun.* (1984) **120** 607.
112. S. Wolfe, D. Westlake and S. Jensen, *Eur. Pat. Appl.* 83304687.3 (1984).
113. S. Usui and C.-A. Yu, *Biochem. Biophys. Acta* (1989) **999** 78.
114. N. Esaki and C.T. Walsh, *Biochemistry* (1986) **25** 3261.
115. S. Kovacevic, M.B. Tobin and J.R. Miller, *J. Bacteriol* (1990) **172** 3952.
116. M. Koshaka and A.L. Demain, *Biochem. Biophys. Res. Commun.* (1976) **70**, 465.
117. M. Yoshida, T. Konomi, M. Kohsaka, J.E. Baldwin, S. Herchen, P. Singh, N.A. Hunt and A.L. Demain, *Proc. Natl. Acad. Sci. USA* (1978) **75** 6253.

118. J.E. Baldwin, P.D. Singh, M. Yoshida, Y. Sawada and A.L. Demain, *Biochem. J.* (1980) **186** 889.
119. J.E. Baldwin, S.R. Herchen and P.D. Singh, *ibid.* (1980) **186** 881.
120. S.J. Brewer, T.T. Boyle and M.K. Turner, *Biochem. Soc. Trans.* (1977) **5** 1026.
121. M.K. Turner, J.E. Farthing and S.J. Brewer, *Biochem. J.* (1978) **173** 839.
122. Y. Fujisawa, H. Shirafuji, M. Kida, K. Nara, M. Yoneda and T. Kanazaki, *Nature (London), New Biol.* (1973) **246** 154.
123. *Idem, Agric. Biol. Chem.* (1975) **39** 1295.
124. Y. Fujisawa, H. Shirafuji and T. Kanazaki, *ibid.* (1975) **39** 1303.
125. D.J. Hook, L.T. Chang, R.P. Elander and R.B. Morin, *Biochem. Biophys. Res. Commun.* (1979) **87** 258.
126. J. Kupka, Y.-Q. Shen, S. Wofe and A.L. Demain, *FEMS Microbiol. Lett.* (1983) **16** 1.
127. C.M. Stevens, E.P. Abraham, F.C. Huang and C.J. Sih, *FASEB* (1975) **34** 625.
128. A. Scheidegger, M.T. Kuenzi and J. Nuesch, *J. Antibiot.* (1984) **37** 518.
129. S.E. Jensen, D.W.S. Westlake, R. Bowers and A. Wolfe, *J. Antibiot.* (1984) **35** 1351.
130. J.E. Baldwin, H.-H. Ting and C.J. Schofield, unpublished results.
131. J.E. Baldwin, R.M. Adlington, J.B. Coates, M.J.C. Crabbe, N.P. Crouch, J.W. Keeping, G.C. Knight, C.J. Schofield, H.-H. Ting, C.A. Vallejo, M. Thorniley and E.P. Abraham, *Biochem. J.* (1987) **245** 831.
132. J.E. Dotzlaf and W.-K. Yeh, *J. Bacteriol.* (1987) **169** 1611.
133. J.E. Baldwin, K.-C. Goh, R.T. Aplin and C.J. Schofield, unpublished results.
134. S.E. Jensen, D.W.S. Westlake and S. Wolfe, *Antimicrob. Agents Chemotherap.* (1983) **24** 307.
135. *Idem, J. Antibiot.* (1985) **38** 263.
136. J. Cortes, J.F. Martin, J.M. Castro, L. Laizi and P. Liras, *J. Gen. Microbiol.* (1987) **133** 3165.
137. S.M. Samson, J.E. Dotzlaf, M.L. Slisz, G.W. Becker, R.M. van Frank, L.E. Veal, W.-K. Yeh, J.R. Miller, S.W. Queener and T.D. Ingolia, *Bio/technology* (1987) **5** 1207.
138. S. Kovacevic, B.J. Weigel, M.B. Tobin, T.D. Ingolia and J.R. Miller, *J. Bacteriol.* (1989) **171** 754.
139. J.E. Baldwin, R.M. Adlington, N.P. Crouch, J.B. Coates, J.W. Keeping, C.J. Schofield, W.A. Shuttleworth and J.D. Sutherland, *J. Antibiot.* (1988) **41** 1694.
140. J.E. Baldwin, J.D. Sutherland and R. Heath, unpublished results.
141. B.J. Baker, J.E. Dotzlaf and W.-K. Yeh, *J. Biol. Chem.* (1991) **266** 5087.
142. J.E. Baldwin and R.D.G. Cooper, unpublished results.
143. K.I. Kivirikko and R. Myllyla, in *The Enzymology of Post-transitional Modification of Proteins* (eds R.B. Freeman and C.H. Hawkins), Academic Press, London (1980) pp 53–104; J.S. Blanchard and S. Englard, *Biochemistry* (1983) **22** 5922.
144. E. Holme, *ibid.* (1975) **14** 4999.
145. P. Ververidis and P. John, *Phytochemistry* (1991) **30** 725; A.J. Hamilton, G.W. Lycett and D. Grieson, *Nature* (1990) **346** 284.
146. H. Kluender, C.H. Bradley, C.J. Sih, P. Fawcett and E.P. Abraham, *J. Am. Chem. Soc.* (1973) **95** 6149.
147. N. Neuss, C.H. Nash, J.E. Baldwin, P.A. Lemke and J.B. Grutzner, *J. Am. Chem. Soc.* (1973) **95** 3797.
148. J.E. Baldwin, R.M. Adlington, N.P. Crouch, N.J. Turner and C.J. Schofield, *J. Chem. Soc., Chem. Commun.* (1989) 1141.
149. R.B. Morin, B.G. Jackson, R.A. Mueller, E.R. Lavagnino, W.B. Scanlon and S.L. Andrews, *J. Am. Chem. Soc.* (1969) **91** 1401.
150. J.E. Baldwin, S.R. Herchen, J.C. Clardy, K. Hirotsu and T.S. Chou, *J. Org. Chem.* (1978) **43** 1342.
151. G. Stork and H.T. Cheung, *J. Am. Chem. Soc.* (1965) **87** 3783.
152. J.E. Baldwin and P.D. Singh, *Tetrahedron Lett.* (1979) 3757.
153. C.A. Townsend, A.B. Theis, A.S. Neese, E.B. Barrabee and D. Poland, *J. Am. Chem. Soc.* (1985) **107** 4760.
154. C.A. Townsend, *J. Nat. Prod.* (1985) **48** 708.
155. C.-P. Pang, R.L. White, E.P. Abraham, D.H.G. Crout, M. Lutstoft, P.J. Morgan and A.E. Derome, *Biochem. J.* (1984) **222** 777.
156. C.A. Townsend and E.B. Barrabee, *J. Chem. Soc., Chem. Commun.* (1984) 1586.

157. J.E. Baldwin, R.M. Adlington, T.W. Kang, E. Lee and C.J. Schofield, *J. Chem. Soc., Chem. Commun.* (1987) 104.
158. *Idem, Tetrahedron* (1988) **44** 5953.
159. J.E. Baldwin, R.M. Adlington, R.T. Aplin, N.P. Crouch, C.J. Schofield and H.-H. Ting, *J. Chem. Soc., Chem. Commun.* (1987) 1654.
160. J.E. Baldwin, R.M. Adlington, R.T. Aplin, N.P. Crouch, G.C. Knight and C.J. Schofield, *ibid.* (1987) 1651.
161. R.D. Miller, L.L. Huckstep, J.P. McDermott, S.W. Queener, S. Kukolja, D.O. Spry, T.K. Elzey, S.M. Lawrence and N. Neuss, *J. Antibiot.* (1981) **34** 984.
162. J.E. Baldwin, R.M. Adlington, C.J. Schofield, W.J. Sobey and M.E. Wood, *J. Chem. Soc., Chem. Commun.* (1989) 1012.
163. J.E. Baldwin, R.M. Adlington, M.A. Russell, C.J. Schofield and M.E. Wood, *J. Labelled Compounds and Radiopharmaceuticals* (1989) **27** 1091.
164. J.E. Baldwin, R.M. Adlington, M.E. Wood and C.J. Schofield, unpublished results.
165. R. Myllyla, K. Majamaa, V. Gunzler, H.M. Hanauske-Abel and K.I. Kivirikko, *J. Biol. Chem.* (1984) **259** 5403.
166. J.E. Baldwin, R.M. Adlington, M.J.C. Crabbe, G. Knight, T. Nomoto, C.J. Schofield and H.-H. Ting, *Tetrahedron* (1987) **43** 3009.
167. J.E. Baldwin, R.M. Adlington, J.B. Coates, M.J.C. Crabbe, J.W. Keeping, G. Knight, T. Nomoto, C.J. Schofield and H.-H. Ting, *J. Chem. Soc., Chem. Commun.* (1987) 374.
168. J.E. Baldwin and C.J. Schofield, unpublished results.
169. J.E. Baldwin and S. Sukamaran, unpublished results.
170. J.E. Baldwin, R.M. Adlington, M.E. Wood and C.J. Schofield, unpublished results.
171. J.E. Baldwin, R.M. Adlington, C.J. Schofield, N.P. Crouch and H.-H. Ting, *J. Chem. Soc., Chem. Commun.* (1987) 1556.
172. J.E. Baldwin, R.M. Adlington, N.P. Crouch and C.J. Schofield, *Tetrahedron* (1988) **44** 643.
173. J.E. Baldwin, N.P. Crouch and I. Perrera, unpublished results.
174. J.E. Baldwin, R.M. Adlington, N.P. Crouch, J.W. Keeping, S.W. Leppard, J. Pitlik, C.J. Schofield and M.E. Wood, *J. Chem. Soc., Chem. Commun.*, submitted for publication.
175. B. Siegel, *Biorg. Chem.* (1979) **8** 219.
176. Y. Fujisawa and T. Kanzaki, *J. Antibiot.* (1975) **38** 373.
177. J. Liersch, J. Neusch and H.J. Treichler, in *Second International Symposium on the Genetics of Industrial Microorganisms*, (Ed. K.D. MacDonald), Academic Press, New York (1976), pp. 179–195.
178. Y. Fujisawa and T. Kanzaki, *Agric. Biol. Chem.* (1975) **39** 2043.
179. A. Hinnen and J. Neusch, *Antimicrob. Agents Chemotherap.* (1976) **9** 824.
180. D.K. Brannon, D.S. Fakuda, J.A. Mabe, F.N. Huber and J.G. Whitney, *ibid.* (1972) **1** 237.
181. R. Nagarajan, L.D. Boeck, M. Gorman, R.L. Hamill, C.R. Higgens, M.M. Hoehn, W.M. Stark and J.G. Whitney, *J. Am. Chem. Soc.* (1971) **93** 2308.
182. S.J. Brewer, J.E. Farthing and M.K. Turner, *Biochem. Soc. Trans.* (1977) **5** 1024.
183. S.J. Brewer, P.M. Taylor and M.K. Turner, *Biochem. J.* (1980) **185** 555.
184. E.O. Stapley, M. Jackson, S. Hernandez, S.B. Zimmerman, S.A. Currie, S. Mochales, J.M. Mata, H.B. Woodruff and D. Hendlin, *Antimicrob. Agents. Chemotherap.* (1972) **2** 122.
185. G.F. Gauze, V.A. Chugosava, L.P. Terekhova, N.N. Lomakina and G.B. Fedorova, *Antibiotiki (Moscow)* (1976) **21** 1059.
186. P.D. Singh, M.G. Young, J.H. Johnson, C.M. Cimarusti and R.B. Sykes, *J. Antibiot.* (1984) **37** 773.
187. J. O'Sullivan, R.T. Aplin, C.M. Stevens and E.P. Abraham., *Biochem. J.* (1979) **179** 47.
188. S.W. Queener and N. Neuss, in *The Chemistry and Biology of β-Lactam Antibiotics*, Vol 3 (Eds R.B. Morris and M. Gorman), Academic Press Inc., London (1982), Chapter 7.
189. M. Cole, *Process Biochemistry* (1966) 334.
190. A.L. Demain, *Advances in Applied Microbiology* (1953) **1** 23.
191. D.L. Pruess and M.J. Johnson, *J. Bacteriol.* (1967) **94** 1502.

192. B. Spencer, *Biochem. Biophys. Res. Commun.* (1968) **31** 170.
193. V.R. Brunner, M. Rohr and M. Zinner, *Hoppe-Seyler's, Z. Physiol. Chem.* (1968) **349** 95.
194. S. Gatenbeck and U. Brunsberg, *Acta Chem. Scand.* (1968) **22** 1059.
195. B. Spencer and C. Maung, *Proc. of Biochem. J.* (1970) **118** 29P–30P.
196. P.B. Loder, *Postepy. Hig. Med. Dos.* (1972) **26** 493.
197. P.A. Fawcett, J.J. Usher and E.P. Abraham, *Biochem. J.* (1975) **151** 741.
198. J.J. Usher, P.B. Loder and E.P. Abraham, *ibid.* (1975) **151** 729.
199. E.P. Abraham, *Jpn. J. Antibiot. Supplement* (1977) **30** S1–S26.
200. E.P. Abraham, in *Antibiotics and Other Secondary Metabolites*, Academic Press, London (1978), pp. 141–164.
201. S.P. Brundidge, F.C.A. Gaeta, D.J. Hook, C. Sapino, Jr., R.P. Elander and R.B. Morin, *J. Antibiot.* (1980) **33** 1348.
202. R.G. Kogekar and V.N. Deshpande, *Ind. J. Biochem. Biophys.* (1983) **20** 208.
203. *Idem, ibid.* (1982) **19** 257; see also R. Brunner and M. Rohr, *Methods Enzymol.* (1975) **43** 476.
204. J.M. Luengo, G. Revilla, J.R. Villanueva and J.F. Martin, *J. Gen. Microbiol.* (1979) **115** 207.
205. G. Revilla, M.J. Lopez-Nieto, J.M. Luengo and J.F. Martin, *J. Antibiot.* (1984) **37** 781.
206. G. Revilla, F.R. Ramos, M.J. Lopez-Nieto, E. Alvarez and J.F. Martin, *J. Bacteriol.* (1986) **168** 947.
207. J.M. Luengo, J.L. Iriso and M.J. Lopez-Nieto, *J. Antibiot.* (1986) **39** 1565.
208. *Idem, ibid.* (1986) **39** 1754.
209. M.J. Alonso, F. Bermejo, A. Reglero, J.M. Fernandez-Canõn, G. Gonzales de Buitrago and J.M. Luengo, *J. Antibiot.* (1985) **41** 1074.
210. E. Alvarez, J.M. Cantoral, J.L. Barredo, B. Diez and J.F. Martin, *Antimicrob. Agents Chemotherap.* (1987) **31** 1675.
211. P.A. Whiteman, E.P. Abraham, J.E. Baldwin, M.D. Fleming, C.J. Schofield, J.D. Sutherland and A.C. Willis, *FEBS Lett.* (1990) **262** 342.
212. J.L. Barredo, P. Van Solingen, B. Diez, E. Alvarez, J.M. Cantoral, A. Kattevilder, E.B. Smaal, M.A.M. Groenen, A.E. Veenstra and J.F. Martin, *Gene* (1989) **83** 291.
213. M.B. Tobin, M.D. Fleming, P.L. Skatrud and J.R. Miller, *J. Bacteriol.* (1990) **172** 5908.
214. J.M. Fernandez-Canõn, A. Reglero, H. Martinez-Blanco and J.M. Luengo, *J. Antibiot.* (1989) **42** 1398.
215. J.M. Fernandez-Cañon, A. Reglero, J. Martinez-Blanco, M.A. Ferrero and J.H. Luengo, *ibid.* (1989) **42** 1410.
216. J.M. Fernandez-Cañon, M. Fernandez-Cañon, A. Reglero, J. Martinez-Blanco, M.A. Ferrero and J.H. Luengo, *ibid.* (1989) **42** 1416.
217. A.G. Brown, D.F. Corbett, J. Goodacre, J.B. Harbridge, T.T. Howarth, R.J. Ponsford, T.J. King and I. Stirling, *J. Chem. Soc. Perkin Trans. 1* (1984) 635; T.T. Howarth, A.G. Brown and T.J. King, *J. Chem. Soc., Chem. Commun.* (1976) 266.
218. S.W. Elson and R.S. Oliver, *J. Antibiot.* (1978) **31** 586.
219. I. Sterling and S.W. Elson, *ibid.* (1979) **32** 1135.
220. C.A. Townsend and M.-F. Ho, *J. Am. Chem., Soc.* (1985) **107** 1066.
221. A.L. Gutman, V. Ribon and A. Boltanski, *J. Chem. Soc., Chem. Commun.* (1985) 1627; A.L. Gutman, E. Meyer and A. Boltanski, *Synth. Commun.* (1988) **18** 1311.
222. A. Basak, S.P. Salowe and C.A. Townsend, *J. Am. Chem., Soc.* (1990) **112** 1654.
223. S.W. Elson, in *Recent Advances in the Chemistry of β*-Lactam Antibiotics, Vol 3 (1988) Royal Society of Chemistry, London, pp. 303–320.
224. C.A. Townsend and S.-S. Mao, *J. Chem. Soc., Chem. Commun.* (1987) 86.
225. S.W. Elson, R.S. Oliver, B.W. Bycroft and E.A. Faruk, *J. Antibiot.* (1982) **35** 81.
226. C.A. Townsend and M.-F. Ho, *J. Am. Chem. Soc.* (1985) **107** 1065.
227. C.A. Townsend, M.-F. Ho and S.-S. Mao, *J. Chem. Soc., Chem. Commun.* (1986) 638.
228. B.W. Bycroft, A. Penrose, J. Gillet and S.W. Elson, *J.Chem. Soc., Chem. Commun.* (1988) 980.
229. C.A. Townsend and W.J. Krol, *J. Chem. Soc., Chem. Commun.* (1988) 1234.
230. S.W. Elson, K.H. Baggaley, J. Gillet, S. Holland, N.H. Nicholson, J.T. Sime and S.R. Woroniecki, *J. Chem. Soc., Chem. Commun.* (1987) 1736.
231. *Idem, ibid.* (1987) 1738; K.H. Baggaley, S.W. Elson, N.H. Nicholson and J.T. Sime,

J. Chem. Soc., Perkin Trans. 1 (1990) 1513, For subsequent studies see W.J. Krol, S.-S. Mao, D.L. Steele and C.A. Townsend, *J. Org. Chem.* (1991) **56** 728.
232. K.H. Baggaley, N.H. Nicholson and J.T. Sime, *J. Chem. Soc., Chem. Commun.* (1988) 567; K.H. Baggaley, S.W. Elson, J.T. Sime and N.H. Nicholson, *J. Chem. Soc. Perkin Trans. 1* (1990) 1521.
233. S.W. Elson, K.H. Baggaley, J. Gillet, S. Holland, N.H. Nicholson, J.T. Sime and S.R. Woroniecki, *J. Chem. Soc., Chem. Commun.* (1987) 1739.
234. S.W. Elson, J. Gillet, N.H. Nicholson, J.T. Sime and J.W. Tyler, *J. Chem. Soc., Chem. Commun.* (1988) 979.
235. S.W. Elson, S.R. Woroniecki and K. Baggaley, *Chem. Abs.* (1988) **108** 204405U (*Eur. Pat. Appl.* EP 213, 914).
236. S.P. Salowe, E.N. Marsh and C.A. Townsend, *Biochemistry* (1990) **29** 6499.
237. W.J. Krol, A. Basak, S.P. Salowe and C.A. Townsend, *J. Am. Chem., Soc.* (1989) **111** 7625.
238. C.A. Townsend and A. Basak, *Tetrahedron* (1991) **47** 2591.
239. D. Hoppe and T. Hilpert, *Tetrahedron* (1987) **43** 2467.
240. J.E. Baldwin, R.M. Adlington, J.S. Bryans, A.O. Bringhen, J.B. Coates, N.P. Crouch, M.D. Lloyd, C.J. Schofield, S.W. Elson, K.H. Baggaley, R. Cassells and N.H. Nicholson, *J. Chem. Soc., Chem. Commun.* (1990) 617.
241. J.E. Hodgson and A.J. Earl, *Eur. Pat. Appl.* 349121 (1990).
242. C.R. Bailey, M.J. Butter, I.D. Normansell, R.T. Rowlands and D.J. Winstanley, *Bio/technology* (1984) **2** 808.
243. G. Albers-Schönberg, B.H. Arison, O.D. Hensens, J. Hirshfield, K. Hoogsteen, E.A. Kaczka, R.E. Rhodes, J.S. Kahan, F.M. Kahan, R.W. Ratcliffe, E. Walton, L.J. Ruswinkle, R.B. Morin and B.G. Christensen, *J. Am. Chem. Soc.* (1978) **100** 6491; A.G. Brown, D.F. Corbett, A.J. Eglington and T.T. Howarth, *J. Chem. Soc., Chem. Commun.* (1977) 523; D.F. Corbett, A.J. Eglington and T.T. Howarth, *J. Chem. Soc., Chem. Commun.* (1977) 953.
244. W.L. Parker, M.L. Rathnum, J.S. Wells, Jr., W.H. Trejo, P.A. Principe and R.B. Sykes, *J. Antibiot.* (1982) **35** 653; M. Miyashita, N. Chida and A. Yoshikoshi, *J. Chem. Soc., Chem. Commun.* (1984) 195.
245. G. Albers-Schönberg, B.H. Arison, E. Kaczka, F.M. Kahan, J.S. Kahan, B. Lago, W.M. Maiese, R.E. Rhodes and J.L. Smith, Abstract 229, *16th Interscience Conference on Antimicrobial Agents and Chemotherapeutics*, Chicago, 1976; J.M. Williamson, E. Inamine, K.E. Wilson, A.W. Douglas, J.M. Liesch and G.A. Albers-Schönberg, *J. Biol. Chem.* (1985) **260** 4637.
246. B.W. Bycroft, C. Maslen, S.J. Box, A. Brown and J.W. Tyler, *J. Antibiot.* (1988) **41** 1231.
247. *Idem, J. Chem. Soc., Chem. Commun.* (1987) 1623.
248. Y. Ueda, C.E. Damus and V. Vinet, *Can. J. Chem.* (1983) **61** 2257; M. Miyashita, N. Chida and A. Yoshikoshi, *J. Chem. Soc., Chem. Commun.* (1984) 195.
249. B.W. Bycroft and S.R. Chhabra, *ibid.* (1989) 423.
250. J.M. Williamson, *CRC Crit. Rev. in Biotechnol.* (1986) **4** 111.
251. K.E. Wilson, A.J. Kempf, J.M. Liesch and B.H. Arison, *J. Antibiot.* (1983) **36** 1109.
252. K.E. Wilson, A.J. Kempf, L.M. Liesch and B.H. Arison, *J. Antibiot.* (1983) **36** 1103.
253. D.R. Houck, K. Kobayashi, J.M. Williamson and H.G. Floss, *J. Am. Chem. Soc.* (1986) **108** 5365.
254. R.W. Woodard, M.-D. Tasi, H.G. Floss, P.A. Crooks and J.K. Coward, *J. Biol. Chem.* (1980) **255** 9124.
255 T.M. Zydowsky, L.F. Courtney, V. Frasca, K. Kobayashi, H. Shimizu, L.-D. Yuen, R.G. Mathews, S.J. Benkovic and H.G. Floss, *J. Am. Chem. Soc.* (1986) **108** 3152.
256 Y. Nozaki, K. Kitano and A. Inada, *Agric. Biol. Chem.* (1984) **48** 37.
257. N. Shibamato, A. Koki, M. Nishino, K. Nakamura, K. Kiyoshima, K. Okamura, M. Okabe, R. Okamato, Y. Fukagawa, Y. Shimauchi and T. Ishikura, *J. Antibiot.* (1980) **33** 1128; K. Yamamoto, T. Yoshioka, Y. Kato, N. Shibamoto, K. Okamura, Y. Shimauchi and T. Ishikura, *J. Antiobiot.* (1980) **33** 796.
258. N. Tsuji, K. Nagashima, H. Kobayashi, Y. Terui, K. Matsumoto and E. Konda, *J. Antibiot.* (1982) **35** 536 and references therein.

259. N. Tsuji, M. Kobayashi, Y. Terui, K. Matsumoto, Y. Takahashi and E. Kondo, *ibid.* (1985) **38** 270 and references therein.
260. S.J. Box, G. Hanscomb and S.R. Spear, *ibid.* (1981) **34** 600 and references therin.
261. P.J. Cassidy, G. Albers-Schönberg, R.T. Goegelman, T.W. Miller, B.H. Arison, E.O. Stapley and J. Birnbaum, *ibid.* (1981) **34** 637 and references therein.
262. Y. Kawamura, Y. Yasuda, M. Mayama and K. Tanaka, *J. Antibiot.* (1982) **35** 10; J. Shoji, H. Hinoo, R. Sakazaki, N. Tsuji, K. Nagashima, K. Matsumoto, Y. Takahashi, S. Kozuki, T. Hattori, E. Kondo and K. Tanaka, *ibid.* (1982) **35** 15; S. Tanabe, M. Okuchi, M. Nakayama, S. Kimura, A. Iwasaki, T. Mizoguchi, A. Murakami, H. Itoh and T. Mori, *ibid.* (1982) **35** 1237.
263. N. Shibamoto, M. Nishino, K. Okamura, Y. Fukagawa and T. Ishikura, *ibid.* (1982) **35** 763.
264. M. Nakayama, S. Kimura, S. Tanabe, T. Mizoguchi, I. Watanabe and T. Mori, *ibid.* (1981) **34** 818; M. Nakayama, S. Kimura, T. Mizoguchi, S. Tanabe, A. Iwasaki, A. Murakami, M. Okuchi, H. Itoh and T. Mori, *ibid.* (1983) **36** 943.
265. S.J. Box, J.D. Hood and S.R. Spear, *ibid.* (1979) **32** 1239.
266. D. Butterworth, J.D. Hood and M.S. Vernall, in *Advances in Biotechnological Processes 1*, (Ed. A. Mizraki) Alan R. Liss, New York (1983) p. 264.
267. I. Kojima, Y. Fukagawa, M. Okabe, T. Ishikura and N. Shibamoto, *J. Antibiot.* (1988) **41** 899.
268. Y. Nakamura, K. Ishi, E. Ono, M. Ishihara, T. Kohda, Y. Yokogawa and H. Shibai, *J. Antibiot.* (1988) **41** 707.
269. Y. Fukagawa, M. Okabe, S. Azuma, I. Kojima, T. Ishikura and K. Kubo, *ibid.* (1984) **37** 1388.
270. K. Kubo, T. Ishikura and Y. Fukagawa, *ibid.* (1984) **37** 1394.
271. *Idem, ibid.* (1985) **38** 622.
272. *Idem, ibid.* (1985) **38** 333.
273. *Idem, ibid.* (1985) **38** 904.
274. J.H. Bateson, R.I. Hickling, T.C. Smale and R. Southgate, *J. Chem. Soc., Perkin Trans. 1* (1990) 1793.
275. A. Imada, K. Kitano, K. Kintaka, M. Muori and M. Asai, *Nature (London)* (1981) **289** 590.
276. R.B. Sykes, C.M. Cimarusti, D.P. Bonner, K. Bush, D.M. Floyd, N.H. Georgopapadoakou, W.H. Koster, W.C. Liu, W.L. Parker, P.A. Principe, M.L. Rathnum, W.A. Slusarchyk, W.H. Trejo and J.S. Wells, *Nature (London)* (1981) **291** 489.
277. J. O'Sullivan, A.M. Gillum, C.A. Aklonis, M.L. Souser and R.B. Sykes, *Antimicrob. Agents Chemother.*, (1982) **21** 558.
278. J. O'Sullivan, M.L. Souser, C.C. Kao and C. Aklonis, *Antimicrob. Agents Chemother.* (1983) **23** 598.
279. W.L. Parker, J. O'Sullivan and R.B. Sykes, *Adv. Appl. Microbiol.* (1986) **31** 181.
280. M. Hashimoto, T. Komori and T. Kamiya, *J. Am. Chem. Soc.* (1976) **98** 3023; J. Hosoda, T.-A. Konomi, N. Tani and H. Imanaka, *Agric. Biol. Chem.* (1977) **41** 2013; J. Hosada, N. Tani, T. Konomi, S. Ohsawa, H. Aoki and H. Imanaka, *Agric. Biol. Chem.* (1977) **41** 2007.
281. C.A. Townsend and A.M. Brown, *J. Am. Chem. Soc.* (1981) **103** 2873; *idem, ibid.* (1983) **105** 913.
282. *Idem, ibid.* (1982) **104** 1748.
283. C.A. Townsend, A.M. Brown and L.T. Bguyen, *J. Am. Chem. Soc.* (1983) **105** 919.
284. C.A. Townsend and L.T. Nguyen, *J. Am. Chem. Soc.* (1981) **103** 4582; C.A. Townsend, G.M. Salituro, L.T. Nguyen and M.J. DiNovi, *Tetrahedron Lett.* (1986) 3819.
285. C.A. Townsend, A. McE. Reeve and G.M. Salituro, *J. Chem. Soc., Chem. Commun.* (1988) 1579.
286. B.T. Golding, I.K. Nassereddin and D.C. Billington, *J. Chem. Soc., Perkin Trans. 1.* (1985) 2007; B.T. Golding and I.K. Nassereddin, *ibid.* 2017 and references therein.
287. H.G. Floss in *Molecular Mechanisms in Bioorganic Processes*. (Eds C. Bleasdale and B.T. Golding) (1990) Proceedings Royal Society of Chemistry Joint Perkin Division Biorganic Group Meeting, 1989. See also references therein.

288. C.A. Townsend and G.M. Salituro, *J. Chem. Soc., Chem. Commun.* (1984) 1631.
289. C.A. Townsend and B.A. Wilson, *J. Am. Chem. Soc.* (1988) **110** 3320.
290. B.A. Wilson, S. Bantia, G.M. Salituro, A. McE. Reeve and C.A. Townsend, *J. Am. Chem. Soc.* (1988) **110** 8238.
291. D.W. Woolley, R.B. Pringle and A.C. Braun, *J. Biol. Chem.* (1952) **197** 409; W.J. Stewart, *Nature* (*London*) (1971) **229** 174.
292. P.J. Langeston-Unkefer, A.C. Robinson, T.J. Knight and R.D. Durbin, *J. Biol. Chem.* (1987) **262** 1608.
293. Comment in B. Müller, A. Hadener and C. Tamm, *Helv. Chim. Acta* (1987) **70** 412.
294. P. Roth, A. Hadener and C. Tamm, *Helv. Chim. Acta* (1990) **73** 476.
295. C.J. Unkefer, R.E. London, R.D. Durbin, T.F. Uchytil and P. J. Langston-Unkefer, *J. Biol. Chem.* (1987) **262** 4994.
296. H. Wehrli, *Helv. Chim. Acta* (1980) **63** 1915.

2 Structure–activity relationships: chemical

M.I. PAGE

2.1 The reactivity of the β-lactam

β-Lactams occur relatively rarely in nature, therefore it is not surprising that the biological activity of these compounds should be attributed to the chemical reactivity of the β-lactam ring. Shortly after the introduction of penicillin to the medical world it was suggested that the antibiotic's activity was due to the inherent strain of the four-membered ring[1] or to reduced amide resonance.[2] Amide resonance is responsible for the lower susceptibility of the carbonyl group to nucleophilic attack, and a reduction of this resonance will lead to increased reactivity. In a normal amide the planar arrangement of the O, C and N atoms is generally assumed to be necessary for the effective delocalisation of the nitrogen lone pair, and therefore the non-planar butterfly shape of the penicillin molecule (**1**) could reduce amide resonance. For several decades these two proposals dominated the thoughts of synthetic chemists who were, and to some extent still are, convinced that more effective antibiotics may be made by making the β-lactam system more strained or non-planar. However, the evidence to support an unusually strained or an amide-resonance inhibited β-lactam in penicillin is ambiguous.

(**1**) (**2**)

It is estimated[3] that resonance stabilises amides by about 18 kcal mol^{-1}. The reason for the greater stability of amides compared with other carbonyl-containing functionalities is attributed to the unfavourable loss of resonance when nucleophiles attack the amide carbonyl carbon to form a tetrahedral intermediate (Scheme 2.1). If delocalisation is completely inhibited in an amide then the rate of nucleophilic attack could occur up to 10^{13}-fold faster than in the analogous resonance-stabilised system. The strain energy of a four-membered ring[4] is 26–29 kcal mol^{-1} and a reaction involving opening of the β-lactam ring could therefore take place faster than the analogous bond fission process in a strain-free amide by a factor of up to 10^{20}. If strain or resonance inhibition is even slightly significant in penicillins and cephalosporins their effects should therefore be easily observable.[4,5]

Scheme 2.1

The treatment of amide resonance as a result of delocalisation of the nitrogen lone pair by overlap of the 2*p*-orbitals on the participating atoms (**2**) has been used to predict that a pyramidal amide nitrogen will cause loss of resonance energy.[2] Pyramidalisation of the nitrogen will, however, not necessarily produce the same effect as that caused by rotation about the C–N bond, i.e. a change in the dihedral angle between the *p*-orbitals with the O, C and N atoms remaining coplanar. These contrasting effects are illustrated, respectively, by the Newman projection formulae (**3**) and (**4**) obtained by looking along the C–N bond. Despite these differences, the assumption that amide resonance in penicillins and cephalosporins is inhibited has been generally accepted and several experimental observations have been used to support this suggestion.

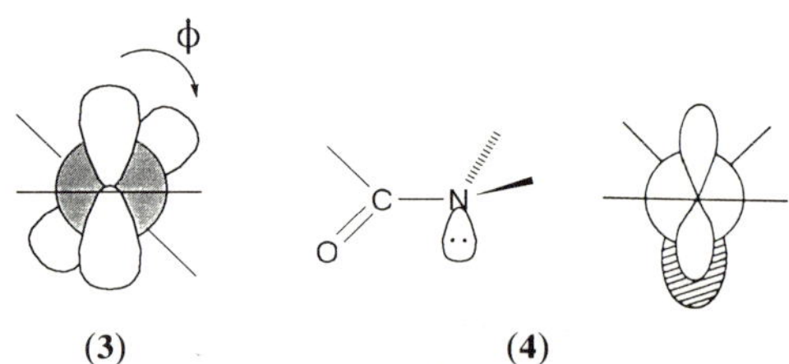

(**3**) (**4**)

2.2 Structural and ground-state effects

Amide resonance is usually depicted by the canonical forms (**5**) and (**6**). Inhibition of amide resonance should make the amide resemble (**5**) at the

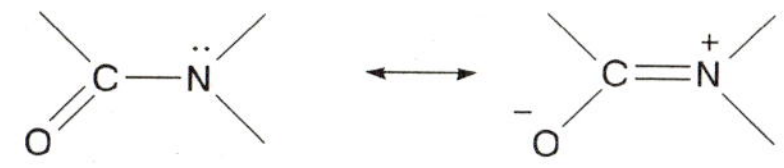

(5) (6)

expense of (**6**). The reasonable conclusion would be that, compared with a normal amide, resonance inhibition would:

(i) increase the C–N bond length and decrease the C–N bond strength;
(ii) decrease the C–O bond length and increase the C–O bond strength;
(iii) decrease the positive charge density on nitrogen; and
(iv) decrease the negative charge density on oxygen.

2.2.1 *Planarity of the nitrogen and bond lengths*

Although acetamide has C_s symmetry in the gas phase and solution, the carbonyl carbon is pyramidalised in the crystalline state.[6] The degree of coplanarity of the β-lactam nitrogen in β-lactam antibiotics can be expressed either by the perpendicular distance, h, of the nitrogen from the plane of its three substituents or by the sum of the bond angles about nitrogen. The former is easier to visualise and the nitrogen ranges from being essentially in the plane of its three substituents in monocyclic β-lactams to being 0.5 Å out of the plane in bicyclic systems. Examples of h-values are given in Table 2.1[7] and, until recently, it had been generally assumed that a more pyramidal nitrogen decreased amide resonance and increased biological activity. Therefore great effort has been discharged in making non-planar β-lactams. Some 1-carba-1-penems show very high h-values of up to 0.54 Å and yet may be biologically inactive. Furthermore, there is no direct correlation between h-values and chemical reactivity. The C–N bond length in planar monocyclic β-lactams (1.35 Å) is generally longer than that in amides (1.33 Å). The converse is true for C=O bond lengths — 1.24 Å for amides compared with 1.21 Å for monocyclic β-lactams. In non-planar penicillins and cephalosporins there is a general trend for the C–N bond length to increase as the C=O bond length decreases. However, this trend is by no means linear. Bond lengths for C=O vary from 1.17 to 1.24 Å and for C–N from 1.33 to 1.46 Å. There is also a tendency for the C–N bond length to increase with h (Table 2.1).

It is difficult to discern reasons and reactivity consequences of these differences in bond length. Penicillin V ((**1**): R = $PhOCH_2$) shows the longest C–N bond length of 1.46 Å and yet the C=O bond length 1.21 Å is identical to that commonly found in planar monocyclic β-lactams. In monocyclic β-lactams the nitrogen is coplanar with its three substituents and yet the bond length differences are also in the direction predicted by inhibition of amide resonance. The degrees of non-planarity in penicillin V ((**1**): R = $PhOCH_2$) and ampicillin ((**1**): R = $PhCH(NH_2)$) are similar (h = 0.40

Table 2.1 Structural parameters of some β-lactams.

Compound	C=O stretch (cm^{-1})	Distance of N from plane *h* (Å)	β-lactam C=O bond length (Å)	β-lactam C–N bond length (Å)
Penicillins[a]	1770–1790			
Ampicillin[b]		0.38	1.20	1.36
Benzylpenicillin[c]		0.40	1.17	1.34
Phenoxymethyl-penicillin[d]		0.40	1.21	1.46
Δ^3-Cephalosporins[e]	1760–1790			
Cephaloridine[f]		0.24	1.21	1.38
Δ^2-Cephalosporins[e]	1750–1780			
Phenoxymethyl Δ^2-cephalosporin[f]		0.06	1.22	1.34
Anhydropenicillins[g]	1810			
Phenoxymethyl-anhydropenicillin[h]		0.41	1.18	1.42
Monocyclic β-lactams	1730–1760	0	1.21	1.35
Amides	1600–1680	0	1.24	1.33

[a] R.B. Morris and B.G. Jackson, *Fortsch. Chem. Org. Naturst* (1970) **28** 343.
[b] M.N.G. James, D. Hall and D.C. Hodgkin, *Nature (London)* (1968) **220** 168.
[c] G.J. Pitt, *Acta Cryst.* (1952) **5** 770.
[d] S. Abrahamsson, D.C. Hodgkin and E.N. Maslen, *Biochem. J.* (1963) **86** 514.
[e] G.G.F.H. Green, J.E. Page and S.E. Straniforth, *J. Chem. Soc.* (1965) 1595.
[f] R.M. Sweet and L.F. Dahl, *J. Am. Chem. Soc.* (1970) **92** 5489.
[g] S. Wolfe and W.S. Lee, *J. Chem. Soc., Chem. Commun.* (1968) 242.
[h] G.L. Simon, R.B. Morin and L.F. Dahl, *J. Am. Chem. Soc.* (1972) **94** 855.

and 0.38 Å, respectively) and yet the C–N bond length in the former is 0.10 Å longer than in the latter.

Structural data have also been used to support the suggestion that enamine resonance is important in cephalosporins and that this also reduces amide resonance.[8] However, there is no significant difference in the C–O and C–N bond lengths in cephalosporins from the general trend exhibited by penicillins. Furthermore, the C-4–N-5 bond length of 1.51 Å in the Δ^3-cephalosporin, cephaloglycin, is longer than that of 1.45 Å in Δ^2-cephems and that of 1.46 Å in cephams where enamine resonance cannot occur.[9]

It would seem logical to conclude that variations in bond lengths within penicillin and cephalosporin derivatives are caused by the nature of substituents and the minimisation of unfavourable strain energies caused by the geometry of the molecule. To attribute these differences to the inhibition of amide resonance seems speculative and is only supported by the selection of examples.

2.2.2 *Nmr chemical shifts*

The conformation of substituents on nitrogen relative to the carbonyl group has a significant effect on the carbonyl ^{13}C chemical shift in amides. For example, a difference of 4 ppm is observed in the carbonyl ^{13}C resonances of the *E*- and *Z*-isomers of N-methylformamide.[10] There does not appear to be a simple relationship between ^{13}C chemical shifts and the local electron density distribution. However, there have been several attempts to correlate antibacterial activity with chemical shifts.

The β-lactam carbonyl carbon usually resonates between 160 and 167 ppm in a ^{13}C nmr spectrum.[11] This is the same region in which the carbonyl resonances of formamide and its N-methylated derivatives also appear.[10] It is interesting to note that the carbonyl resonances of γ- and larger-membered lactams appear between 170 and 180 ppm.[12]

There is little variation in the chemical shifts of the β-lactam carbonyl carbon of penicillins and cephalosporins.[7] The carbonyl carbon of the β-lactam in penicillins resonates about 10 ppm to lower field than that in cephalosporins. Surprisingly, the shifts in the biologically active Δ^3- and the inactive Δ^2-cephalosporins are similar. Inhibition of amide resonance may be expected to make the carbonyl carbon more electron deficient. Although the difference in chemical shifts between penicillins and Δ^3-cephalosporins support this proposal, it is not apparent from the Δ^2-/Δ^3-cephalosporin comparison. In penicillins the nitrogen is 0.4 Å from the plane of its substituents compared with 0.2 Å in Δ^3-cephalosporins, whereas the ceph-2-em systems are planar. The similarity of the values of the ^{13}C shifts found for the β-lactam carbonyl carbons in ceph-3-ems and ceph-2-ems indicates that the charge density and bond order at the carbonyl carbons in both systems are approximately the same.

^{15}N Chemical shifts of the β-lactam nitrogen in ceph-3-ems are almost invariant ($< \pm 1$ ppm) with the nature of the substituent at C-3 and therefore also do not indicate significant enamine-type resonance in these systems. Not surprisingly, there is a large difference of 15 ppm in the ^{15}N chemical shifts of the β-lactam nitrogen in ceph-2-ems and ceph-3-ems. Interestingly, there is an upfield shift of 30 ppm in the β-lactam nitrogen on going from non-planar penicillins to planar ceph-2-ems.[13] Increased amide conjugation in the planar system would be expected to have induced a downfield shift.

2.2.3 *Infrared carbonyl stretching frequency*

The β-lactam infrared stretching frequency ($\nu_{C=O}$) has been regarded as an important index both for inhibition of amide resonance and for investigating structure–activity relationships of the β-lactam antibiotics.[14]

In normal penams the β-lactam carbonyl stretching frequency occurs in the 1770–1790 cm^{-1} range compared with 1730–1760 cm^{-1} for monocyclic unfused β-lactams and about 1600–1680 cm^{-1} for amides (Table 2.1). In general, the non-planar 3-cephems show higher stretching frequencies (1786–1790 cm^{-1}) than the planar 2-cephems, which absorb at 1750–1780 cm^{-1}. The frequency in cephalosporins increases by *ca.* 5 cm^{-1} when the ring sulphur is replaced by oxygen but decreases by a similar amount when the 7-α-hydrogen is substituted by a methoxy group. It is difficult to make generalisations about the observed β-lactam frequency, since different conditions of measurement (KBr, film, solution, etc.) may cause variations comparable with those produced by structural changes. There is a tendency for a high carbonyl stretching frequency to be associated with a shorter β-lactam C=O bond length and a more pyramidal nitrogen. Furthermore, it has been tempting to associate a high carbonyl stretching frequency with increased strain, increased double-bond character and reduced amide resonance. However, the evidence is again ambiguous. Although selected examples may show some of these interrelationships, there are many exceptions; for example, the carbonyl stretching frequency for some penems decreases 20 cm^{-1} whilst the β-lactam nitrogen becomes more pyramidal by 0.12 Å.

The direct interpretation of carbonyl stretching frequencies in terms of bond strengths or electron density distributions is not straightforward. Many subtle effects can alter the frequency, even if the force constant for C=O stretching — which is presumably the best indicator of bond strength — remains constant. For example, in the system X–C=O the carbonyl stretching frequency can be increased by decreasing the C–X bond length, by increasing the C–X stretching force constant or by increasing the XCO bending force constant.[15]

2.2.4 *Theoretical calculations*

Theoretical geometry optimisation of β-lactams at a semi-empirical level, and a limited *ab initio* study using minimal basis STO-3G calculations at fixed geometries have been reported.[16] The calculated STO-3G energy of formamide in a penicillin-like geometry is only 2.8 kcal mol^{-1} higher than the planar geometry.[17] Furthermore, in general, the geometrical parameters associated with the β-lactam ring vary only slightly with changes in the hybridisation at nitrogen. An exception is the C–N bond length, which becomes longer as the nitrogen becomes pyramidal.

The barrier to inversion at nitrogen in ammonia is 5.8 kcal mol^{-1}, which is much greater than that in molecules like formamide.[18] Formamide lies in a potential well, which is very flat with respect to inversion at nitrogen. The inversion barrier is lower for molecules favouring a large angle at nitrogen (amides) and higher for systems adopting a small angle at nitrogen (e.g.

aziridine). It appears that the nitrogen in amides can be made pyramidal without severe changes in energy.

2.2.5 *Basicity of β-lactam nitrogen*

Inhibition of amide resonance in bicyclic β-lactams will make the amide resemble canonical form (5). An expected consequence of increased localisation of the lone pair on nitrogen would be to increase the basicity of nitrogen.[5,7]

It is well known that torsional strain in amides can increase the basicity of nitrogen. For example, 6,6-dimethyl-1-azabicyclo[2.2.2]octan-2-one (**7**) presumably has the nitrogen lone pair almost orthogonal to the carbonyl π system, and amide resonance is consequently inhibited.[19]

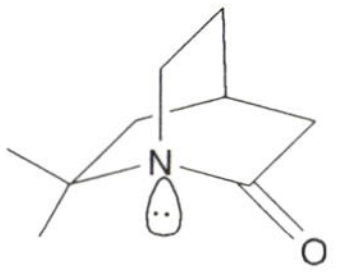

N O

(**7**)

Amides are normally only very weakly basic and the pK_a-values of their conjugate acids are around zero. By contrast, (**7**) is half protonated at pH 5.3, consistent with the increased basicity of the amide nitrogen.

If amide resonance in penicillins is inhibited because of the pyramidal nature of the β-lactam nitrogen, penicillins should also show enhanced basicity compared with normal amides. There is no evidence to suggest that this is the case. In fact, penicillins appear to show reduced basicity and cannot be detectably protonated even in 12 M hydrochloric acid.[20]

Another indication of increased nitrogen basicity would be a large binding constant of penicillin to metal ions. However, the equilibrium constant for metal-ion coordination between the carboxyl group and β-lactam nitrogen (**8**) is only about 100–200 M^{-1} for a variety of metal ions.[21] This is the order of magnitude expected for coordination between a normal amide and a carboxyl group.

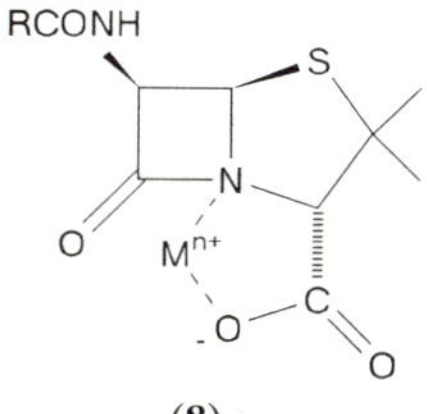

RCONH S N O M^{n+} ^-O C O

(**8**)

There appears, therefore, to be no evidence of substantial inhibited amide resonance in penicillins, and the β-lactam nitrogen shows no enhanced electron pair donating ability either to a proton or to metal ions.

2.3 Kinetic effects

Nucleophilic substitution at the carbonyl group of an amide invariably occurs in a stepwise manner by initial formation of a tetrahedral intermediate (Scheme 2.1). Conversion of the three-coordinate, sp^2-hybridised carbonyl carbon to a four-coordinate sp^3-hybridised carbon in the intermediate must be accompanied by the loss of amide resonance. This contribution to the activation energy will be reduced if amide resonance is inhibited and, in such cases, a rate enhancement is expected. Similarly, the release of strain energy will increase the rate if the four-membered ring is opened in the transition state. The total strain energy of four-membered rings is probably not released until there is significant bond extension.

A simple way to see if either of these effects is apparent is by examining the rates of hydrolysis of the β-lactam antibiotics.[5,7] The alkaline hydrolysis of benzylpenicillin opens the β-lactam ring to give benzylpenicilloate (**9**) and occurs at a rate similar to that for ethyl acetate. The pK_a-value of the protonated amine in the thiazolidine derivate (**9**) is 5.2 and, because of this weakly basic nitrogen, the leaving group ability of the amine is expected to be improved. Consequently, in order to assess any special reactivity of the β-lactam antibiotics, the dependence of the rate of hydrolysis of simple amides and β-lactams upon substituents must be known.

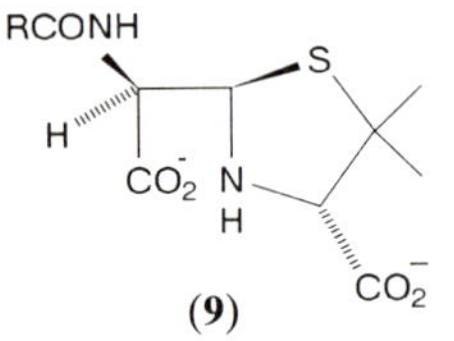

(9) 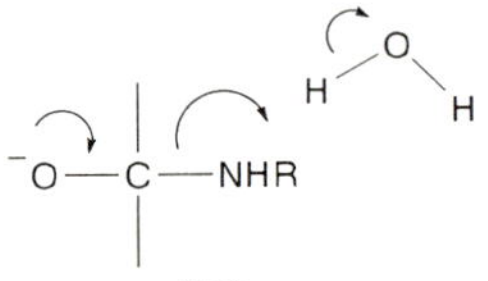 **(10)**

A Bronsted plot of the second-order rate constants for the hydroxide-ion catalysed hydrolysis of acyclic amides, monocyclic β-lactams and bicyclic β-lactams is shown in Figure 2.1.[20,22]

The Bronsted β_{1g} value for N-substituted acyclic amides and anilides is -0.07, which is compatible with rate-limiting breakdown of the tetrahedral intermediate in which the nitrogen has a partial positive charge. The observations are consistent with water acting as a general acid catalyst in the breakdown of the tetrahedral intermediate (**10**). The rates of alkaline hydrolysis of β-lactams exhibit a first-order dependence on hydroxide ion concentration and show a Bronsted β_{1g} value of -0.44, which is indicative of

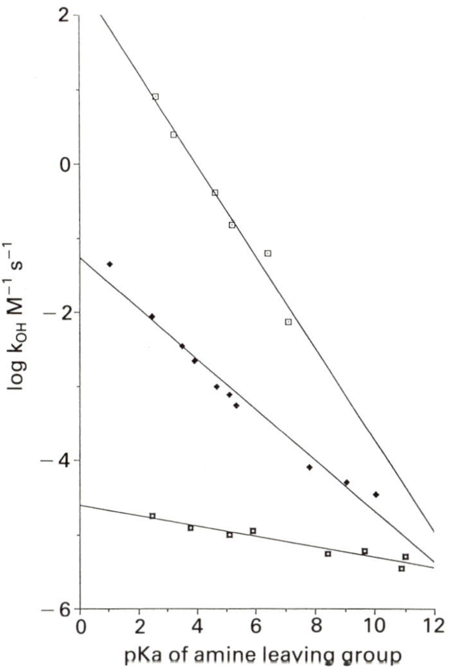

Figure 2.1 Bronsted plot of the second-order rate constants for the hydroxide-ion catalysed hydrolysis of acyclic amides (□), monocyclic β-lactams (◆), and bicyclic β-lactams (⊡) against pK_a for the leaving group amine. Data refer to 30°C and are taken from P. Proctor, N.P. Gensmantel and M.I. Page, *J. Chem. Soc.* (1982) **2** 1185.

rate-limiting formation of the tetrahedral intermediate. β-Lactams of basic amines show very similar reactivity to that of analogous non-cyclic amides. It is only β-lactams of weakly basic amines which show enhanced reactivity. The rate enhancement of β-lactams compared with acyclic amides thus depends upon the basicity of the leaving group amine. β-Lactams of weakly basic amines (pK_a of conjugate acid less than 4) are *ca.* 500-fold more reactive than an acyclic amide of the same amine. It is worth noting that β-lactones are also only about 10-fold more reactive than analogous esters towards alkaline hydrolysis.

Crystallographic and spectroscopic evidence shows that N-substituted β-lactams are planar and resonance-stabilised as in amides. The rate enhancement of 30 to 500-fold shown by β-lactams of weakly basic amines may be rationalised by the change in coordination number/hybridisation of the carbonyl carbon as the tetrahedral intermediate is formed in the four-membered ring. The magnitude is similar to the 500-fold faster rate of reduction of cyclobutanone by borohydride compared with that of acetone, which is due to the release of strain energy upon converting three-co-ordinated carbon to four.[23]

As the rate-limiting steps for the alkaline hydrolysis of amides and β-lactams are different, the relatively small rate enhancement shown by β-lactams indicates that the energy of the transition state for breakdown of the tetrahedral intermediate in amide hydrolysis is probably not significantly greater than that for formation of the intermediate.

Fusing the β-lactam ring to a five-membered ring to make 1-aza-bicyclo-[3.2.0]heptan-2-ones increases the reactivity by *ca*. 100-fold but does not significantly change the Bronsted β_{1g} value, which is -0.55 for the bicyclic system. Although the rate enhancement is substantial, it is hardly of the magnitude expected from the release of strain energy in opening a four-membered ring, or from a system in which amide resonance is significantly inhibited. Ring opening does not lower the activation energy because the rate-limiting step for the alkaline hydrolysis of penicillins is formation of the tetrahedral intermediate. The Bronsted β_{1g} of -0.55 indicates that the nitrogen behaves as if it has no charge in the transition state and has lost all of the expected 0.6 positive charge present in the resonance-stabilised β-lactam. This is compatible with a transition state that very much resembles the tetrahedral intermediate. Furthermore, the β_{1g} of -0.55 indicates that the positive charge density on the β-lactam nitrogen in penicillins is similar to that in monocyclic β-lactams and amides where resonance is established.

It is worth noting that monocyclic β-lactams of weakly basic amines can be as chemically reactive as penicillins and cephalosporins. It is not necessary to make the β-lactam part of a bicyclic system to have a reactive amide. The second-order rate constants for the alkaline hydrolysis of some β- and γ-lactam derivatives and some amides are given in Table 2.2. It can be seen that there are potentially many amide and lactam derivatives that are as chemically reactive as the penicillins and cephalosporins. The molecular recognition between the lactam or amide and its host protein is essential for biological activity. Nonetheless, the chemical reactivity shown by many non-β-lactam compounds indicates their potential acylating power and possible antibiotic activity with suitably placed substituents (see chapter 9).

2.4 Summary of kinetic and ground-state effects[5,7]

Both kinetic and ground-state effects do not indicate a significant degree of inhibition of amide resonance in penicillins and cephalosporins. The bicyclic β-lactam antibiotics do not exhibit exceptional chemical reactivity. Monocyclic β-lactams with suitable electron-withdrawing substituents may be as reactive as the bicyclic systems. A pyramidal geometry of the β-lactam nitrogen does not necessarily give a chemically more reactive β-lactam. Strained β-lactams are not necessarily better antibiotics and biological activity is not directly related to chemical reactivity.

Table 2.2 Second-order rate constants ($dm^3\ mol^{-1}\ s^{-1}$) for the hydroxide-ion catalysed hydrolysis of penicillins, cephalosporins, β-lactams and amides at 30°C.

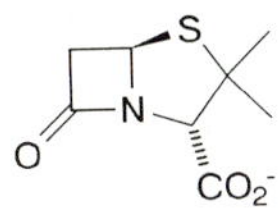 7.4×10^{-3}	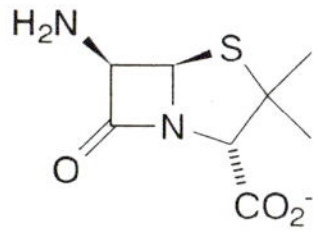 6.31×10^{-2}	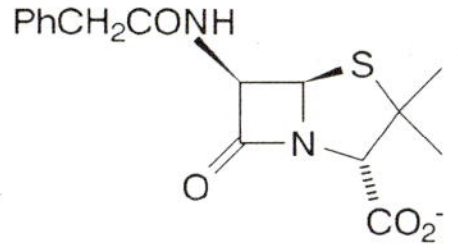 1.54×10^{-1}
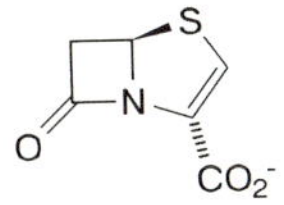 5.10×10^{-1}	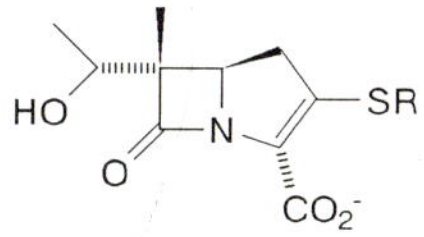 1.59	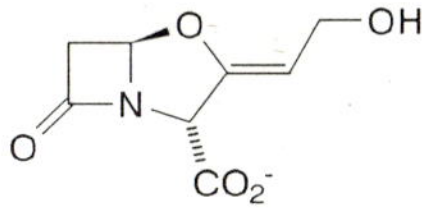 2.39
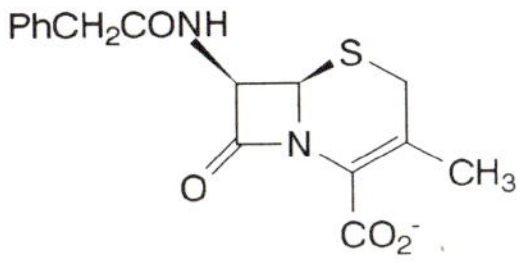 2.90×10^{-2}	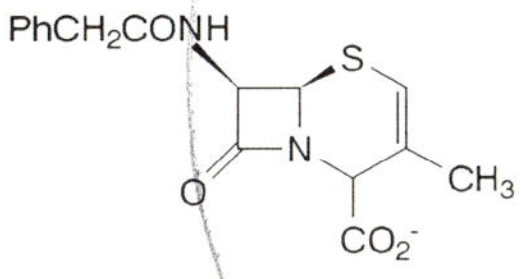 1.10×10^{-2}	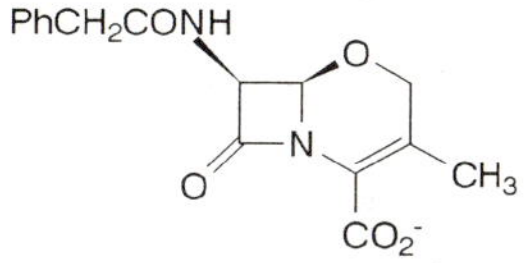 1.78×10^{-1}
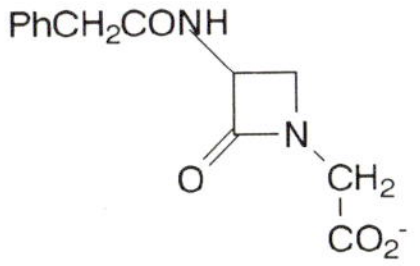 1.02×10^{-4}	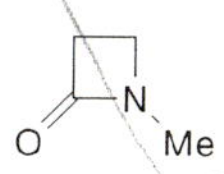 6.1×10^{-5}	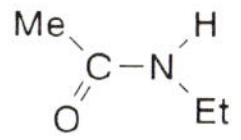 5.4×10^{-6}
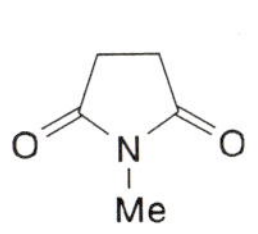 0.93	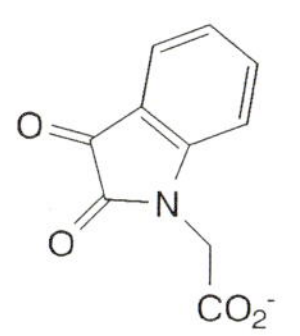 0.15	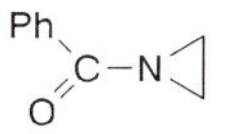 1.9

2.5 Structure–chemical reactivity relationships

It is well known that minor substituent changes in β-lactam antibiotics can have a dramatic effect on antibacterial activity and susceptibility to β-lactamase catalysed hydrolysis. In order to identify the molecular recognition between β-lactams and their host enzymes, the effect of substituents on their chemical reactivity must be known.

The relative rates of enzyme catalysed reactions of a series of substrates may be due to a combination of differences in intrinsic 'chemical' reactivity of the substrates and their binding interactions with the enzyme. The effect of changes in substrate structure on enzyme catalytic activity are often used directly to identify specific binding sites between parts of the substrate and the enzyme. However, before this is done, allowance should be make for the changes substituents can make to the intrinsic 'chemical' reactivity by inductive, resonance and steric effects. A simple method to estimate intrinsic chemical reactivity is from the rates of alkaline hydrolysis of β-lactams. The ratio of the second-order rate constant, k_{cat}/K_m, for the enzyme catalysed reaction to that for the hydroxide-ion catalysed hydrolysis, k_{OH}, gives the 'enzyme rate enhancement factor' (EREF).[24] This factor normalises intrinsic chemical reactivity and relative values, and gives a much better way of estimating binding interactions between the substrate and enzyme than do the simple relative values of k_{cat}/K_m.

2.5.1 *Hydrolysis*

The effects of structural changes on the rates of alkaline hydrolysis of penicillin and cephalosporin derivatives are summarised in Table 2.2. Decreasing the basicity of the leaving group amine increases the rate of alkaline hydrolysis of *penicillins*. The Bronsted β_{lg} of -0.55 is indicative of rate-limiting formation of a tetrahedral intermediate (**11**).[20]

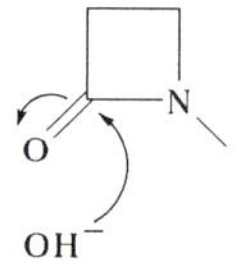

(11)

Electron-withdrawing substituents at C-6 also increase the rate of hydroxide ion hydrolysis and give a Hammett ϱ_I-value of 2.0, which is slightly less than the value of 2.7 for acyclic amides.[25] Lactams are cyclic, therefore substituents at C-6 affect the rate of nucleophilic substitution by their effect on both the electrophilicity of the carbonyl carbon and the leaving ability of the amine. Although the acylamido side chain at C-6 is important for biological activity, and increases the rate of alkaline hydrolysis 20-fold relative to penicillanic acid, its effect on chemical reactivity is purely inductive.

The replacement of the thiazolidine S by CH_2 to give a carbapenam increases the rate by a factor of three, whereas substitution by O as in the oxapenams increases the rate *ca.* five-fold (Table 2.2). The effect of the replacement by oxygen is that expected on the basis of an inductive effect.

The C–S–C bond angle is relatively small and the C–S bond length relatively long, so that the carbapenam is expected to be more strained. However, the replacement of S by CH_2 will also decrease the leaving group ability of the β-lactam amine by making it more basic. Presumably these effects must cancel so that there is little difference in reactivity between the penams and carbapenams.

The incorporation of a double bond into the thiazolidine ring of a penam to give the corresponding penem system increases the rate of hydrolysis by *ca.* 25-fold. This is the order of magnitude expected from the decrease in basicity of the leaving group amine brought about by the introduction of a conjugated amine in the tetrahedral intermediate. Conversion of a Δ^2-carbapenem to a Δ^1-carbapenem similarly decreases the reactivity 25-fold. Esterification of the C-3 carboxyl group also reduces the basicity of the leaving group amine so that the rate of alkaline hydrolysis is increased 16-fold.[24]

The major structural differences between *cephalosporins* and penicillins are that the five-membered thiazolidine ring of penicillins is replaced by a six-membered dihydrothiazine ring in cephalosproins and that the degree of pyramidalisation of the β-lactam nitrogen is generally smaller in cephalosporins. In addition, many of the cephalosporins have a leaving group, e.g. acetate, pyridine or thiol, at C-3′ and expulsion of these groups occurs during the hydrolysis of the β-lactam as shown in Scheme 2.2. There is no evidence[7] other than that suggested by theoretical calculations[26] that nucleophilic attack on the β-lactam carbonyl carbon is concerted with departure of the leaving group at C-3′.

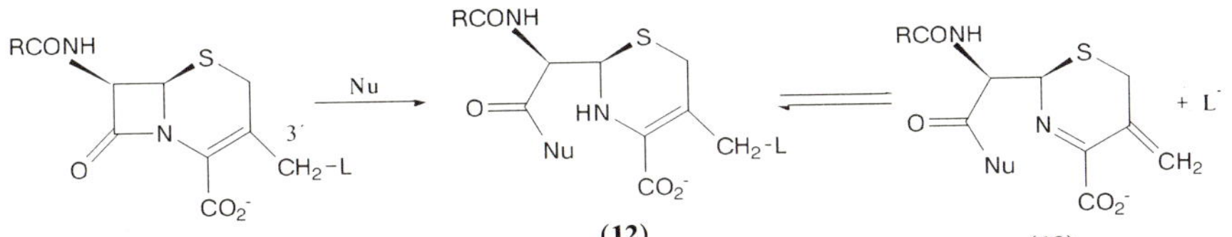

(12) (13)

Scheme 2.2

In general, the second-order rate constants for the hydroxide-ion catalysed hydrolysis of cephalosporins are similar to those of penicillins. This similarity indicates that the non-planarity of the β-lactam nitrogen does not significantly affect amide resonance since the nitrogen is pyramidal by 0.4 Å in penicillins, whereas in the cephalosporins it deviates by only 0.2–0.3 Å. The similarity in rates also indicates that the presence of a leaving group at C-3′ does not significantly affect the reactivity of cephalosporins. The stepwise mechanism of hydrolysis involves the breakdown of the tetrahedral intermediates to generate the enamine **(12)** followed by expulsion of the leaving group at C-3′ to reversibly[27] form the conjugated imine **(13)**.

The second-order rate constants for the hydroxide-ion catalysed hydrolysis of cephalosporins are correlated with σ_I for C-3 substituents and give a Hammett ϱ of 2.5 for CH_2L. Several substituents at C-3, e.g. CH_3, H, CH_2CO_2Et, are not expelled during hydrolysis and yet are controlled by the same, linear, free-energy relationships. Leaving groups of different nucleofugalities influence the rate of reaction only by their inductive effect, and there is little or no change in the effective charge on the leaving group on going from the ground to the transition state.

Esterification of the C-4 carboxylate group or its conversion to a lactone makes the β-lactam carbonyl carbon more electrophilic and facilitates β-lactam C–N bond cleavage so that the rates of hydrolysis increase up to 130-fold.[24]

In addition to the leaving group at C-3′, many other structural parameters within cephalosporins have been varied.[7] Although the change from a Δ^3- to a Δ^2-cephem system causes the β-lactam nitrogen to become planar, there is little difference, only two- to three-fold, in the chemical reactivity.

Other effects of structural changes are given in Table 2.2. Replacement of the dihydrothiazine S by O increases the rate of alkaline hydrolysis about six-fold whereas that by CH_2 decreases the rate by up to 30-fold.[28]

The introduction of a 7-α-methoxy group has an almost insignificant effect, less than two-fold, on the susceptibility of cephalosporins to alkaline hydrolysis. Inductively, a 7-α-methoxy group should slightly increase the rate, but unfavourable steric interactions in the tetrahedral intermediate must lower the rate. This steric effect is supported by the 10-fold reduction in the rate of alkaline hydrolysis of penicillins and cephalosporins by the introduction of a 6-α or 7-α methyl group, respectively.

The β-lactam ring of cephalosporins has a reactivity comparable with that of ethyl acetate, therefore it is not surprising that hydrolysis of an acetoxy ester side chain at C-3 is competitive with hydrolysis of the β-lactam. The second-order rate constant for the base catalysed conversion of the C-3 ester to the 3-hydroxylmethylcephalosporin is similar to that for β-lactam hydrolysis.

The pH-rate profile for the hydrolysis of benzylpenicillin shows no significant spontaneous hydrolysis, but the β-lactam does undergo an acid catalysed degradation. By contrast, the hydrolysis of cephaloridine exhibits a spontaneous, pH-independent hydrolysis and is less reactive towards acid than are the penicillins — by a factor of about 10^4.

In addition to the expected hydrolysis product, benzylpenicilloic acid (**9**), the acid catalysed degradation of benzylpenicillin gives benzylpenicillenic acid (**14**), benzylpenamaldic acid (**15**), benzylpenillic acid (**16**) and benzylpenilloic acid (**17**). The proportion of each product formed depends on the pH.[29] Although several kinetic studies have been reported on the degradation of penicillins in acidic media, there is still considerable uncertainty about the details of the reaction pathway.

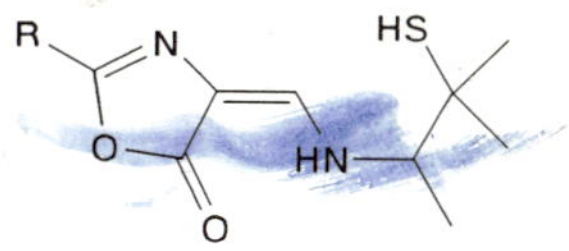

(14) 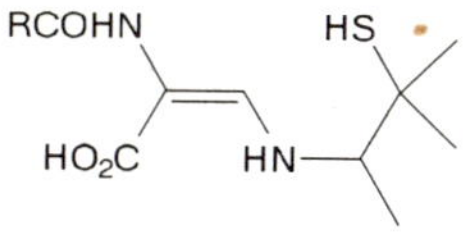 (15)

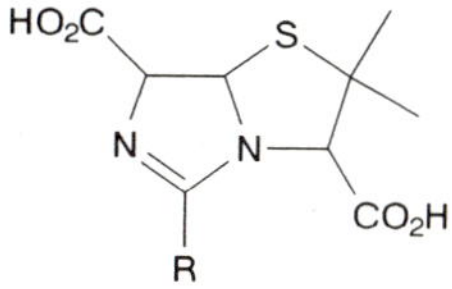

(16) 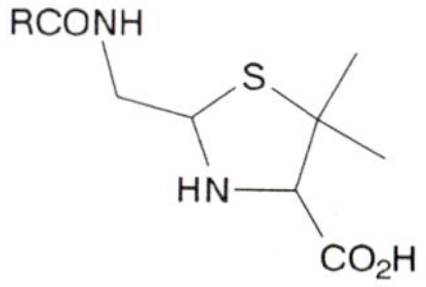 (17)

Remarkably, the logarithms of the pseudo first-order rate constants for the acid catalysed hydrolysis of some β-lactam antibiotics and derivatives increase linearly with decreasing H_o values up to -5. This is quite unlike the behaviour of other amides, for which the rate of hydrolysis passes through a maximum, attributed both to complete conversion of the amide into its O-conjugate acid, and to decreasing water activity. This indicates that the β-lactams are far less basic then normal amides for O-protonation and that a different mechanism of hydrolysis is operating. Neither the nitrogen nor the oxygen of the bicyclic β-lactams is sufficiently basic for substantial conversion to the conjugate acid; the pK_a for O- or N-protonation must be < -5. This behaviour is not peculiar to bicyclic β-lactams, since monocyclic β-lactams show similar reactivities and behaviour. The mechanism of the acid catalysed hydrolysis of β-lactams appears to be a unimolecular A-1 type process, with N-protonation of the β-lactam[20] (Scheme 2.3). N-protonation

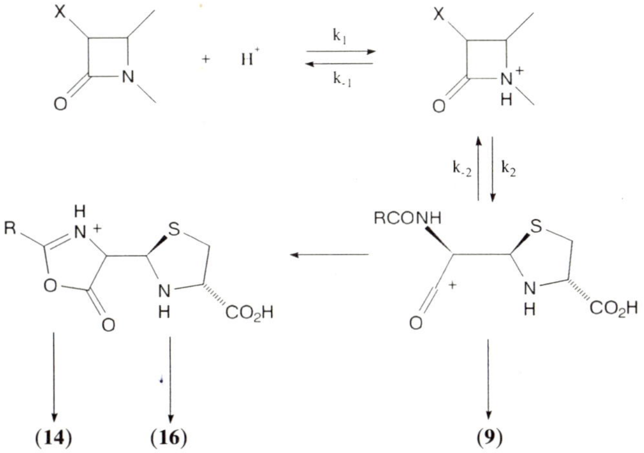

(14) (16) (9)

Scheme 2.3

takes place because of an intrinsic property of β-lactams, and is not the result of reduced amide resonance in penicillins and cephalosporins. The introduction of the A-1 mechanism could result because the normal A-2 mechanism is retarded or because the A-1 pathway is favoured. The most likely explanation is the enhanced rate of C–N bond fission that occurs in β-lactams as a result of the relief of ring strain and the reduced O-basicity of β-lactams.

Substituents at C-6 in penicillins affect both the carbonyl carbon and the nitrogen of the β-lactam inductively, but their effect on C–N bond cleavage will be predominantly that of an acyl substituent. Electron-withdrawing substituents that cannot be involved in neighbouring group participation greatly retard the rate of hydrolysis with a Hammett ϱ_I-value of *ca.* -4.0 to -5.0, depending on the acidity.[20] By contrast, the effect of acyl substituents on the rate of acid catalysed hydrolysis of acyclic amides is small, with electron-withdrawing substituents producing either a small increase or decrease.

Electron-withdrawing substituents in the amine portion of the β-lactam decrease the rate of acid catalysed degradation of penicillins. The Bronsted β-value is *ca.* 0.35 compared with -0.26 for acyclic anilides and amides. Although the effects of substituents are not large, they are significant and in the opposite direction for β-lactams compared with other amides, which again is indicative of a different mechanism.

There is a large dependence of the rate of the acid catalysed degradation of penicillins upon the nature of the acylamido side chain. The acid catalysed degradation of C-6 acylamido penicillins shows a rate of enhancement of *ca.* 10^3 compared with that predicted from the Hammett plot for C-6 substituents. The mechanism of degradation must therefore incorporate the acylamido group in the rate-limiting step or in a pre-equilibrium step.

Cephalosporins are about 10^4-fold less-reactive than penicillins towards acid catalysed hydrolysis.[20] Electron-withdrawing substituents at C-7 in cephalosporins decrease the rate of acid hydrolysis and, as for penicillins, the Hammett ϱ_I-value is *ca.* -5. However, there is no evidence for neighbouring group participation by the 7-acylamido group as postulated for the penicillins. There seems to be no obvious explanation for this different behaviour between the cephalosporins and penicillins.

Penicillins undergo an acid and a base catalysed hydrolysis, but there is no significant uncatalysed reaction. The pH minimum is *ca.* 7 and k_0, the apparent first-order rate constant for spontaneous or water catalysed degradation, is $1 \times 10^{-7}\,s^{-1}$ at 30°C. By contrast, cephalosporins often show a significant pH-independnent reaction between pH 3 and 7, with k_0 in the range 5×10^{-7} to $3 \times 10^{-6}\,s^{-1}$ at 30°C. It has been suggested that this pH-independent reaction involves intramolecular nucleophilic attack on the β-lactam by the 7-amido side chain but, as cephalosporins do not show neighbouring group participation in their acid catalysed degradation, it is difficult to understand why it should occur in the uncatalysed reaction. Furthermore, 7-amido cephalosporins show a similar reactivity to 7-aminocephalosporanic acid in their spontaneous degradation. However, the deuterium solvent isotope effect, $k_0^{H_2O}/k_0^{D_2O}$ of 0.93 is not typical of a water catalysed hydrolysis. Unlike the spontaneous degradation of cephalosporins, that of penicillins shows a significant solvent isotope effect — $k_0^{H_2O}/k_0^{D_2O}$ is 4.5.

2.5.2 Alcoholysis

The alcoholysis of penicillin is thought to be the first chemical step in the reaction of β-lactam antibiotics with transpeptidase and β-lactamase enzymes (chapters 5 and 6). The first of these enzymes is the primary killing site for the lethal action of β-lactam antibiotics against bacteria. The second is the primary method of defence used by bacteria to resist this bactericide. Both enzymes share a common method of opening the β-lactam ring, which involves the attack of a serine hydroxy group on the carbonyl carbon to give a penicilloyl enzyme intermediate, an ester of penicilloic acid (Scheme 2.4). The difference between the two enzymes is that, with transpeptidase, the acyl enzyme is thought either to be relatively stable or to undergo another chemical step to generate the actual inhibited enzyme complex, whereas, with β-lactamase, the acyl enzyme is rapidly hydrolysed to regenerate the enzyme and penicilloic acid (Scheme 2.4).

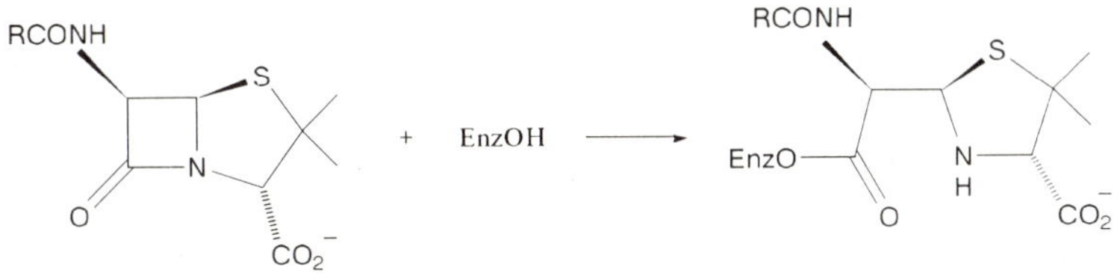

Scheme 2.4

The reaction of alcohols and other oxygen nucleophiles with penicillin is therefore of obvious interest. In the absence of enzymes, do penicilloyl esters undergo reactions at a rate that is competitive with or faster than hydrolysis? This is discussed in chapter 4. The alcoholysis of β-lactam antibiotics represents a model for the first step of the enzyme catalysed reactions (Scheme 2.4).

The second-order rate constants for the hydrolysis of benzylpenicillin catalysed by alkoxide ions and other oxygen bases generate a non-linear Bronsted plot when plotted against the pK_a of the conjugate acid of the base.[30] There are two slopes; one for weak oxygen bases with a Bronsted β-value of 0.38 is indicative of general base catalysis, whereas the steeper slope for more basic alkoxide ions corresponds to a Bronsted β-value of 0.97, indicative of nucleophilic catalysis. Basic alkoxide ions catalyse hydrolysis by initial alcoholysis and the intermediate formation of a penicilloyl ester. This ester is, in turn, formed from the tetrahedral intermediate generated by nucleophilic attack of the alkoxide ion on the β-lactam carbonyl. The alcoholysis of benzylpenicillin with trifluoroethoxide shows a kinetic solvent isotope effect k_{H_2O}/k_{D_2O} of 3.16, which suggests rate-limiting, general acid catalysed breakdown of the intermediate by proton transfer from water to nitrogen (**18**).[30]

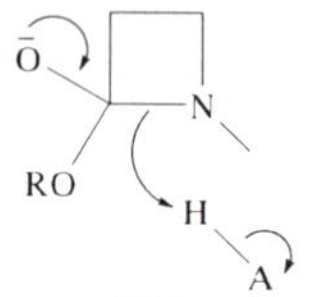

(**18**)

The traditional chemical view of penicillin as an effective antibiotic was that it was a good acylating agent because of enhanced reactivity, due either to strain in the four-membered ring, or to reduced amide resonance. Table 2.3 shows the rate constants for the acylation of trifluoroethoxide ion by a variety of acylating agents. Penicillin is seen not to be particularly reactive towards this oxygen nucleophile. This is yet another simple demonstration that the high rate of reaction between penicillin and β-lactamase or transpeptidase enzymes is not due to the intrinsic reactivity of penicillin, but is a result of the favourable, non-bonded interactions between the substrate and enzyme, which provide molecular recognition and binding energy to lower the activation energy.

2.6 Ease of C–N bond fission in β-lactams

Many observations indicate that opening the four-membered ring is not the facile process that is normally assumed to result from this strained system. Some of these are described in chapter 4; here is presented some particularly unambiguous evidence.[31]

The hydrolysis of azetidin-2-ylideneammonium salts (**19**) can give two products depending on which C–N bond is broken. The alkaline hydrolysis proceeds by formation of the tetrahedral intermediate, T^0, which can undergo exocyclic C–N bond fission to give the β-lactam (**20**) and an amine, or endocyclic C–N bond fission to give the β-amino amide (**21**) (Scheme 2.5).

The breakdown of the tetrahedral intermediate T^0 is general acid catalysed, and both exocyclic and endocyclic C–N bond fission require

Table 2.3 A comparison of the second-order rate constant for the reaction of 2,2,2-trifluoroethoxide ion with various acylating agents in water at 25°C, unless stated otherwise.

Acylating agent	k (dm^3 mol^{-1} s^{-1})	Reference
CH_3CO—N^+(pyridinium)—CH_3	1.2×10^7	*a*
CH_3CO—N(imidazolium, $\overset{+}{N}H$)	6.5×10^5	*b*
CH_3CO—Cl	8.9×10^4*	*c*
CH_3CO—O—N^+(pyridinium)—OMe	4.6×10^4	*d*
CH_3CO—N(imidazole, N)	4.3×10^3	*b*
CH_3CO—O—C_6H_3(NO_2)—NO_2	4.0×10^2	*d*
CH_3CO—O—C_6H_4—NO_2	64	*e*
CH_3CO—OCH_2CF_3	20	*f*
CH_3CO—O—Ph	6.8	*d*
Benzylpenicillin	0.223	*g*

* at 200°C.
[a] A.R. Fersht and W.P. Jencks, *J. Am. Chem. Soc.* (1970) **92** 5442.
[b] D.G. Oakenfall and W.P. Jencks, *J. Am. Chem. Soc.* (1971) **93** 178.
[c] D.J. Palling and W.P. Jencks, *J. Am. Chem. Soc.* (1984) **106** 4869.
[d] W.P. Jencks and M. Gilchrist, *J. Am. Chem. Soc.* (1968) **90** 2622.
[e] R.P. Bell and W.C.E. Higginson, *Proc. R. Soc. London (A)* (1949) **197** 141; W.P. Jencks and M. Gilchrist, *J. Am. Chem. Soc.* (1962) **84** 2910.
[f] Estimated from reference a.
[g] A.M. Davis, D. Proctor and M.I. Page, *J. Chem. Soc., Perkin Trans. 2* (1991) 1213.

proton transfer to the departing amine nitrogen. Despite the release of strain on opening the four-membered ring, the endocyclic nitrogen still needs protonation to aid C–N bond fission.[31]

The major product of the hydrolysis of azetidinyl amidinium salts is the

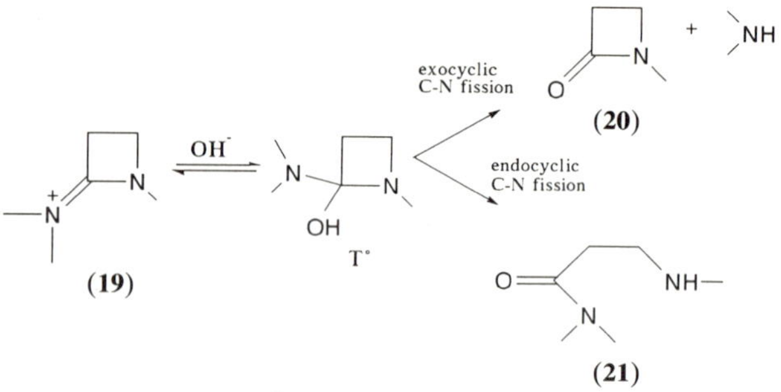

Scheme 2.5

β-lactam. Partitioning of the tetrahedral intermediate T^0 favours exocyclic C–N bond fission rather than four-membered ring opening by endocyclic C–N bond fission, which would be expected to be facilitated by the release of strain energy. The apparent reluctance of the four-membered ring to open is not the result of differential basicities of the two nitrogens, entropic factors, or stereoelectronic effects.[31]

Other reports have suggested that the four-membered ring does not open as readily as would be suspected from its strain energy. It is conceivable that there is either an increase in strain energy as four-membered rings open, or a non-linear relationship between strain energy and bond length so that an early transition state with little bond fission is still highly strained. One problem with this type of explanation is that although four-membered rings are often formed slowly relative to other ring sizes, ring closure is not exceptionally slow and is often only about 100 times slower in four-membered rings than in six-membered rings despite a predicted maximum difference of 10^{20} based solely on strain-energy differences.[4,32]

The rate of ring opening of some four-membered rings is sometimes enhanced. For example, the hydrolysis of oxetidinyl and ketals is about 10^5 times faster than that of an analogous acyclic system.[33] However, the rates of both ring opening and ring closure are not exceptional compared with the total strain energy involved.[34] Despite several attempts to rationalise this anomaly, an acceptable theory remains to be developed.

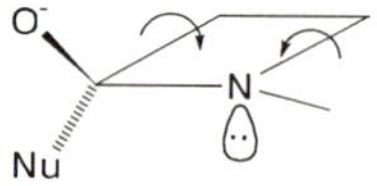

(22)

The C–N bond in acyclic peptides and larger lactams probably breaks by a C–N stretching motion. This method of bond fission is unlikely to occur in four-membered rings because it would involve an increase in strain energy due to a decrease in bond angles.[31] A more favourable reaction coordinate involves rotation about the single bonds (**22**). One consequence of this motion for bond breaking would be that any proton donor on the enzyme which facilitates carbon–nitrogen bond cleavage would be situated in a different position to that required in the reactions of acyclic peptides. Furthermore, substituents attached to the β-lactam would move considerably upon ring opening. It would then, presumably, be inappropriate to have these substituents tightly anchored to the enzyme.[35]

References

1. J.L. Strominger, *Antibiotics* (1967) **1** 706.
2. R.B. Woodward, in *The Chemistry of Penicillin* (Eds H.T. Clarke, J.R. Johnson and R. Robinson), Princeton University Press, Princeton, New Jersey (1949), p. 443.
3. A.R. Fersht and Y. Requena, *J. Am. Chem. Soc.* (1971) **93** 3499, 3502.
4. M.I. Page, *Chem. Soc. Rev.* (1973) 295.
5. M.I. Page, *Accounts Chem. Res.* (1984) **17** 144.
6. G.A. Jeffrey, J.R. Ruble, R.K. McMullen, D.J. DeFrees, J.S. Binkley and J.A. Pople, *Acta Cryst.* (1980) **B36** 2242.
7. M.I. Page, *Adv. Phys. Org. Chem.* (1987) **23** 165.
8. R.M. Sweet, in *Cephalosporins and Penicillins: Chemistry and Biology* (Ed. E.H. Flynn), Academic Press, New York (1973), p. 280.
9. K. Vijayan, B.F. Anderson and D.C. Hodgkin, *J. Chem. Soc., Perkin Trans. 1* (1973), 484; R.M. Sweet and L.F. Dahl, *J. Am. Chem. Soc.* (1970) **92** 5489.
10. G.C. Levy and G.L. Nelson, *J. Am. Chem. Soc.* (1972) **94** 4897.
11. A.K. Bose and P.R. Srinivasan, *Org. Mag. Res.* (1979) **12** 34.
12. K.L. Williamson and J.D. Roberts, *J. Am. Chem. Soc.* (1976) **98** 5082.
13. R.L. Lichter and D.E. Dorman, *J. Org. Chem.* (1976) **41** 582.
14. M. Takasuka, J. Nishikawa and K. Tori, *J. Antibiot.* (1982) **35** 1729; J. Nishikawa, K. Tori, M. Takasuka, H. Onoue and M. Narisada, *J. Antibiot.* (1982) **35** 1724.
15. A.J. Collings, P.F. Jackson and K.J. Morgan, *J. Chem. Soc. B* (1970) 581.
16. J. Frau, M. Coll, J. Donoso, F. Munoz and F.G. Blanro, *J. Mol. Struct. (Theochem)* (1991) **231** 109; D.B. Boyd, *J. Med. Chem.* (1983) **26** 1010; C. Petrongolo, G. Ranghino and R. Scordamaglia, *Chem. Phys.* (1980) **45** 279.
17. S. Vishveshwara and V.S.R. Rao, *J. Mol. Struct.* (1983) **92** 19.
18. L. Radom and N.V. Riggs, *Austral. J. Chem.* (1980) **33** 249; Y. Li, R.L. Garrell and K.N. Houk, *J. Am. Chem. Soc.* (1991) **113** 5895; C.M. Brenneman and K.B. Wiberg, *J. Comp. Chem.* (1990) **11** 361.
19. H.Y. Pracejus, M. Kehlen, H. Kehlen and H. Matschiner, *Tetrahedron* (1965) **21** 2257; Q-P. Wang, A.J. Bennet, R.S. Brown and B.D. Santarsiero, *J. Am. Chem. Soc.* (1991) **113** 5757.
20. P. Proctor, N.P. Gensmantel and M.I. Page, *J. Chem. Soc., Perkin Trans. 2* (1982) 1185.
21. N.P. Gensmantel, E.W. Gowling and M.I. Page, *J. Chem. Soc., Perkin Trans. 2* (1978) 235; N.P. Gensmantel, P. Proctor and M.I. Page. *J. Chem. Soc., Perkin Trans. 2* (1980) 1225.
22. S.N. Rao, R.A. More O'Ferrall, *J. Am. Chem. Soc.* (1990) **112** 2729.
23. H.C. Brown and K. Ichikawa, *Tetrahedron* (1957) **1** 221.
24. A.P. Laws and M.I. Page, *J. Chem. Soc., Perkin Trans. 2* (1989) 1577.
25. A. Bruylants and F. Kezdy, *Rec. Chem. Progr.* (1960) **21** 213.
26. D.B. Boyd, D.K. Herron, W.H.W. Lunn and W.A. Spitzer, *J. Am. Chem. Soc.* (1980) **102** 1812.
27. S.C. Buckwell, M.I. Page and J.L. Longridge, *J. Chem. Soc., Chem. Commun.* (1986) 1039; S.C. Buckwell, M.I. Page, J.L. Longridge and S.G. Waley, *J. Chem. Soc., Perkin Trans. 2* (1988) 1823; R.F. Pratt and W.S. Faraci, *J. Am. Chem. Soc.* (1986) **108** 5328.

28. L.C. Blaszczak, R.F. Brown, G.K. Cook, W.J. Hornback, R.C. Hoying, J.M. Indeliciato, C.L. Jordan, A.S. Katner, M.D. Kinnick, J.H. McDonald III, J.M. Morin Jr., J.E. Munroe and C.E. Pasini, *J. Med. Chem.* (1990) **33** 1656.
29. D.P. Kessler, M. Chushman, I. Ghebre-Sellassie, A.M. Knevel and S.L. Hem, *J. Chem. Soc., Perkin Trans. 2* (1983) 1699; J. Degelaen, S.L. Loukas, J. Feeney, G.C.K. Roberts and A.S.V. Burgen, *J. Chem. Soc., Perkin Trans. 2* (1979) 86.
30. A.M. Davis, P. Proctor and M.I. Page, *J. Chem. Soc., Perkins Trans. 2* (1991) 1213.
31. M.I. Page, P.S. Webster and L. Ghosey, *J. Chem. Soc., Perkin Trans. 2* (1990) 805, 813.
32. A. Dr. Markino, G. Galli, P. Gargano and L. Mandolini, *J. Chem. Soc., Perkin Trans. 2* (1985) 1345; L. Mandolini, *Adv. Phys. Org. Chem.* (1986) **22** 1.
33. R.F. Atkinson and T.C. Bruice, *J. Am. Chem. Soc.* (1974) **96** 819.
34. M.A. Casadei, A. di Martino, G. Galli and L. Mandolini, *Gazz. Chim. Ital.* (1986) **116** 659.
35. M.I. Page, *Phil. Trans. R. Soc. Lond. (B)* (1991) **332** 149.

3 Structure–activity relationships: biological

H.C. NEU

3.1 Introduction

In the last fifty years a large number of β-lactam compounds have been produced, and our understanding of the structure–activity of β-lactams has increased to a remarkable degree. β-Lactams interfere with cell-wall synthesis by acylating enzymes — penicillin binding proteins (PBPs) — critical in the production of the bacterial cell wall.[1] The activity of β-lactam agents is also markedly influenced by the ability of a β-lactam compound to reach the PBP receptors, by the stability of the compound to attack by β-lactamases present in many microorganisms, which will acylate the β-lactam antibiotic and render it inactive, and by the interaction of entry, PBP binding, β-lactamase induction and stability.[2–4]

This chapter will analyze compounds that are in clinical use and mention a few agents which, although not used clinically, provide insights into the effect of structural changes on antimicrobial activity, pharmacological and/ or toxicological properties.

3.2 General aspects

β-Lactams share a common structural similarity, the four-membered lactam ring. In most β-lactams the β-lactam ring is fused through the nitrogen and the adjacent tetrahedral carbon atoms to a secondary ring structure. A thiazolidine ring is present in penicillins, and a dihydrothiazine ring in cephalosporins. Today there are many other structures such as penems, oxacephams, oxacephems, carbacephams, carbacephems and monocyclic structures. Penicillins have asymmetric centers at carbons C-5 and C-6, which correspond to the asymmetric centers present at C-6 and C-7 in cephalosporins. The configuration of the carbon atoms C-6 and C-7, which bear the amide, is the same in both cases. Thus, the hydrogen atoms attached to the C-5, C-6 and C-6, C-7 carbons are in a *cis*-configuration on the α-side of the ring system. Carbapenems, by contrast, usually have

Table 3.1 Activity[a] of β-lactam compounds against Enterobacteriaceae.

Agent	*Escherichia coli*	*Citrobacter freundii*	*Enterobacter aerogenes*	*Enterobacter cloacae*	*Klebsiella pneumoniae*
Azlocillin	16/>128	16/128	8/>128	8/>128	>128
Carbenicillin	16/>128	16/128	8/>128	8/>128	>128
Mezlocillin	16/>128	8/128	8/>128	8/>128	>128
Piperacillin	16/>128	8/128	8/>128	8/>128	32/>128
Ticarcillin	16/>128	16/128	8/>128	8/>128	>128
Cefazolin	4/>128	>128	>128	>128	4/>128
Cefuroxime	1/32	>128	4/>128	4/>128	64/>128
Cefotiam	2/32	8/>128	16/>128	8/>128	8/>128
Cefoxitin	2/8	16/>128	>128	>128	2/16
Cefotetan	1/4	8/>128	8/>128	8/>128	2/16
Cefoperazone	0.12/32	0.5/16	0.25/16	0.5/64	0.25/16
Cefotaxime	0.06/0.12	0.25/128	0.12/0.5	0.25/128	≤0.12
Ceftizoxime	0.06/0.12	0.25/128	0.12/1	0.25/128	≤0.12
Cefmenoxime	0.06/0.12	0.25/128	0.25/1	0.25/128	≤0.12
Ceftriaxone	0.06/0.12	0.25/128	0.25/1	0.25/128	≤0.12
Ceftazidime	0.12/0.25	0.25/128	0.25/1	0.25/128	≤0.12
Cefpirome	0.03/0.06	0.12/16	0.12/0.5	0.25/16	≤0.12
Cefepime	0.03/0.06	0.12/16	0.12/0.5	0.25/16	0.06
Aztreonam	0.06/0.12	0.25/128	0.12/0.5	0.25/128	0.12/0.25
GR 69153	0.06/0.12	0.25/8	0.12/0.5	0.25/16	0.12/0.25
Imipenem	0.12/0.5	0.5/1	0.5/2	1/2	0.12/0.5
Meropenem	0.25/0.5	0.5/1	1/4	0.25/2	0.12/0.25
FCE 21101	0.25/0.5	2/4	1/4	0.5/8	0.5/1
Moxalactam	0.12/0.25	0.5/2	0.5/2	0.12/8	0.12/0.25

[a] Activity measured as minimal inhibitory concentration (MIC) 50%/90%; a single number denotes MIC_{50} equal to MIC_{90}.

a *trans*-configuration, with the C-6 substituent in the β-position. Most β-lactams have a carboxyl group on the carbon atom attached to the β-lactam nitrogen. In the case of penams, clavams and cephems, the carbon atom carrying the carboxyl group is tetrahedral, so that the carboxyl group is on the α-side of the ring system. A variety of different substitutions are possible on the C-6 position of penicillins, carbapenems and clavams, and on the C-7 position of oxacephems and cephamycins. For most penicillins, cephalosporins and monobactams, a β-acylamino group is necessary to achieve antimicrobial activity; this group is not required for the penems or carbapenems, in which the chemical reactivity of the β-lactam ring appears adequate to achieve high antibacterial activity. Activities of the compounds discussed in this chapter are provided in Tables 3.1, 3.2 and 3.3.

3.3 Natural penicillins

The structural relations of two natural penicillins, penicillin G and penicillin V, were reviewed by Florey and colleagues at the end of the 1940s.[5] Work in

Proteus mirabilis	*Morganella morganii*	*Proteus stuartii*	*Serratia marcescens*	*Salmonella*
<1	8/128	8/>128	32/>128	4/>128
<1	8/128	8/>128	32/>128	8/>128
<1	4/128	4/>128	16/>128	4/>128
<1	4/128	4/>128	8/>128	2/>128
<1	8/128	8/>128	32/>128	4/>128
0.5/2	>128	>128	>128	4/32
0.5/2	32/128	64/>128	>128	2/16
0.5/2	32/>128	32/>128	>128	2/32
2/8	8/64	8/64	32/64	4/16
0.5/1	4/32	8/32	16/64	1/8
≤0.5	1/8	2/64	8/128	0.5/16
≤0.01	0.06/2	0.06/0.5	0.5/128	≤0.12
≤0.01	0.06/16	0.06/1	0.5/128	≤0.12
≤0.01	0.06/2	0.06/1	0.5/128	≤0.12
≤0.01	0.06/4	0.06/0.1	0.5/128	≤0.12
≤0.01	0.06/1	0.06/1	0.5/128	0.25
≤0.01	0.06/1	0.06/1	0.5/16	0.12
≤0.01	0.06/5	0.06/1	0.5/16	≤0.06
≤0.01	<0.06/0.5	<0.12	0.12/32	≤0.12
≤0.01	≤0.12	<0.12	0.12/8	≤0.12
0.5/2	1/4	1/4	1/4	0.5
0.5/2	1/2	1/2	2/8	0.5
2/4	4	2/4	8/16	0.5
≤0.1	0.25/1	0.5/1	0.5/4	0.12

the 1960s by Strominger and colleagues established the transpeptidase target.[6] There have not been any developments in the natural penicillins, and all of our current agents are semisynthetic or synthetic compounds.

3.4 Penicillinase-resistant antistaphylococcal penicillins

Antistaphylococcal, penicillinase-resistant penicillins have been available since 1960. The basis of the activity of the antistaphylococcal penicillins against β-lactamase-producing isolates is steric hindrance around the carbon atom attached to the C-6 side chain amide carbonyl group, as is found in methicillin and the isoxazolyl penicillins such as oxacillin, cloxacillin, dicloxacillin, and flucloxacillin[7] (see Editorial Introduction, Figure 1). Methicillin is a phenyl penicillin in which both of the *ortho* positions of the phenyl group are occupied by a methoxy group.[8] Use of larger moieties in this position results in penicillinase resistance, but markedly lowers antimicrobial activity due to poor binding to penicillin-binding proteins. Nafcillin has only one *ortho* position filled, since a naphthylene nucleus is

Table 3.2 Activity[a] of β-lactam compounds against gram-positive bacteria.

Agent	*Streptococcus pneumoniae*	*Streptococcus pyogenes*	*Streptococcus faecalis*	*Staphylococcus aureus*		*Staphylococcus epidermidis*[b]	*Listeria*
				Meth-S	Meth-R		
Carbenicillin	0.25	0.25/1	16/64	32/128	>128	32/128	2/8
Mezlocillin	≤0.02	≤0.02	1/4	32/128	>128	32/128	1/2
Piperacillin	≤0.02	≤0.02	1/4	32/128	>128	32/128	1/2
Ticarcillin	0.25/1	0.25/1	32/64	32/128	>128	32/128	2/8
Cefazolin	≤0.25	≤0.25	>256	0.5/4	16/>128	1/32	>128
Cefotiam	≤0.12	≤0.12	64/256	0.5/2	16/>128	1/32	>128
Cefuroxime	≤0.25	≤0.25	256	0.5/2	16/>128	1/32	>128
Cefoxitin	1/2	0.5/1	>256	2/4	>128	4/128	>128
Cephalexin	0.5/2	0.5/1	>256	2/8	>128	4/128	>128
Cefotaxime	0.01/0.06	≤0.03	32/256	2/4	16/128	2/128	>128
Ceftizoxime	0.01/0.06	≤0.03	32/256	2/4	16/>128	2/128	>128
Ceftriaxone	0.01/0.06	≤0.03	32/256	2/4	16/>128	4/128	>128
Cefmenoxime	0.01/0.06	0.01/0.03	64/256	2/4	16/>128	2/128	>128
Ceftazidime	0.25/0.5	0.12	>256	8/16	128	32/128	>128
Cefoperazone	≤0.25	0.12	32/128	2/4	16/>128	4/128	>128
Cefaclor	1/2	0.25/0.5	>128	0.5/8	>128	>128	>128
Moxalactam	1/2	1/2	>128	1/4	128	2/32	>128
Cefpodoxime	0.5/4	<0.12	>128	4/16	>128	8/>128	>128
Cefpirome	0.02/0.25	0.02/0.25	8/32	0.5/1	4/64	0.5/8	16/64
Cefepime	≤0.02/0.06	≤0.02/0.06	16/64	1/4	4/16	1/8	16
Cefixime	0.25/2	0.05/0.25	>128	16/32	>128	16/64	>128
Aztreonam	>128	>128	>128	>128	>128	>128	>128
Imipenem	≤0.02/0.06	≤0.02	0.5/1	0.06/0.12	0.5/8	0.12/8	0.06
Meropenem	≤0.02	≤0.02	4/8	≤0.06	0.5/64	0.12/8	0.06
FCE 21101	≤0.02	≤0.02	4/8	≤0.06	0.5/64	0.12/8	0.06

[a] Activity measured as minimal inhibitory concentration (MIC) 50%/90%; a single number denotes MIC_{50} equal to MIC_{90}.
[b] Includes methicillin-resistant.

Table 3.3 *In vitro* susceptibility of *Pseudomonas* and *Acinetobacter* to β-lactam agents.[a]

Agent	*Pseudomonas aeruginosa*	*Pseudomonas cepacia*	*Acinetobacter anitratus*
Azlocillin	16/128	8/64	64
Carbenicillin	128/>128	128/>128	16/32
Mezlocillin	16/128	16/64	64/>128
Piperacillin	8/64	16/64	16/64
Ticarcillin	16/>128	>128	8/64
Cefotaxime	32/128	16/256	16
Ceftizoxime	32/128	16/256	8/16
Cefmenoxime	32/128	16/128	32/64
Ceftriaxone	32/128	16/256	8/16
Ceftazidime	4/8	8/64	8/16
Cefoperazone	8/16	32/128	64/128
Cefsulodin	4/16	64/>256	32/64
Moxalactam	16/64	32/128	>32/64
Cefpirome	4/16	8/64	16/128
Cefepime	4/16	8/64	16/128
E-1040	2/8	>64/>128	4/>128
Aztreonam	2/8	32/128	16/64
Carumonam	2/8	32/128	8/32
Imipenem	1/2	0.5/8	0.25/0.25
Meropenem	1/16	1/2	0.5/1
GR 69153	2/8	16/128	16/128

[a] Activity measured as MIC 50%/90%.

utilized.[9] Combination of an isoxazole ring with a phenyl group on the acyl side chain produces β-lactamase stability for oxacillin. The orally absorbed agent cloxacillin contains a chlorine on the phenyl group, as do dicloxacillin and flucloxacillin. This improves the oral absorption of the compound, but protein binding increases as the number of chlorine atoms are increased.[7]

The isoxazolyl penicillins are not as β-lactamase-stable as is methicillin, but this difference in penicillinase stability is not felt to be of clinical significance. Many β-lactamase-stable penicillins have been synthesized, and have shown that when a five-membered heterocyclic ring is attached to the amide carbonyl group, larger substituents are needed to provide resistance to penicillinase attack. Two *ortho* methyl groups or a single *ortho* phenyl group will not make an isoxazolyl-penicillin resistant to penicillinase. Larger groups, such as a naphthyl or polar, 4-pyridyl rings, lower *in vitro* activity. There have been no new antistaphylococcal penicillins for decades, since the problem with staphylococci is no longer β-lactamase, but resistance due to failure to bind to PBP2a — namely methicillin-resistance.

3.5 Amino penicillins

Ampicillin, a D-α-aminobenzyl penicillin available since 1960, has the antibacterial activity of penicillin G against gram-positive species. Since it is not β-lactamase stable, many of the formerly susceptible microorganisms, such as *Escherichia coli*, *Salmonella* and *Shigella* species, and *Proteus mirabilis*, are now resistant.[10] Many lipophilic and hydrophilic α-substituents have been prepared, but none of them is appreciably more active than ampicillin.[11] Analogues of ampicillin, such as epicillin (1,4-cyclohexadienyl) and cyclacillin (1-aminocyclohexylpenicillin) have *in vitro* activity similar to ampicillin. Structural changes increased the oral absorption from the gastrointestinal tract. The major advance in amino penicillins was amoxicillin, a *p*-hydroxy derivative, which has the same *in vitro* activity as ampicillin but much-improved oral absorption. Introduction of hydroxyl groups at *meta* or *para* positions on the phenyl ring reduce antibacterial activity.

A number of pro-drugs of ampicillin were developed to improve the absorption of the drug when administered orally. These involved carbonyl adducts involving the α-amino group and esters of the carboxyl group. The adducts include hetacillin, which released acetone after ingestion, and metampicillin, which releases formaldehyde. These had no advantage over ampicillin, and more careful analysis showed that they were not better absorbed. Pivolyl esters, bacampicillin and talampicillin produce serum levels of ampicillin comparable to those of amoxicillin.[12] Since aminopenicillins are not β-lactamase-stable and there has been a general increase worldwide in β-lactamase-producing organisms, there is unlikely to be further work on these agents.

3.6 Carboxy and sulfo penicillins

The addition of a carboxylic, sulfamic or sulfonic acid to the α-carbon atom of the C-6 acyl side chain alters the *in vitro* activity. Carbenicillin, the prototype of the carboxy penicillins, shows reduced gram-positive activity, particularly against streptococcal and enterococcal species. This is due to decreased binding to PBPs. However, organisms such as *Pseudomonas aeruginosa*, some *Enterobacter* species, *Proteus vulgaris*, *Morganella morganii* and *Providencia* species, which contain chromosomal β-lactamases — cephalosporinases — are inhibited.[13] Replacement of the phenyl group with a 3-thienyl group has increased the activity of ticarcillin against *P. aeruginosa* but not against gram-positive or gram-negative species.[14] Replacement of the carboxyl group with a sulfonic acid group, as in sulbenicillin, produces better chemical stability but no improvement in *in vitro* antibacterial activity or pharmacological properties.[14] Introduction of an α-sulfamic acid substituent provides activity against *P. aeruginosa* but results in loss of activity against many gram-negative species.[15] Benzyl penicillin derivatives, which contain strongly acidic α-carbon side chain functions, bind to an ADP receptor site on platelets, resulting in prolongation of bleeding time and alteration of platelet aggregation.[16] There has not been further interest in carboxy penicillins, and they are no longer used.

3.7 Acyl-ureido penicillins

Ampicillin derivatives that contain a complex acyl moiety attached to the α-amino group have been synthesized. Most of the agents have an α-ureido grouping and an additional carbonyl group or a guanidyl on the N-3 atom. Azlocillin, mezlocillin, and piperacillin have activity against streptococci similar to that of ampicillin.[17] These agents inhibit *P. aeruginosa* at concentrations lower than carbenicillin or ticarcillin. Although they are less stable against attack by β-lactamases, they have greater affinity for PBPs.[1] The improved ability of the ureido penicillins to penetrate the outer membrane of *P. aeruginosa* and *K. pneumoniae* is a major factor in the activity of the agents against these organisms.[18,19] The side chain changes do not increase β-lactamase stability. There is a marked inoculum effect with these penicillins, and difference between minimum inhibitory concentrations (MICs) and minimum bactericidal concentrations (MBCs). Many other ureido side chains have been added, but toxicological problems have resulted. Azlocillin, mezlocillin and piperacillin are safe and effective agents. Piperacillin is the most active penicillin against *P. aeruginosa*.

3.8 Amidino penicillanic acid penicillins

Lund[20] synthesized the compound mecillinam, 6-β-[(hexahydro-1H-azepin-1-yl)-methyleneamino]penicillanic acid. An *n*-alkyl group on the amidino nitrogen atom yielded optimal gram-negative activity, but alkyl group changes had less-predictible effect on activity against gram-positive species. Total ring size was important, since small alkyl-substituted rings with the same number of carbons were less active. Polar groups could not be attached without loss of activity, and any modification of the basic nucleus reduced biological activity.[21]

Mecillinam inhibits *E. coli*, including a number of plasmid-containing, β-lactamase strains, and many *Klebsiella*, *Enterobacter* and *Citrobacter* species,[22] but has variable activity against *Proteus* species and does not inhibit *P. aeruginosa* or *Bacteroides fragilis*. It binds almost exclusively to PBP 2 of *E. coli*[1] and does not bind to PBPs of gram-positive species. Its activity against some β-lactamase-producing *E. coli* and *Klebsiella* is due to a poor affinity for β-lactamases, and thus although it is hydrolyzed at a much lower rate than are aminopenicillins or acylureido penicillins.[23] Mecillinam acts synergistically with penicillins and cephalosporins which bind to PBP 3.[24]

A pivolyl ester of mecillinam, which is orally absorbed to yield mecillinam, has been used to treat urinary tract infections and diarrhoea. In view of increases in plasmid β-lactamase organisms, mecillinam has little clinical use.

3.9 6-α-Substituted penicillins

A 6-α-methoxy penicillin — temocillin, a derivative of ticarcillin — is highly active against the Enterobacteriaceae, including β-lactamase-producing strains, but has lost activity against gram-positive species and has poor activity against *P. aeruginosa*.[25,26] The 6-α-methoxy group and acyl carboxyl group virtually eliminate gram-positive activity. The presence of the methoxy group on the 6-α-position results in stability of staphylococcal β-lactamases and classes I, III, IV, and V of the Richmond–Sykes classification.[26]

The poor activity of temocillin against *Enterobacter* species and *P. aeruginosa* is due to a lesser ability to cross the outer cell wall, since altering cell-wall permeability increases the activity temocillin against these organisms.[26] The C-6 methyl group markedly reduced the affinity of temocillin for cephalosporinases.[27,28] This may explain why temocillin inhibits some *E. cloacae* resistant to ticarcillin.

The methoxy group has effect on the pharmacological properties of temocillin. The half-life of temocillin is 4 h and tubular secretion is minimal, whereas the half-life of ticarcillin is 1 h.[29] Temocillin consists of diastereoisomers, with markedly different pharmacokinetic behavior, the *R*-epimer being more active, but in individuals with normal renal function this is of no consequence.[29]

Compound BRL 36650 — foramidocillin — has a formamido (NHCHO) substituent in the 6-α-position. On the acyl side chain the phenyl group contains *ortho* and *para* hydroxyl groups, and the α-carbon atom contains a ureido substituent, which consists of a piperazine ring analogous to the structure of piperacillin.[30] Foramidocillin does not inhibit gram-positive species, MIC $>128\ \mu g\ ml^{-1}$, or *Bacteroides* species. The activity of foramidocillin against the Enterobacteriaceae and *P. aeruginosa* is similar to the activity of the third-generation cephalosporins.[30–32] The formamido group prevents hydrolysis by plasmid and chromosomal β-lactamases.[31] The acyl ureido component probably contributes to its excellent activity against *P. aeruginosa*, which is superior to that of piperacillin since it inhibits many piperacillin-resistant organisms. Foramidocillin inhibits *P. cepacia* and other *Pseudomonas* species, such as *P. putida*, *P. stutzeri* and *P. fluorescens*. Unlike the ureidopenicillins, foramidocillin has a minimal inoculum effect due to β-lactamase stability. The formamido moiety is not an inducer of β-lactamase activity, as is the methoxy group in this position. Foramidocillin has poor affinity for plasmid and chromosomal β-lactamases, which explains the inhibition of *Enterobacter* and *Citrobacter* resistant to third-generation cephalosporins. It utilizes the *tonB* iron transport system to enter gram-negative bacteria such as *Pseudomonas*. Unfortunately, this agent did not survive the toxicological studies. It is questionable whether further penicillins will be synthesized, since improvement in activity has been associated with renal toxicity.

3.10 Cephalosporins

The cephalosporin structure permits greater modification of biological and pharmacological properties (see Editorial Introduction, Figure 1). It is possible to vary the 3′-substituent and the 7-α-acylamino groups, and to make substitution at the 7-α-position to alter β-lactamase stability. It is also possible to replace the sulfur atom in the ring system with an oxygen to increase the reactivity of the molecule and produce oxacephems.

3.11 7-β-Acylamino group modifications

The first cephalosporins such as cephalothin and cephaloridine contained no substitutions on the α-carbon of the C-7-acyl side chain and possessed a thienyl ring or a tetrazole structure, such as in cefazolin (see Editorial Introduction, Figure 1). These compounds inhibited staphylococci and streptococci, with the exception of enterococci and the gram-negative species *E. coli*, *K. pneumoniae*, and *P. mirabilis*. When the α-carbon was multiply substituted or incorporated into a ring, there was a marked reduction in activity, and when the lipophilicity of the side chain was increased, gram-negative activity was decreased.[33] Presence of an amino group or hydroxyl group on the α-carbon slightly enhanced activity against gram-negative species. Such a change was used in the oral compounds cephalexin and cephradine. A hydroxyl function is present in two parenteral cephalosporins, cefamandole and cefonicid.

A major improvement in antibacterial activity was produced by the introduction of a 2-aminothiazolyl side chain. This moiety has been used in a large number of parenteral and new oral cephalosporins. The 2-aminothiazolyl moiety provides high affinity for PBPs both both gram-positive and gram-negative bacteria.[19] Replacement of the amino group at position 2 of the thiazole ring with either –H, –OH, or $-NHCH_3$ groups has resulted in lowered activity against *Staphylococcus aureus*, Enterobacteriaceae and *P. aeruginosa*.[34,35] Cefotiam, which has no more β-lactamase stability than cefazolin, inhibits a number of *E. coli* and *Klebsiella* resistant to cefazolin due to its high affinity for PBPs 1b and 3 of these species. However, cefotiam will be destroyed by organisms that produce a large amount of β-lactamase.

Since the aminothiazolyl group did not provide any β-lactamase stability, a number of other modifications of the 7-β-acyl side chain were made. Introduction of an α-iminomethoxy group provided β-lactamase stability without significant loss of *in vitro* activity. Cefuroxime was the first widely used agent to possess an α-oxyimino grouping. It has β-lactamase stability against many plasmid and chromosomal β-lactamases with the exception of the β-lactamases present in *Bacteroides* and the cephalosporinases of the type Ic Richmond–Sykes classification found in *Pseudomonas* and *Enterobacter*.[36,37] Presence of a hydroxyl group on the 10-α-carbon does not provide significant β-lactamase stability, as is seen by the compounds cefamandole and cefonicid.[38,39] Both of these agents are more active than cephalothin against gram-negative bacteria, but this is due to better binding to the PBPs and to passage across the outer cell wall. Cefamandole is more stable to attack by some of the *S. aureus* β-lactamases, but the structural reason for this is not apparent. Cefonicid is less active against staphylococci, but this is related to protein binding due to its C-3 substituent, and is unrelated to the β-acyl side chain.

Cefuroxime demonstrates another aspect of changes in the 7-β-acyl side chain. A furyl ring replaces the phenyl and thienyl groups found in most of the first- and second-generation cephalosporins. Whether this is the reason for its greater activity or the smaller $-CH_2OCONH_2$ at position 3 is not established.

The aminothiazoyl group and iminomethoxy groups have been utilized in what have been referred to as third-generation cephalosporins — cefotaxime, ceftizoxime, cefmenoxime, ceftriaxone and cefodizime. All of these agents have very similar activity against gram-negative species, streptococci and staphylococci (Tables 3.2 and 3.3). Different C-3 substituents do not appear to make major changes in the activity of these compounds against gram-positive species such as beta hemolytic streptococci and *S. pneumoniae* or organisms such as *Haemophilus influenzae*, *Neisseria* species and *Moraxella catarhallis*. Most of the newer parenteral cephems such as ME-1228,[40] E-1040,[41] Ro 09-1428[42] and GR 69153[43,44] possess a 2-aminothiazoyl moiety. Indeed, all of the new, highly active oral cephalosporins possess a 2-aminothiazolyl side chain.

The oximino group that provides β-lactamase stability must be in a *syn* position.[45,46] The most common plasmid β-lactamase was the TEM-1 enzyme, which accounted for the resistance of *Neisseria gonorrhoeae*, *H. influenzae*, *E. coli*, *Salmonella* and *Shigella* species to penicillins, and to first- and some second-generation cephalosporins.[47] The *syn*-positioned iminomethoxy group prevented hydrolysis of the cephalosporins synthesized in the late 1970s and early 1980s. In 1987, however, new β-lactamases were found in clinical specimens.[48] These new enzymes differ by one or two amino acids from the TEM-1 or SHV-1 enzymes,[49] and can hydrolyze iminomethoxy-containing cephalosporins.[49–52] The substitution at C-3 may alter the MICs of compounds against different organisms that possess these enzymes.[50,52] The MICs of cefotaxime and ceftazidime prototype agents are shown in Table 3.4. Although these new β-lactamases are important enzymes, they fortunately are not as well expressed as is TEM-1 and do not seem to have the ability to be transferred as readily as does TEM-1. These new β-lactamases are inhibited by β-lactamase inhibitors such as clavulanate, and they do not hydrolyze compounds that have a 7-α-methoxy group (see chapters 6 and 7).

Interestingly, ceftibuten, an oral cephalosporin that has a $C{=}CHCH_2COOH$ moiety instead of the iminomethoxy group, inhibits organisms containing the new TEM and SHV enzymes[53] (Table 3.5). However, this moiety markedly reduces ceftibuten's activity against *S. pneumoniae* and *S. agalactiae*.

The spatial structure of the C-7 side chain of these cephalosporins has had a major influence on their activity. The amide group forms an angle of 60° to the plane of the β-lactam ring, and the *syn* methoxy group and aminothiazolyl ring form a plane that is virtually perpendicular to the plane of the amide group.[54] The N atoms of the oxime group and thiazolyl ring are vicinal. There is no interference between the oxime group and the β-lactam ring, and the aminothiazolyl cephalosporins are fairly rigid molecules.

Table 3.4 Effect of extended spectrum β-lactamases in *E. coli* on antimicrobial susceptibility of cephalosporins, carbapenems and monobactams.

	MIC (μg ml^{-1}) of			
Enzyme	Cefotaxime	Ceftazidime	Aztreonam	Imipenem
TEM-1	0.12	0.25	0.12	0.12
TEM-2	0.12	0.5	0.25	0.12
TEM-3 (CTX-1)	32	64	16	0.25
TEM-4	32	32	16	0.12
TEM-5 (CAZ-1)	4	128	8	0.25
TEM-6	1	128	64	0.5
TEM-7	0.5	64	2	0.5
TEM-9	2	128	128	0.25
TEM-10	1	64	32	0.25
TEM-11	0.06	4	0.25	0.5
TEM-12	0.06	4	0.25	0.5
SHV-1	0.12	1	0.5	0.5
SHV-2	64	32	32	0.5
SHV-3	64	32	32	0.25
SHV-4	128	128	256	0.25
SHV-5	64	128	256	0.25

Table 3.5 MICs of oral cephalosporins against *E. coli* containing new β-lactamases.

Drug	MIC (μg ml^{-1}) against *E. coli* transconjugants containing										
	No β-lactamase	TEM-1	TEM-2	TEM-3	TEM-4	TEM-7	SHV-1	SHV-2	SHV-3	SHV-4	SHV-5
Cephalexin	4	8	8	16	16	8	8	32	256	32	128
Cefuroxime	4	4	4	64	64	4	16	16	32	8	32
Cefaclor	1	4	8	32	32	2	64	32	128	32	128
Cefproxil	2	4	64	64	64	4	64	>256	256	64	256
Cefcanel	2	16	32	64	32	8	16	128	64	32	128
Cefixime	0.25	0.25	0.25	4	8	0.25	0.25	1	1	8	32
Cefetamet	0.25	0.25	0.25	4	8	0.25	1	1	2	1	1
Cefpodoxime	0.25	0.5	0.5	64	32	1	8	8	32	8	32
Loracarbef	0.5	2	2	8	16	1	8	32	64	8	64
Ceftibuten	0.12	0.12	0.12	0.25	0.25	<0.06	0.5	0.5	0.25	1	8

The iminomethoxy-containing cephalosporins vary in their ability to induce cephalosporinases in species such as *Citrobacter freundii*, *Enterobacter cloacae*, *P. aeruginosa*, and *Serratia marcescens*. None of the agents induces β-lactamase at the rate of that of the methoxy compounds, such as cefoxitin or the carbapenems. The large amount of β-lactamase present in the periplasmic space of *Enterobacter* species, when induced or derepressed for β-lactamase production, allows the enzyme to form an acyl–enzyme complex with iminomethoxy cephalosporins, and to hydrolyze the small number of molecules that have reached the periplasm before they bind to PBPs. This has resulted in clinical failures.[55–57]

The 2-aminothiazolyl group may provide some activity against enterococci in the presence of blood.[58] Clinically fewer enterococcal superinfections have been encountered with the 2′-aminothiazolyl cephalosporins than with cephalosporins that lack the moiety.

Among the clinically used cephalosporins there have been other C-7-β-acyl side chain modifications that have been successful. Cefoperazone contains a 2,3-dioxopiperazine group on the α-carbon, similar to piperacillin.[59] Cefoperazone does not have increased β-lactamase stability.[60] It has the β-lactamase stability of all cephalosporins for staphylococcal β-lactamases, and the *p*-hydroxyphenylglycol group provides some stability against cephalosporinases.[59] The piperazine side chain also provides some stability against cephalosporinases, in contrast to the lack of such stability seen in compounds possessing a hydrogen or hydroxyl on this carbon. The piperazine modification provides improved binding to PBPs and improved entry through gram-negative cell-wall porin channels.[59] Cefoperazone inhibits *P. aeruginosa* provided that the organisms do not contain the TEM, SHV, OXA and PSE β-lactamases. Cefoperazone does not achieve activity against enterococci, illustrating that acyl side chain modifications cannot overcome the lack of enterococcal activity caused by the cephem nucleus. This is due to poor PBP affinity.

Other 7-β substituted cephalosporins have been synthesized. Cefpiramide has a 4-hydroxy-6-methylpyridine group and a 4-hydroxyphenyl group, and is actually less β-lactamase stable than is cefoperazone. The hydroxyphenyl group produces a longer serum half-life and higher serum levels. A 6,7-dihydroxy-4-oxo-4H-1-benzopyran group was used in compound E-0702 to increase activity against many Enterobacteriaceae, but it lacked β-lactamase stability and anti-*Pseudomonas* activity was not increased. Cefpimazole has a 4(S) carboxy imidazole moiety, is less active than cefoperazone against *Pseudomonas*, and is not β-lactamase-stable.

The effect of 7-β-side chain structure on activity against *P. aeruginosa* is illustrated by cefsulodin, which contains a 7-D(β-sulfo) phenylacetamido side chain similar to sulbenicillin. Cefsulodin has an extremely narrow spectrum of activity, which includes only *P. aeruginosa* and *S. aureus*.[61] This is due to poor affinity for PBPs in the Enterobacteriaceae, and poor penetration into many gram-negative organisms while being able to pass through porins of *P. aeruginosa*.[18]

There has been great interest in modifying the imino group on the acyl side chain. Addition of a propyl carboxy group as in ceftazidime provided excellent anti-*Pseudomonas* activity.[62] Some other non-fermenting gram-negative organisms, such as *P. cepacia* and Acinetobacter, were inhibited. *Xanthomonas maltophilia* is not inhibited.[63] The iminopropyl carboxy group caused some decrease in gram-positive activity. The cefotiam and ceftazidime MICs for *S. pneumoniae* are $0.12\,\mu g\,ml^{-1}$ and $0.5\,\mu g\,ml^{-1}$ respectively. Activity of ceftazidime against *S. aureus* is also less than that of the iminomethoxy aminothiazolyl cephalosporins. The iminopropyl carboxy group does not protect against attack by the new TEM enzymes. One of these enzymes was called CAZ-1 because of its greater attack on ceftazidime compared to cefotaxime.[51] This enzyme is now called TEM-5. As noted, the *syn* position $C{=}CHCH_2COOH$ group of ceftibuten provides stability against new enzymes.

There has been a major effort to utilize the *tonB* iron transport pathway to increase the activity of cephalosporins against Enterobacteriaceae and *Pseudomonas*.[64,65] Compound M-14659 showed enhanced activity against *Pseudomonas* and *S. aureus*, but has failed to achieve clinical use. Another agent, GR 69153, showed similar excellent activity[43,44] which was further enhanced by an iron-deficient medium.[66] GR 69153 was also resistant to most of the new TEM β-lactamases except TEM-3 and TEM-9.[67] Unfortunately, the increased activity and entry into the periplasmic space did not protect this agent against the cephalosporinases of *Enterobacter*. An isoxazolidine-containing cephalosporin, SPD-391, has been shown to be 30-fold more active than ceftazidime against *P. aeruginosa*.[68] Specific proteins, Fiv and Cir, from the *tonB*-dependent iron transport locus involved in the uptake of catechol- and hydroxypyridone-containing cephalosporins have been identified.[69] The function of these proteins is the recapture of hydrolytic products resulting from the breakdown of the siderophore enterobactin.[70] Utilizing the *tonB*-dependent iron transport, catechol structures that can inhibit bacteria lacking both *OmpC* and *OmpF* are being constructed.[71,72] In the mouse protection tests, Nakagawa *et al.*[73] showed that catechol cephalosporins are effective due to the iron-limited environment. However, *tonB*-lacking mutants can easily arise.[74]

3.12 Substitutions on the 7-α-position

Cephalosporins that contain a 7-α-methoxy group are referred to as cephamycins. This grouping provides β-lactamase stability. Other substitutions, such as ethoxy, or larger moieties, have been tried at position 7, but there has not been improvement in gram-positive activity, and gram-negative activity has been reduced as the side chain is lengthened.[75] Cefmetazole and cefotetan also have a 7-α-methoxy group. Cefbuperazone is similar to cefoperazone except for the 7-α-methoxy group. Cefbuperazone has less gram-positive activity and no activity against *P. aeruginosa*.[76] The activity of cephamycin compounds is modulated by the 7-β acyl side chain substitutions and substitutions at the C-3 position of the cephem nucleus. Cefotetan, which contains a 7-β-4(carbamoyl carboxylatomethylene)-1,3-dithietan-2-yl moiety, is more active against Enterobacteriaceae than is cefoxitin, but is less active than cefoxitin against a number of gram-positive species.[77] In addition, it does not inhibit some *Bacteroides* species inhibited by cefoxitin.[78]

3.13 C-3 Substituent modifications

The first parenteral cephalosporin contained an acetoxy side chain, CH_2OCOCH_3 at C-3. The acetoxy group is subject to metabolism in serum and liver, producing a desacetyl derivative. In the case of cephalothin, this derivative is markedly less active than the parent compound. Conversely, desacetyl cefotaxime, although less active than the parent compound, is more active than many other cephalosporins.[79] Desacetyl cefotaxime acts synergistically with the parent cefotaxime against many Enterobacteriaceae, but lacks activity against *P. aeruginosa* and *Morganella morganii*. Other moieties used at C-3 include hydrogen, as is the case for ceftizoxime, alkoxy in cefroxadime, and chloride in cefaclor. A pyridinium methyl group provided cephaloridine with metabolic stability, low serum binding, water solubility, and excellent serum levels.[80] The pyridinium methyl group caused cephaloridine to be concentrated in the proximal tubular cells of the kidney, producing nephrotoxicity at high doses. The pyridinium methyl group present in ceftazidime has not caused renal toxicity.

Cefpirome contains a 2,3-cyclopentenopyridine at position 3 and has better activity against *S. aureus* than cefotaxime.[81] Cefepime contains a methyl-pyrrolidinio group at position 3. This structure does not provide the same increase in anti-staphylococcal activity, but causes a two- to four-fold increase in activity against Enterobacteriaceae compared to other aminothiazolyl cephalosporins.[82,83]

C-3-Quaternary ammonium cephalosporins such as cefpirome and cefepime have a markedly reduced affinity for cephalosporinases of the *Enterobacter* P99 type, which explains the activity of these compounds against the hyperproducing β-lactamase strains resistant to the other aminothiazolyl agents.[81,84] Compound E-1040 has a 4-carbamoyl-1-quinuclidino methyl at C-3, a low affinity for *Citrobacter freundii* β-lactamase, and is more stable against attack at low concentrations.[84] The presence of nitrogen nucleophiles at position 3 has significant effects on the biological properties, and on the pharmacokinetic properties of the cephalosporins.

Replacement of the acetoxy group with sulfur nucleophiles, in general, does not increase activity of cephalosporins against gram-negative organisms.[12] Heterocyclic thioyl structures may have slightly improved antibacterial activity and pharmacokinetic properties. For example, cefazolin has increased activity against *E. coli* and *Klebsiella pneumoniae* compared to its acetoxy analogue cephalothin, and much higher serum levels and longer half-life.

A methylthiotetrazole group is present in cefamandole, cefotetan, ceftmetazole, cefoperazone, cefpiramide, cefmenoxime, and the oxacephem moxalactam. The methylthiotetrazole group probably contributes to the antibacterial activity of cefamandole, cefoperazone, cefpiramide, cefotetan and moxalactam, but it does not improve activity of the aminothiazolyl cephalosporin cefmenoxime in comparison with aminothiazolyl compounds that have acetoxy, H or heterocyclic acid functions at C-3.[59]

The thiomethyl tetrazole group has been associated with hematological reactions by interfering with vitamin K metabolism and with reactions due to alcohol caused by interference with alcohol dehydrogenase.[85,86]

Cephalosporins that have an acidic heterocyclic thiomethyl grouping at the 3 position have high and prolonged serum levels and are highly bound to serum proteins. Examples include ceforanide, cefonicid and ceftriaxone. Furthermore, the replacement of the methyl group with an alkyl group carrying an acidic function (CH_2COOH, CH_2SO_3H, $CH_2CH_2NHSO_3H$) is not associated with the aforementioned adverse hematological side effects. Variation in the substituents on the thiotetrazole ring may affect antibacterial activity since the CH_2COOH of ceforanide provides less antibacterial activity than the CH_2SO_3H of cefonicid. The 1,2,4-triazine and 1,3,4-triazole systems found in ceftriaxone and related compounds probably impart a long half-life due to the enolate anion of the six-membered heterocycle, and there is increased biliary excretion due to this grouping. This can result in sludge in the gallbladder.[87]

3.14 Orally absorbed cephalosporins

Most of the early cephalosporins that had adequate antibacterial activity and good oral absorption contained a 7-β-arylglycine side chain. In general, oral absorption has been related to the presence of an α-amino group on the 7-β-acyl substituent, and a small uncharged group at the 3-position.[88] The presence of a *para* hydroxyl group on the phenyl side chain provides cefadroxil with increased absorption and longer half-life. Cefaclor, which contains a chlorine at position 3, is metabolized. An axetil ester of cefuroxime provides useful serum concentrations and a half-life similar to the parenteral form.

Many esters of aminothiazolyl iminomethoxy cephalosporins have been synthesized. Most of these compounds lack anti-staphylococcal activity, and do not have as good activity against *Enterobacter*, *Serratia* and *Morganella* as do the parenteral agents. These oral compounds also lack activity against *P. aeruginosa*. Cefixime contains a vinyl moiety at C-3 and the 7-β side chain is aminothiazolyl with a carboxyl group affixed to the iminomethoxy portion. It has excellent β-lactamase stability, but is destroyed by some *Enterobacter*, *K. oxytoca* K-1, and by *Bacteroides* β-lactamases and the new TEM and SHV β-lactamases. The vinyl group is not the cause of the reduced activity against staphylococci, since cefdinar — which also has a vinyl group — has some antistaphylococcal activity.[89] It appears that the combination of the moiety at C-3 and the moiety at C-7 on the acyl side chain affect gram-positive activity.

Ceftetrame contains an ester in the carboxylic acid function at position 4, and a 5-methyl-2H-tetrazole group at position C-3. It does not inhibit *S. aureus* but has excellent activity against hemolytic streptococci and *S. pneumoniae*.[89–91] It has lower activity against *M. morganii* and *P. vulgaris*, and is hydrolyzed by β-lactamases of these species. Similar to other oral cephalosporins, it lacks activity against *Bacteroides* and *P. aeruginosa*. Ceftamet, which contains $CH_2OCOC(CH_3)_3$ on the carboxyl at C-4 and a $-CH_3$ at C-3, has less activity than the comparable parenteral agents, and has poor activity against *E. cloacae*, *C. freundii* and *M. morganii*.[89–91] The presence of a $-CH_3$ group on C-3 generally seems to be associated with lowered activity against staphylococci.[92]

Another oral cephalosporin that is an aminothiazolyl compound is cefpodoxime, which has an OCH_2OCH_3 at position 3.[93,94] Cefpodoxime has anti-staphylococcal activity, but less than that of older agents. Its activity against gram-negative species is similar to that of cefixime. Cefitbuten, which contains a group for β-lactamase stability and $C{=}CHCH_2COOH$ at C-3, has lower activity against *S. pnuemoniae* and does not inhibit group B streptococci. Compound BMY 28232 contains a $C{=}C_2H_5$ at C-3 and $=N{-}OH$ at C-7. This agent is not hydrolyzed by TEM-1 but is hydrolyzed by the new TEM and SHV enzymes.[95] It has activity similar to other amino-thiazolyl cephalosporins.[95,96] Synthesis of cyclopropyl-amido cephems has decreased the activity of the agents.

Other cephalosporins, which structurally resemble earlier agents in that they lack aminothiazyl side chains or groups to protect the β-lactam ring, have been synthesized.[97–99] These agents inhibit streptococci, pneumococci, *Haemophilus* and *Moraxella* at reasonable concentrations, but do not provide activity against β-lactamase-producing Enterobacteriaceae and are of minimal importance microbiologically — although their pharmacokinetic properties are usually superior to the esters of the aminothiazolyl compounds.

A novel oral compound, loracarbef, is a carbacephem otherwise similar to cefaclor in antibacterial properties,[100] but with longer half-life and better oral absorption. This illustrates that carbacephems are of activity similar to their prototype dihydrothiazine agents. Table 3.5 shows the effect of new β-lactamases on the activity of oral cephalosporins.

3.15 Oxacephalosporins

Replacement of the sulfur of the bicyclic ring structure has a number of important biological effects. The S→O shift results in increased antibacterial activity against gram-negative bacteria, particularly microorganisms lacking β-lactamases. β-Lactamase stability in the oxacephem compounds is less, and there is decrease in activity against gram-positive organisms.[101] In general, the *in vitro* structure–activity relationships of oxacephems are similar to those of cephems with respect to substituent variation at the 3,7-α and 7-β positions. Introduction of a methoxy group at the 7-α position produces less loss of activity except against *S. aureus* and *S. pneumoniae*. The presence of an acidic function, a carboxyl group, on the α-carbon of the 7-acyl side chain provides stability against cephalosporinases of the chromosomal, inducible type. The presence of the 7-α-methoxy and the carboxyl group on the α-carbon of the acyl side chain provides excellent activity against *Bacteroides* species since they contain primarily cephalosporinases.[101] Moxalactam contains a methylthiotetrazole group, which is associated with abnormalities of clotting due to decreased vitamin K synthesis.[102] The carboxyl group similar to the carboxy penicillins causes decreased platelet aggregation. The pharmacological properties of moxalactam, with its long half-life and high blood levels, appear related to the hydroxyl group in the *para* position of the phenyl side chain, which apparently changes excretion from tubular secretion to glomerular filtration.

Compound flomoxel has a CH_2OH attached to the thiotetrazole ring at C-3, and an F_2CHCH_2–SCH_2CO acyl side chain. It has excellent anti-anaerobic activity and does not have the prothrombin problems of moxalactam.[103–106]

3.16 Carbapenems

Carbapenems basically are 1-carbapen-2-em-3-carboxylic acid with substituents at the 2 and 6 positions[107] (see Introduction, Figure 1). Instead of a C-7 acyl amino substituent on the β-lactam ring, the compounds contain a hydroxyethyl side chain or other substituent. The original compound of this series was called thienamycin, the name derived from the exocyclic sulfur to the ene-lactam system of the molecule.

The major differences between the carbapenems, penicillins and cephalosporins are the absence of the ring sulfur atom and a more strained pyrroline–azetidinone ring system. The β-lactam is highly reactive due to ring strain and the electron-withdrawing effects associated with the adjacent double bond. Carbapenems possess a 6-alkyl, or substituted alkyl substituents that can be in a *cis* or *trans* relationship to the substituents about the azetidinone ring. β-Lactamase stability is due to the *trans* configuration of the side chain, since the epithienamycins, which have a similar configuration, except that they contain the side chain the the C-6-β-configuration, are not β-lactamase stable.[107] The basic alkylthio side chain in imipenem is the source of its activity against *P. aeruginosa*.[108] The side chain is critical in preventing metabolism of the structure by dipeptidases present in the kidney, as shown by newer compounds such as meropenem, SM-7338 and LJC 10,627.[109–111]

Thienamycin hasextremely broad antibacterial activity, but is unstable in an aqueous environment at high concentration. Imipenem, the carbapenem in clinical use, is not only highly β-lactamase stable, but also an effective inhibitor of β-lactamases.[112–114] Imipenem is an effective inducer of cephalosporinases, but inhibits bacteria that contain these enzymes and will inhibit β-lactamase hyperproducing *E. cloacae* and *C. freundii*.[4]

An important factor in the activity of the imipenem is its ability to rapidly diffuse through the porin channels of bacteria.[115] Recently it has been shown that imipenem-resistant *E. cloacae* and *P. rettgeri* lack porins.[116] Furthermore, it is the D2 protein of *P. aeruginosa* that facilitates diffusion of penems and carbapenems through the outer wall of *P. aeruginosa*. Imipenem and other carbapenems bind to PBP2 in gram-negative species.[114] Carbapenems produce a lytic reaction and rapid cell death when they reach their receptor site, which is related to binding to other PBPs as well as to PBP 2. Imipenem and other carbapenems produce a marked post-antibiotic effect not only with gram-positive species, but also with Enterobacteriaceae and *P. aeruginosa*.[110] *Xanthomonas maltophilia* produces a β-lactamase that destroys all carbapenems, and none of the structural changes has protected the carbapenems.[110,111] The instability of imipenem to the renal dihydropeptidase has been overcome by using the renal dipeptidase inhibitor, cilastatin.[118]

Other carbapenems that have been synthesized in the past few years include RS-533 which has excellent β-lactamase stability and inhibitory activity.[119,120] In general, heterocyclic derivatives with a variety of groups attached to the sulfur yield some increase in gram-negative activity compared to imipenem, and increased renal dehydropeptidase stability; however gram-positive activity is slightly less.[110,111,120]

3.17 Penems

Although penems were constructed by Woodward before the discovery of the carbapenems,[121] their structure–activity relationships were determined later. In penems the bridgehead carbon must be *R*, since the corresponding 5-*S* epimers are inactive.[121] Substituents at the 6 position significantly affect the level of activity and provide stability against the β-lactamases of gram-negative organisms. As has been observed with the carbapenems, a C5–C6 *trans* configuration is necessary to achieve β-lactamase stability. Substitution on the 6-α position provides greater β-lactamase stability, since the unsubstituted 6-α is less active against β-lactamase-producing strains.[121] A 2-cysteamino group does not improve the spectrum of activity over a 2-ethylthio group as is seen with carbapenems.[122] The compound SCH 29482 has a hydroxyethyl group at position 6 and a 2-ethylthio group. The lack of a basic alkylthio side chain, which prevents metabolism, results in loss of anti-*Pseudomonas* activity.

In penems, depending on the side chain components, activity against certain of the Enterobacteriaceae is strikingly different. This is seen in particular with *Proteus* species, *Serratia marcescens* and *Enterobacter*.[123–125] Penems are effective inducers of β-lactamase in *Enterobacter* spp., *C. freundii* and *P. aeruginosa*, but organisms of these species resistant to cephalosporins and penicillins are often inhibited. The penems bind poorly to the β-lactamases that they induce, and are not hydrolyzed by them. The small compact size of the penems contributes to their ability to reach the PBPs, and their affinity for PBP 2 contributes to their activity.

The compound SCH 29482, which has an ethylthio group at position 2, undergoes metabolism in urine with the production of a thiol compound that has a foul-smelling odor. Another compound, SCH 38342, can only be used parenterally. It inhibits *S. aureus* and other gram-positive species but has no activity against *P. aeruginosa*.[126] Anti-anaerobic activity, however, is extremely high, comparable or superior to that of the carbapenems such as imipenem.

Compound FCE 22101 is a penem that contains a 2-carba-moyloxy-methyl group, and has its ester esterified on C-3 with CH_2COCH_3. It has activity similar to other penems.[125,127]

A number of other penems have been studied, in particular CGP 31608.[128,129] It was less potent than imipenem except against *Pseudomonas* and methicillin-resistant *S. aureus*. It did not inhibit enterococci and, due to a very short half-life, it was abandoned.

In general, penems have not yet proved clinically useful due to poor pharmacokinetic properties or toxic properties.

3.18 Monobactams

The term monobactam was used by Sykes *et al.*[130] to describe a group of monocyclic, bacterially-produced β-lactam antibiotics isolated from soil bacteria such as *Chromobacterium*, *Agrobacterium* and *Flexibacter*.[76,77]

Naturally occurring monobactams contain a 2-oxoazetidine-1-sulfonic acid moiety with variation in the 3-α- or 3-β side chain.[131] Monobactams can undergo side chain substitution and, depending on the side chain, compounds may exhibit activity against a variety of different microorganisms. Production of monobactams that have side chains similar to the first-generation cephalosporins, containing an acetyl side chain with a thienyl derivative, have been produced. These homologs were less active than the cephalosporins with a ureido acyl side chain. A compound with an acyl piperazine side chain has activity similar to piperacillin.[131]

Aztreonam contains a carboxypropyl and a 2-aminothiazol-4-yl-acetyl side chain. The methoxyimino group does not provide the β-lactamase stability seen with cephalosporins, and 4-methylation is necessary to achieve β-lactamase stability.[132] The sulfonic acid group attached to the nitrogen of the β-lactam ring is responsible for 'activating' the β-lactam carbonyl. The aminothiazolyl oxime moiety on the acyl side chain is responsible for gram-negative bacterial activity in the cephalosporins. Activity against *Pseudomonas* is achieved by the addition of methyl groups and a carboxylic acid function on the oxime side chain, similar to that of ceftazidime. Placement of a β-methyl group at position 4 enhances the stability of the ring to attack by the K1 β-lactamase and also increases anti-*Pseudomonas* activity.[132] Conversely, a 4-fluoromethyl monobactam loses *Pseudomonas* activity.[133] Replacement of the methyl group at the 4 position with other groups results in increased activity against *P. aeruginosa*, as is seen with the α-carbamyloxymethyl, carumonam.[134] Carumonam inhibits *K. oxytoca* resistant to aztreonam, indicating that variation in position 4 can alter β-lactamase stability to specific β-lactamases. This is seen with Bo-1165, a 1-carboxy-1-cycloproxyamino 4-fluoromethyl monobactam.[135]

Replacement of the $-SO_3H$ of monobactams with an acidic moiety yields activity of appreciable nature. For example, $-CON-SO_2N-R_1R_2$, where R_1 and R_2 are H, alkyl, aryl, amino and methoxy groups, are all β-lactamase-stable and highly active against aerobic gram-negative bacteria. Monophosphams have more β-lactamase stability but less intrinsic antibacterial activity. Increasing the size and lipophilicity of phosphonate esters at position-1 decreases intrinsic activity. Conversely, β-lactam activation by use of a carboxyl group is not useful, since the carboxylates are unstable, undergoing decarboxylation. Incorporating a hydroxypyridine moiety into the N-1 activating group produces a monocarbam, which is more active against non-fermenting gram-negative species.[136]

An acetic acid derivative of the monobactams — SQ 82,291 — is a [2-oxo-1-(azetidinyl)oxy] acetic acid, which inhibits aerobic gram-negative bacteria. The glycolate-activated monobactam, which contains an amino-thiazolyl oxime side chain, has excellent β-lactamase stability but lost *Pseudomonas* activity and, like other monobactams, does not inhibit gram-positive species.[137,138] This derivative was made into an orally absorbed ester — tigemonam – but, due to pharmacological problems it was abandoned.

A most interesting aspect of the monobactams is that they do not cross-react allergenicly with penicillins or cephalosporins; consequently, they can be used to treat penicillin-allergic patients.[139]

3.19 β-Lactamase inhibitors

Compounds such as methicillin and the isoxazolyl penicillins inhibit staphylococcal β-lactamases, some of the common plasmic β-lactamases and some of the Richmond–Sykes Ia and Id type β-lactamases, which are often called cephalosporinases.[3] Penicillins themselves have not proved successful as β-lactamase inhibitors of gram-negative organisms, since their penetration through the porin channels of gram-negative organisms is exceedingly poor. A substance that is attacked by β-lactamases may, in principle, also function as an inhibitor. This has been possible both with natural compounds such as clavulanic acid, and with the penicillanic acid sulfone derivatives. These compounds function as progressive β-lactamase inhibitors and fit into the characteristics of suicide-enzyme inhibitors[139] (see chapter 7).

The β-lactamase-inhibiting compounds usually show a number of basic criteria. These include possession of a β-lactam ring, which can form an acyl enzyme intermediate that hydrolyzes at a relatively slow rate. They often contain an acidic 6-α-proton to participate in an elimination reaction to generate a relatively stable α,β-unsaturated acyl compound.[140] Alternatively, this vinylogous carbamate intermediate, which is sufficiently stable to permit a virtually irreversible inactivation, can be generated by having a good leaving group at C-6 (see chapter 7).

Critical to the activity of any β-lactamase inhibitor is its ability to pass readily through porin channels in gram-negative organisms to reach an adequate concentration in the periplasmic space.

In clavulanate, the sulfur in position 1 has been replaced by an oxygen, increasing the reactivity of the molecule. At position 2, is an unsaturated bond to an ethyl group. Substitution on the ethyl group with a variety of different groups is possible,[141] and in clavulanic acid itself there is a hydroxyl group. However, NH_2, NHCHO, $NHCOCH_3$ and a variety of other substitutions of the hydroxyl group all produce β-lactamase inhibition of

Richmond–Sykes type III and type IV β-lactamases, plasmid enzymes, and the penicillinases of *S. aureus*. Many modifications of the allylic hydroxymethyl substituent result in retention or improvement of inhibitory activity.

Other inhibitors are penicillanic acid sulfone derivatives. Sulbactam has a similar, although somewhat different inhibition of β-lactamases compared to clavulanate.[142] It also has a slightly better inhibition of cephalosporinases compared to clavulanate, but it would not be clinically useful. The ability of sulbactam to pass through cell walls is considerably less than that of clavulanate.

Alkyl derivatives of penicillanic acid in which the side chain is varied at position 6 of the penicillanic acid vary to an extreme degree in their ability to inhibit different β-lactamases,[143] but many of these β-lactamase inhibitors do not cross the bacterial cell wall. A 6-acetylmethylene penicillanic acid (Ro 15-1903) has also been shown to function like sulbactam and clavulanic acid.[144]

A triazolyl derivative of penicillanic acid, tazobactam, has β-lactamase activity similar to sulbactam and, like sulbactam, is slightly less effective with some *Klebsiella* than is clavulanate.[145–147] It is in clinical use combined with piperacillin. It inhibits some type I β-lactamases.

Another approach to β-lactamase inhibition was use of a penem, BRL 42715. This triazolyl methylene penem is a potent inhibitor not only of plasmid β-lactamases of the TEM and SHV type, but also of chromosomal cephalosporinases of *Enterobacter* and *Citrobacter*.[148] All of these studies seem to indicate that bulky groups at C-6 of a sulfone or penem created better inhibition of chromosomal cephalosporinases.

3.20 Conclusions

This chapter has reviewed some of the structure–activity relationships that influence *in vitro* activity and pharmacological properties of β-lactams. Although there has been progress in understanding how additions to the basic β-lactam nucleus will be reflected in marked changes in antibacterial activity, there is much to be learned. Understanding the structural aspects of β-lactams has made it possible to synthesize compounds that have the ability to pass through the walls of gram-negative organisms in a more rapid fashion, to provide agents that have an increased affinity for the β-lactam receptor sites, and to increase stability against β-lactamase attack.

In some cases, β-lactamase stability has been achieved at a significant loss of overall *in vitro* activity whereas, in others, it has been possible to achieve high binding to β-lactam receptor sites, and to achieve a compound that is not excluded from the receptor sites by an inability to cross the cell wall.

Advances in the chemistry of carbapenems, monobactams and catechol cephalosporins may produce molecules that will meet the changing needs of bacterial resistance.

References

1. B.G. Spratt, *Phil. Trans. Royal Soc. London (B)* (1980) **289** 273–283.
2. H. Nakaido, in *Antibiotic Inhibitors of Bacterial Cell Wall Synthesis* (Ed. D.J. Tipper), Pergamon Press, Oxford (1987), pp. 203–240.
3. H.C. Neu, *Am. J. Med.* (1985) **79** (Suppl. 5B) 2–12.
4. C.C. Sanders, *Ann. Rev. Microbiol.* (1987) **41** 573–593.
5. H.W. Florey, E. Chain, M.G. Heatley, M.A. Jennings, A.G. Saunders, E.P. Abraham and M.E. Florey, *Antibiotics*, Oxford University Press, London (1949).
6. D.J. Tipper and J.L. Strominger, *Proc. Natl. Acad. Sci. USA* (1965) **54** 1113–1141.
7. J.H.C. Nayler, in *Advances in Drug Research*, Vol. 7, (Eds N.J. Harper and A.B. Simmonds), Academic Press, New York (1973), pp. 1–105.
8. F.P. Doyle, A.A.W. Long, J.H.C. Naylor and E.R. Stove, *Nature* (1961) **192** 1183–1184.
9. E.G. Brain, F.P. Doyle, M.D. Mehta, D. Miller, J.H.C. Naylor and E.R. Stove, *J. Chem. Soc.* (1962) 491–497.
10. H.C. Neu, *Rev. Infect. Dis.* (1983) **5** (Suppl. 1) 9–20.
11. H.C. Neu, *Int. J. Clin. Pharmacol. Biopharm.* (1975) **11** 132–144.
12. J.R.E. Hoover and G.L. Dunn, in *Burgers Medicinal Chemistry*, 4th edition, (Ed. M. Wolff), Wiley, New York (1979), pp. 83–172.
13. P. Acred, D.M. Brown, E.T. Knudsen, G.N. Rolinson and R. Sutherland, *Nature* (1967) **215** 66–70.
14. H.C. Neu and E.B. Winshell, *Appl. Microbiol.* (1971) **21** 66–70.
15. K. Tsuchiya, M. Kondo and H. Nagatomo, *Antimicrob. Agents Chemother.* (1978) **13** 137–145.
16. K.E. Price, D.R. Chisholm, F. Leitner, M. Misiek and A. Gourevitch, *Appl. Environ. Microbiol.* (1969) **17** 881–887.
17. H.C. Neu, *Rev. Infect. Dis.* (1983) **5** (Suppl. 2) 319–336.
18. W. Zimmerman, *Antimicrob. Agents Chemother.* (1980) **18** 94–100.
19. N.A.C. Curtis, D. Orr, G.W. Ross and M.G. Boulton, *Phil. Trans. Royal Soc. London (B)* (1980) **289** 368–370.
20. F. Lund and L. Tybring, *Nature* (1977) **236** 135–137.
21. F.J. Lund, in *Recent Advances in the Chemistry of Beta-lactam Antibiotics* (Ed. J. Elks), Royal Society of Chemistry, London (1977), pp. 24–45.
22. H.C. Neu, *Antimicrob. Agents Chemother.* (1976) **9** 793–799.
23. H.C. Neu, *Am. J. Med.* (1983) **75** (2A) 9–20.
24. H.C. Neu, *Antimicrob. Agents Chemother.* (1976) **10** 535–542.
25. B. Slocombe, M.J. Basker, P.H. Bentley, J.P. Clayton, M. Cole, K.R. Comber, R.A. Dixon, R.A. Edmonson, D. Jackson, D.J. Merriken and R. Sutherland, *Antimicrob. Agents Chemother.* (1981) **20** 38–46.
26. K. Jules and H.C. Neu, *Antimicrob. Agents Chemother.* (1982) **22** 453–460.
27. W. Cullman and W. Dick, *Zbl. Bakt. Hyg.* (1983) **A256** 109–118.
28. K.L. Then and P. Angehrn, *Chemiotherapia* (1985) **4** 83–89.
29. E.A. Guest, R. Horton, G. Mellows, B. Slocombe, A.J. Swaisland and T.C.G. Tasker, *J. Antimicrob. Chemother.* (1985) **15** 327–336.
30. M.J. Basker, R.A. Edmondson, S.J. Knott, R.J. Ponsford, B. Slocombe and S.J. White, *Antimicrob. Agents Chemother.* (1984) **26** 734–740.
31. W. Mandell and H.C. Neu, *Antimicrob. Agents Chemother.* (1986) **29** 769–773.
32. H.W. VanLanduyt, A. Lambert, J. Boelad and B. Gorts, *Antimicrob. Agents Chemother.* (1986) **29** 362–366.
33. M.L. Sassiver and A. Lewis, *Adv. Appl. Microbiol.* (1970) **13** 163–236.
34. H. Nomura, I. Minami, T. Hitaka and T. Fuguno, *J. Antibiot.* (1976) **29** 928–936.

35. J.A. Webber and W.J. Wheeler, in *Chemistry and Biology of Beta-lactam Antibiotics*, (Eds R.B. Morin and M. Gorman), Academic Press, New York (1982), pp. 372–436.
36. C.H. O'Callaghan, R.B. Sykes, A. Giffith and J.E. Thornton, *Antimicrob. Agents Chemother.* (1976) **9** 511–519.
37. H.C. Neu and K.P. Fu, *Antimicrob. Agents Chemother.* (1978) **13** 657–664.
38. S. Normark, T. Grundstrom and S. Bergstrom, *Scand. J. Infect. Dis.* (1980) **25** 23–29.
39. A.L. Barry, R.N. Jones and J. Thornsberry, *Clin. Microbiol.* (1983) **17** 232–239.
40. H.C. Neu, G. Saha and N.X. Chin, *Antimicrob. Agents Chemother.* (1989) **33** 1260–1267.
41. N. Watanabe, K. Katsu, M. Moriyaama and K. Kitoh, *Antimicrob. Agents Chemother.* (1988) **32** 693–701.
42. M. Arisawa, Y. Sekine, S. Himizu, H. Takano, P. Angehrn and R.L. Then, *Antimicrob. Agents Chemother.* (1991) **35** 653–659.
43. N.X. Chin, J.W. Gu, W. Fang and H.C. Neu, *Antimicrob. Agents Chemother.* (1991) **35** 259–262.
44. R. Wise, J.M. Andrews, J.P. Ashby and D. Thornberey, *Antimicrob. Agents Chemother.* (1991) **35** 329–334.
45. Y. Shigi, H. Kojo, M. Wakasugi and N. Nishida, *Antimicrob. Agents Chemother.* (1981) **19** 393–396.
46. R. Bucourt, D. Borman, R. Heymes and M. Perronett, *J. Antimicrob. Chemother.* (1981) **8**(B) 15–21.
47. H.C. Neu, *Pharm. Ther.* (1985) **30** 1–18.
48. A. Philipon, G. Paul, G. Vedel and P. Nerot, *Presse Med.* (1988) **17** 1833–1889.
49. C. Mabilat and P. Courvalin, *Antimicrob. Agents Chemother.* (1990) **34** 2210–2216.
50. M.D. Kitzis, N. Liassine, B. Fere, L. Gutmann, J.F. Acar and F. Goldstein, *Antimicrob. Agents Chemother.* (1990) **34** 1783–1786.
51. G.A. Jacoby and A.A. Medeiros. *Antimicrob. Agents Chemother.* (1991) **35** 1697–1704.
52. G.A. Jacoby and I. Carreras, *Antimicrob. Agents Chemother.* (1990) **34** 856–862.
53. R.N. Jones and A.L. Barry, *Eur. J. Clin. Microbiol. Infect. Dis.* (1988) **7** 802–806.
54. W. Durckheimer, J. Blumbach, R. Heymes, E. Schrinner and K. Seeger, in *New Beta-lactam Antibiotics* (Ed. H.C. Neu), Am. Coll. Phys., Philadelphia (1981), pp. 3–22.
55. W.E. Sanders, Jr. and C.C. Sanders, *Rev. Infect. Dis.* (1988) **10** 830–838.
56. C.S. Bryan, J.F. Johns, Jr., M.S. Pai and T.L. Austin, *Am. J. Dis. Child.* (1985) **159** 1086–1089.
57. J.W. Chow, M.J. Fine, D.M. Shales, J.P. Quinn, D.C. Hooper, M.P. Johnson, R. Ramphal, M.W. Wagener, D. Minashiro and V.L. Yu, *Ann. Intern. Med.* (1991) **115** 584–590.
58. G.M. Eliopoulos, E. Reiszner, S. Willey, W.J. Novick, Jr. and R.C. Moellering, Jr., *Diagn. Microbiol. Infect. Dis.* (1989) **12** 149–156.
59. I. Saikawa, N. Matsubara, S. Minami and S. Mitsuhashi, in *Beta-Lactam Antibiotics* (Eds M. Salton and G.D. Shockman), Academic Press, New York (1981), pp. 353–360.
60. H.C. Neu, F.P. Fu, N. Asawapokee, P. Asawapokee and K. Kung, *Antimicrob. Agents Chemother.* (1979) **16** 150–157.
61. H.C. Neu and B.E. Scully, *Rev. Infect. Dis.* (1984) **6**(S3) 667–677.
62. C.H. O'Callaghan, P. Acred, P.B. Harper, D.M. Ryan, K. Kirby and S.M. Harding, *Antimicrob. Agents Chemother.* (1980) **17** 876–883.
63. H.C. Neu and P. Labthavikul, *Antimicrob. Agents Chemother.* (1982) **21** 11–18.
64. N. Watanabe, T. Nagasu, K. Katsu and K. Kitoh, *Antimicrob. Agents Chemother.* (1987) **31** 497.
65. H. Mochizuki, Y. Oikawa, H. Yamada, S. Kusababe, T. Shirikawa, K. Murakami, K. Kato, J. Ishiguro and H. Kosuzume, *J. Antibiot.* (1988) **41** 377–382.
66. M.E. Erwin, R.N. Jones, M.S. Barrett, B.M. Briggs and D.N. Johnson, *Antimicrob. Agents Chemother.* (1990) **35** 929–937.
67. N.X. Chin, J.W. Gu, W. Fang and H.C. Neu, *Antimicrob. Agents. Chemother.* (1991) **35** 259–266.
68. H. Ogino, K. Iwamatsu, K. Katano, S. Nakabayasha, Y. Yoshida, S. Shibahara, T. Tsurorka, S. Inouye and S. Kondo, *J. Antibiot.* (1990) **43** 189–195.
69. H. Nikaido and E.Y. Rosenberg, *J. Bacteriol.* (1990) **172** 1361–1369.
70. K. Hantke, *FEMS Microbiol. Lett.* (1990) **67** 5–9.

71. T. Hashizume, M. Sanada, S. Nakagawa and N. Tanaka, *J. Antibiot.* (1990) **43** 1617–1621.
72. I.A. Critchley, *J. Antimicrob. Chemother.* (1990) **26** 733.
73. S. Nakagawa, M. Sanada, K. Matsuda, T. Hashizume, T. Asahi and R. Ushijima, *Antimicrob. Agents Chemother.* (1989) **33** 1423–1427.
74. N.A.C. Curtis, R.L. Eisenstadt, S.J. East, R.J. Cornfeld, L.A. Walker and A.J. White, *Antimicrob. Agents Chemother.* (1988) **32** 1879–1886.
75. E.D. Stapley and J. Birnbaum, in *Beta-lactam Antibiotics* (Eds M.R.J. Salton and G.D. Shockman), Academic Press, New York (1981), pp. 327–351.
76. H.C. Neu, N.X. Chin and P. Labthavikul, *Chemiotherapia* (1985) **4** 415–423.
77. I. Phillips, A. King, K. Shannon and C. Warren, *J. Antimicrob. Chemother.* (1983) **11** 1–9.
78. H.M. Wexler and S.M. Finegold, *Antimicrob. Agents Chemother.* (1988) **32** 601–604.
79. H.C. Neu, *Rev. Infect. Dis.* (1982) **4**(S) 374–378.
80. J.L. Spencer, F.Y. Siu, E.H. Flynn, B.G. Jackson, M.V. Sigal, H.M. Higgins, E.R. Chauvette, S.L. Andrews and E.D. Bloch, *Antimicrob. Agents Chemother.* (1966) **XX** 573–580.
81. H.C. Neu, N.X. Chin and P. Labthavikul, *Infection* (1985) **13** 146–155.
82. R.N. Jones, C. Thornsberry and A.L. Barry, *Antimicrob. Agents Chemother.* (1984) **25** 710–718.
83. H. Goosens, P. DelMol, H. Coignau, J. Levy, O. Grados, G. Ghysels, H. Innocent and J.P. Butzler, *Antimicrob. Agents Chemother.* (1985) **27** 388–392.
84. E. Ingue and S. Mitsuhashi, *Antimicrob. Agents Chemother.* (1989) **33** 2157–2159.
85. G. Agnelli, A. DelFavero and P. Parise, *Antimicrob. Agents Chemother.* (1986) **29** 1108–1109.
86. B.R. Meyers, *Am. J. Med.* (1985) **79**(2A) 96–103.
87. R.F. Jacobs, *Pediat. Infect. Dis. J.* (1986) **5** 708–710.
88. C.H. O'Callaghan, *J. Antimicrob. Chemother.* (1975) **1**(S) 1–12.
89. H.C. Neu, N.X. Chin and P. Lavthabikul, *Antimicrob. Agents Chemother.* (1986) **30** 423–428.
90. R. Wise, J.M. Andrews and L.J.V. Piddock, *Antimicrob. Agents Chemother.* (1986) **29** 1067–1072.
91. R.J. Fass and V.L. Heisel, *Antimicrob. Agents Chemother.* (1986) **30** 429–434.
92. H.C. Neu, G. Saha and N.X. Chin, *Antimicrob. Agents Chemother.* (1989) **33** 1795–1800.
93. N.X. Chin and H.C. Neu, *Antimicrob. Agents Chemother.* (1988) **32** 671–677.
94. M. Sheppard, A. King and I. Phillips, *Eur. J. Clin. Microbiol. Infect. Dis.* (1991) **10** 573–580.
95. N.X. Chin, K.W. Yu and H.C. Neu, *Eur. J. Clin. Microbiol. Infect. Dis.* (1990) **9** 841–845.
96. K. Tomatsu, S. Masuyoshi, M. Hirano, H. Kawguchi, T. Oli, J. Fung-Tomc, J.V. Desiderio and R.E. Kessler, *Antimicrob. Agents Chemother.* (1989) **33** 489–487.
97. N.X. Chin and H.C. Neu, *Antimicrob. Agents Chemother.* (1987) **31** 480–483.
98. C.E. Nord, A. Lindmark and I. Pearson, *Eur. J. Clin. Microbiol. Infect. Dis.* (1989) **8** 550–551.
99. R. Wise, J.M. Andrews, J.P. Asbby and D. Thornber, *Antimicrob. Agents Chemother.* (1990) **34** 813–818.
100. R.N. Jones and A.L. Barry, *Eur. J. Clin. Microbiol. Infect. Dis.* (1987) **6** 570–571.
101. Y. Yoshida, *Phil. Trans. Royal Soc. London (B)* (1980) **289** 231–237.
102. J.J. Lipsky, J.C. Lewis and W.J. Novick, *Antimicrob. Agents Chemother.* (1985) **25** 380–381.
103. H.C. Neu and N.X. Chin, *Antimicrob. Agents Chemother.* (1986) **30** 638–644.
104. S. Iyobe, T. Kubota and S. Mitsuhashi, *Chemotherapy (Tokyo)* (1987) **35**(Suppl. 1) 33–43.
105. T. Yokota, E. Suzuki, K. Azai and N. Kato, *Chemotherapy (Tokyo)* (1987) **35**(Suppl. 1) 57–69.
106. K. Sawa, M. Asoki, H. Kohmo, T. Kobayashi, K. Watanabe and K. Uemo, *Chemotherapy (Tokyo)* (1987) **35**(Suppl. 1) 44–56.

107. R.W. Radcliffe and G. Albers-Schonberg, in *Chemistry and Biology of Beta-lactam Compounds*, Vol. 2, (Eds R.B. Morin and M. Gorman), Academic Press, New York (1982), pp. 227–313.
108. B.G. Christensen, in *Beta-lactam Antibiotics* (Eds M.E.J. Salton and G.D. Schockman), Academic Press, New York (1981), pp. 101–125.
109. J.R. Edwards, P.J. Turner, C. Wannop, E.S. Withnell, A.J. Grindeg and K. Nairn, *Antimicrob. Agents Chemother.* (1989) **33** 215–222.
110. H.C. Neu, A. Novelli and N.X. Chin, *Antimicrob. Agents Chemother.* (1989) **33** 1009–1018.
111. K. Ubukata, M. Hikida, M. Yoshida, K. Nishiki, Y. Furukawa, K. Tashiro, M. Konno and S. Mitsuhashi, *Antimicrob. Agents Chemother.* (1990) **34** 994–1000.
112. H. Kroop, J.G. Sundelof, J.S. Kahan, F.M. Kahan and J. Birnbaum, *Antimicrob. Agents Chemother.* (1980) **17** 993–1000.
113. H.C. Neu and P. Labthavikul, *Antimicrob. Agents Chemother.* (1982) **21** 180–187.
114. S. Mitsuhashi, *J. Antimicrob. Chemother.* (1983) **12**(Suppl. D) 53–64.
115. F. Yoshimura and H. Nikaido, *Antimicrob. Agents Chemother.* (1985) **27** 84–92.
116. A. Raimondi, A. Traveras and H. Nikaido, *Antimicrob. Agents Chemother.* (1991) **35** 1174–1180.
117. J. Trias and H. Nikaido, *Antimicrob. Agents Chemother.* (1990) **34** 52–57.
118. F.M. Kahan, H. Kropp, J.G. Sundelof and J. Birnbaum, *J. Antimicrob. Chemother.* (1983) **12**(Suppl. D) 1–35.
119. T. Shibata, K. Lino and Y. Sugimura, *Heterocycles* (1986) **24** 1331–1333.
120. H.C. Neu, N.X. Chin, S. Saha and P. Labthavikul, *Antimicrob. Agents Chemother.* (1986) **30** 828–834.
121. R.B. Woodward, *Phil. Trans. Royal Soc. London (B)* (1980) **289** 239–250.
122. A.K. Ganguly, V.M. Girjavallaban, S. McComble, P. Pinto, R. Rizvi, P.D. Jeffrey and S. Lin, *J. Antimicrob. Chemother.* (1982) **9**(Suppl. C) 1–6.
123. W. Cullmann and M. Steiglitz, *Antimicrob. Agents Chemother.* (1988) **32** 1090–1093.
124. T. Gootz, J. Retsema, A. Girard, E. Hamanaka, M. Anderson and S. Solowski, *Antimicrob. Agents Chemother.* (1989) **33** 1160–1166.
125. H.C. Neu, N.X. Chin and P. Labthavikul, *Antimicrob. Agents Chemother.* (1985) **28** 305–313.
126. K. Matsuda, K. Sasaki, T. Inoue, H. Kondo, M. Inoue and S. Mitsuhashi, *Antimicrob. Agents Chemother.* (1985) **28** 684–688.
127. R. Wise, J.M. Andrews and G. Danks, *Antimicrob. Agents Chemother.* (1983) **24** 909–914.
128. H.C. Neu, N.X. Chin and N.M. Neu, *Antimicrob. Agents Chemother.* (1987) **31** 558–569.
129. R. Wise, J.M. Andrews and L.J.V. Piddock, *Antimicrob. Agents Chemother.* (1987) **31** 267–272.
130. R.B. Sykes, C.M. Cimarusti, D.P. Bonner, K. Bush, D.M. Floyd, N.M. Georgopapdakou, N.H. Koster, W.C. Liu, W.L. Parker, P.A. Principe, M.L. Rathnum, W.A. Slusarchyk, W.H. Trejo and J.S. Wells, *Nature* (1981) **291** 489–491.
131. C.M. Cimaruski, R.B. Sykes and M.A. Appelgate, in *Beta-lactam Antibiotics* (Ed. H.C. Neu), Am. Coll. Phys., Philadelphia (1982), pp. 35–44.
132. H.C. Neu and P. Labthavikul, *Antimicrob. Agents Chemother.* (1983) **24** 227–232.
133. J.S. Skotnicki, T.J. Commons, R.W. Rees and J.L. Speth, *J. Antibiot.* (1983) **36** 1201–1204.
134. R.J. Fass and V.L. Helsel, *Antimicrob. Agents Chemother.* (1985) **28** 834–836.
135. H.C. Neu and N.X. Chin, *Antimicrob. Agents Chemother.* (1987) **31** 505–511.
136. G.E. Zurenko, S.E. Truedell, B. Yagi, R.J. Mourey and A.L. Laborde, *Antimicrob. Agents Chemother.* (1990) **34** 884–888.
137. N.X. Chin and H.C. Neu, *Antimicrob. Agents Chemother.* (1988) **32** 84–91.
138. R.C. Fuchs, R.N. Jones and A.L. Barry, *Antimicrob. Agents Chemother.* (1988) **32** 346–349.
139. M. Cole, *Phil. Trans. Royal Soc. London (B)* (1980) **289** 207–223.
140. R.L. Charnas, J. Fisher and J.R. Knowles, *Biochemistry* (1978) **17** 2185–2189.
141. P.C. Cherry and C.E. Newall, in *Chemistry and Biology of Beta-lactam Antibiotics* (Eds R.B. Morin and M. Gorman), Academic Press, New York (1982), pp. 361–402.

142. A.R. English, J.A. Retsema, V.A. Ray, J.E. Lunch and W.E. Barth, *Antimicrob. Agents Chemother.* (1978) **14** 414–419.
143. M.J. Calverley and M. Begtrup, *J. Antibiot.* (1983) **36** 1507–1515.
144. M. Arisawa and R.L. Then, *J. Antibiot.* (1982) **35** 1578–1583.
145. R.G. Micetich, S.N. Maiti, P. Spevak, T.W. Hall, Y. Yamabe, N. Ishida, M. Tanaka, Y. Yamazaki, A. Nakai and K. Ogawa, *J. Med. Chem.* (1987) **30** 1469–1471.
146. L. Gutman, M.D. Kizis, S. Yamabe and J.F. Acar, *Antimicrob. Agents Chemother.* (1986) **29** 955–957.
147. N.A. Kuck, N.V. Jacobus, P.J. Peterson, W.J. Weiss and R.T. Testa, *Antimicrob. Agents Chemother.* (1989) **33** 1964–1969.
148. K. Coleman, R.D.J. Griffin, J.W.J. Page and P.A. Upshaw, *Antimicrob. Agents Chemother.* (1989) **33** 1580–1587.

4 The mechanisms of reactions of β-lactams

M.I. PAGE

4.1 Introduction

The reactions of β-lactam antibiotics and their derivatives have been extensively studied. In this chapter an attempt is made to review those reactions that may be of relevance to the biological activity of β-lactams, and those reactions for which some reasonable quantitative observations have been made.

4.2 The aminolysis of β-lactam antibiotics

The reaction of amines with penicillins (**1**) to give penicilloyl amides (**2**) (Scheme 4.1) is of interest because the major antigenic determinant of penicillin allergy is the penicilloyl group bound by an amide linkage to ε-amino-groups of lysine residues in proteins.[1] The aminolysis of penicillin is an amide exchange reaction, a normally difficult process because of the small free-energy change, but one which occurs readily with β-lactams. Carbon–nitrogen bond fission in β-lactams is accompanied by a large release of strain energy, which makes the reaction thermodynamically feasible. The aminolysis reaction is also of interest because the bacterial enzyme inhibited by β-lactam antibiotics, catalyses a transpeptidation reaction.

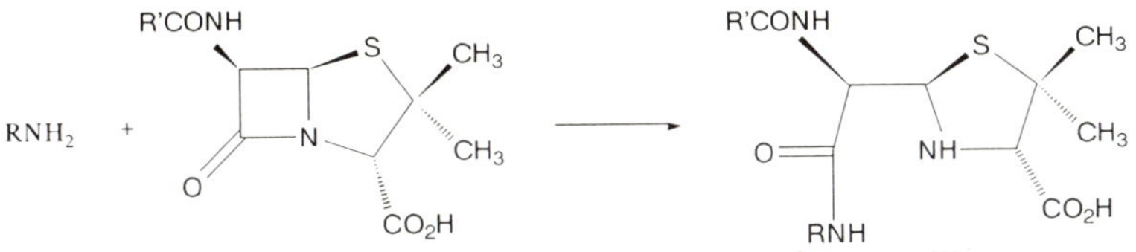

(**1**) (**2**)

Scheme 4.1

The aminolysis of penicillin is a substitution reaction in which an acyl group is transferred from one amino group to another. This reaction requires at least two proton transfers: (i) proton removal from the attacking amine; and (ii) proton addition to the leaving amino group. These proton transfers are facilitated by buffers,[2] and the kinetic importance of such catalysis is usually related to the observation that 'catalysis occurs where it is most needed'. Buffer catalysis is needed in the aminolysis of penicillin because covalent bond formation and fission between heavy atoms is accompanied by large changes in the acidity and basicity of the reacting groups.

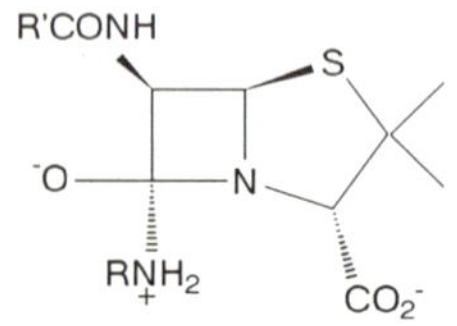

(3)

Acyl-transfer reactions generally involve the intermediate formation of a tetrahedral addition compound, such as (**3**), i.e. bond formation to the attacking group occurs before bond fission to the leaving group. The aminolysis of penicillins and cephalosporins is a stepwise process catalysed predominantly by bases, which remove a proton from the attacking amine. The evidence for the reversible formation of the tetrahedral intermediate is kinetic and based on linear free-energy relationships.[3]

The aminolysis of benzylpenicillin in aqueous solutions of the amine follows the rate law (equation (4.1)), where k_{obs} is the observed pseudo first-order rate constant for the disappearance of penicillin and k_0 is the second-order rate constant for the hydrolysis reaction.[4,5]

$$k_{obs} = k_0[OH^-] + k_u[RNH_2] + k_b[RNH_2]^2 + k_{OH}[RNH_2][OH^-] \quad (4.1)$$

The general acid catalysed aminolysis of penicillin makes a negligible contribution to the observed rate. The dominant form of buffer catalysis in the aminolysis reaction is general base catalysis. The relative importance of the terms in equation (4.1) depends on the basicity and the concentration of the amine and the pH. For strongly basic amines, the amine catalysed (k_b) and the hydroxide-ion catalysed (k_{OH}) terms contribute most to the observed rate, with the k_b term being more important with increasing concentration of amine. Consequently, the rate constants k_u for the uncatalysed reactions of basic monoamines are difficult to determine accurately. For the more weakly basic amines, aminolysis occurs mainly through the uncatalysed (k_u) and amine catalysed (k_b) pathways because of the low concentration of hydroxide ion. The hydroxide-ion catalysed term (k_{OH}) makes a negligible contribution to the observed rate of aminolysis in buffers of amines with $pK_a < ca.\ 9$, and can only be determined in solutions of sodium hydroxide.[5]

Linear free-energy relationships have been determined by varying, independently, the reactivity of the amine nucleophile and the catalyst. A plot of k_b for the general base catalysed aminolysis of benzylpenicillin for a series of primary monoamines against the pK_a values of the amines is a straight line, the slope of which gives the Bronsted β-value of 1.09.[5] This means that the reaction behaves as if approximately a unit positive charge is developed in the transition state and is distributed between the nucleophilic and the catalysing amine molecules. The Bronsted β-value of *ca.* unity is indicative of a transition state in which full covalent bond formation has taken place between the nitrogen of the attacking amine and the β-lactam's carbonyl carbon, and which carries the positive charge on either the nitrogen of the nucleophilic amine or on the catalytic amine molecule. The simplest mechanism that is consistent with this observation is shown in Scheme 4.2. The first step involves nucleophilic attack of the amine to reversibly form the tetrahedral intermediate, $T^{\pm}$, for which there is independent kinetic evidence. However, the intermediate $T^{\pm}$ breaks down rapidly to regenerate starting materials by expulsion of the attacking amine (k_{-1}). Catalysis of the reaction occurs by the formation of an encounter complex between $T^{\pm}$ and the basic catalyst, B (Scheme 4.2). Subsequent proton transfer from $T^{\pm}$ to B forms T^{-}, which then breaks down to products.

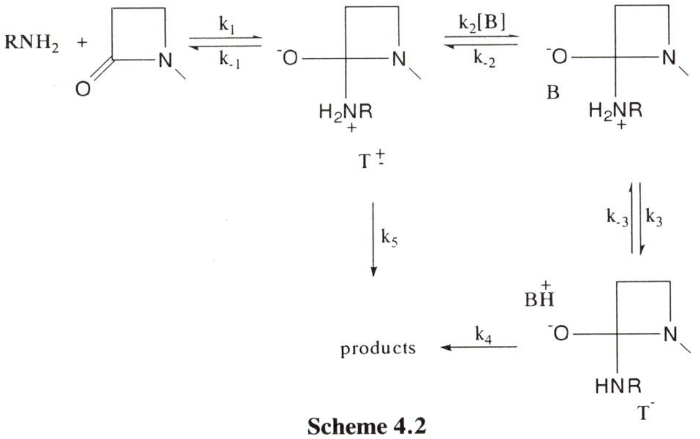

Scheme 4.2

The Bronsted β_{nuc}-value for the hydroxide-ion catalysed aminolysis of benzylpenicillin (k_{OH}) for a series of primary monoamines is 0.96.[5] This value also indicates that the reaction behaves as if a unit positive charge is developed on the attacking amine in the transition state. In this case the assignment of charge density is unambiguous, and the simplest interpretation of the β_{nuc}-value is that the attacking amine resembles its conjugate acid, i.e. is fully protonated, in the transition state. This is compatible with the mechanism of Scheme 4.2, in which the base B is a hydroxide ion, and is consistent with rate-limiting diffusion-controlled encounter of the tetrahedral intermediate $T^{\pm}$ and the hydroxide ion, k_2, to give T^- in a subsequent proton transfer step.

Hydrazine shows an enhanced nucleophilic reactivity towards penicillin compared with amines of similar basicity, which is attributed to the α-effect. This has allowed a study of the effect of varying the basicity of the catalyst with a constant nucleophile, even in the presence of strongly basic catalysts. For example, catalysis of the reaction of hydrazine with benzylpenicillin occurs even with the strongly basic amine, propylamine.[6] These is a non-linear dependence of the rate of hydrazinolysis of benzylpenicillin upon the basicity of both oxygen and nitrogen base catalysts. For strongly basic catalysts there is little dependence of the rate constants upon basicity, and the Bronsted β-value is <0.2. However, catalysis by weak bases shows a much stronger dependence upon the basicity of the catalyst, with $\beta > 0.8$. A curved or non-linear Bronsted plot is required to describe the behaviour of both oxygen and nitrogen bases.

The large sensitivity of the rate constants to base strength for weakly basic catalysts indicates that the catalyst resembles its conjugate acid in the transition state, i.e. there is a large amount of, or complete, proton transfer to the catalyst. For strongly basic catalysts the small sensitivity of the rate constants upon base strength suggests that the catalyst resembles its free, unprotonated basic form in the transition state.

These observations are also compatible with Scheme 4.2. Proton transfer between electronegative atoms is thought to occur by a stepwise process involving the diffusion-controlled encounter of the proton donor and acceptor, followed by proton transfer itself, and then by diffusion apart. Proton transfer itself, k_3, is not usually rate-limiting. The application of these suggestions to the mechanism of aminolysis of penicillin provides an explanation for the non-linear Bronsted plot. When the tetrahedral intermediate, $T^{\pm}$, is a stronger acid than the conjugate acid of the basic catalyst, proton transfer is thermodynamically favourable. The rate-limiting step will therefore be the diffusion-controlled encounter of $T^{\pm}$ and the catalyst (k_2), and the observed rate will be independent of the basicity of the catalyst. However, for weakly basic catalysts, proton transfer is thermodynamically unfavourable and the rate-limiting step changes to the diffusion apart of the deprotonated intermediate, T^-, and the protonated catalyst (k_4). This kinetic scheme explains the large dependence of the rate upon the basicity of the catalyst for weakly basic catalysts, and its insensitivity for strongly basic catalysts.[6]

In addition to this change in rate-limiting step deduced from non-linear free-energy relationships by changing the basicity of the catalyst, another change has been observed directly from the kinetics of the hydroxide-ion catalysed aminolysis of benzylpenicillin.[7] In aqueous sodium hydroxide the aminolysis occurs largely by the k_{OH} pathway (equation (4.1)). There is a non-linear dependence of the apparent second-order rate constants upon the concentration of hydroxide ion. At low concentrations of hydroxide ion the rate is first-order in hydroxide ion, and the initial slopes give values of k_{OH} that agree well with those determined at lower pH in buffer solutions.[6,7] At high concentrations of hydroxide ion the rate becomes independent of the concentration of hydroxide ion. This change in the kinetic dependence on the hydroxide ion is indicative of a change in the rate-limiting step of the reaction which, in turn, requires that there be at least two sequential steps in the reaction. One of these steps is rate-limiting at low concentrations of hydroxide ion, and the transition state for this step contains hydroxide ion, or its kinetic equivalent. The other step is rate-limiting at high concentrations of hydroxide ion, but the transition state for this step does not contain hydroxide ion. The existence of two sequential steps demands that there be an intermediate in the reaction, which is probably the tetrahedral intermediate $T^{\pm}$.

The mechanism of Scheme 4.2 is also compatible with this observation. At low concentrations of hydroxide ion the rate of collapse of the tetrahedral intermediate to reactants must be faster than its reaction with hydroxide ion ($k_{-1} > k_2[OH^-]$); the observed rate constant is dependent upon the concentration of hydroxide ion, with k_2, the diffusion-controlled step, being rate-limiting. Proton transfer from the tetrahedral intermediate to the hydroxide ion is in the thermodynamically favourable direction, and it is to be expected that the rate-limiting step for this process is the diffusion-controlled encounter of the proton donor and acceptor.

At high concentrations of hydroxide ion the tetrahedral intermediate and hydroxide ion diffuse together faster than the intermediate collapses back to reactants ($k_2[OH^-] > k_{-1}$). Under these conditions the observed rate constant is independent of hydroxide ion concentration, and k_1, the rate of formation of the tetrahedral intermediate, is rate-limiting. Values of k_1 thus determined for a series of amines yield a Bronsted β_{nuc} of 0.3. This indicates that the reaction *behaves as if* there is a development of a charge of *ca.* $+0.3$ on the attacking amine nitrogen in the transition state, which must therefore occur relatively early along the reaction coordinate with little C–N bond formation.

Assuming that the diffusion-controlled step, k_2, has a value of $10^{10}\,M^{-1}\,s^{-1}$, values of k_{-1} and the equilibrium constants for the formation of the tetrahedral intermediates have been obtained.[7] The rates of expulsion of the attacking amine from the tetrahedral intermediate to regenerate the reactants (k_{-1}) are very rapid, *ca.* 10^9–$10^{10}\,s^{-1}$. Although these rate constants are very large, they are of the order of magnitude that has been postulated for the breakdown of tetrahedral intermediates formed in acyl-transfer reactions.

The partitioning of the tetrahedral intermediate ($T^{\pm}$, Scheme 4.2) formed by nucleophilic attack on a β-lactam is controlled by the ease of exocyclic versus endocyclic bond fission. Endocyclic C–N bond fission is expected to be favoured by the release of the strain energy of the four-membered ring. Expulsion of the attacking nucleophile by exocyclic bond cleavage is accompanied by a relatively favourable entropy change because two molecules are generated from one. That expulsion of the attacking amine nucleophile from **(3)** occurs more readily than fission of the β-lactam C–N bond, is confirmed by the observation that 2-azetidinylideneammonium salts **(4)** react with hydroxide to give β-lactams **(5)** through the formation of the intermediate **(6)**.[8] Similarly, several synthetic reactions in which the exocyclic β-lactam oxygen is exchanged without fission of the β-lactam ring have been reported.[9] These observations substantiate those outlined in chapter 2, where it was shown that the β-lactam ring is not particularly reactive and that the four-membered ring opens much more slowly than is generally assumed.

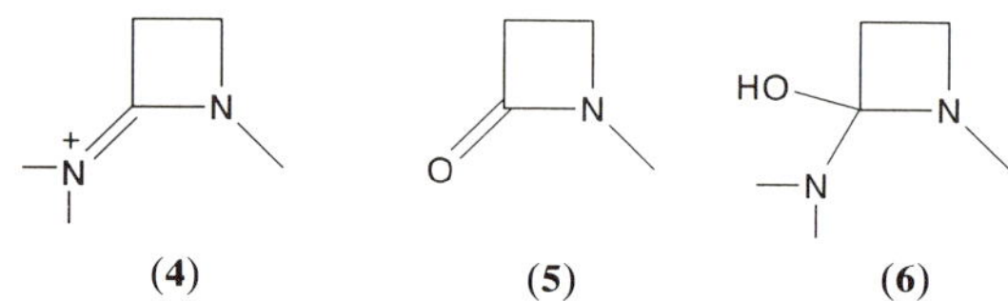

Intermolecular general base catalysis in the aminolysis of penicillins by monoamines is a major pathway for product formation. It is not surprising, therefore, that intramolecular general base catalysed aminolysis has been observed[10,11] with diamines. The rate constant, k_u, for the reaction of 1,2-diaminoethane with benzylpenicillin is *ca.* 30-fold greater than that predicted for a monoamine of the same basicity from the Bronsted plot.[11] The rate enhancement is interpreted as evidence for intramolecular general base catalysis of aminolysis by the second amino group in 1,2-diaminoethane. Proton transfer occurs from the amino group that acts as the nucleophile to the terminal group acting as a general base. Intramolecular catalysis is observed because of the importance of general base catalysis in these reactions compared with uncatalysed aminolysis.

4.3 Metal-ion catalysed hydrolysis

The possible interaction of β-lactam antibiotics with metal ions *in vivo* is of obvious interest, but it is also of importance as a model for metallo-enzymes, such as β-lactamase II, which act as catalysts for the reactions of the β-lactams.

Transition-metal ions cause an enormous increase in the rate of hydrolysis of penicillins and cephalosporins.[12,13,14] For example, copper(II) ions enhance the rate of hydrolysis of benzylpenicillin 10^8-fold, a change in the half-life from 11 weeks to 0.1 seconds at pH 7. In the presence of excess metal ions, the observed apparent first-order rate constants for the hydrolysis of the β-lactam derivatives are first-order in hydroxide ion but show a saturation phenomenon with respect to the concentration of metal ion. This is indicative of the formation of an antibiotic–metal-ion complex. A kinetic scheme is shown in equation (4.2):

$$M + L \overset{K_1}{\rightleftharpoons} ML \xrightarrow{k_2(OH)} \text{products} \quad (4.2)$$

where M is the metal ion and L is the β-lactam. The rate of hydroxide-ion catalysed hydrolysis of benzylpenicillin bound to metal ion shows the following rate enhancements compared with the uncoordinated substrate:[13] Cu(II), 8×10^7; Zn(II), 4×10^5; Ni(II), 4×10^4; Co(II), 3×10^4. The analogous data for cephaloridine are: Cu(II), 3×10^4; Zn(II), 2×10^3.

The copper(II) ion is thought to coordinate to the carboxylate group and the β-lactam nitrogen of benzylpenicillin as shown in (**7**). Coordination to the carboxylate group is indicated because esterification of this group decreases the rate enhancement by a factor of *ca.* 5×10^3. Although it has been suggested[14] that copper(II) ions coordinate to the 6-acylamino side chain and the β-lactam carbonyl group, replacement of the side chain by the more basic amino group has little effect upon the binding constant, and the rate enhancement for the hydroxide-ion catalysed hydrolysis for 6-aminopenicillanic acid is very similar to that for benzylpenicillin. Furthermore, penicillanic acid, in which the amido side chain has been removed, also give similar binding constants and rate enhancements. It is apparent from these observations that copper(II) ions do not bind to the amido side chain in penicillins, and that coordination probably occurs between the carboxylate oxygen and the β-lactam nitrogen (**7**).[13,15]

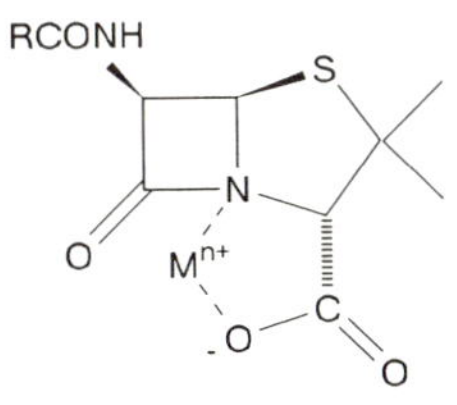

(**7**)

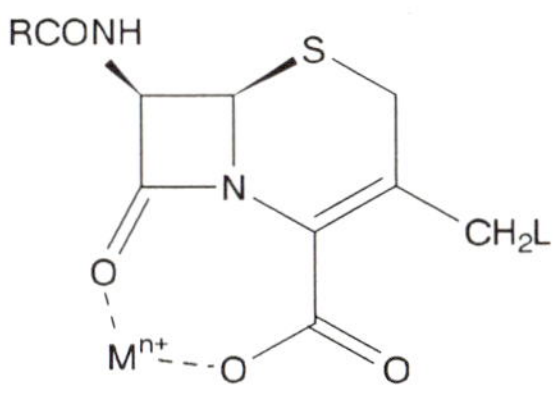

(**8**)

Copper(II) ions bind 10-fold more tightly to cephalosporins than to penicillins, which would be surprising if the sites of coordination were similar. Molecular models indicate that one of the conformations of cephalosporins would be very suitable for metal-ion coordination between the carboxylate group and the β-lactam carbonyl oxygen (**8**). The shortest distance between the carboxylate oxygen and the β-lactam nitrogen is similar (~2.7 Å) in penicillins and cephalosporins. However, the carboxylate oxygen–β-lactam oxygen shortest distance is much smaller (~2.7 Å) in cephalosporins than in penicillins (~4.6 Å). Precipitation of the β-lactam/metal-ion complex in the presence of excess ligand gives solids with interestingly different characteristics. Benzylpenicillin forms a 1 : 1 complex with both copper(II) and zinc(II), in which the asymmetric stretching frequencies of the β-lactam carbonyl and the carboxylate are decreased by *ca.* 30 cm^{-1} compared with uncoordinated penicillin. The nmr spectrum of the zinc(II)/benzylpenicillin complex shows a downfield shift for the C-3 hydrogen, consistent with the proposed mode of binding.[16] However, solid ML_2 complexes of Mn(II), Pb(II) and penicillins have been reported.[17]

Cephalothin and 3-methyl-7β-phenylacetamidoceph-3-em-4-carboxylic acid form solid 2 : 1 complexes with transition-metal ions in which not only is the asymmetric stretching frequency of the carboxylate decreased, but also the β-lactam carbonyl stretching frequency by 10–30 cm^{-1}, depending on the nature of the metal ion. The nmr spectrum of the zinc(II)/cephalothin complex shows a downfield shift of the C-7 hydrogen. The site of metal-ion coordination could thus be different for cephalosporins, and may involve the β-lactam carbonyl oxygen. The kinetic data indicate only a 1 : 1 complex and, of course, the thermodynamically favoured binding site in the solid is not necessarily the kinetically important one in solution.

The hydroxide-ion catalysed hydrolysis of benzylpenicillin probably proceeds by the formation of the tetrahedral intermediate (**9**). The pK_a-value of the bridgehead nitrogen in (**9**) is estimated to be about 8, so there is an enormous change in the basicity of the β-lactam nitrogen as the intermediate is formed. The role of the metal ion in the hydroxide-ion catalysed hydrolysis is to stabilise the tetrahedral intermediate.

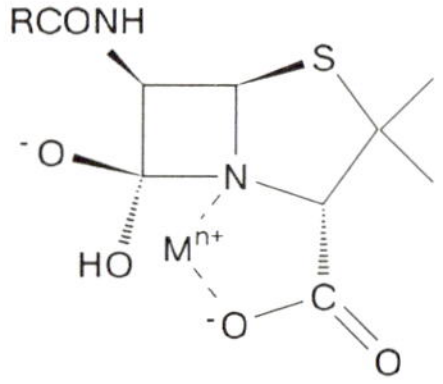

(**9**)

The rate of the hydroxide-ion catalysed hydrolysis of copper(II)-bound benzylpenicillin is 8×10^7 faster than that of uncoordinated benzylpenicillin.[12,13] An estimate of the stabilisation of the transition state by the metal ion can be made from the comparison of the third-order rate constant, $k_2 K_1$, for the metal and hydroxide-ion catalysed hydrolysis with the second-order rate constant for the hydroxide-ion catalysed hydrolysis. For copper(II) ions and benzylpenicillin this ratio is 1.2×10^{10} M. The copper(II) ion thus stabilises the transition state for hydroxide ion catalysis by 13.9 kcal mol^{-1} at 30°C.

The rate enhancements brought about by various metal ions for the hydrolysis of penicillins and cephalosporins are in the order of reactivity of the Irving–Williams series: Co(II) < Ni(II) < Cu(II) > Zn(II).[13]

Copper(II) ions bind *ca.* 10-fold more tightly to cephalosporins than to penicillins.[13] This is at first surprising in view of the greater non-planarity of the penicillin molecule and the expected greater basicity of the β-lactam nirogen. In cephalosporins the possibility of enamine-type conjugation, and the less-favourable geometry for metal-ion coordination to the β-lactam nitrogen and the carboxylate group would also be expected to hinder coordination. Nonetheless, the rate of hydroxide-ion catalysed hydrolysis of copper(II)-bound cephaloridine is *ca.* 3×10^4-fold faster than that for the uncoordinated compound. This may be compared with a rate enhancement of 8×10^7 for benzylpenicillin. The ratio of the third-order rate constant, $k_2 K_1$, for the copper(II) ion plus hydroxide-ion catalysed hydrolysis of cephaloridine, to the second-order rate constant for hydroxide-ion catalysed hydrolysis of the same substrate is 1.6×10^8 M. The corresponding ratio for benzylpenicillin is 1.2×10^{10} M. The transition state for cephaloridine hydrolysis is therefore stabilised by copper(II) ions *ca.* 100-fold less than that for penicillin hydrolysis, but both transition states are greatly stabilised by the metal ion. Again, *ad hoc* explanations for this difference may be found in the lower basicity of the ring nitrogen in the tetrahedral intermediate formed from cephaloridine and/or a less favourable geometry.

Whether or not the group at the C-3′ of the cephalosporin is expelled, makes little difference to the rate enhancement brought about by the metal ion. The 3-methyl derivative has a similar association constant for binding of the copper(II) ion to that for cephaloridine, and the rate enhancement brought about by the copper(II) ion is the same within a factor of 2.[13]

It has been suggested that a ternary complex is formed between benzylpenicillin, zinc(II) and tris buffers, and that hydrolysis occurs by intramolecular nucleophilic attack of one of the coordinated buffer hydroxyl groups on the β-lactam.[18]

4.4 Micelle catalysed hydrolysis

Micelles are of particular interest with respect to the hydrolysis of penicillins because they can provide different microenvironments for different parts of the reactant molecule. In the micelle there is a non-polar, hydrophobic core that can provide binding energy for similar groups on penicillin, and a polar, usually charged, outer shell that can interact with the penicillin's polar groups.

Hydrophobic substrates and counterions are attracted to the micelle, therefore a cationic micelle should assist the reaction between a neutral molecule and an anionic nucleophile, while anionic micelles will inhibit such reactions. The micelle catalysed hydrolysis of penicillins in alkaline solution is unusual because it involves the reaction between two anions, the hydroxide ion and the negatively charged benzylpenicillin.[19] The acid catalysed degradation of penicillins is inhibited in cationic micelles of cetyltrimethylammonium bromide[20] and, as expected, neither anionic micelles of sodium dodecylsulphate nor polyoxyethylene lauryl ether promote the hydroxide-ion catalysed hydrolysis of benzylpenicillin.[19]

In the presence of cetyltrimethylammonium bromide (CTAB) the pseudo first-order rate constants for the alkaline hydrolysis of penicillins[19] and cephalosporins[21] increase rapidly with surfactant concentration once above the critical micelle concentration (cmc) of the surfactant. Increasing the surfactant concentration eventually leads to a slow decrease in the observed rate. This general shape of surfactant-rate profile has been found for many bimolecular reactions catalysed by cationic micelles. However, unusually, the observed pseudo first-order rate constant is not independent of penicillin concentration. The binding constant between the micelle and substrate is unlikely to change significantly with concentration, and yet the lower the concentration of benzylpenicillin, the faster the rate increases and the greater the maximal rate obtained — the rate maximum shifting to a lower surfactant concentration. This observation could be explained if both hydroxide ion and benzylpenicillin compete for the same type of sites in the micelle, and if benzylpenicillin binds better than the hydroxide ion. Increasing the hydroxide ion concentration thus eventually inhibits the rate of the micellar catalysed reaction.

The observed pseudo first-order rate constant for the micelle catalysed hydrolysis does not increase linearly with increasing hydroxide-ion concentration at constant surfactant concentration, but reaches a maximum value.[19]

The kinetic evidence implies that there must be some binding between the benzylpenicillin anion and the micelles of CTAB, and this has been shown spectroscopically.[22] The maximum rate acceleration in the alkaline hydrolysis of benzylpenicillin by CTAB micelles is about 50.

Catalysis by micelles of the hydroxide-ion catalysed hydrolysis of substrates appears to be qualitatively understood on the basis of a concentration effect on the reactants on, or around, the micelle surface and need not necessarily involve a difference in the free energies of activation in the micelle and bulk phase. That is not to say that the cationic micelles do not cause electrostatic stabilisation of the transition state. The cationic micelle surface can act as an electrostatic sink for the anionic intermediate leading to its stabilisation, but a rate enhancement requires preferential stabilisation of this intermediate compared with the reactant. The small rate enhancement of the micelle catalysed reaction, about 50-fold, is equally well explained by considering that the increased concentration of reactants at the micelle surface leads to a higher observed rate. Incorporation of the reactants into a limited volume, decreases the entropy loss that is associated with bringing reactants together in the transition state, and leads to an increase in the pseudo first-order rate constants in the presence of surfactant micelles.

Added salts decrease the rate of the CTAB micelle catalysed alkaline hydrolysis of benzylpenicillin.[19] The salt effect can be considered to be due to competitive binding of the anions with the micelle. Increasing the unreactive anion concentration displaces hydroxide ion bound in the Stern layer, leading to a reduction in the observed rate.

The contribution of hydrophobic effects to micellar catalysis has been demonstrated by modifying the 6-β-side chain of penicillin to increase the substrate lipophilicity and hence the micelle–substrate hydrophobic interaction.[23] The second-order rate constants for the simple hydroxide-ion catalysed hydrolysis of 6-substituted penicillins are independent of the alkyl substituent. The rate maximum in the rate-surfactant concentration profiles for the base catalysed hydrolysis of alkylpenicillins in the presence of CTAB moves to a lower surfactant concentration with increasing substrate lipophilicity, and the rate 'maximum' is dependent on the penicillin concentration. Increasing the 6-β-acylamino chain length increases the lipophilic character of the substrate, and increases the binding constant and the rate enhancement.[23]

Increasing the 6-acylamino chain length of the penicillin substrate not only decreases the surfactant concentration at which the maximum rate is observed, but also results in a slightly increased maximal rate. The first effect may be rationalised on the basis of increasing affinity of the substrate for the micelle phase, brought about by the increased hydrophobic interaction when the 6-acylamino side chain is increased in length. The second aspect, that of different rate maxima, must be more subtle. Increasing the surfactant concentration should eventually lead to all the substrate being associated with the micelle and, since the substrates hydrolyse in water with similar second-order rate constants, then the same rate maximum would be expected for each substrate if, as generally accepted, the rate constant within the micelle is similar to that in the aqueous phase. For compounds with lower affinities for the micelle, it is necessary to use higher concentrations of surfactant to incorporate all the penicillin substrate. Increasing the surfactant concentration also increases the concentration of unreactive counterion, and it is probably the displacement of reactants from the micelle surface by bromide ion that causes different rate maxima to be achieved for different substrates.

The CTAB catalysed hydrolysis of penicillin derivatives appears to exhibit some degree of specificity. Increasing the hydrophobicity of the 6-β-side chain increases micellar catalysis. The association of the penicillin substrate with the micelle is presumably the result of interactions similar to those that give micelles stability relative to their monomeric form in aqueous solution;

hence the not unexpected increase in substrate binding with increased lipophilicity of the molecule. It appears that once the 6-β-side chain has been extended to $CH_3(CH_2)_4CONH$–, further extension does not significantly increase the binding constant. It is interesting to note that there is no evidence of the longer chain compounds pulling the whole penicillin molecule into the interior of the micelle. Conversion of the C-3 carboxylate of benzylpenicillin to its methyl ester leads to only a small reduction in the micelle–substrate binding constant, which suggests that the carboxylate anion points away from the micelle surface and has only a weak electrostatic interaction with the micelle surface. The 6-β-acylamino side chain is probably located in the hydrophobic core of the micelle, with the β-lactam carbon suitably situated for reaction with the hydroxide ion in the micelle surface.

The logarithm of the binding constants shows a non-linear dependence upon the Hansch π-substituent constant for the 6-alkyl side chain.[19] This non-linear relationship is reflected in an apparent decrease in the free energy of transfer of a methylene unit from water to the micelle, with a maximum value of 0.48 kcal mol^{-1} for transfer in the ground state and a maximum of 0.71 kcal mol^{-1} for transfer in the transition state.[23] It has been estimated that the free-energy change for the complete transfer of a single methylene unit from water to the micellar phase is 0.65 kcal mol^{-1}, which corresponds to a maximum rate or equilibrium difference of 3 at 25°C. The free energy of transfer of a methylene group from water to a non-polar liquid is about 1.0 kcal mol^{-1}, and that to an enzyme from 2.1 to 3.8 kcal mol^{-1}. The smaller value for transfer to micelles compared with enzymes presumably results from the 'loose' interactions between the micelle — composed of several molecules of surfactant separated by their van der Waals radii — and the substrate, compared with the 'tight' interactions available from the substrate molecule and one molecule of enzyme — composed of many atoms closely packed together.[24]

4.5 The direction of nucleophilic attack

According to the theory of stereoelectronic control of Deslongchamps,[25] the breakdown of tetrahedral intermediates is facilitated by the lone pairs of the heteroatoms attached to the incipient carbonyl carbon being *anti*periplanar to the leaving group. Application of this theory to the microscopic reverse steps predicts that the direction of nucleophilic attack on the carbonyl carbon be such that the lone pairs on the heteroatoms will be antiperiplanar to the attacking group. Penicillins have a fairly rigid structure because of the fusion of the β-lactam and the thiazolidine rings giving a V-shaped molecule. A consequence of the non-planarity of the fused bicyclic ring system is that the electron density of the lone pair of the β-lactam nitrogen will be concentrated heavily on the α-face of the penicillin molecule, and particularly in the tetrahedral intermediate resulting from nucleophilic attack on the β-lactam carbonyl carbon. According to the theory of stereoelectronic control, nucleophilic attack on penicillins should, therefore, take place from the β-side.[26] However, this face is sterically hindered and it has been suggested that nucleophilic attack may therefore take place from the less hindered α-side to give the stereoisomer (**10**).[7,11]

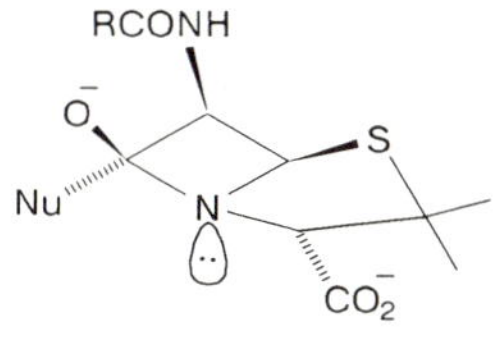

(**10**) 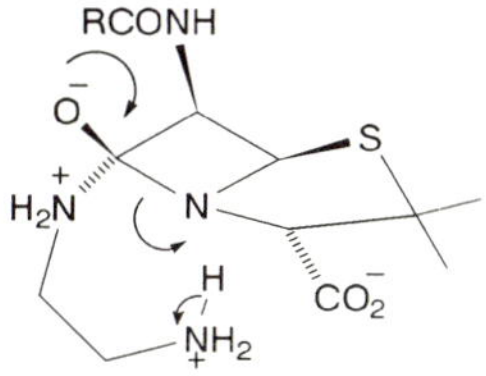 (**11**)

The rate constant for the reaction of penicillin with the monocation of 1,2-diaminoethane is *ca.* 100-fold greater than that predicted from the Bronsted plot for a monoamine of the same basicity. The rate enhancement is attributed to intramolecular general acid catalysis of aminolysis by the protonated amine.[2,11] Breakdown of the tetrahedral intermediate is facilitated by proton donation from the terminal protonated amino group to the β-lactam nitrogen (**11**).

The observation of intramolecular general acid catalysis in the reaction with the monocation of 1,2-diaminoethane gives an indication of the direction of nucleophilic attack on penicillin. In order that ready proton transfer takes place from the protonated amine to the β-lactam nitrogen, it is essential that the tetrahedral intermediate has the geometry shown (**11**) resulting from α-attack. Although intramolecular general acid catalysis could conceivably take place if the amine attacked from the β-face, this would involve considerable non-bonded interactions and/or the proton transfer taking place through one or more water molecules. *Intramolecular* aminolysis, however, involving nucleophilic attack of an amino group (in the C-7 or C-6 side chain of cephalosporins or penicillins, respectively) on the β-lactam carbonyl must occur for the formation of γ- or δ-lactams from the β-face.[27]

Further evidence for nucleophilic attack taking place from the α-face comes from the *absence* of intramolecular general base catalysis in the aminolysis of 6-β-aminopenicillanic acid. That the lone pair on the β-lactam nitrogen takes up the geometry with respect to the carboxy-group shown in (**10**) is also supported by the observation that copper(II) ions catalyse the aminolysis of penicillin by coordination to the β-lactam nitrogen and the carboxy-group, thus stabilising the tetrahedral intermediate.[19] Thus, nucleophilic attack on penicillins, at least by amines, appears to take place from the least hindered α-side, in disagreement with the prediction of the theory of stereoelectronic control. It is unlikely that α-attack would give the stereoisomer predicted by stereoelectronic control, because this would introduce a highly strained *trans*-fused bicyclic system.

4.6 Thiazolidine ring opening

The initial product of alkaline hydrolysis of benzylpenicillin is 5*R*, 6*R*-benzylpenicilloic acid (**12**). However, epimerisation then occurs at C-5 to give a mixture of the 5*R*, 6*R*- and 5*S*, 6*R*-penicilloic acids (**13**).[28,29] This change in configuration at C-5 is accompanied by a decrease in pK_a of the protonated thiazolidine from 5.3 to 4.8, a change in the nmr chemical shifts of the protons at C-5 and C-6 and those in the α- and β-methyl groups at C-2, a decrease in the coupling constants between the C-5–C-6 hydrogens, and a change in specific rotation. The rate of epimerisation at C-6 is negligible compared with that at C-5, and with that of other degradation pathways.

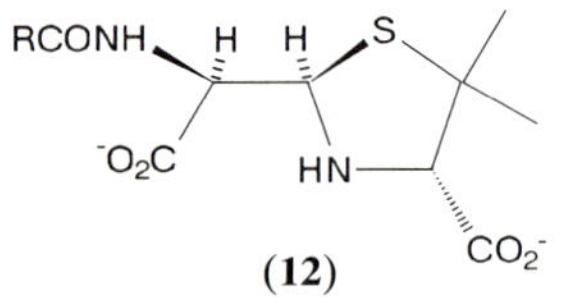

(**12**) 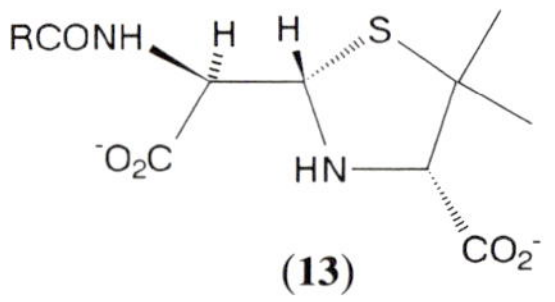 (**13**)

The equilibrium constant for the ratio of the 5*S*, 6*R*-epimer to the 5*R*, 6*R*-benzylpenicilloate at neutral pH is 5.7 (85 : 15). The observed polarimetric pseudo first-order rate constants for epimerisation are pH- and buffer-independent from pH 6 to pH 12.5 but become first order in hydroxide ion at higher pH and acid-dependent at low pH.[30] Epimerisation in D_2O occurs without significant D-incorporation at C-5 or C-6 over most of the pH range. The pH-independent epimerisation at C-5 probably occurs, therefore, by unimolecular ring opening of the thiazolidine to give the iminium ion (**14**) (Scheme 4.3). Intramolecular thiolate-ion attack on the iminium ion recloses the ring to give either the *R* or *S* epimer at C-5. No hydrolysis products of the iminium ion are observed during epimerisation, therefore intramolecular thiolate addition occurs faster than attack by hydroxide ion. Similarly, ring closure to the thiazolidine occurs faster than intramolecular attack by the C-3 carboxylate to form an oxazolidinone.[30]

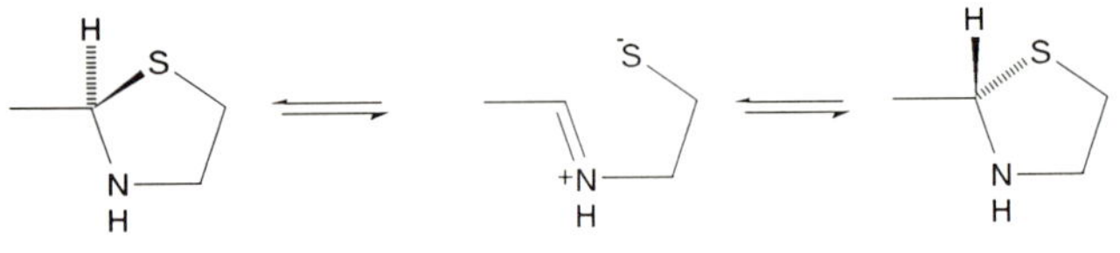

5R (**14**) 5S

Scheme 4.3

The rates of C-5 epimerisation of the mono- and di-methyl esters of (5*R*, 6*R*)-benzylpenicilloate (**15**) and (**16**) also show a pH-independent reaction. However, esterification of the carboxy groups reduces the rate of spontaneous thiazolidine ring opening. The observed pH-independent first-order rate constant for epimerisation of the dimethyl ester (**16**) is 1.7×10^3 times smaller than that observed for the dicarboxylate anion (**12**). Esterification of the carboxylate groups makes them electron-withdrawing and it is therefore expected that thiazolidine ring opening, and the generation of positive charge at C-5 and N-4 in the iminium ion (**14**) (Scheme 4.3), should become unfavourable. This is also manifested in the reduced basicity of the thiazolidine nitrogen, the conjugate acid (**16**) of which shows a pK_a of 1.2 for the 5*R* epimer. A similar, but less marked, reduction in the rate of epimerisation and basicity of the thiazolidine nitrogen is observed for the mono-methyl ester (**15**). The rate of C-5 epimerisation of (**15**) is 21 times less than that of the dianion (**12**) whereas the pK_a of the N-conjugate acid of (**15**) is 3.8. Below pH 8 the rate of C-5 epimerisation of (**15**) is faster than hydrolysis of the methyl ester.

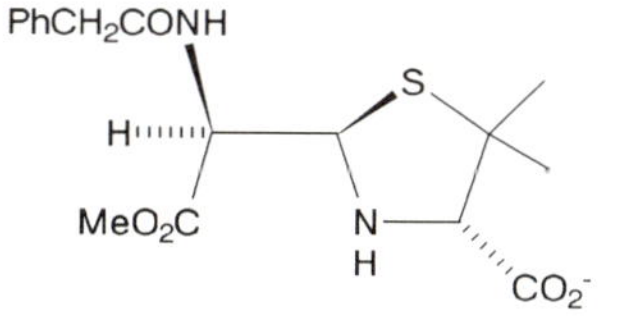

(**15**) 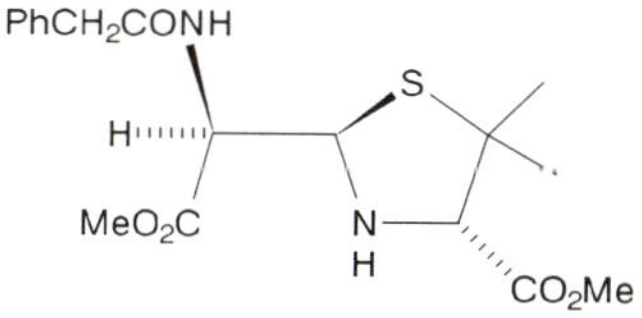 (**16**)

Base catalysed epimerisation of (5*R*, 6*R*)-benzylpenicilloate (**12**) at C-5 occurs above pH 12.5 because the rate becomes dependent on the hydroxide ion concentration. The most likely mechanism for this process is a concerted E2-type mechanism (**17**) to form the neutral imine. The re-closure of the thiazolidine ring occurs by the microscopic reverse process involving general acid catalysis by water (**18**).[30]

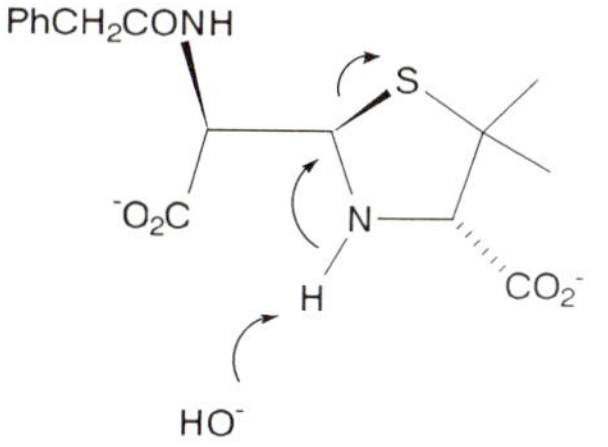

(**17**) 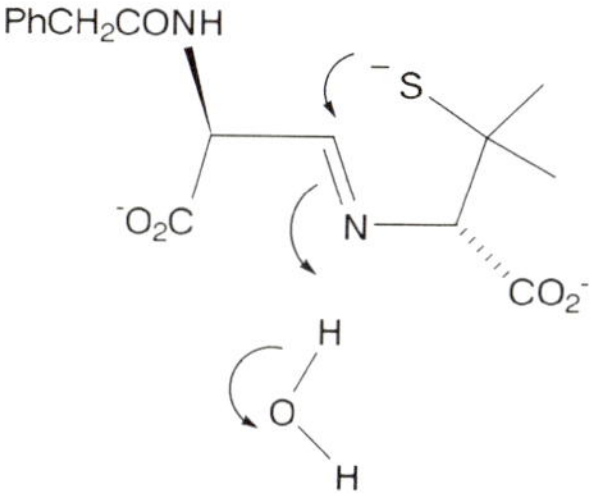 (**18**)

The observed pseudo first-order rate constant for epimerisation of benzylpenicilloate (**12**) at C-5 increases at low pH and follows a sigmoidal curve. The increase in rate passes through the pK_a of the protonated thiazolidine nitrogen. The 5*R* and 5*S* epimers have a different pK_a and the observed pseudo first-order rate constant becomes a complicated function of forward and reverse rates.[30]

The increase in the rate of epimerisation at low pH is interesting. It is difficult to believe that the mechanism involves the unimolecular ring opening of the N-protonated thiazolidine. Epimerisation could occur via either the kinetically equivalent form of the unprotonated, thiazolidine-undissociated carboxylic acid, or the S-protonated thiazolidine. Both the monomethyl ester of (3*S*, 5*R*, 6*R*)-benzylpenicilloate (**15**) and the dimethyl ester (**16**) show a similar pH-dependent rate of epimerisation. It is therefore unlikely that the higher rate of epimerisation of benzylpenicilloate at low pH is due to intramolecular general acid catalysis. In fact, the pH-independent rate at low pH is similar for all three derivatives. This presumably arises from the similar electronic effects of the ester and undissociated carboxy groups. The rate of C-5 epimerisation of the dimethyl ester (**16**) increases markedly at low pH, therefore the pH-independent rate at pH 1 is 10^4 times greater than that at pH 7. The mechanisms of the acid catalysed reaction presumably occurs by thiazolidine ring opening of the S-conjugate acid to generate the thiol-iminium ion, which can re-close to give either epimer at C-5.[30]

The first chemical step in the antibacterial activity of penicillins (**1**) is thought to be the opening of the β-lactam ring by a serine hydroxy group of a transpeptidase enzyme to form a penicilloyl enzyme intermediate (**19**), which is an ester of penicilloic acid (Scheme 4.4).[31,32] It is not known if it is the resistance of this intermediate towards hydrolysis, and the consequential lack of regeneration of the enzyme, which causes the inhibition. It is conceivable that a reaction of the penicilloyl ester produces an electrophilic entity, which is ultimately responsible for enzyme inhibition by irreversibly reacting with another nucleophilic group on the enzyme.

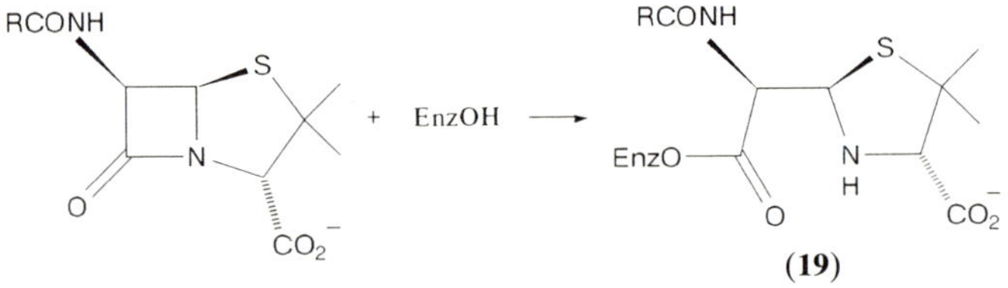

Scheme 4.4

Methyl 5*R*, 6*R*-benzylpenicilloate (**15**) in water undergoes reactions other than simple ester hydrolysis, and produces intermediates that may be of relevance to enzyme inhibition. The alkaline hydrolysis of methyl 5*R*, 6*R*-benzylpenicilloate shows an optical density increase followed by a decrease

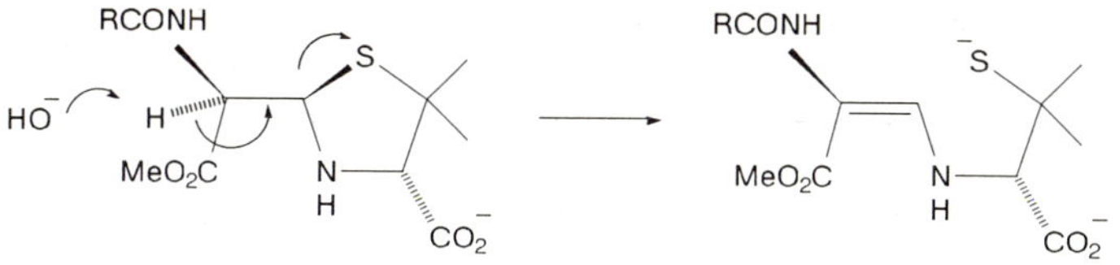

(20)

Scheme 4.5

at 280 nm, with the absorbance maximum increasing with concentration of hydroxide ion.[33] The observed pseudo first-order rate coefficients for both of these phases show non-linear dependences upon the hydroxide ion.

The simplest interpretation of these observations is that methyl 5*R*, 6*R*-benzylpenicilloate (**15**) undergoes reversible elimination across C-6–C-5, and ring opening of the thiazolidine to give the enamine tautomer of methyl penamaldate (**20**) (Scheme 4.5). The amount of enamine formed increases

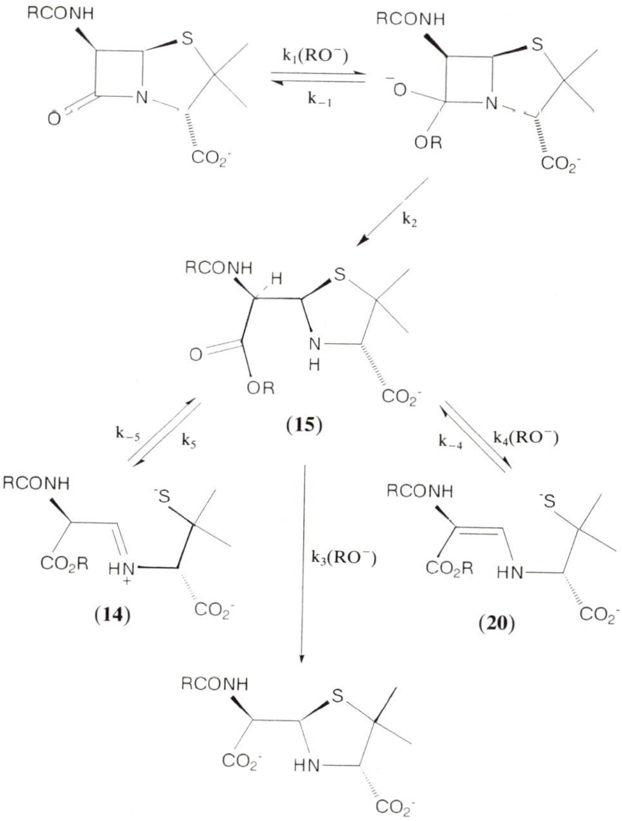

$k_1(RO^-)$, k_{-1}, k_2, (15), k_{-5}, k_5, k_{-4}, $k_4(RO^-)$, $k_3(RO^-)$, (14), (20)

Scheme 4.6

with increasing hydroxide concentration, which explains the increasing absorbance at 280 nm, the wavelength of the absorption maxima of penamaldates. The thiol anion can be trapped with Ellman's reagent and there is a significant deuterium kinetic isotope effect on the ring closure reaction involving reprotonation at C-6.[33]

The reactions of alcohols with penicillins are summarised in Scheme 4.6.[33] The initial alcoholysis proceeds by reversible formation of a tetrahedral intermediate, which generates, in the rate-limiting step, the penicilloyl ester (**15**).[34] At high pH the monoester (**15**) then undergoes thiazolidine ring opening by reversible, base catalysed elimination across C-5–C6 to give the enamine (**20**). This reaction is competitive with the hydrolysis of the ester function. At neutral pH the penicilloyl ester (**15**) undergoes slow, unimolecular thiazolidine ring opening to give the iminium ion (**14**) at a rate which is faster than that for hydrolysis of the ester function.

References

1. B.B. Levine, *Arch. Biochem. Biophys.* (1961) **93** 50; A.L. De Weck and G. Bulm, *Int. Arch. Allergy Appl. Immunol.* (1965) **27** 221; C.W. Parker, J. Shapiro, M. Kern and H.N. Eisen, *J. Exp. Med.* (1962) **115** 821.
2. J.J. Morris and M.I. Page, *J. Chem. Soc., Perkin Trans. 2* (1980) 212.
3. M.I. Page, *Accounts Chem. Res.* (1984) **17** 144.
4. A. Tsuji, T. Yamana, E. Miyamoto and E. Kiya, *J. Pharm. Pharmacol.* (1975) **27** 580; H. Bundgaard, *Arch. Pharm. Chemi. Sci. Ed.* (1976) **4** 25.
5. M.I. Page, *Adv. Phys. Org. Chem.* (1987) **23** 165.
6. J.J. Morris and M.I. Page, *J. Chem. Soc., Perkin Trans. 2* (1980) 220.
7. N.P. Gensmantel and M.I. Page, *J. Chem. Soc., Perkin Trans. 2* (1979) 137.
8. M.I. Page, P.S. Webster and L. Ghosez, *J. Chem. Soc., Perkin Trans. 2* (1990) 805, 813.
9. M.L. Gilpin, J.B. Harbridge, T.F. Howarth and T.J. King, *J. Chem. Soc., Chem. Commun.* (1981) 929; P.W. Wojtkowski, J.E. Dolfini, O. Kocy and C.M. Cimarusti, *J. Am. Chem. Soc.* (1975) **97** 5628.
10. M.A. Schwartz, *J. Pharm. Sci.* (1968) **57** 1209.
11. A.F. Martin, J.J. Morris and M.I. Page, *J. Chem. Soc., Chem. Commun.* (1979) 298.
12. N.P. Gensmantel, E.W. Gowling and M.I. Page, *J. Chem. Soc., Perkin Trans. 2* (1978) 235.
13. N.P. Gensmantel, P. Proctor and M.I. Page, *J. Chem. Soc., Perkin Trans. 2* (1980) 1725.
14. W.A. Cressman, E.T. Sugita, J.T. Coluisio and P.J. Niebergall, *J. Pharm. Sci.* (1969) **58** 1471.
15. G.V. Fazakerley, G.E. Jackson and P.W. Linder, *J. Inorg. Nucl. Chem.* (1976) **38** 1397; G.V. Fazakerley and G.E. Jackson, *J. Pharm. Sci.* (1977) **66** 533.
16. N.P. Gensmantel, D. McLellan, J.J. Morris, M.I. Page, P. Proctor and G. Randahawa, in *Recent Advances in the Chemistry of β-Lactam Antibiotics* (Ed. G. Gregory), Royal Society of Chemistry, London (1981), p. 227.
17. P.B. Chakrawarti, C.P. Tiwari, A. Tiwari and H.N. Sharma, *J. Indian Chem. Soc.* (1984) **61** 705.
18. H. Tomida and M.A. Schwartz, *J. Pharm. Sci.* (1983) **72** 331; H. Tomida, K. Kohashi, Y. Tsuruta, S. Kiryu and M.A. Schwartz, *Pharm. Res.* (1987) **4** 214.
19. N.P. Gensmantel and M.I. Page, *J. Chem. Soc., Perkin Trans. 2* (1982) 147.
20. A. Tsuji, E. Miyamoto, M. Matsuda, K. Nishimura and T. Yamana, *J. Pharm. Sci.* (1982) **71** 1313.
21. M. Yatsuhara, F. Sato, T. Kimura, S. Muranishi and H. Sezaki, *J. Pharm. Pharmacol.* (1977) **29** 638.

22. H. Chaimovich, V.R. Correia, P.S. Araujo, R.M.V. Aleixo and I.O.M. Caccovia, *J. Chem. Soc., Perkin Trans. 2* (1985) 925.
23. N.P. Gensmantel and M.I. Page, *J. Chem. Soc., Perkin Trans. 2* (1982) 155.
24. M.I. Page, in *The Chemistry of Enzyme Action* (Ed. M.I. Page), Elsevier, Amsterdam (1984), p. 1.
25. P. Deslongchamps, *Tetrahedron* (1975) **31** 2463.
26. P. Deslongchamps, *Stereoelectron Effects in Organic Chemistry*, Pergamon Press, Oxford (1983).
27. A.G. Oliveira, M.S. Nothenberg, I.M. Caccovia and H. Chaimovich, *J. Phys. Org. Chem.* (1991) **4** 19.
28. A.M. Davis and M.I. Page, *J. Chem. Soc., Chem. Commun.* (1985) 1702.
29. D.P. Kessler, M. Cushman, I. Ghebre-Sellassie, A.M. Knevel and S.L. Hem, *J. Chem. Soc., Perkin Trans. 2* (1983) 1699; A.E. Bird, E.A. Cutmore, K.R. Jennings and A.C. Marshall, *J. Pharm. Pharmacol.* (1983) **35** 138.
30. A.M. Davis, M. Jones and M.I. Page, *J. Chem. Soc., Perkin Trans. 2* (1991) 1219.
31. J.M. Frère and B. Joris, *CRC Crit. Revs. Microbiol.* (1985) **11** 299.
32. D.J. Waxman and J.L. Strominger, *Ann. Rev. Biochem.* (1983) **52** 825.
33. A.M. Davis, N.J. Layland, M.I. Page, F. Martin and R. More O'Ferrall, *J. Chem. Soc., Perkin Trans. 2* (1991) 1225.
34. A.M. Davis, P. Proctor and M.I. Page, *J. Chem. Soc., Perkin Trans. 2* (1991) 1213.

5 Mode of action: interaction with the penicillin binding proteins

J.M. FRÈRE, M. NGUYEN-DISTÈCHE, J. COYETTE and B. JORIS

5.1 Introduction

The introduction of penicillins as antibacterial agents fifty years ago was one of the major breakthroughs in chemotherapy. The heroic efforts of the Oxford group to prepare and elucidate the structure of these molecules paved the way for a tremendous number of research projects designed to understand their mode of action, and to obtain new compounds exhibiting similar properties and capable of circumventing the different resistance mechanisms used by bacteria to escape their lethal effects. These latter studies resulted in the synthesis or isolation, from natural sources, of a large number of molecules; these became more and more different from the original penicillins, giving rise to a vast family of chemicals whose sole common structural feature was the four-membered β-lactam ring (see Editorial Introduction, Figure 1). Recently, non-β-lactam structures that appear to inactivate the β-lactam-specific targets have been described. These targets are members of a vast family of β-lactam-recognizing proteins, which comprises the β-lactamases and the penicillin binding proteins (PBPs). The *β-lactamases* hydrolyse the amide bond of the β-lactam ring and represent the most efficient mechanism currently responsible for bacterial resistance phenomena. The *PBPs* form rather stable covalent adducts with β-lactams. When that inactivation results in cell death, the PBP is considered as 'essential'. In most cases, the physiological function of non-essential PBPs remains mysterious. A few of them probably act as detectors in the induction of β-lactamase synthesis. *In vitro*, some PBPs exhibit a DD-peptidase activity.

The β-lactam and non β-lactam structures, and the β-lactamases are thoroughly discussed in other chapters of this book. The present contribution is centred on the DD-peptidases and penicillin binding proteins.

Addition of penicillin to a growing population of a sensitive strain of bacteria usually results in cell lysis. In the 1950s, it was generally accepted that penicillin interfered with the formation of the bacterial cell wall — more precisely with the biosynthesis of the peptidoglycan, a cross-linked polymer which completely surrounds the cell. The structure and biosynthesis of this giant macromolecule has been reviewed[1,2] and here only the points that are of major importance for the understanding of β-lactam-induced cell death will be described.

5.2 Structure and biosynthesis of peptidoglycan

Peptidoglycan is a mesh-like structure in which short peptides cross-link linear strands of glycan, consisting of alternating N-acetylglucosamine and N-acetylmuramic acid residues. The structure of the peptides varies with the bacterial species and, on this basis, four different chemotypes have been defined.[3]

The peptidoglycan of all gram-negative and of several gram-positive bacteria is of chemotype I, that of *Streptomyces* sp. and *Staphylococcus aureus* is of chemotype II (Figure 5.1(b)). Bacteria elegantly solved the problem of synthesizing a polymer larger than themselves in the extra-cellular space by (i) building activated precursors inside the cell; (ii) exporting them via a C_{55}, membrane-soluble isoprenoid carrier; and (iii) assembling the translocated pieces with the help of membrane-bound enzymes. The first, intracellular, stage of the process results in the formation of UDP-N-acetylglucosamine and UDP-N-acetylmuramyl-peptide. When compared to the mature peptidoglycan 'unit peptide', the precursor contains one additional D-alanine residue, exhibiting a D-alanyl–D-alanine C-terminus. The second stage involves the successive transfer of the N-acetylmuramyl-pentapeptide-phosphate and N-acetylglucosamine residues on the mono-phosphorylated C_{55} isoprenoid alcohol, and secondary modifications of the peptide. At this stage, the five glycine residues are added on the ε-NH_2 group of the lysine in *S. aureus*, and the α-carboxylate of the D-Glu residue is eventually amidated. The disaccharide–peptide unit is then translocated across the cytoplasmic membrane. The third and final stage in the process takes place in the extracellular space and consists of two distinct reactions; (i) a transglycosylation, which lengthens the saccharidic strands (Figure 5.1(a)); and (ii) a transpeptidation, which closes the peptide cross-bridges (Figure 5.1(b)). Both reactions are catalysed by membrane-bound enzymes.

In 1964, it was proposed[4] that the enzyme responsible for the catalysis of the latter reaction, a DD-transpeptidase (Figures 5.1(b), 5.2(a)), was the target of penicillins. Indeed, it was rapidly demonstrated that a large number of uncross-linked units accumulated in the peptidoglycan of *S. aureus* grown in the presence of sublethal concentrations of penicillin[5,6] and it was proposed[6] that inactivation of the transpeptidase resulted from an

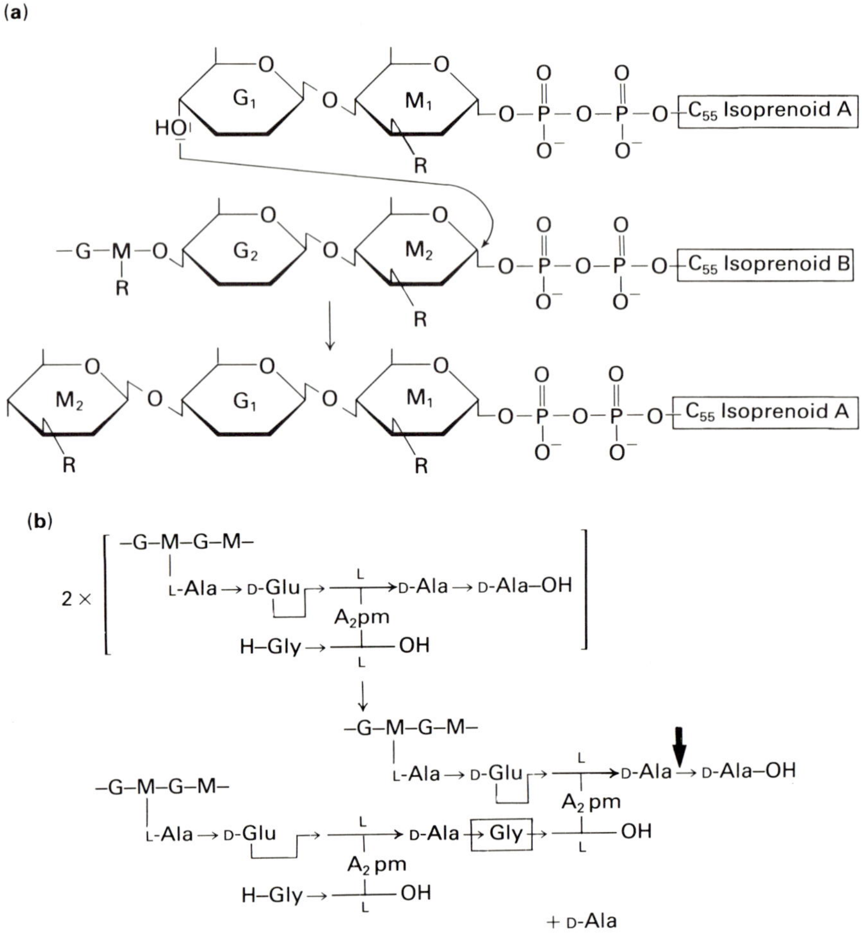

Figure 5.1 The final steps in peptidoglycan biosynthesis. (a) Transglycosylation reaction. The growing chain is transferred on a new disaccharide–peptide unit linked to the membrane-bound isoprenoid lipid carrier. The second lipid carrier (B) is released as a pyrophosphoryl derivative. (b) Transpeptidation reaction in *Streptomyces* sp. (peptidoglycan of type II). A_2pm is diaminopimelic acid, the LL isomer in this case. The 'bridging group', a glycyl residue, is boxed. In *S. aureus*, the bridging group is a pentaglycyl peptide and A_2pm is replaced by L-lysine. In chemotype I, there is no bridging group and A_2pm is the *meso* isomer (*m*-A_2pm). The cross-link is thus a direct peptide bond between D-alanine and the carboxyl group on the D centre of *m*-A_2pm. The heavy arrow indicates the site of action of the DD-carboxypeptidase. Both reactions (transpeptidation and carboxypeptidation) release D-alanine.

acylation by penicillin, acting as a structural analogue of the D-alanyl–D-alanine C-terminus of the substrate.

All the peptide units do not form cross-bridges, and the degree of cross-linking varies according to the species and the growth conditions. In *Escherichia coli*, 75% of the D-alanine residues usually exhibit a free (**a**): R → D-Ala → D-Ala + R′NH_2 ⟶ R → D-Ala–NH–R′ + D-Ala

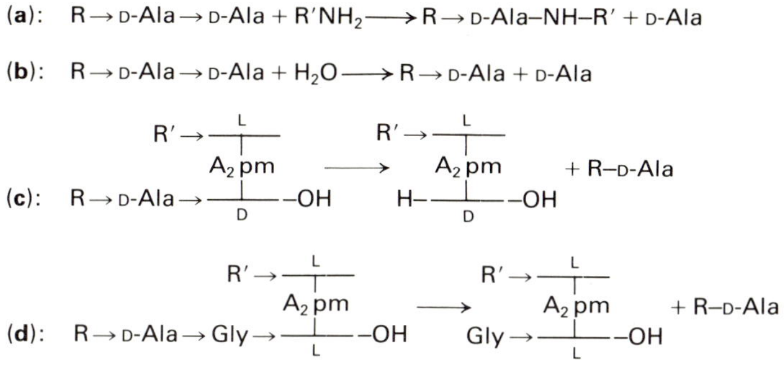

(**b**): R → D-Ala → D-Ala + H_2O ⟶ R → D-Ala + D-Ala

(**c**):
R′ → (L) A_2pm (D) –OH, with R → D-Ala → attached at D ⟶ R′ → (L) A_2pm (D) –OH, with H– at D + R–D-Ala

(**d**):
R′ → (L) A_2pm (L) –OH, with R → D-Ala → Gly → attached at L ⟶ R′ → (L) A_2pm (L) –OH, with Gly → at L + R–D-Ala

Figure 5.2 Penicillin-sensitive enzymatic activities. (a) Transpeptidase; (b), (c) DD-carboxypeptidase but the activity depicted in (c) is often mischaracterized as 'endopeptidase'; (d) true endopeptidase, generally *not* penicillin-sensitive.

carboxyl group but in *S. aureus* the value is only 10%.[7] The fact that D-alanyl–D-alanine C-termini are seldom observed in normal peptidoglycan implies the existence of a D-alanyl–D-alanine carboxypeptidase activity (Figure 5.2(b)), which was also found to be penicillin-sensitive. Peptidoglycan is continuously remodelled to allow cell growth and division. Endopeptidases, which hydrolyse the peptidoglycan cross-bridges, might thus be involved in the creation of new growing sites by making new aminated acceptor groups available. Some of these 'endopeptidases' were also reported to be penicillin-sensitive. However, in the latter case, the hydrolysed peptide bond is always α of a free carboxylate (Figure 5.2(c)) and, in fact, the activity should be considered as a DD-carboxypeptidase activity. True endopeptidases (Figure 5.2(d)) are usually not penicillin-sensitive.

There are thus two clearly identifiable groups of penicillin-sensitive activities: (i) DD-transpeptidation; and (ii) DD-carboxypeptidation. These activities can sometimes be carried out by the same enzymes.

It can easily be realized that inhibition of the transpeptidation reaction results in the formation of a defective peptidoglycan, which induces the appearance of osmotically fragile cells. However, that inhibition can have some more unexpected consequences, such as a triggering of the autolytic system, leading to rapid cell lysis or a decrease of RNA and protein synthesis.[8] The mechanisms that relate the primary interaction with the DD-peptidase(s), with these secondary phenomena remain unknown, as does that which might explain the existence of penicillin-tolerant bacteria that stop dividing in the presence of the antibiotic but resume growth as soon as the penicillin is removed from the medium.

Two complementary approaches have been used in the study of the primary events in penicillin 'recognition' by the bacterial cell.

(i) A search for penicillin-sensitive DD-peptidase activities, followed by the purification of the enzymes and study of the purified proteins.

(ii) The detection of penicillin binding proteins (PBPs) in the bacterial membranes, followed by attempts to identify their physiological role by indirect methods, such as the obtaining of thermosensitive mutants.

In some cases, the two approaches converge when DD-peptidase activities can be attributed to PBPs but, at the present time, these represent a minority and many PBPs do not exhibit a detectable activity *in vitro*. The most widely studied PBPs are those of *E. coli*.[9] Seven proteins have been identified and the corresponding genes sequenced. They have been numbered in order of decreasing M_r. PBPs 1a, 1b, 2 and 3 appear to act as transpeptidases. PBPs 1a and 1b are involved in cell elongation, PBP 2 in the determination of cell shape, and PBP 3 in septation. PBP 4 is probably a DD-carboxypeptidase/transpeptidase, and PBPs 5 and 6, DD-carboxypeptidases. The physiological function of the three low-M_r PBPs is less clear (see section 5.5).

5.3 Penicillin-recognizing proteins as members of an 'active serine' enzyme family

The present understanding of the interaction between DD-peptidases and β-lactam antibiotics is largely based on data obtained with soluble enzymes. Although their physiologically important DD-transpeptidases are membrane-bound, some filamentous bacteria of the Actinomycetales order secrete soluble DD-peptidases in the extracellular medium. Moreover, these enzymes utilize short, synthetic peptides as substrates, which greatly facili-

Table 5.1 General properties of the soluble DD-peptidases.

	Actinomadura R39	*Streptomyces* R61	*Streptomyces albus* G
Molecular weight	50 000	37 500	22 200
Number of residues	489	349	213
Number of polypeptide chains	1	1	1
Number of disulfide bridges	1	1	3
Isoelectric pH	<5	4.8	8.5
Maximum of fluor spectrum (nm)	339	319	350
Catalysis of carboxypeptidation			
— turnover numbers at 37° (min^{-1}) on			
Ac_2–L-Lys–D-Ala–D-Ala	1050	3300	180
αAc–L-Lys–D-Ala–D-Ala	1900	15	3
benzoyl–D-Ala–SCH_2COO^-	360	2400	ND
Catalysis of transpeptidation	Yes	Yes	No
Essential catalytic group	Ser	Ser	Zn^{2+}
Acylation parameter (k_2/K') by benzylpenicillin ($M^{-1}s^{-1}$)	300 000 (37°C)	13 000 (25°C)	0.01 (37°C)

tates the study of their catalytic properties. The three best-characterized enzymes are those of *Streptomyces albus* G, *Actinomadura* R39 and *Streptomyces* R61.[10] Their general properties are summarized in Table 5.1.[11–13] These enzymes are not lethal targets in the bacteria that produce them, and their exact physiological function remains mysterious. Neverthless, they have been extremely valuable models and, qualitatively, most of the principles deduced from their study have been shown to be valid for the membrane-bound penicillin-sensitive enzymes.

5.3.1 The 'active serine' model of interaction with β-lactams

The kinetic analysis demonstrated that the interaction obeyed a three-step model (Model 5.1)

$$\mathrm{E} + \mathrm{C} \underset{k_{-1}}{\overset{k_{+1}}{\rightleftharpoons}} \mathrm{EC} \xrightarrow{k_2} \mathrm{EC}^* \xrightarrow{k_3} \mathrm{E} + \mathrm{P(s)}$$

Model 5.1

where E is the enzyme, C the antibiotic, EC a non-covalent complex, EC* a covalent acyl-enzyme and P(s) the inactive product(s) of degradation of the antibiotic. Efficient inactivation of the enzyme depends on a rapid and nearly quantitative accumulation of the EC* complex, which is the result both of its stability (low k_3), and of its rapid formation — generally due to high k_2 values.

In 1976, the enzyme group involved in the formation of the acyl-enzyme was identified as a serine side chain[14] in the *Streptomyces* R61 DD-peptidase (Figure 5.3). The same result has now been obtained with all penicillin-sensitive DD-peptidases. The resistance of the *Streptomyces albus* G DD-carboxypeptidase was easily explained by the fact that it is not an active-site serine enzyme, but a metallo-protein containing an essential Zn^{2+} ion. When non-covalently bound to the active-site, the β-lactam reacts very slowly with a side chain that has not been identified. The Zn^{2+} enzyme is also characterized by the fact that it is unable to catalyse transpeptidation reactions. It is not a penicillin target and will not be further discussed here.

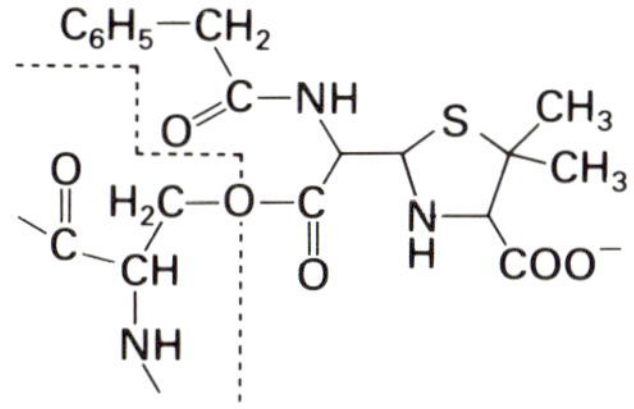

Figure 5.3 Structure of the inactive acylenzyme EC* formed when benzylpenicillin reacts with a penicillin-sensitive enzyme. The discontinuous line delimits the serine residue of the enzyme.

5.3.2 *The family of penicillin-recognizing proteins*

In subsequent years, it became clear that Model 5.1 depicted how β-lactams interacted not only with PBPs but also with a large number of β-lactamases, involving in all cases the acylation of a serine side chain.[15] This suggested the existence of a vast family of 'penicillin-recognizing enzymes' (PREs or PRPs), in which the position of the active serine allowed the distinction between low molecular weight (LMW) and high molecular weight (HMW) proteins. The LMW proteins ($M_r < 55\,000$) include β-lactamases and the LMW-PBPs, where the active serine is 30 to 60 residues from the N-terminus. In the HMW-PBPs the corresponding residue is situated early in the second half of a significantly longer polypeptide chain.[16] These observations, together with the three-dimensional structural data (see section 5.3.3) allow the distinction in these larger proteins of a C-terminal, penicillin-binding part — which is expected to be the site of the transpeptidase activity — and an N-terminal portion — responsible for anchoring the protein in the membrane. In addition, the N-terminal extension might fulfil other functions that vary with the protein, such as transglycosylase activity for *E. coli* PBP 1b or signal transduction by the *blaR* gene product, a protein which is probably the sensor for β-lactamase induction in *Bacillus licheniformis*.

While all HMW-PBPs are strongly membrane-bound via their N-terminal part, the LMW proteins exhibit three distinct features:

(i) most β-lactamases and the *Streptomyces* R61 and *Actinomadura* R39 DD-peptidases are soluble, secreted enzymes;
(ii) *E. coli* PBP 4 and the *Streptomyces* K15 DD-transpeptidase are loosely bound to the membrane and can be solubilized by high salt concentrations;[17,18] and
(iii) *E. coli* PBPs 5 and 6 are membrane-bound via a C-terminal peptide and a *B. licheniformis* β-lactamase via fatty acid residues linked to its N-terminal cysteine.[19,20]

5.3.3 *Three-dimensional structural data*

Several refined β-lactamase structures have now been obtained by X-ray diffraction (see chapter 6), but the only PBP for which similar data are available is the *Streptomyces* R61 extracellular DD-peptidase.[21] These results indicate striking similarities in the general architectures of the molecules, sometimes in the absence of significant analogies in the primary structures.[22,23] The molecules are composed of an all-α and an α/β domain (Figure 5.4) , and the active serine is located in a cavity at the hinge between those two domains. The following conserved structural and functional

Figure 5.4 3-D structure of the *Streptomyces albus* G β-lactamase (O. Dideberg and P. Charlier, personal communication). Only the α-carbons are represented. Important residues in the four structural elements are represented: ○ = the S and K residues of the S*XXK tetrad; □ = the three residues of the SDN triad; △ = the three residues of the KTG triad; and ● = the E and N residues of the EXEXN element, which has only been clearly identified in class A β-lactamases. Courtesy of Drs O. Dideberg, P. Charlier and J. Lamotte-Brasseur, University of Liège.

elements are involved in forming the walls of the cavity (Figure 5.4).:

(i) an S*XXK tetrad, where S* is the active serine residue at the N-terminus of a long hydrophobic α helix, the K side chain pointing into the cavity;
(ii) an SDN triad in class A β-lactamases or YXN in the DD-peptidase and class C β-lactamases — on a loop between two helices of the all-α domain; and
(iii) a KT(S)G or HSG sequence in the β-lactamases and DD-peptidase, respectively — on the innermost strand of the β-pleated sheet, facing the SDN/YXN loop.

A fourth element, EXEXN, where the first acid and the amidated residues appear to be of paramount importance for the catalytic machinery (the E carboxylate probably playing the role of a general base), is also found in class A β-lactamases but a similar structure has not been identified with certainty in class C β-lactamases and the DD-peptidase.

5.3.4 *Sequence comparisons and the seven classes of PRPs*

The sequences of all serine β-lactamases and PBPs can be aligned[24,25] and the three major conserved elements localized in the polypeptide chains (Figure

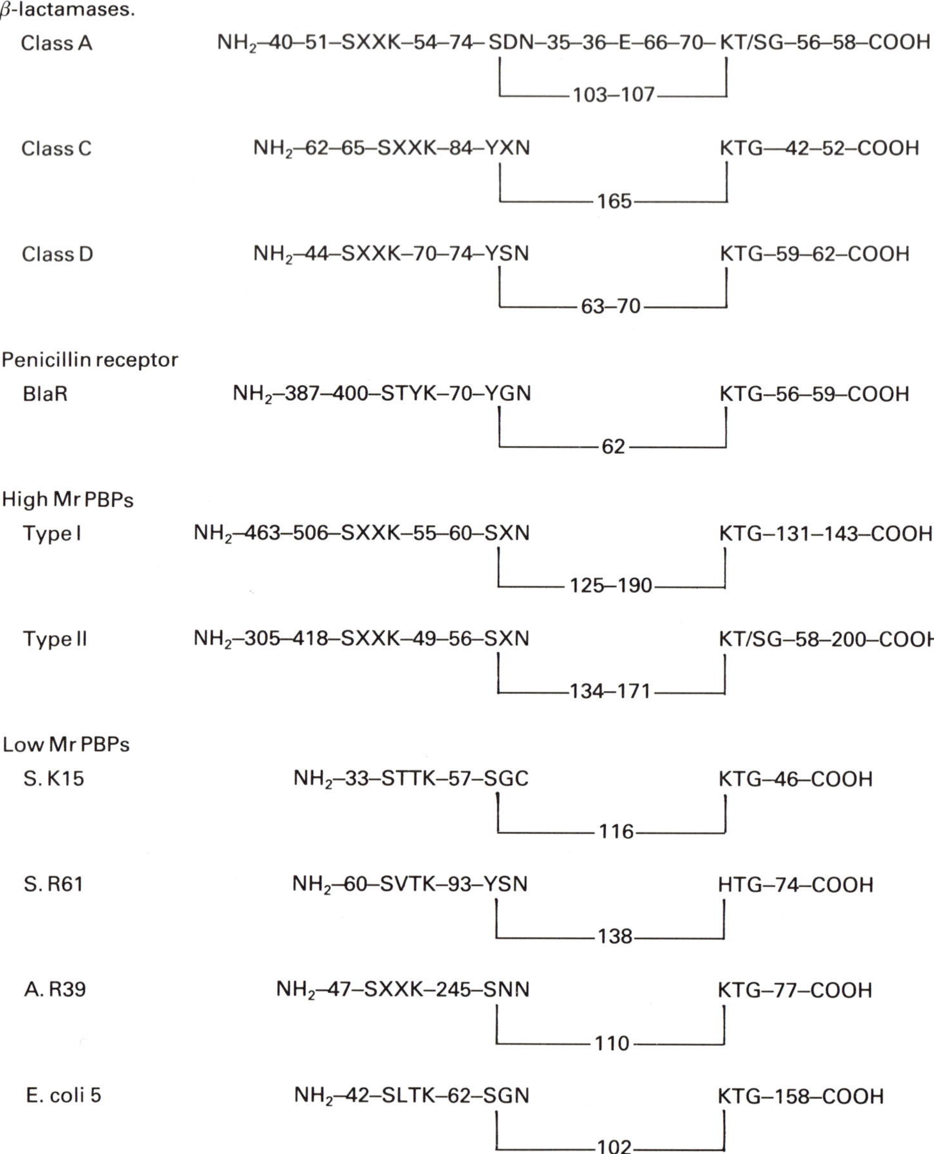

Figure 5.5 Conserved elements in PRPs. The figure shows the number of residues between the N-terminus and the first (SXXK) structural element, between the various structural elements, and between the last (KTG) structural element and the C-terminus. The identity of the 'acidic' structural element is well established only in class A β-lactamases.

5.5). Acidic residues, possibly corresponding to the fourth element of class A β-lactamases are consistently observed between the second and third elements.

The α/β domain is particularly well conserved in the 3-D structures — an N-terminal helix and two pieces of β-strands precede the S*XXK tetrad. Conversely, the K(H)TG triad forms a part of the third β-strand, followed by the two last β-strands and the C-terminal helix. Secondary structure predictions indicate the presence of the same five β-strands and two helices in similar positions in the sequences of class D β-lactamases and LMW-PBPs. Those structures are also predicted in HMW-PBPs, but with a lower degree of certainty. The all-α domain appears to be much more variable, with the striking exception of a hydrophobic helix, starting with the S*XXK sequence, which is consistently predicted.

These comparisons show that penicillin-recognizing proteins can be divided into seven major classes. Among the β-lactamases, classes A, C and D can be easily identified and individualized. The LMW-PBPs include the *Actinomadura* R39, *Streptomyces* R61 and *Streptomyces* K15 DD-peptidases, in addition to PBPs 4, 5 and 6 of *E. coli* and PBP 5 of *Bacillus subtilis*. Of these, the two largest molecules, the *Actinomadura* R39 DD-peptidase and the *E. coli* PBP 4 ($M_r \pm 50\,000$) contain a large insertion in the all α-domain, between elements 1 and 2, when compared to the smaller proteins.[12] Three classes of HMW-PBPs can be distinguished on the basis of the primary structures — *E. coli* PBPs 1a and 1b (type 1) differ from the other HMW-PBPs (type II), which include both 'sensitive' (*E. coli* PBPs 2 and 3, *Neisseria gonorrhoeae* PBP 2 and *Streptococcus pneumoniae* PBPs 2b and 2x) and 'resistant' PBPs (*S. aureus* PBP2′, *Enterococcus hirae* PBPs 3R and 5, and numerous mutants of the sensitive proteins). In the N-terminal part of that latter group, a few conserved elements, indicating a common function, can also be detected[16] but the role of the 300–350 residue extension preceding the active serine has not been clearly established. Finally, the *blaR* gene product seems to be in a class by itself, although its C-terminal part exhibits a surprising similarity with the OXA-2 β-lactamase. It has, however, no detectable enzymatic activity and its N-terminal extension contains several predicted transmembrane helical segments.[26] If β-lactamases are evolutionarily derived from PBPs, one must assume that the various classes evolved independently from distinct PBPs, which were already quite different from each other.

5.4 Kinetics of the β-lactam–PRP interaction

After underlining the similarities between β-lactamases and PBPs, it must be stressed that the differences between the two types of proteins are also important. Firstly, β-lactamases do not recognize the peptide substrates of the PBPs. However, some esters and thiol esters analogues of these peptides can be hydrolysed by the β-lactamases, which can also catalyse the transfer of the acyl moiety of those compounds on a suitable acceptor.[27,28] Secondly, in the interaction with β-lactams, the rate of deacylation is much larger with β-lactamases, for which k_3 values larger than $1000\,s^{-1}$ are not uncommon. By contrast, k_3 values larger than $10^{-3}\,s^{-1}$ are exceptional with PBPs, the usual range being 10^{-4}–$10^{-5}\,s^{-1}$. The rate of acylation is also generally larger with β-lactamases, sometimes reaching the diffusion-limited values of 10^7–$10^8\,M^{-1}\,s^{-1}$. Thus, purely quantitative factors translate into gross qualitative differences—β-lactamases inactivate penicillins, while penicillins inactivate PBPs. Although the equations discussed in the following section are valid whenever Model 5.1 applies, the main emphasis here will be on their implications for the PBP systems, where the k_3 values are low.

5.4.1 Characteristic equations

When the initial concentration of antibiotic (C_0) is much larger than that of the enzyme (E_0), a steady-state identified by the ss subscript is eventually reached where

$$\frac{[EC^*]_{ss}}{E_0} = \frac{k_f}{k_3 + k_f} \tag{5.1}$$

$$\frac{[EC]_{ss}}{E_0} = \frac{C_0}{C_0 + K'} \cdot \frac{k_3}{k_f + k_3} = \frac{C_0 k_3}{k_2 C_0 + k_3(K' + C_0)} \tag{5.2}$$

$$\frac{[E]_{ss}}{E_0} = \frac{K'}{C_0 + K'} \cdot \frac{k_3}{k_f + k_3} \tag{5.3}$$

where

$$k_f = \frac{k_2 C_0}{C_0 + K'} \tag{5.4}$$

$$K' = \frac{k_{-1} + k_2}{k_{+1}} \tag{5.5}$$

and

$$k_a = k_3 + k_f \tag{5.6}$$

When the interaction is analysed, k_2 is found to be negligible when compared to k_{-1} and $K' = k_{-1}/k_{+1} = K$, the dissociation constant of EC.

From equation (5.1), it follows that a large proportion of enzyme is immobilized as EC* if $k_f \gg k_3$. In most cases, this occurs when $C_0 \ll K'$ because k_2 is much larger than k_3. Under these conditions, $k_a = k_f = k_2 C_0/K'$ and the concentration of the non-covalent complex remains negligible. For instance, if $K' = K = 1\,mM$, $k_2 = 10\,s^{-1}$ and $k_3 = 10^{-4}\,s^{-1}$, a 0.1 μM concentration of β-lactam yields at the steady-state

$$[\mathrm{EC}^*] = 0.9\,\mathrm{E}_0$$

and

$$[\mathrm{E}] = 0.1\,\mathrm{E}_0$$

The important factor governing the rate of acyl-enzyme formation is then k_2/K', a second-order rate constant (10 000 $\mathrm{M}^{-1}\,\mathrm{s}^{-1}$ in the present example).

The immobilization of a large proportion of enzyme as the EC* complex is not, however, a sufficient condition for the antibiotic to be efficient. This immobilization must also occur within a short time — a few minutes if the growth of the bacterium is rapid — and new targets are continuously produced at an elevated rate. *In vitro*, the total enzyme concentration remains equal to E_0 and

$$(\mathrm{EC}^*) = \frac{k_\mathrm{f}}{k_\mathrm{a}} E_0(1 - \mathrm{e}^{-k_\mathrm{a}t}) \tag{5.7}$$

In fact, k_3 is often so small ($\leqslant 10^{-4}\,\mathrm{s}^{-1}$) that the rate of formation of EC* (k_f) is the only significant factor. *In vivo*, the problem might be summarized by stating that the killing targets must be inactivated faster than the growing bacteria can synthesize new ones, although the exact number of active transpeptidase molecules per cell necessary for cell survival is not known. In the present example, EC* = 0.38 E_0 after 500 s, 0.61 E_0 after 1000 s, and 0.74 E_0 after 1500 s.

Table 5.2 shows the values of the characteristic parameters for the *Streptomyces* R61 and *Actinomadura* R39 DD-peptidases, and some typical compounds. From these data, several conclusions can be drawn.

(i) Nearly all the k_3 values are very low, of the order of $10^{-4}\,\mathrm{s}^{-1}$ or below. The penem supplies the striking exception with both enzymes, and 6-amino-penicillanate with R39.

(ii) Acylation of the R39 enzyme is always faster than that of R61. However, the ratio of the k_2/K values is very far from constant, varying from 1.5 with clavulanate, to 10 000, with β-iodopenicillanate. This is a clear indication of the different structures of the enzyme active sites, an observation which will be confirmed when the acylation rates of other penicillin-recognizing enzymes are considered.

(iii) At the steady-state, the major form of inactive enzyme is always the covalent adduct EC*. Even with R61 and ampicillin, where both k_2 and K are relatively low, a 10 μM concentration of antibiotic yields the following values at the steady-state:

$$[\mathrm{EC}^*] = 98.17\%$$
$$[\mathrm{EC}] = 0.15\%$$
$$[\mathrm{E}] = 1.68\%$$

Table 5.2 Kinetic parameters for the interaction between the R61 and R39 DD-peptidases and various β-lactam antibiotics (at 37°C, unless otherwise stated).

	R61				R39	
	K (mM)	k_2 (s^{-1})	k_2/K ($M^{-1}s^{-1}$)	k_3 (s^{-1})	k_2/K ($M^{-1}s^{-1}$)	k_3 (s^{-1})
6-Aminopenicillanate	0.8	2×10^{-4}	0.25	$<6 \times 10^{-5}$	1200	6×10^{-3}
Benzylpenicillin	13[a]	180[a]	13 000[a]	1.4×10^{-4}	300 000	3×10^{-6}
Carbenicillin	0.11	0.09	830	1.4×10^{-4}	6000	5×10^{-6}
Ampicillin	7	0.8	110	1.4×10^{-4}	70 000	4×10^{-6}
β-Iodopenicillanate	4	3×10^{-3}	0.7	$<3 \times 10^{-5}$	7600	$<10^{-5}$
Cephalosporin C	>1	>1	1500	1×10^{-6}	65 000	0.3×10^{-6}
Cephaloglycin	0.4	9×10^{-3}	22	3×10^{-6}	70 000	0.8×10^{-6}
Nitrocefin	ND	ND	460[b]	3×10^{-4}	2 600 000[b]	1.5×10^{-6}
Cefotaxime	ND	ND	16	$<4 \times 10^{-6}$	2600	$<2 \times 10^{-6}$
Cefuroxime	ND	ND	350	4×10^{-6}	3900	2×10^{-6}
Cefoxitin	ND	ND	1500	5×10^{-5}	7000	$<3 \times 10^{-5}$
Imipenem	ND	ND	1000	7×10^{-6}	10 000	2×10^{-6}
Clavulanate	ND	ND	21	5×10^{-6}	32	$<10^{-6}$
(penem) [structure: S, N, O, COO^-]	ND	ND	670	0.05	1750	0.008
Aztreonam	12	$<2 \times 10^{-4}$	$<2 \times 10^{-3}$	ND	15	$<5 \times 10^{-6}$

[a] at 25°C; [b] at 10°C; ND = not determined.

and the time necessary for [EC*] to reach 50% of that steady-state value is 90 s. Under these conditions, and although the antibiotic concentration is well below the K value, more than 90% of the enzyme activity is lost in less than 6 min. If this represented an *in vivo* situation with an essential PBP, the synthesis of new enzyme by the cell would not be able to compensate for such a rapid inactivation, and cell death would certainly be observed.

5.4.2 *Nature of the reaction products*

With β-lactamases, the primary products are penicilloic or cephalosporoic acids, which can eventually undergo further spontaneous degradation. With the DD-peptidases, the k_3-step can represent two distinct reactions. Penicilloic or cephalosporoic acids are often obtained but, in some cases, an unexpected hydrolysis of the C-5–C-6 bond of penicillins has been observed prior to acyl-enzyme hydrolysis.[29] With the R61 enzyme and benzylpenicillin, the main reaction product at pH 7.0 is thus phenylacetylglycine. At pH 9 and 10, benzylpenicilloic acid is also formed, probably resulting from a direct attack of the penicilloyl-enzyme by OH^- ions.[30] Formation of phenylacetylglycine has been well documented with various other DD-peptidases and PBPs[10] but the reaction mechanism remains obscure and there is presently no way to predict if hydrolysis of the C-5–C-6 bond will occur upon degradation of a given acyl-enzyme. This reaction does not appear to be of major physiological importance, since there does not seem to be a direct relationship between the half-life of the EC* complex and the deacylation pathway.

5.4.3 *Interaction with substrates*

Three categories of penicillin binding proteins can be distinguished on the basis of their *in vitro* enzymatic properties.

(i) Proteins in the first group, which represent a majority of the known HMW-PBPs, have not been found to catalyse any reaction except their own acylation by β-lactams. They can only be identified on the basis of their ability to covalently bind radioactively labelled β-lactam compounds.[9] Most of the physiologically important PBPs appear to belong to this category. Their physiological role can be approached by genetic methods (thermosensitive mutations) or by the utilization of β-lactams that specifically inactivate one of them, for example aztreonam and *E. coli* PBP 3, or mecillinam and *E. coli* PBP 2.

(ii) Proteins whose transpeptidase activity can be demonstrated with complex systems, involving the true peptidoglycan precursors such

Table 5.3 Enzymatic properties of some penicillin binding proteins.

			Interaction with benzylpenicillin		
			k_2/K ($M^{-1} s^{-1}$)		
	M_r	Enzyme activities[a]	Activity decrease	Penicillin binding	k_3 (s^{-1})
Streptomyces R61 D-peptidase	37 500	Cbase Tpase 1[c]	13 000 (25°C)	13 000	1.4×10^{-4}
Actinomadura R39 DD-peptidase	50 000	Cbase Tpase 1[c]	3×10^5	3×10^5	3×10^{-6}
Streptomyces K15 DD-transpeptidase	26 000	Tpase 1[b]	150		1.1×10^{-4}
Salmonella typhimurium PBP 4	52 000	Cbase Tpase 1[c]	40 000	6000	1.3×10^{-4}
PBP 5	38 000	Cbase Tpase 1[c]	7800	2000	1.2×10^{-3}
Proteus mirabilis PBP 4	46 000	Cbase Tpase 1[c]	$2–3 \times 10^5$		4×10^{-5}
PBP 5[d]	43 000	Cbase Tpase 1[c]	$2–8 \times 10^4$		1.6×10^{-3}
Staphylococcus aureus PBP 4	46 000	Cbase Tpase 2	2000[e]		8×10^{-3}
Enterococcus hirae PBP 6	43 000	Cbase Tpase 2	1045	445	$2.8–4.4 \times 10^{-5}$
Bacillus subtilis PBP 5	50 000	Cbase Tpase 2	1200 (4°C)		6×10^{-5}

Bacillus megaterium KM PBP 5	45 000	Cbase Tpase 2	200	500	3×10^{-4}
PBP 1	123 000			110 000	1.2×10^{-3}
Bacillus licheniformis PBP 4	46 000	Cbase Tpase 2	80		$<6 \times 10^{-5}$
Escherichia coli PBP 6	32 000	Cbase Tpase 2	175		7×10^{-4}
PBP 5	34 000	Cbase	175		2.3×10^{-3}
PBP 4	49 000	Cbase Tpase 1[f]		7000	4×10^{-4}
PBP 3	60 000	Tpase 1a[g]		300[i]	1.8×10^{-4}
PBP 2	66 000	Tpase 1a[g]		130	1.4×10^{-4}
PBP 1a[h]		Tglase			
	90 000				
PBP 1b[h]		Tpase 1a[c] Tglase		200	3×10^{-5}

[a] Cbase = DD-carboxypeptidase; Tpase 1 = DD-transpeptidase mimicking the natural reaction, e.g. dimer formation; Tpase 1a = transpeptidation only with natural precursor; Tpase 2 = DD-transpeptidase only with simple acceptors (glycine, hydroxylamine); Tglase = transglycosylase.
[b] A DD-carboxypeptidase activity can be detected, mainly with ester substrates (Figure 5.7, compound I) but it disappears in the presence of a suitable acceptor.
[c] Dimer formation with donor–acceptor substrates.
[d] From L-forms.
[e] Obtained with a different strain.
[f] Observed with an impure preparation in which PBP 4 was the sole PBP that could catalyse the reactions.
[g] The observation of Tpase activity with natural precursors remains doubtful and extremely low. A DD-Cbase and Tpase activity has, however, been demonstrated with PBP 3 and the thiol esters IV and V (Figure 5.7).
[h] Many results obtained for the mixture PBP 1a + PBP 1b.
[i] Recent data indicate the real value to be closer to 3000 $M^{-1} s^{-1}$.

as UDP-N-acetylmuramyl pentapeptide and UDP-N-acetylglucosamine. A good representative of this group is *E. coli* PBP 1b, which synthesizes cross-linked peptidoglycan when supplied with the isoprenoid-pyrophosphate disaccharide peptide.[31,32] The complexity of these systems precludes their utilization for detailed kinetic analysis, or for routinely monitoring the activity of the proteins.

(iii) Proteins known as DD-carboxypeptidases or DD-peptidases, which recognize simple peptides, identical or similar to those undergoing transpeptidation *in vivo*. They catalyse the hydrolysis of the D-alanyl–D-alanine peptide bond, and a concomitant transpeptidation reaction if supplied with a suitable aminated acceptor (R′NH2, Model 5.2).

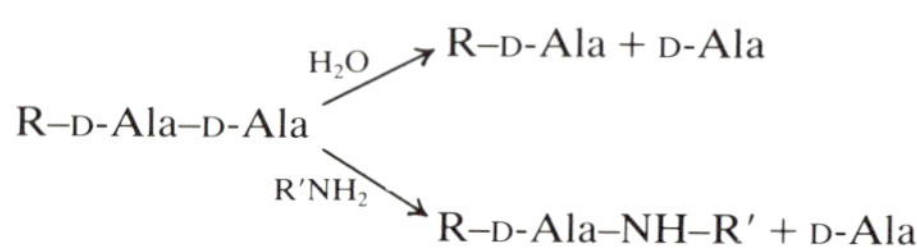

Model 5.2 Concomitant hydrolysis and transpeptidation reactions catalysed by DD-carboxypeptidases. R–D-ala–D-ala and R′–NH_2 are the 'donor' and the 'acceptor' substrates, respectively.

Quite often, only very simple acceptor molecules such as glycine or hydroxylamine are utilized, and with poor efficiency. However, a few enzymes behave as good transpeptidases, and catalyse the formation of peptide products very similar to those found in cross-linked peptidoglycan (Table 5.3). All the enzymes in this third group are members of the LMW-PBPs class described previously. Among these, the *Streptomyces* R61 and *Actinomadura* R39 DD-peptidases have been particularly studied, and exhibit interesting transpeptidation properties. Both enzymes catalyse polymerization reactions with substrates that contain both a D-alanyl–D-alanine C-terminus, and an adequately positioned amino group, as shown in Figure 5.6,[33,34] thus closely mimicking the physiological transpeptidation reactions. A third interesting enzyme in this group is the membrane-bound transpeptidase of *Streptomyces* K15, which behaves mainly as a transpeptidase; the properties of this enzyme will be discussed in section 5.5.2.1.

The R61 and R39 soluble enzymes also utilize simple esters and thiol esters (Figure 5.7) as donor substrates.[28] The recent introduction of the thiol esters represents a major breakthrough in the field, both for technical and conceptual reasons. Firstly, it is possible to directly monitor the disappearance of thiol esters by spectrophotometry, which greatly facilitates the kinetic analyses and increases the accuracy of the results. Secondly, thiol esters have allowed the visualization of the accumulation of catalytically competent acyl-enzymes, and supplied a completely new approach to the

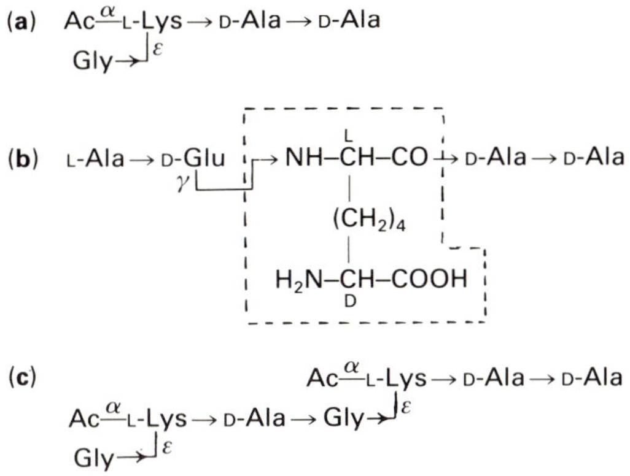

(a) Ac$\overset{\alpha}{—}$L-Lys → D-Ala → D-Ala
Gly → ⌋ε

(b) L-Ala → D-Glu ⟶ NH–CH(L)–CO ⟶ D-Ala → D-Ala
γ
(CH₂)₄ → $(CH_2)_4$
H_2N–CH(D)–COOH

(c) Ac$\overset{\alpha}{—}$L-Lys → D-Ala → D-Ala
Ac$\overset{\alpha}{—}$L-Lys → D-Ala → Gly → ⌋ε
Gly → ⌋ε

Figure 5.6 Structures of substrates that can be polymerized by the R61 (a) and R39 (b) DD-peptidases. In the first case, the structure of the dimer product (c) is shown, but trimers and tetramers have also been obtained. The boxed residue in (b) is *meso*-diaminopimelic acid (m-A_2pm).

study of the transpeptidation mechanism. Thirdly, the thiol esters are recognized by several high molecular weight PBPs.

It seems that the kinetic pathway for the hydrolysis reaction catalysed by both R61 and R39 DD-peptidases can be represented by Model 5.1, thus involving the three steps shown in Figure 5.8. With the most widely-used tripeptide (Ac_2–L-Lys–D-Ala–D-Ala, where Ac = acetyl), acylation is rate-limiting at substrate saturation, and the kinetic parameters k_{cat} and K_m represent k_2 and K', respectively.[30] For instance, with the R61 enzyme, $k_2 = 50\,s^{-1}$ and $K' = 12$ mM. There is no acyl-enzyme accumulation, and the value of k_3 can be safely assumed to be large; however, it cannot be determined. In contrast, with the thiol ester benzoyl–Gly–SCH_2–COO^-, deacylation is rate-limiting and $k_{cat} = k_3$.[35] However, the individual values of k_2 and K' can be obtained by measuring the rate of accumulation of acyl-enzyme,

$$R_1\text{–NH–CHR}_2\text{–CO–X–CHR}_3\text{–COO}^-$$
↑
HO H

R_1	R_2	X	R_3	
Ac_2–L-Lys	CH_3	O	CH_3	I
C_6H_5–CO	H	O	C_6H_5	II
C_6H_5–CO	H	O	C_6H_5–CH_2	III
C_6H_5–CO	H	S	H	IV
C_6H_5–CO	CH_3	S	H	V

Figure 5.7 Structures of some esters and thiol esters that behave as substrates for the DD-peptidases. When R_2 and R_3 are not H, the asymmetric carbon is D. The arrow indicates the site of enzyme action.

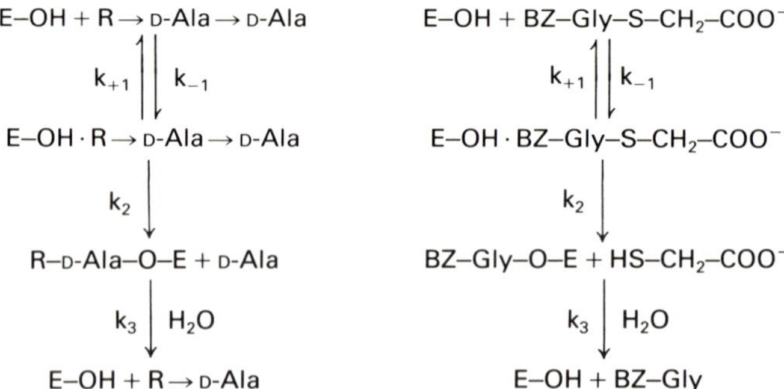

Figure 5.8 Detailed mechanism of the hydrolysis of an R–D-Ala–D-Ala peptide, and of a simple thiol ester by a DD-peptidase (E–OH). If E–OH is the R61 enzyme and R = Ac_2–L-Lys, k_3 is much larger than k_2 (k_2 = 50 s^{-1}). With the thiol ester and the same enzyme, k_2 = 700 s^{-1} and k_3 = 5–6 s^{-1}. BZ = benzoyl.

which is easily monitored by fluorescence spectroscopy. In the presence of an acceptor, the rate of disappearance of the tripeptide is generally not strongly modified, and the sum of transpeptidation (T) and hydrolysis (H) is often similar to the hydrolysis observed in the absence of acceptor. In contrast, the acceptor increases the k_{cat} value for the utilization of the thiol esters, in a concentration-dependent manner (Figure 5.9). The maximum k_{cat} value observed at a saturating acceptor concentration depends upon the structure of the substrate.[35] These results are in agreement with a simple partition model, represented by Model 5.3.

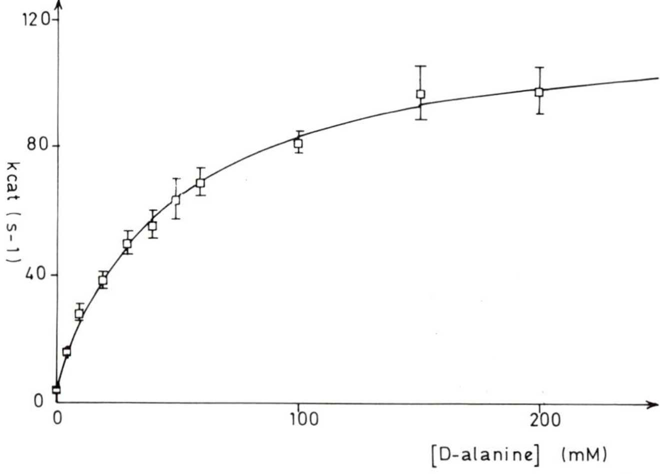

Figure 5.9 Influence of the acceptor (D-alanine) concentration of the k_{cat} value for the utilization of thiol ester IV by the *Streptomyces* R61 DD-peptidase.

$$E + C \underset{k_{-1}}{\overset{k_{+1}}{\rightleftharpoons}} EC \xrightarrow{k_2} EC^* \begin{cases} \xrightarrow[k_3]{H_2O} E + H \\ \underset{k_{-4}}{\overset{k_4 A}{\rightleftharpoons}} EC^*A \xrightarrow{k_5} E + T \end{cases}$$

Model 5.3 Simple partition model accounting for concomitant transfer and hydrolysis reactions. C = donor substrate, A = acceptor substrate, EC* = acyl-enzyme, H = hydrolysis product, T = transpeptidation product. If $k_2 < k_3$, A does not accelerate the disappearance of C. If $k_2 > k_3$, the k_{cat} value for the utilization of C increases with acceptor concentration if $k_5 > k_3$. The formation of ternary EC*A complex is demonstrated by the fact that different acceptors yield different maximum k_{cat} values at saturation. If transpeptidation occurred through a direct aminolysis (EC* + A → E + T), the same maximum k_{cat} value, equal to k_2, would be reached with all acceptors.

This simple model, however, fails to explain various experimental observations:

(i) The model predicts that, at the steady-state, the T/H ratio must be strictly proportional to the acceptor concentration. In fact, the ratio often exhibits a hyperbolic dependency upon [A].
(ii) Similarly, the model predicts that the T/H ratio should be independent of [C], and this is again in contradiction with most results.
(iii) With large concentrations of some acceptors, an inhibition of the total donor utilization is observed.[34] This could be explained by assuming k_5 to be $< k_3$ but, in some cases, the inhibition is disproportionate and a complete 'freezing' of the system has been reported, although a significant proportion of transpeptidation product was detected at low [A].

To account for these data, more complicated models are needed, probably involving multiple binding sites for donor and acceptor substrates, possibly similar to that proposed for the catalysis of transacylation reactions by a class C β-lactamase,[36] where the acceptor binds to the non-covalent ES complex. The authors' own data indicate, however, that with the *Streptomyces* R61 DD-peptidase, no binding of the acceptor occurs before the acyl-enzyme is formed.

5.4.4 *Extension of the model to membrane-bound PBPs*

The general model of interaction with penicillin that has been established for the R61 DD-peptidase has also been recognized as valid for all other PBPs. In fact, the demonstration of the involvement of the same serine residue in the interactions with both substrates and penicillin was first performed with *Bacillus stearothermophilus* PBP 5, a protein that exhibited a DD-carboxypeptidase activity.[37] The molecular properties of the PBPs of numerous bacteria have been extensively reviewed.[10]

The interaction between a PBP and a β-lactam is best described by the kinetic parameters k_2/K' and k_3. With enzymatically active proteins, these values can be determined by monitoring the time-dependent loss of activity after addition of the antibiotic, and the recovery of the catalytic properties after isolation of the acyl-enzyme.

With the proteins devoid of *in vitro* enzymatic activity, one must directly measure the rate of penicillin binding. Most often, an ID_{50} value (representing the penicillin concentration necessary to label 50% of a given PBP) is determined by incubating the membrane or purified PBP preparation with various concentrations of antibiotic and, after a standard contact time, terminating the reaction by addition of sodium dodecylsulphate and heating the mixture to 100°C. The PBPs are subsequently separated by PAGE and the extent of labelling quantified by fluorography.[9,38] The ID_{50} value is then considered as characteristic of the sensitivity of the PBP to the tested β-lactam.

This quite laborious procedure only applies to labelled β-lactams, and the number of such compounds is rather limited. The interaction with unlabelled compounds must be studied by even more complicated counter-labelling procedures. In fact, deducing the values of the kinetic parameters from ID_{50} measurements is not always a straightforward task.

In the following sections the major pitfalls that should be avoided will be briefly summarized, and rigorous methods for the analysis of β-lactam–PBP interactions will be suggested.

5.4.4.1 The presence of a β-lactamase Minute amounts of β-lactamases can be found in some membrane, or even in purified PBP preparations. This can significantly increase the apparent resistance of the studied PBP(s),[39] even if the quantity of β-lactamase is too small to allow its detection as a protein. This is due to the enormous difference underlined previously between the k_3 values for the two types of proteins.

The presence of a β-lactamase is clearly suggested by the following results:

(i) the ID_{50} seems to increase when an increasing concentration of membrane protein or purified PBP is utilized; and
(ii) the rate of penicillin disappearance is larger than that expected from the PBP concentration and its k_3 value(s).

This was observed with a purified preparation of *S. aureus* PBP 4, which exhibited a k_{cat} value with benzylpenicillin of $0.42\,s^{-1}$ while the k_3 value, as directly measured by monitoring the degradation of the PBP–penicillin complex, was only $7.7 \times 10^{-3}\,s^{-1}$.[40,41]

5.4.4.2 The titration effect It can be easily deduced from Model 1 and equations (5.6) and (5.7) that the value might vary with the time of contact between the reagents before the ID_{50} reaction is terminated. When the PBP is very sensitive, the reaction becomes quite rapid, and the ID_{50} might represent only 50% of the PBP concentration in the assay mixture.[10,42]

5.4.4.3 The value of k_3 If the value of k_3 is not small when compared with k_f, the maximum proportion of PBP labelled at the steady-state is $k_f/(k_3 + k_f)$. This underlines the fact that a careful determination of the k_3-values is a prerequisite for meaningful analyses of penicillin–PBP interactions. This factor is too often neglected in many studies, although it is not difficult to measure — the excess of penicillin can be rapidly destroyed simply by the addition of a suitable β-lactamase, and the deacylation rate measured by following the time-dependent loss of label by the various PBPs.

5.4.4.4 The value of k_2/K' The relationship between the k_2/K' and ID_{50} values has been extensively discussed.[43] The time-dependence of the accumulation of the labelled PBP is given by equation (5.6), which is valid if C_0 is much larger than E_0. With a mixture of PBPs, C_0 should be larger than ΣE_0, unless the studied PBP reacts significantly faster than all the others. When k_3 is much smaller than k_f, the quation simplifies to

$$(EC^*) = E_0(1 - e^{-k_f . t}) \quad (5.8)$$

where k_f is given by equation (5.4).

In most cases, $C_0 \ll K'$ and

$$k_f = \frac{k_2}{K'} C_0 \quad (5.9)$$

The value of k_f can then be directly deduced from that of the ID_{50}:

$$\frac{k_2}{K'} = -\frac{\ln 0.5}{t \,.\, ID_{50}} = \frac{0.69}{t \,.\, ID_{50}} \quad (5.10)$$

where t is the contact time in the ID_{50} determination.

When this simple procedure is utilized, it is necessary to verify that both simplifying assumptions are fulfilled (i.e. $k_f > k_3$ and $C_0 < K'$). To do so, theoretical k_f values must be computed (i.e. $k_f = k_2 \,.\, C_0/K'$), compared to the k_3 value, and utilized to compute theoretical EC* values with the help of equation (5.7). If C_0 is not much smaller than K', the computed (EC*) values will be consistently larger than the measured ones for $C_0 > ID_{50}$.

It is thus much more rigorous to directly and individually determine each k_f value by monitoring the time-dependent accumulation of EC* for each C_0 value. If k_f does not vary linearly with C_0, the general equation (5.4) applies, and the individual values of k_2 and k' can be obtained by a non-linear regression or a linearization of the rectangular hyperbola. For example:

$$\frac{C_0}{k_f} = \frac{K'}{k_2} + \frac{C_0}{k_2} \quad (5.11)$$

Unfortunately, this laborious procedure has been seldom used, which probably explains why very few reliable k_2/K' values can be deduced from the numerous published ID_{50} data.

5.4.4.5 Counter-labelling and competition methods When the tested compound is not radioactively labelled, one usually measures the residual amount of free PBP by performing a second incubation with a saturating concentration of a labelled antibiotic, most often benzylpenicillin. Although this procedure is generally safe, it is again important to first determine the k_3 value for the tested compound, because an unstable acyl-enzyme will decay during the second incubation with radioactive penicillin and this will result in an underestimation of the (EC*) value for the first β-lactam.

There is, however, a much more simple method to determine the k_2/K' values for an unlabelled compound, which has surprisingly never been used. Assume that the k_2/K' and k_3 values are known for the reference labelled compound. The k_3 value of the tested compound can be readily obtained by the counter-labelling method. A true competition can then be performed between the labelled and unlabelled β-lactams, i.e. they are incubated together with the PBP for a period of time which results in complete immobilization in the covalent complexes. If the period of time is much shorter than the half-lives of the acyl-enzymes, it can be shown that

$$\frac{[EC_u^*]}{[EC_l^*]} = \frac{[(k_2/K')\,C_0]_u}{[(-k_2/K')\,C_0]_l} \tag{5.12}$$

where the u and l subscripts refer to the unlabelled and labelled compounds, respectively. The value of $[EC_l^*]$ can be readily determined. In a reference experiment, the value $[EC_l^*]$ is also measured in the absence of unlabelled compound, i.e. $[EC_l^*]_0$. The value of $[EC_u^*)$ is then easily obtained as $[EC_l^*]_0 - [EC_l^*]$.

In the example presented in Figure 5.10(a), the k_f values for the disappearance of free PBP are $0.005\,s^{-1}$ in the presence of the sole labelled β-lactam and $0.01\,s^{-1}$ when both compounds are present. However, under the latter conditions, equal concentrations of the PBP are immobilized as labelled and unlabelled acyl-enzymes, and only 50% of the PBP can be detected as a radioactive species. According to equation (5.12), this shows that

$$\left(\frac{k_2}{K'}\cdot C_0\right)_u = \left(\frac{k_2}{K'}\cdot C_0\right)_l$$

and, since $(C_0)_u$ and $(C_0)_l$ are known, the correct value for $(k_2/K')_u$ can be deduced.

It is evident that the same $(EC_u^*)/(EC_l^*)$ ratio will be obtained if the initial concentrations of labelled and unlabelled β-lactams are modified by the same factor. This can aid in circumventing the k_3 problem. Indeed, in the

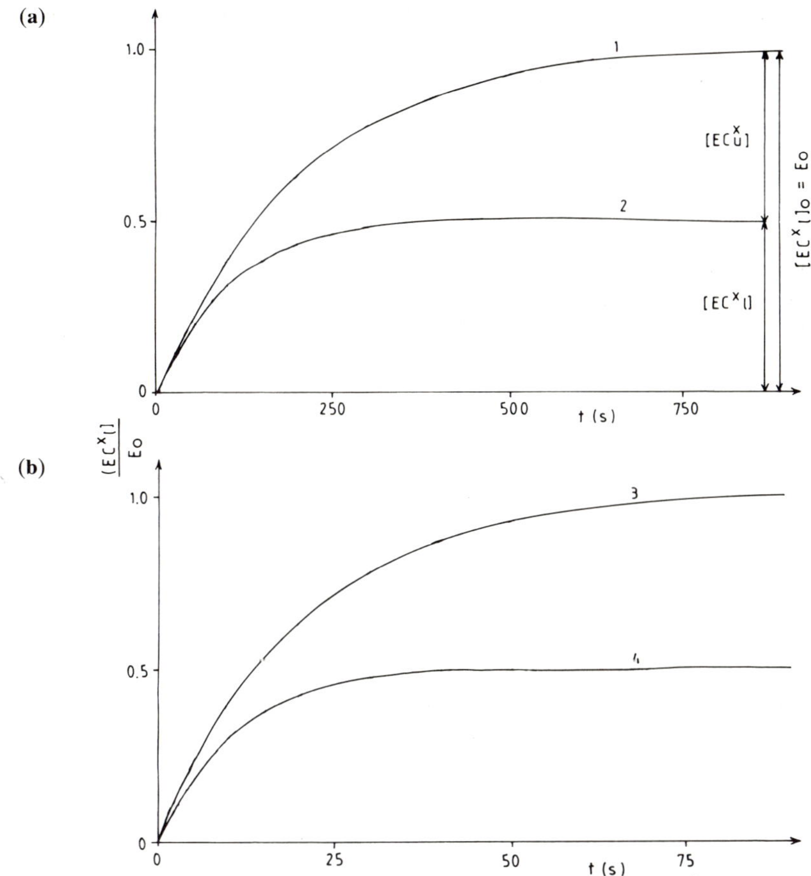

Figure 5.10 Accumulation of the labelled acyl-enzyme (EC_l^*) in the absence (curves 1 and 3) and presence (curves 2 and 4) of a competitor. For both labelled (l) and unlabelled (u) compounds, the k_3 values are less than $10^{-3}\,s^{-1}$ in (a) and less than $10^{-2}\,s^{-1}$ in (b). For the labelled compound, $k_2 = 10\,s^{-1}$ and $K = 1\,mM$, so that $(k_2/K)_l = 10\,000\,M^{-1}\,s^{-1}$. For the unlabelled compound, $k_2 = 2\,s^{-1}$ and $K = 2\,mM$, so that $(k_2/K)_u = 1000\,M^{-1}\,s^{-1}$. (a): $(C_0)_l = 0.5\,\mu M$; curve 1: $(C_0)_u = 0$ and $k_f = 5 \times 10^{-3}\,s^{-1}$; curve 2: $(C_0)_u = 5\,\mu M$ and $k_f = 10 \times 10^{-3}\,s^{-1}$. (b): $(C_0)_l = 5\,\mu M$; curve 3: $(C_0)_u = 0$ and $k_f = 5 \times 10^{-2}\,s^{-1}$; curve 4: $(C_0)_u = 50\,\mu M$ and $k_f = 9.7 \times 10^{-2}\,s^{-1}$.

example described by Figure 5.10(a), and in the presence of both β-lactams, (EC_l^*) reaches 99% of its final value after 450 s. If the value of k_3 for one of the compounds is larger than $10^{-3}\,s^{-1}$ (i.e. 10% of k_f), equation (5.12) is no longer valid and the (EC_u^*)/(EC_l^*) ratio at the steady-state can be significantly different from 1. However, if both C_0 values are increased by a factor of 10, the k_f value for the disappearance of the free PBP increases to $0.097\,s^{-1}$ and equation (5.12) approximately holds for k_3 values up to $10^{-2}\,s^{-1}$ (Figure 5.10(b)).

5.4.4.6 Compounds that rapidly reach the steady-state A poor understanding of the kinetics involved in the 'counter-labelling' method might also lead to gross misinterpretations of the properties of compounds with which a steady-state is rapidly reached.

$$E + X \underset{k'_{-1}}{\overset{k'_{+1}}{\rightleftharpoons}} EX \overset{k'_2}{\longrightarrow} E + Y$$

For such a compound, the apparent K_i value is:

$$\frac{k'_{-1} + k'_2}{k'_{+1}}$$

If the counter-labelling method is used to study such a compound, and the values of k'_{-1} and k'_2 are such that the half-life of EX is small compared to the duration of incubation with the labelling β-lactam, saturation by the β-lactam will be observed even if, at the steady-state during the first incubation, a large proportion of the enzyme is immobilized in the EX complex. This is illustrated by Figure 5.11(b), where a long incubation with the labelled β-lactam would lead to the false conclusion that no interaction occurred between E and X. Thus, when using the counter-labelling method for assessing the inhibitory capacity of a new molecule, it might be sensible to utilize the labelling β-lactam in non-saturating conditions (Figure 5.11(a)).

5.4.4.7 Examples of measured kinetic parameters Table 5.3 shows the values of the kinetic parameters for the interaction of benzylpenicillin with some PBPs. The data for the soluble *Streptomyces* R61 and *Actinomadura* R39 enzymes are included as reference values. The table presents only a very small proportion of the PBPs that have been described and studied. The selection was based on the proteins for which an enzymatic activity has been tentatively or convincingly demonstrated. In some cases (e.g. for the HMW-PBPs of *E. coli*), the k_2/K' values were calculated by the authors on the basis of ID_{50} values found in the literature; these figures are probably less reliable than those obtained by measuring the rate of enzyme activity decrease.[10]

The most striking feature of the figures given in Table 5.3 is the wide dispersion of the k_3 and k_2/K' values, which span at least three orders of magnitude. The highest k_3 value was found for *S. aureus* PBP 4 ($8 \times 10^{-3} s^{-1}$), which yields a 1.5 min half-life for the benzylpenicilloyl–enzyme covalent adduct. Since this value is significantly shorter than the generation time of the bacterium, it would contribute to the resistance of the strain if PBP 4 were the lethal target in the organism. However, this does not seem to be the case. By contrast, k_3 values lower than $10^{-4} s^{-1}$ probably have less physiological meaning, at least for bacteria with 'short' generation times

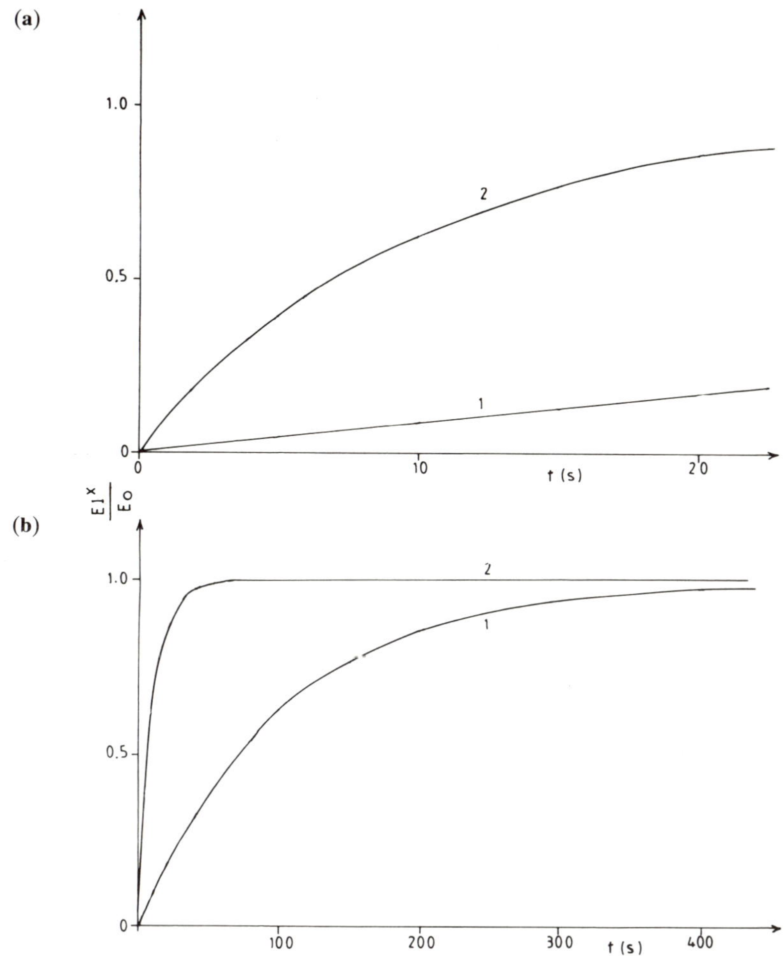

Figure 5.11 Influence of a compound with which the steady-state is reached rapidly, on the rate of counter-labelling by a second β-lactam. The curves were calculated by assuming that $[X] = 9K_i$ and that either k_2' or k_{-1}' was very much greater than $0.1\,s^{-1}$. For the labelling β-lactam, k_3 was negligible and k_f was $0.1\,s^{-1}$ in the absence of X (curve 2) but decreased to $0.01\,s^{-1}$ (i.e. $k_f/(1 + [X]/K_i)$) in its presence (curve 1). If the counter-labelling incubation is short (10 or 20 s (a)) the effect of X is readily detected. If it is too long (300–400 s (b)), the effect of X becomes undetectable.

(20–60 min). The k_2/K' values range from 10^2 to $3 \times 10^5\,M^{-1}s^{-1}$, which illustrates the contribution of the intrinsic properties of the PBPs to the resistance phenomenon. Indeed, if PBP 1 is the lethal target in *Bacillus megaterium*, it is not surprising that this bacterium is much less resistant than *E. coli*, even in the absence of β-lactamase or permeability barrier, since the essential HMW-PBPs of *E. coli* bind penicillin at a much lower rate. If the gene coding for PBP 1b or 1a of *E. coli* could be expressed in a *Bacillus* cell and, assuming that the newly acquired enzyme could catalyse the synthesis of cross-linked peptidoglycan in the host cell, the resistance of the recipient *Bacillus* would be increased by two to three orders of magnitude. In this perspective, the appearance in the 1960s of the famous methicillin-resistant *S. aureus*,[44] which owed its resistance to the acquisition of a very resistant PBP ($k_2/K' \simeq 10\,M^{-1}s^{-1}$) might not appear as very surprising, since this value is only one order of magnitude lower than that observed for several HMW-PBPs in 'normal' strains.

5.4.4.8 Enzymatic activity of HMW-PBPs Recently, the authors demonstrated that various HMW-PBPs from *Enterococcus hirae*, *S. pneumoniae* and *E. coli* would catalyse hydrolysis and transfer reactions[45] with the thiol esters IV or V (Figure 5.7). The efficiencies of the various enzymes were very variable (Table 5.4). In all cases, in the presence of D-alanine, a new peptide

$$C_6H_5\text{–CO–NH–CHR}_2\text{–CO–NH–CH(CH}_3\text{)–COO}^-$$

where R_2 was H or CH_3 depending on the structure of the donor substrates, was formed. In some cases, the presence of D-alanine also increased the rate of donor utilization indicating that, as with the R61 DD-peptidase, the deacylation was rate-limiting in the hydrolysis reaction.

Since the disappearance of the thiol ester can be monitored in the spectrophotometer, such substrates supply an easy method to detect the presence of the relevant HMW-PBPs, and will allow the dissection of the kinetic mechanism of the proteins. Their introduction represents a real breakthrough in the field, although they can only be utilized with solubilized proteins.

5.5 The physiological functions of PBPs

The role of the various PBPs in peptidoglycan biosynthesis and cell division is presently best understood in *E. coli*. A complex picture has recently emerged, in which the LMW-PBPs, initially considered as dispensable, might play a non-negligible role.

5.5.1 The situation in E. coli

5.5.1.1 The HMW-PBPs 1a and 1b PBPs 1a and 1b, encoded by the *ponA* and *ponB* genes, respectively, are not synthesized as pre-proteins.[46] The N-terminus of PBP 1b acts as an uncleaved signal-like sequence, which is responsible for both the translocation of the protein into the periplasmic space, and its anchoring to the cytoplasmic membrane.[47] When compared to

Table 5.4 Acyltransferase activities of HMW-PBPs.

PBP	Hydrolysis		Transpeptidation			
			Substrates			
	Substrate	k_{cat}/K_m ($M^{-1} s^{-1}$)	Donor	Acceptor	Acceleration factor	T/H
Enterococcus hirae PBP 3	A	3200	A, 1 mM	D-alanine, 10 mM	1.25	0.42
	B	3900	B, 0.25 mM	D-alanine, 10 mM	ND	1.5
Enterococcus hirae PBP 1	A	1600	A, 1 mM	D-alanine, 50 mM	2.2	0.67
Streptococcus pneumoniae PBP 2B	A	300	A, 1 mM	D-alanine, 10 mM	1.0	0.24
	B	1100	B, 0.25 mM	D-alanine, 10 mM	1.0	0.95
Escherichia coli PBP 3	A	80	A, 1.2 mM	D-alanine, 5 mM	1.0	>5

Substrates A and B are C_6H_5–CO–NH–CH(CH_3)–CO–S–CH_2–COO^- and C_6H_5–CO–NH–CH(CH_3)–CO–S–CH(CH_3)–COO^-, respectively. The PBPs were obtained as soluble proteins by proteolysis of membrane preparations (*Enterococcus hirae*) or modification of the corresponding gene (*Streptococcus pneumoniae* and *Escherichia coli*). Results are from reference 45.

that of PBP 1b, the sequence of PBP 1a suggests a similar membrane insertion mode, although the small cytoplasmic N-terminal domain (residues 1 to 63, see Figure 5.12) seems to be absent. PBPs 1a and 1b are essential for cell growth, and their inactivation by β-lactam antibiotics results in cell lysis.[9] Mutant studies show that the absence of either PBP 1a or 1b is not lethal under laboratory conditions, but that a double mutant lacking both of them is not viable. The two proteins may thus have similar

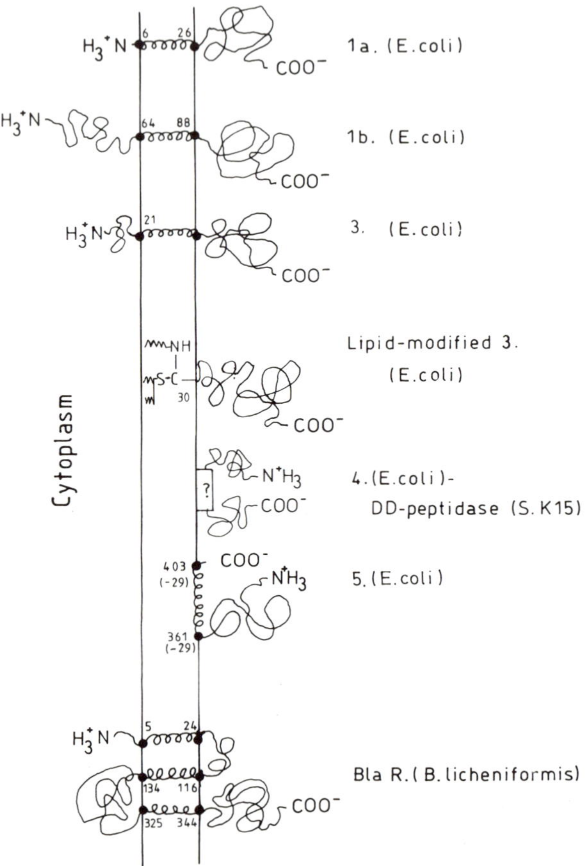

Figure 5.12 Anchoring of the various PBPs in the cytoplasmic membrane. The situation for *E. coli* PBP 2 is similar to that described for PBP 3, with one single N-terminal trans-membrane helix. In addition, an undetermined proportion of PBP 3 might be lipid-modified, and is represented here on the basis of the model established for the *E. coli* lipoprotein and the *Bacillus licheniformis* β-lactamase—the jagged lines representing fatty acid chains. The residues involved in the attachment of *E. coli* PBP 4 and of the *Streptomyces* K15 DD-peptidase have not been identified. Overproduction of both proteins results in a partial solubilization. The mode of attachment of *E. coli* PBP 6 and of *B. subtilis* PBP 5 is similar to that of *E. coli* PBP 5. In these cases, the anchoring helices are amphiphilic.

functions and can replace each other in peptidoglycan biosynthesis.[48–50] They behave as bifunctional enzymes, exhibiting both transglycosylase and transpeptidase activities. Indeed, they catalyse the formation of cross-linked peptidoglycan from the undecaprenol diphosphate-linked disaccharide peptide. The cross-linking activity is higher with PBP 1a, 39% vs. 14%.[31,32] The C-terminal parts of the proteins contain the structural elements found in other PRPs, particularly the active serine residue, and are responsible for both the transpeptidase activity and the penicillin sensitivity. The N-terminal portion, which is absent in monofunctional LMW-PBPs, catalyses the transglycosylase reaction.

Inhibition of the formation of cross-linked peptidoglycan by low concentrations of penicillin or by various monoclonal antibodies[51] seems to increase the rate of polymerization of glycan chains. By contrast, inhibition of the transglycosylase activity by moenomycin also inhibits the transpeptidation. Deletion of the C-terminal part of PBP 1b (residues 424–844 or 405–844) eliminates the transpeptidase activity as expected, but partially conserves the transglycosylation (reference 32; Wierenga and Keck, personal communication). Conversely, attempts to isolate a functional penicillin-binding domain have failed; although truncated proteins consisting of residues 413 to 844 or 272 to 844 recognize antibodies directed against the membrane-bound enzyme, they exhibit no penicillin-binding properties (Wierenga and Nguyen-Distèche, unpublished results). Consequently, it appears that the N-terminal domain can function and fold into an active conformation independently of its C-terminal counterpart but that the reverse is not true; according to the present results, transpeptidation cannot occur without transglycosylation, and the transpeptidase domain cannot by itself fold into a functional conformation.

Analysis of cross-linked peptidoglycan synthesized by PBP 1b[51] shows that its preferred acceptor is the tetrapeptide and not the pentapeptide (Figure 5.13). This suggests that optimum activity of PBP 1b is dependnent on the preliminary action of a DD-carboxypeptidase (PBP 5), which controls the proportion of tetrapeptides in the nascent peptidoglycan.

5.5.1.2 The HMW-PBPs 2 and 3 PBP 2 encoded by the *pbpA* gene appears to be essentially a periplasmic protein attached to the cytoplasmic membrane by an N-terminal, uncleaved signal-like sequence.[52] PBP 3 is the product of the *ftsI* gene, synthesized as a precursor and processed to a mature protein by removal of residues 578 to 588 at the C-terminus.[53] Near the N-terminus, the sequence Leu–Leu–Cys–Gly–Cys-30 is similar to the bacterial consensus for the modification and processing of the lipoproteins. However, only 15% of the molecules are lipid-modified on Cys-30 in a strain overproducing PBP 3.[54] Attachment to the membrane of the non-modified molecules again seems mainly due to the N-terminus, acting as an uncleaved lipoprotein signal peptide. Starting with residue 36, the whole polypeptide

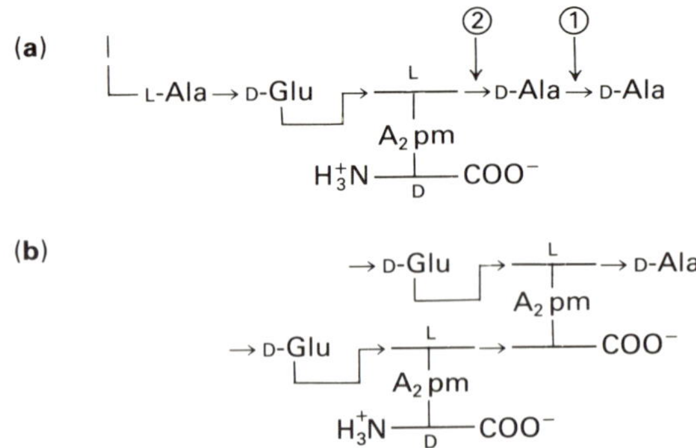

Figure 5.13 (a) The precursor pentapeptide unit in *E. coli*. The DD-carboxypeptidase (arrow 1) liberates the C-terminal D-alanine, yielding the tetrapeptide, which is the substrate for the LD *trans*- or carboxypeptidase (arrow 2). The product of the former is shown in (b), and that of the latter is the tripeptide. (b) Product formed by the action of the LD-transpeptidase on two tetrapeptides.

chain is in the periplasm.[55] A signal peptide containing a similar consensus sequence Val–Leu–Phe–Ala–Cys-41 for lipid modification is also found in PBP 2 of *Neisseria gonorrheoae*, which corresponds to PBP 3 in *E. coli*. In *Neisseria meningitidis* PBP 2, the Cys-41 residue is replaced by a glycine[55] making lipid modification impossible.

Replacement of the putative lipoprotein signal peptide by that of PBP 5 or that of the OmpA protein resulted in the production, in the periplasm of *E. coli*, of a water-soluble truncated form of PBP 3 (reference 55; Fraipont and Nguyen-Distèche, unpublished data). A simple removal of the signal peptide sequences of PBPs 2 and 3 leads to the synthesis of water-soluble proteins in the cytoplasm. When overexpressed intracellularly, PBP 3 forms inclusion bodies, which can be solubilized with guanidinium chloride and refolded. All of these various derivatives appear to bind penicillin as efficiently as does the intact protein. They are only water-soluble in rather high salt concentrations.[56,57]

Fusion of the C-terminal part of PBP 3 (residues 240 to 588) to β-galactosidase or removal of the 240 N-terminal residues do not appear to modify the affinity of the protein for penicillin, but have a profound effect on its stability (reference 58; Nguyen-Distèche, unpublished data). This part of the protein thus represents a functional penicillin-binding domain. The role of the N-terminal part is not yet clearly established.

PBP 2 is the lethal target of mecillinam, and its transpeptidase activity is very sensitive to this antibiotic. In collaboration with the RodA protein, it is responsible for carrying out lateral cell-wall elongation and maintaining the bacillary cell shape. In the presence of mecillinam, wild-type cells become spherical.[9] Thermosensitive mutants are mecillinam resistant, have no functional PBP 2, and are spherical even in the absence of mecillinam.[59]

Conversely, *rodA* mutants are spherical and mecillinam resistant, but produce a normal PBP 2.[59] The gene coding for PBP2 (*pbpA*), and the *rodA* gene are located in the same cluster, at 14 min on the chromosome.[60] The RodA protein seems to be essential for the activity of PBP 2. Membranes prepared from a strain that overproduces PBP 2 and RodA, and lacks PBP 1b, catalyse the synthesis of cross-linked peptidoglycan from the lipid intermediate in the presence of cefmetazole, which inactivates all the other PBPs. When the gene coding for either PBP 2 or RodA is inactivated by insertion of a transposon, there is no peptidoglycan synthesis.[61] Which of the two proteins is responsible for the transglycosylation activity has not yet been determined. RodA may regulate the activity of PBP 2, or the two proteins may form a complex functioning as a peptidoglycan synthetase.

PBP 3 is involved in septum formation. Mutations affecting PBP 3, or selective inactivation by β-lactam antibiotics such as ampicillin, lead to the formation of filamentous cells.[9] PBP 3 also seems to require the presence of another protein, FtsW, to catalyse transglycosylation and transpeptidation *in vitro*, but the degree of cross-linking is very low.[62] Mutations in FtsW, which is homologous to RodA, cause inhibition of the cell division.[63]

Finally, in related Enterobacteria, PBP 2 might also play a role in the induction of the chromosomal β-lactamase.[64]

5.5.1.3 The LMW-PBPs PBPs 4, 5 and 6 are synthesized as pre-proteins with typical amino-terminal signal peptides, which are cleaved upon maturation.[65,66] Their mode of membrane anchoring is described in Figure 5.12. They are responsible for most of the DD-carboxypeptidase activity, and the first results indicated that, under laboratory conditions, they seemed to be dispensable for growth and division. Strains in which the chromosome contains single or double deletions for PBP 5 (*dacA*) and PBP 6 (*dacC*) grow with normal morphology.[67,68] However, *dacA* mutants appear to be more sensitive to penicillin than the parent strain.[69] Point mutations in *dacB*, which abolish the PBP 4 penicillin-binding capacity and reduce the total DD-carboxypeptidase activity, do not alter cell morphology, thus suggesting that PBP 4 is not essential under these experimental conditions.[70] However, deletion mutants for PBP 4 must be constructed before its role can be definitively considered as accessory, and it should be noted that no mutant completely devoid of DD-carboxypeptidase activity has yet been obtained. Moreover, PBP 4 could play an important role in cells growing in their natural environment, since a *dacB* mutant was found to be sensitive to bile acids.[71]

PBP 4 also cleaves the D-Ala–(D)-*meso*-A_2pm (Figure 5.2(c)) peptide bonds in the peptide cross-links, thus exhibiting an ‘apparent endopeptidase’ activity. It was thus proposed to be involved in the turnover of cross-bridges[72] and in the maturation process of newly synthesized peptidoglycan, acting *in vivo* as a secondary transpeptidase.[73,74] However, a later study of a PBP 4 overproducer indicated that PBP 4 catalysed *in vivo* as well *in vitro* DD-carboxypeptidase and 'endopeptidase' activities,[75] resulting in a decrease of the D-Ala–(D)-A_2pm cross-bridges, partially compensated by an increase of the L–D cross-links involving two A_2pm residues (Figure 5.13(b)).

Although it has often been considered that the three LMW-PBPs were dispensable, the most recent investigations indicate that these enzymes are required to maintain a correct balance between the various precursors in such a way that cells can elongate or divide. The interplay of the transpeptidase and carboxypeptidase activities in the cell cycle is discussed in the next section.

5.5.1.4 Integration and regulation of the enzymatic activites The activity of PBPs has been studied *in vivo*, and in ether-permeabilized cells, by monitoring the effects of specific β-lactams or mutations on the rate of peptidoglycan synthesis — measured as the incorporation of [^{3}H]-*m*-A_2pm.[76–79] Two distinct peptidoglycan synthesis systems appear to be involved in elongation or division. In contrast to the previously accepted hypotheses, it has been suggested that PBPs 1a and 1b do not have a specific function, but supply 'primers', i.e. newly cross-linked peptidoglycan, inserted in new initiation sites; PBPs 2 and 3, behaving as specialized enzymes, utilize these sites to synthesize peptidoglycan during longitudinal wall growth or cross-wall formation, respectively.[79] Some analyses of either longitudinal wall or septum peptidoglycan appeared to indicate that the former contained only peptide dimers, while the latter exhibited a higher degree of cross-linking; this suggested that peptidoglycan would be inserted as single strands during elongation, while a multi-stranded incorporation would occur during septum formation. PBP 3 can use exogenously-added, soluble, tripeptide-containing substrates as acceptors, with the newly insolubilized pentapeptide as donor.[77,80] Initiation of the division appears to require a penicillin-insensitive peptidoglycan synthesis.[81] One such activity might be due to the LD-transpeptidase, which remains to be identified. This enzyme could be responsible for the formation of L–D peptide bonds between two *m*-A_2pm residues (Figure 5.13(b)), in a reaction where the tetrapeptide would act as a donor. The number of such L–D cross-bridges increases in the presence of penicillins, in the stationary phase and during amino-acid starvation.[82,83] Other candidates might be a monofunctional glycan polymerase, thus different from PBPs 1a and 1b,[84] or the *mepA* 'endopeptidase',[85] which is neither penicillin-sensitive nor homologous to the PREs.

The cell cycle consists of alternating phases of elongation and septation and thus depends on a periodically shifting balance between the activities of the two competing morphogenetic systems. This balance can be altered by modifying the amounts either of the PBPs themselves, or of their preferred substrates. For instance, expression of PBP 3 is under the negative control of the *mreB* gene product — a 37 000 M_r protein; overexpression of this gene causes filamentation, while a deletion or mutations induce the formation of round cells that overproduce PBP 3 and exhibit a somewhat increased PBP 1b content.[86,87]

Similarly, elevated levels of PBP 5 or PBP 6 can restore division in *fts123* mutants, which produce a thermolabile PBP 3.[88] Indeed, larger amounts of DD-carboxypeptidase increase the number of tetra- and tripeptides preferentially utilized as acceptors by PBP 3. A similar correction of the mutation can be performed by the addition of low concentrations of D-cycloserine, which inhibits the alanine racemase and the D-Ala–D-Ala ligase and, consequently, decreases the concentration of the pentapeptide.[88] In addition, overproduction of PBP 5 causes conversion of the rod-shaped *E. coli* into round cells, probably by displacing the elongation/septation balance towards that latter reaction.[89] As suggested by Botta and Park,[76] the proportion of tri-, tetra- and pentapeptides might be an important factor in determining the shape of the cells by controlling the relative rates of division and elongation. The racemase, the D-Ala–D-Ala ligase, and the adding enzyme are all necessary for the transformation of tri- into pentapeptides and might also, and for obvious reasons, be involved in this mechanism. Further data corroborate this hypothesis:

(i) The level of LD-carboxypeptidase (Figure 5.13(a)) fluctuates during the cell cycle and is maximum at the time of septation.[90]
(ii) Production of PBP 6 in the transition to the stationary phase is induced by the *bolA* gene product, which is not required during exponential growth. Overexpression of *bolA* induces the formation of spherical cells.[91]

The control system is indeed very complex. Many proteins are involved in septum formation, and the interplay of the PBPs with the various division gene products is not completely understood. Most of the evidence relies on genetic data, and the biochemical activity of the proteins remains undetermined. Several division genes (*ftsI*, *ftsW*, *ftsQ*, *ftsA*, *ftsZ*) are located in a large cluster of genes, involved in peptidoglycan biosynthesis, at 2 min on the chromosome.[92]

Together with PBP 3, some of the corresponding proteins (FtsA, Q, W, Z) might associate to form a cell-wall biosynthetic complex called a septator or divisome.[81,93] For instance, a mutation in *ftsA* affects the binding of penicillin to PBP 3.[94] Overproduction of the *ftsZ* gene product stimulates cell division, inducing the formation of mini-cells.[95] Properties of some *fts* gene products are summarized in Table 5.5. The translocation of cell division proteins across the cytoplasmic membrane is also an essential step of cell division, as shown by the effects of mutations in the *ftsH*, *ftsY* and *secA* genes, which are located elsewhere on the chromosome.[100,101]

Finally, this phenomenon would also be controlled by the cyclic AMP–CAP protein complex[102] and the FicA protein,[103] while the product of the *lov*

Table 5.5 Properties of the *fts* gene products directly involved in cell division.

Gene product	Assumed function	Homologous proteins and assumed functions
FtsZ	Initiation of septation Target of cell division inhibitors: products of *sulA* (SOS response) and *min C* and *D* genes[95]	FtsZ in *B. subtilis*: symmetric septation in vegetative growth and asymmetric septation in sporulation[95]
FtsA	Represses the transcription of cell division genes Involved in the late events of cell division Interacts with PBP 3[94]	MreB: repression of cell division Heat shock proteins—Hsp70 and DnaK—protein kinases[92,96] FicA: folate metabolism Cell division proteins in yeast[97]
FtsW	Interacts with PBP 3	RodA SpoVE: spore formation in *B. subtilis*[98,99]
FtsQ	Involved in the late events of cell division[100]	

gene seems to connect peptidoglycan and protein biosynthesis, and growth rate.[104] Mutations affecting the first complex and the *lov* gene can confer high resistance to mecillinam.

5.5.2 *LMW-PBPs in other bacteria*

Most, if not all bacteria possess one or several LMW-PBPs, often exhibiting DD-carboxypeptidase activities, and which are usually assumed to control the extent of peptidoglycan cross-linking by limiting the availability of donor D-alanyl–D-alanine-containing precursors. However, as seen previously, they might play a more subtle role in the control of the cell cycle; the *Streptomyces* K15 transpeptidase seems to represent an exception.

5.5.2.1 The DD-transpeptidase of Streptomyces *K15* This enzyme is the smallest PBP described so far. Although its substrate specificity reflects the structural features of the *Streptomyces* peptidoglycan, its physiological function remains undetermined. It is synthesized with a signal peptide, which is cleaved in the mature form of the protein.[105] In the wild-type strain, it remains membrane-bound but, when overexpressed, 30% of the enzyme activity is found in the culture medium.[17]

Sequence alignments and hydrophobic cluster analysis (HCA) show clear similarities with *E. coli* PBPs 5 and 6, *B. subtilis* PBP 5 and the *spoIIA* gene product, a putative PBP involved in the sporulation of *B. subtilis*. These proteins might thus form a distinct structural subgroup in the LMW-PBPs. Moreover, the HCA and secondary structure predictions indicate a folding similar to that of class A β-lactamases.[105] In the K15 enzyme, predicted helices $\alpha 1$ and $\alpha 11$ are shorter, leaving one face of the β-sheet partially uncovered. This would expose a hydrophobic area, which might help binding the protein to the cytoplasmic membrane, possibly via an interaction with an intrinsic membrane protein.

When assayed on a peptide substrate such as Ac_2–L-Lys–D-Ala–D-Ala, the enzyme behaves as a very poor hydrolase — a low proportion of dipeptide is rapidly produced and the enzyme seems to stop working.[106] In fact, it efficiently utilizes the small amount of released D-alanine as an acceptor, and catalyses a 'silent', homogeneous interchange between the tripeptide C-terminal D-alanine and the free amino acid. The elimination of the free D-alanine with D-amino acid oxidase allows the hydrolysis reaction to proceed to completion. In the presence of dipeptides, such as glycyl–glycine or glycyl–L-alanine, which mimic the structure of the Gly–LL-A_2pm acceptor in nascent peptidoglycan, virtually no hydrolysis product is formed and the enzyme becomes a strict, efficient transpeptidase (Table 5.6).

By contrast, the ester Ac_2–L-Lys–D-Ala–D-lactate is efficiently and completely hydrolysed, D-lactate being a much poorer acceptor than D-alanine. But, as shown in Table 5.6, the presence of a dipeptide acceptor also increases the rate of utilization of this donor substrate.

In agreement with the results obtained with the *Streptomyces* R61 DD-peptidase, a simple model involving the partition of the acyl-enzyme between hydrolysis and aminolysis fails to explain various experimental observations.

(i) The presence of the acceptor dipeptide appears to modify the k_{cat}/K_m ratio.
(ii) The concentration of the donor substrate influences the transpeptidation/hydrolysis ratio.[106]

5.5.2.2 S. aureus *PBP 4* *In vitro*, this protein behaves as a DD-carboxypeptidase[40] but, *in vivo*, it seems to only catalyse a transpeptidation involved in secondary cross-linking reactions.[107] A decreased cross-linking is observed in cells treated with low concentrations of cefoxitin or in a mutant lacking PBP 4.[108] However, PBP 4 is dispensable since the mutant appears to be viable.[109]

5.5.2.3 The DD*-carboxypeptidase of* Gaffkya homari In this bacterium, a transpeptidase of low penicillin sensitivity incorporates linear, non-cross-linked peptidoglycan into existing cell wall.[110] The enzyme, however, exhibits a high specificity for tetrapeptides as acceptors. Those tetrapeptides are generated by a penicillin-sensitive DD-carboxypeptidase. Consequently, the presence of penicillin indirectly inhibits transpeptidation by depriving the enzyme of its preferred substrate — a situation reminiscent of the functioning of the PBP 3 system in *E. coli*. Unfortunately, analysis of PBP saturation by different β-lactams during cell growth inhibition sheds some

Table 5.6 Kinetic constants for the interaction between the *Streptomyces* K15 DD-peptidase and its synthetic substrates.

	Transpeptidation[a]			Hydrolysis[b]					
Substrate	K_m (mM)	k_{cat} (s^{-1})	k_{cat}/K_m ($M^{-1} s^{-1}$)	K_m (mM)	K	k_2 (s^{-1})	k_3 (s^{-1})	k_2/k_3	k_2/K ($M^{-1} s^{-1}$)
Ac_2–L-Lys–D-Ala–D-Ala (peptide)	6	0.55	92	8	9	0.13	1.2	9.5	14
Ac_2–L-Lys–D-Ala–D-lactate (ester)	8.9	5.6	630	1.3	2.2	0.86	1.2	1.4	390

[a] Transpeptidation studied in the presence of 10 mM glycyl–glycine.
[b] With the peptide, steady-state hydrolysis rates are difficult to measure (see text) and the values were determined with a lesser degree of certainty.

shadow on this rather clear and simple picture, the primary target in growing cells would be PBP 6 and not the DD-carboxypeptidase, PBP 9.[111]

5.5.2.4 PBP 4 in Salmonella *and* Proteus The properties of these PBPs have been reviewed previously.[10] The most striking results were the catalysis of dimer formation by *Salmonella typhimurium*[112] and *Proteus vulgaris* PBP 4,[113] as well as by the *P. vulgaris* L-form PBP 4, a soluble protein.[113] Apparently, no significant new data on the enzymtic properties or the physiological role of LMW-PBPs have been recently obtained.

5.5.3 HMW-PBPs in other bacteria

In *S. aureus*, PBP 1 appears to be essential, and capable of supporting cell growth when PBPs 2, 3 and 4 are inhibited. However, small quantities of the PBPs 2, 3 and 4 might be sufficient.[114] In *B. subtilis*, a functional PBP 1 is also necessary but PBPs 2a, 2b and 3 could control the cell morphology.[115]

Most of the recent publications only contain ID_{50} values, and no kinetic or genetic analyses have been presented. However, the emergence of the resistant PBPs has elicited a large number of studies; these are described in section 5.6.

In the gram-positive bacteria *S. aureus*, *S. pneumoniae* and *Micrococcus luteus*, the major proportion of the transglycosylase activity can be separated from the PBPs.[116] By contrast, *B. subtilis* PBP 1 seems to exhibit a transglycosylase activity *in vitro*.[117]

5.5.4 The product of the blaR *gene*

In *B. licheniformis*, this protein acts as a receptor for β-lactams in the induction of β-lactamase synthesis. It consists of an extracellular, C-terminal, penicillin-binding domain strongly attached to the cytoplasmic membrane via an N-terminal extension containing three membrane-spanning helices.[118,119] The C-terminal domain has been expressed in the periplasm of *E. coli* as a fully soluble protein and purified. It is very efficiently acylated by penicillins.[120] One can hypothesize that the complete protein might have arisen from the fusion of a chemoreceptor with a modified class D β-lactamase, and that acylation of its 'active' serine side-chain results in the transmission, via the transmembrane segments, of a cytoplasmic signal inducing the derepression of the structural gene *blaP*, which encodes the β-lactamase. In spite of detailed genetic studies, the molecular mechanisms responsible for that phenomenon remain obscure but, interestingly, the *blaR* and *blaP* genes are in the same locus and acylation of the *blaR* receptor also induces its own expression.

Recently, two new genes coding for receptors homologous to *blaR* have been isolated, both in *S. aureus*. One is on the Tn *552* transposon which also contains the *blaP* gene coding for the PC1 staphylococcal β-lactamse.[121] The second, *mecR*, is on the chromosome[122] near the gene coding for the penicillin-resistant PBP 2′ (see section 5.6.4). Although the similitudes are striking, the spatial and functional organisation of the various genes is different in *S. aureus* and *B. licheniformis*. Finally, a comparison of the two *blaR* genes (Joris, unpublished data) reveals a higher degree of conservation in the C-terminal domains (36% of identical residues) than in the N-terminal portion, containing the membrane-spanning helices (21%).

5.6 PBPs involved in resistance to β-lactams

Resistance to β-lactams can be directly associated, in laboratory mutants and clinical isolates or wild-type strains of various bacteria, to a decrease in PBP affinity or to modification of the PBP 'pattern'.[123–130] This might not be very surprising if one remembers the large variations described previously in the sensitivity of the original PBPs characteristic of different species. The sequencing of the genes coding for resistant PBPs has allowed the identification of some crucial mutations. Unfortunately, since very little is generally known about the enzymatic properties of the sensitive PBPs themselves, it remains impossible to propose a coherent molecular explanation of the decreased rate of acylation.

5.6.1 E. coli *PBP 3*

The study of cephalexin-resistant *E. coli* mutants obtained by chemical mutagenesis and selection shows that single point mutations can only result in a seven- to eight-fold increase in resistance. Some of those mutations occur in the S307XXK (T308→P) or the S359XN (N361→S) conserved motifs (see section 5.3.3). Higher levels of resistance were obtained after a more drastic remodelling of the active site by multiple amino acid substitutions, which further reduced the affinity of PBP 3 for cephalosporins.[131,132] Assuming a similarity with the known 3-D-structures, two of those mutations, E349K and N361S, can be located near the active site. More surprisingly, the V530I and Y541S substitutions are 40 to 50 residues upstream from the C-terminus and more than 35 residues downstream from the KTG motif; they are thus in the putative α/β domain. In some cases, the mutations make the protein more unstable — an effect that can be corrected by additional mutations — and the N361S substitution alters the cell morphology by inducing the formation of pointed cell poles.[133]

The decreased affinity for cephalosporins was not accompanied by a similar effect for penicillins or monobactams, indicating a very specific effect of the acylating agent's structure. Consequently, it was suggested that the mutants' abilities to process their normal peptide substrates remained unimpaired.

5.6.2 Neisseria gonorrhoeae *PBP 2*

Several chromosomal loci (*mtr*, *penA*, *penB*, *pem*, *tem*, *env*) are involved in the resistance of β-lactamase negative *N. gonorrhoeae* strains.[134] PBPs 2 and 1 — determined by the *penA* gene and another so-far unidentified locus, respectively — show reduced affinity for penicillin in resistant clinical isolates.[125] Sensitive strains acquired resistance in a stepwise manner when transformed with DNA isolated from resistant strains.[135,136] The first-step transformants contained an altered form of PBP 2 but showed only a six- to ten-fold increased penicillin resistance, as described in the previous section for *E. coli* PBP 3.[125] More-resistant transformants exhibited an altered PBP 1, and mutations decreasing the outer membrane permeability.[137]

In a more recent analysis, 44 out of 47 resistant strains were discovered to contain a more extensively altered *penA* gene. First, an additional aspartic acid residue was inserted after Arg-345, between the SXXK and the SXN motifs, probably very near the active site. This mutation might represent a key step in developing a resistant PBP2 but, by itself, only results in a moderately decreased affinity. In more-resistant strains, a completely divergent fragment corresponding to the C-terminal sequence was found, in addition to several other point mutations.[138] Spratt and coworkers have suggested that the insertion of the additional Asp residue represented the first step in the alteration of PBP 2, and was followed by further substitutions, which gradually decreased its affinity for penicillin, yielding the class A resistant strains. An additional exchange of the fragment coding for the C-terminal part of PBP 2 for that of the corresponding *penA* gene of a closely related resistant species (presumably *N. flavescens*) produced the more extensively altered and more-resistant class B *penA* genes.[138,139] A similar acquisition of one or two gene fragments from *N. flavescens* appears to be responsible for the increased resistance of PBP 2 of *N. meningitidis* and *N. lactamica*.[140–142] The way in which PBP 1 loses its affinity, and the role of the other genetic loci in highly resistant gonococci remains to be understood.

5.6.3 *Streptococci*

Multiple modifications in PBPs, and quantitative or qualitative alterations in the PBP patterns are also observed in penicillin-resistant *S. pneumoniae* strains.[127,128] Stepwise transformation of sensitive strains was also performed using the DNA from resistant cells.[130,143]

As in the gonococcal PBP 2, reduction of the penicillin affinity of the pneumococcal PBP 2b requires several amino acid substitutions. Again, the most extensively altered region extends between the active serine S192XXK and the second S249XN motifs, from positions 232 to 238. Another mutation was just adjacent to the SXN motif (T252A).[144] These mutants could be considered as corresponding to the class A of gonococcal mutants, although they show no insertion comparable to the additional Asp in these latter proteins.

When the PBP 2b genes from clinical isolates of penicillin-sensitive and resistant strains are compared, about 14% of the nucleotides are different. The 'resistant' genes exhibit a 'mosaic' structure consisting of blocks of nucleotides very similar to or quite different from those of the 'sensitive' genes. Such a gene structure can only arise by the recruitment — by transformation — of a corresponding PBP gene from a closely related, penicillin-resistant species (probably a streptococcal strain), followed by recombinations between the 'resistant' and 'sensitive' genes.[145] This phenomenon is thus reminiscent of that described in the previous section for *Neisseria*, but appears to be more complex. Two classes (A and B) of resistant genes can be distinguished according to the positions of the substitutions.

Isolates of resistant *S. sanguis* and *S. oralis* also contain a PBP 2b gene with a mosaic structure similar to that of the class B PBP 2b gene of resistant pneumococci,[146] suggesting a horizontal transfer of an altered PBP 2b gene from *S. pneumoniae*.

Resistance in pneumococci can also result from modifications of PBP 2x, due to interspecies recombinations as described for PBP 2b.[147]

Interestingly, alterations in the structure of peptidoglycan itself are correlated with penicillin resistance. The peptidoglycan of resistant strains seems to result from the action of a transpeptidase having different specificity requirements for the structure of the acceptor and donor peptides.[148]

Thus, the remodelling of the active site of resistant PBPs not only reduces their affinity for β-lactams, but can also modify their selectivity for peptide substrates.

5.6.4 Staphylococcus aureus

The first clinical problems involving penicillinase-negative, resistant mutants of an usually sensitive species were due to methicillin-resistant *Staphylococcus aureus* (MRSA) strains. They possess a *mecA* gene coding for a low affinity PBP 2a or 2′. This gene was cloned[149] and found in resistant strains of other staphylococcal species.[150,151] It is located on a DNA fragment carrying additional resistance determinants for tetracycline, heavy-metal ions or tobramycin, and four direct repeat sequences nearly identical to the IS257 insertion sequence of *S. aureus*.[152] The nucleotide sequence in the promoter and neighbouring regions of the *mecA* gene is very similar to that of the staphylococcal PC1 β-lactamase gene, suggesting that the *mecA* gene might be the product of an illegitimate recombination between a β-lactamase gene and a PBP gene from another species.[149] Moreover, a *mecR* gene, closely related to the *blaR* gene, is found next to the *mecA* promoter in an arrangement reminiscent of that of the *blaP* and *blaR* genes in the *S. aureus* Tn*552* transposon.[121,122] Amplification of the *mecA*-associated DNA correlated with increased methicillin resistance after step-wise selection of highly resistant strains.[153] In many *S. aureus* strains, *mecA* expression is inducible, but PBP 2′ is produced constitutively in MRSA mutants that have lost a β-lactamase-encoding plasmid. However, the disappearance of the plasmid often results in the subsequent loss of the *mecA* gene.[154] This suggests that the synthesis of PBP 2′ is controlled by a plasmid-encoded protein also needed for the regulation of the staphylococcal *bla* gene.[155,156] If an active *mecA* gene is required for the appearance of a resistant phenotype, other chromosomal factors unlinked to this *mec* locus are also necessary, since different resistant phenotypes are found in MRSA clinical isolates.[157–161]

The sequence of PBP 2′ is clearly related to those of *E. coli* PBPs 2 and 3, and of the other resistant PBPs described here, both in the N-terminal and the penicillin-binding C-terminal domains. Conversely, it differs from those of *E. coli* PBPs 1a and 1b and, most strikingly, in the N-terminal domain, which acts as a transglycosylase in the latter enzymes.

In the absence of PBP 2′, β-lactam resistance in *S. aureus* can also be attributed to other PBPs with decreased penicillin affinity.[162,163]

5.6.5 Enterococci

Most laboratory selected and naturally occurring resistant strains of enterococcal species constitutively produce large amounts of low affinity PBPs to which the resistance could be unequivocally attributed.[164-167] The genes coding for two of these PBPs — PBPs 3R and 5 — have been cloned[168] from two different resistant strains of *E. hirae* (formerly, *Streptococcus faecium*) and were shown to code for very similar amino acid sequences, a result which was confirmed by the immunological cross-reactions of the purified proteins.

Those sequences are also, but less closely, related to that of the *S. aureus* PBP 2′. The sequences of the enterococcal and staphylococcal PBPs are most divergent near their N-termini, probably indicating different origins for the upstream parts of these genes.

It can be assumed that those PBPs have the ability to take over the cross-linking of peptidoglycan after the other, sensitive, PBPs have been inactivated by penicillin.[169] There is a distinct possibility that they might select substrates different from those utilized by the sensitive PBPs, as do the resistant pneumococcal PBPs.

The necessity of overproduction in these resistant strains raises an interesting question. This might indeed be due to the fact that the decreased affinity for β-lactams is accompanied by a parallel decrease in catalytic efficiency, which could be most simply compensated for by a simple increase in the quantity of the enzyme.

5.7 Site-directed mutagenesis results

In contrast to β-lactamases, for which many point mutations have been engineered and studied, only a small number of residue substitutions have been performed in PBPs and DD-peptidases.

Replacement of the active-site Ser-62 residue of the *Streptomyces* R61 DD-peptidase by a Cys yielded an inactive protein that also failed to bind penicillin. The K65R mutant exhibited a 100–200 fold decreased activity vs. the various substrates, but the rate of penicillin inactivation was decreased 20 000-fold, which essentially yielded a poor but penicillin-resistant enzyme.[170] The H298K and H298Q mutant enzymes exhibited reduced penicillin-binding rates and hydrolytic activities, but the transpeptidation activity suffered an even more important decrease, which indicated a possible specific involvement of this residue in the interaction with acceptor molecules.[171] His-298 is the first residue of the HTG triad, which corresponds to the KT(S)G triad of β-lactamases, in some of which the K residue appears to be important for transition-state stabilization.[172] The results thus indicate that corresponding residues might not play exactly the same functional roles in homologous β-lactamases and DD-peptidases. A surprising result was the significantly increased β-lactamase activity of the W233S and W233L mutants, a residue apparently situated close to the binding site of the β-lactam C-6 (or C-7) acylamido side chain.[173] Other mutants, F58L, Y90N, F164A, D225E and D225N, did not exhibit significantly modified properties — if one excepts the markedly decreased stability of F164A.[170]

Replacement of the active-site Ser-330 of *E. coli* PBP 2 by a Cys yielded a protein unable to bind penicillin, and enzymatically inactive according to genetic complementation experiments.[174] Similar results were obtained when Lys-333 was replaced by Arg, His, Glu, Gln or Asp. Substitution of Lys-544 (in the KTG motif) by the same residues gave identical results — if one excepts the slight penicillin-binding activity of the K544R mutant.[174] Residue Asp-447 might correspond to the first Glu of the EXEXN motif in class A β-lactamases. The D447E mutant retained both penicillin-binding and genetic complementation properties, while mutants D447A and D447K lost both of them. Conversely, penicillin-binding capacity was retained, and genetic complementation lost in mutants D447N and D447Q.

Conversion of the active-site Ser-307 of *E. coli* PBP 3 into a Cys caused cell filamentation and was lethal,[175] but the protein retained a slight penicillin-binding activity.[176]

Unfortunately, the absence of accurate kinetic parameters for the penicillin-*E. coli* PBPs interactions makes the analysis of the results somewhat uncertain.

5.8 Conclusions and perspectives

Although the reactions involved in the biosynthesis of peptidoglycan are now well identified, the genetic studies indicate that the shape of the bacterial cell is determined by the interplay between several enzymes and proteins. Interesting hypotheses have recently been proposed, but the mechanism of the transpeptidation reaction(s) — one of the pivotal elements for the analysis of the regulation phenomena — is still, after more than 25 years, a major 'black box' in the understanding of this specific prokaryotic pathway. The contribution of crystallography remains essential and the elucidation of the first structure of a penicillin-sensitive enzyme will represent an impatiently awaited breakthrough.

The only 3-D structural data obtained concerning a PBP — the *Streptomyces* R61 DD-peptidase — have already supplied a wealth of interesting results, among which the close structural relationship with the β-lactamases of classes A and C was the most striking, but a high resolution structure is not yet available. Soluble, functional forms of some HMW-PBPs can now be purified in suitable amounts and, hopefully, their structures will also be solved in the not too distant future. The comparison of the structures of PBPs exhibiting different sensitivities to β-lactam antibiotics, and the study of their catalytic and regulatory properties will also constitute a basis for the rational design of new inhibitory molecules.

In addition, this will require an integrated analysis of the various components of the resistance phenomenon, relying on the accurate determination of the kinetic parameters involved in antibiotic diffusion, β-lactamase action, PBP inactivation and autolysin activity. However, any new progress in chemotherapy is likely to elicit a clever response from bacteria, which have exhibited a seemingly unbridled imagination in escaping the lethal action of the numerous compounds that were progressively introduced in the chemotherapeutic arsenal; mutations in β-lactamases increased their activity against third-generation cephalosporins, resistant PBPs have been assembled, and the outer membrane has been modified. The struggle between scientists and pathogenic bacteria is far from being finished. This should perhaps not be considered as a discouraging situation. A better understanding of the fundamentals of bacterial physiology and biochemistry produces benefits well outside the clinical world and, more egoistically, biochemists and microbiologists can look forward to more interesting experiments to perform, and new fields to explore.

Acknowledgements

The authors wish to thank all their coworkers for access to unpublished data and fruitful discussions, and Mrs A. Labye, Mrs H. Pottier and Miss N. Riga for their invaluable assistance in preparing the manuscript.

The work in Liège was supported, in part, by the Belgian programme on Interuniversity Poles of attraction initiated by the Belgian State, Prime Minister's Office, Science Policy Programming (PAI no. 19), *Actions concertées* with the Belgian Government (conventions 86/91–90, 89/94-130), the *Fonds de la Recherche Scientifique Médicale* (contract no. 3.4537.88), a *Convention tripartite* between the *Région wallonne*, SmithKline Beecham, UK, and the University of Liège, and the *Institut pour l'Encouragement de la Recherche Scientifique dans l'Industrie et l'Agriculture* (IRSIA, Brussels). B. Joris is *Chercheur qualifié* of the *Fonds National de la Recherche Scientifique*, Brussels.

Dedication

The authors wish to dedicate this article to Professor J.-M. Ghuysen for his vast contribution to the field of bacterial cell-wall structure and penicillin-recognizing enzymes.

References

1. G.D. Shockman and J.F. Barrett, *Ann. Rev. Microbiol.* (1983) **37** 501–527.
2. D. Mirelman, in *Bacterial Outer Membrane* (ed. M. Inouye), John Wiley & Sons, New York (1980), pp. 115–166.
3. J.M. Ghuysen, *Bacteriol. Rev.* (1968) **32** 425–464.
4. H.H. Martin, *J. Gen. Microbiol.* (1964) **36** 441–450.
5. E. Wise and T. Park, *Proc. Natl. Acad. Sci. USA* (1965) **54** 75–81.
6. D.J. Tipper and J.L. Strominger, *Proc. Natl. Acad. Sci. USA* (1965) **54** 1133–1141.
7. J.M. Ghuysen and G.D. Shockman, in *Bacterial Membranes and Walls* (Ed. L. Leive), Marcel Dekker, New York (1973), pp. 37–130.
8. A. Tomasz, *Ann. Rev. Microbiol.* (1979) **33**, 113–137.
9. B.G. Spratt, *Proc. Natl. Acad. Sci. USA* (1975) **72**, 2999–3003.
10. J.M. Frère and B. Joris, *CRC Crit. Rev. Microbiol.* (1985) **11** 299–396.
11. C. Duez, C. Piron-Fraipont, B. Joris, J. Dusart, M. Urdea, J. Martial, J.M. Frère and J.M. Ghuysen, *Eur. J. Biochem.* (1987) **162** 509–518.
12. B. Granier, C. Duez, S. Lepage, S. Englebert, J. Dusart, O. Dideberg, J. Van Beeumen, J.M. Frère and J.M. Ghuysen, *Biochem. J.* (1992) **282** 781–788.
13. B. Joris, J. Van Beeumen, F. Casagrande, C. Gerday, J.M. Frère and J.M. Ghuysen, *Eur. J. Biochem.* (1983) **130** 53–69.
14. J.M. Frère, C. Duez, J.M. Ghuysen and J. Vandekerckhove, *FEBS Lett.* (1976) **70** 257–260.
15. S.G. Waley, *Sci. Prog.* (1988) **72** 579–597.
16. J.M. Ghuysen, *Ann. Rev. Microbiol.* (1991) **45** 37–67.
17. P. Palomeque-Messia, V. Quittre, M. Leyh-Bouille, M. Nguyen-Distèche, C.L. Gershater, I.K. Dacey, J. Dusart, J. Van Beeumen and J.M. Ghuysen, *Biochem. J.* (1992) (In Press).
18. H. Mottl and W. Keck, *Eur. J. Biochem.* (1991) (In Press).
19. M.E. Jackson and J.M. Pratt, *Mol. Microbiol.* (1987) **1** 23–28.

20. J.B. Nielsen, M.P. Caulfield and J.O. Lampen, *Proc. Natl. Acad. Sci. USA* (1981) **78** 3511–3515.
21. J.A. Kelly, J.R. Knox, H. Zhao, J.M. Frère and J.M. Ghuysen, *J. Mol. Biol.* (1989) **209** 281–295.
22. J.A. Kelly, O. Dideberg, P. Charlier, J. Wery, M. Libert, P. Moews, J. Knox, C. Duez, C. Fraipont, B. Joris, J. Dusart, J.M. Frère and J.M. Ghuysen, *Science* (1986) **231** 1429–1431.
23. B. Samraoui, B. Sutton, R. Todd, P. Artimyuk, S.G. Waley and D. Phillips, *Nature* (1986) **320** 378–380.
24. B. Joris, J.M. Ghuysen, G. Dive, A. Renard, O. Dideberg, P. Charlier, J.M. Frère, J. Kelly, J. Boyington, P. Moews and J. Knox, *Biochem. J.* (1988) **250** 313–324.
25. B. Joris, P. Ledent, O. Dideberg, E. Fonzé, J. Lamotte-Brasseur, J. A. Kelly, J.M. Ghuysen and J.M. Frère, *Antimicrob. Agents Chemother.* (1991) **35** 2294–2301.
26. Y.F. Zhu, I.H.A. Curran, B. Joris, J.M. Ghuysen and J.O. Lampen, *J. Bacteriol.* (1990) **172** 1137–1141.
27. C. Govardhan and R. Pratt, *Biochemistry* (1987) **26** 3385–3395.
28. M. Adam, C. Damblon, B. Plaitin, L. Chritiaens and J.M. Frère, *Biochem. J.* (1990) **270** 525–529.
29. J. Frère, J.M. Ghuysen, J. Degelaen, A. Loffet and H.R. Perkins, *Nature* (1975) **258** 168–170.
30. L. Varetto, F. De Meester, D. Monnaie, J. Marchand-Brynaert, G. Dive and J.M. Frère, *Biochem. J.* (1991) **278** 801–807.
31. J. Nakagawa, S. Tamaki, S. Tomioka and M. Matsuhashi, *J. Biol. Chem.* (1984) **259** 13937–13946.
32. H. Suzuki, Y. van Heijenoort, T. Tamura, J. Mizoguchi, Y. Hirota and J. van Heijenoort, *FEBS Lett.* (1980) **110** 245–249.
33. A.R. Zeiger, J.M. Frère and J.M. Ghuysen, *FEBS Lett.* (1975) **52** 221–225.
34. J.M. Ghuysen, M. Leyh Bouille, J.N. Campbell, R. Moreno, J.M. Frère, C. Duez, M. Nieto and H.R. Perkins, *Biochemistry* (1973) **12** 1243–1250.
35. M. Jamin, M. Adam, C. Damblon, L. Chritiaens and J.M. Frère, *Biochem. J.* (1991) **280** 499–506.
36. S. Pazhanisamy and R.F. Pratt, *Biochemistry* (1989) **28** 6875–6882.
37. R.R. Yocum, H. Amanuma, T.A. O'Brien, D.J. Waxman and J.L. Strominger, *J. Bacteriol.* (1982) **149** 1150–1153.
38. B.G. Spratt, *Eur. J. Biochem.* (1977) **72** 341–352.
39. P.E. Reynolds and H.A. Chase, in *Beta-lactam antibiotics* (Eds. M.R.J. Salton and G.D. Shockman), Academic Press, New York (1981), pp. 153–168.
40. J.W. Kozarich and J.L. Strominger, *J. Biol. Chem.* (1978) **253** 1272–1278.
41. D.J. Waxman and J.L. Strominger, *J. Biol. Chem.* (1979) **254** 12056–12061.
42. H.A. Chase, P.E. Reynolds and J.B. Ward, *Eur. J. Biochem.* (1978) **88** 275–280.
43. J.M. Ghuysen, J.M. Frère, M. Leyh-Bouille, M. Nguyen-Distèche and J. Coyette, *Biochem. J.* (1986) **235** 159–165.
44. M.P. Jevons, *Brit. Med. J.* (1961) **1** 124–125.
45. M. Adam, C. Damblon, M. Jamin, W. Zorzi, V. Dusart, M. Galleni, A. El Kharroubi, G. Piras, B.G. Spratt, W. Keck, J. Coyette, J.M. Ghuysen, M. Nguyen-Distèche and J.M. Frère, *Biochem. J.* (1991) **279** 601–604.
46. J.K. Broome-Smith, A. Edelman, S. Yousif and B.F. Spratt, *Eur. J. Biochem.* (1985) **147** 437–446.
47. A. Edelman, L. Bowler, J.K. Broome-Smith and B.G. Spratt, *Mol. Microbiol.* (1987) **1** 101–106.
48. S. Tamaki, S. Nakajima and M. Matsuhashi, *Proc. Natl. Acad. Sci. USA* (1977) **74** 5472–5476.
49. H. Suzuki, Y. Nishimura and Y. Hirota, *Proc. Natl. Acad. Sci. USA* (1978) **75** 664–668.
50. J.I. Kato, H. Suzuki and Y. Hirota, *Mol. Gen. Genet.* (1985) **200** 272–277.
51. T. Den Blaauwen, M. Aarsman and N. Nanninga, *J. Bacteriol.* (1990) **172** 63–70.
52. S. Asoh, H. Matsuzawa, F. Ishino, J.L. Strominger, M. Matsuhashi and T. Ohta, *Eur. J. Biochem.* (1986) **160**, 231–238.
53. H. Nagasawa, Y. Sakagami, A. Suzuki, H. Suzuki, H. Hara and Y. Hirota, *J. Bacteriol.* (1989) **171** 5890–5893.

54. S. Hayashi, H. Hara, H. Suzuki and Y. Hirota, *J. Bacteriol.* (1988) **170** 5392–5395.
55. L.D. Bowler and B.G. Spratt, *Mol. Microbiol.* (1989) **3** 1277–1286.
56. J. Bartholomé-De Belder, M. Nguyen-Distèche, N. Houba-Hérin, J.M. Ghuysen, I.N. Maruyama, H. Hara, Y. Hirota and M. Inouye, *Mol. Microbiol.* (1988) **2** 519–525.
57. H. Adachi, T. Ohta and H. Matsuzawa, *FEBS Lett.* (1987) **226** 150–154.
58. P.J. Hedge and B.G. Spratt, *FEBS Lett.* (1984) **176** 179–184.
59. H. Matsuzawa, S. Asoh, T. Ohta, S. Tamaki and M. Matsuhashi, *Agric. Biol. Chem.* (1980) **44** 2937–2941.
60. S. Tamaki, H. Matsuzawa and M. Matsuhashi, *J. Bacteriol.* (1980) **141** 52–57.
61. F. Ishino, W. Park, S. Tomioka, S. Tamaki, I. Takase, K. Kunugite, H. Matsuzawa, S. Asoh, T. Ohta, B.G. Spratt and M. Matsuhashi, *J. Biol. Chem.* (1986) **261** 7024–7031.
62. M. Matsuhashi, M. Wachi, F. Ishino, M. Ideda, Y. Okada, H. Jung, K.M. Ueki, S. Tomioka, A.N. Pankrushina and M.D. Song, in *Fifty Years of Penicillin Application — History and Trends* (eds H. Kleinhauf and H. von Döhren), Video Press, Prague (1992) (In Press).
63. F. Ishino, H.K. Jung, M. Ikeda, M. Doi, M. Wachi and M. Matsuhashi, *J. Bacteriol.* (1989) **171** 5523–5530.
64. B. Oliva, P.M. Bennett and I. Chopra, *Antimicrob. Agents Chemother.* (1989) **33** 1116–1117.
65. H. Mottl, P. Terpstra and W. Keck, *FEMS Microbiol. Lett.* (1991) **78** 213–220.
66. J.K. Broome-Smith, I. Ioannidis, A. Edelman and B.G. Spratt, *Nucl. Ac. Res.* (1988) **16** 1617.
67. J.K. Broome-Smith, *J. Gen. Microbiol.* (1985) **131** 2115–2118.
68. J.K. Broome-Smith and B.G. Spratt, *J. Bacteriol.* (1982) **152** 904–906.
69. S. Tamaki, J. Nakagawa, I.N. Maruyama and M. Matsuhashi, *Agric. Biol. Chem.* (1978) **42** 2147–2150.
70. M. Matsuhashi, Y. Takagaki, I.N. Maruyama, S. Tamaki, Y. Nishimura, H. Suzuki, U. Ogino and Y. Hirota, *Proc. Natl. Acad. Sci. USA* (1977) **74** 2976–2979.
71. M. Iwaya and J.L. Strominger, *Proc. Natl. Acad. Sci. USA* (1977) **74** 2980–2984.
72. E.W. Goodell and U. Schwarz, *J. Bacteriol.* (1983) **156** 136–140.
73. Y. Nishimura and Y. Hirota, *FEMS Microbiol. Lett.* (1980) **9** 219–221.
74. M.A. de Pedro and U. Schwarz, *Proc. Natl. Acad. Sci. USA* (1981) **78** 5856–5860.
75. B. Korat, H. Mottl and W. Keck, *Mol. Microbiol.* (1991) **5** 675–684.
76. G.A. Botta and J.T. Park, *J. Bacteriol.* (1981) **145** 333–340.
77. W. Kraus and J.V. Höltje, *J. Bacteriol.* (1987) **169** 3099–3103.
78. B.L. de Jonge, F.B. Wientjes, I. Jurida, F. Driehuis, J.T. Wouters and N. Nanninga, *J. Bacteriol.* (1989) **171** 5783–5794.
79. F.B. Wientjes and N. Nanninga, *Res. Microbiol.* (1991) **142** 333–334.
80. A.G. Pisabarro, R. Prats, D. Vázquez and A. Rodriguez-Tebar, *J. Bacteriol.* (1986) **168** *199–206.*
81. N. Nanninga, *Mol. Microbiol.* (1991) **5** 791–795.
82. E. Tuomanen, Z. Markiewicz and A. Tomasz, *J. Bacteriol.* (1988) **170** 1373–1376,.
83. J.V. Höltje and B. Glauner, *Res. Microbiol.* (1990) **141** 75–89.
84. H. Hara and H. Suzuki, *FEBS Lett.* (1984) **168** 155–160.
85. W. Keck, A.M. van Leeuwen, M. Huber and E.W. Goodell, *Mol. Microbiol.* (1990) **4** 209–210.
86. M. Wachi, M. Doi, S. Tamaki, W. Park, S. Nakajima-Iijima and M. Matsuhashi, *J. Bacteriol.* (1987) **169** 4935–4940.
87. M. Wachi, M. Doi, Y. Okada and M. Matsuhashi, *J. Bacteriol.* (1989) **171** 6511–6516.
88. K.J. Begg, A. Takasuga, D.H. Edwards, S.J. Dewar, B.G. Spratt, H. Adachi, T. Ohta, H. Matsuzawa and W.D. Donachie, *J. Bacteriol.* (1990) **172** 6697–6703.
89. Z. Markiewicz, J.K. Broome-Smith, U. Schwarz and B.G. Spratt, *Nature* (1982) **297** 702–704.
90. B.D. Beck and J.T. Park, *J. Bacteriol.* (1976) **126** 1250–1260.
91. M. Aldea, T. Garrido, C. Hernandez-Chico, M. Vicente and R. Kushner, *EMBO J.* (1989) **8** 3923–3931.
92. M. Matsuhashi, M. Wachi and F. Ishino, *Res. Microbiol.* (1990) **141** 89–103.
93. M. Vicente, P. Palacios, A. Dopazo, T. Garrido, J. Pla and M. Aldea, *Res. Microbiol.* (1991) **142** 253–257.

94. A. Tormo, J.A. Ayala, M.A. de Pedro, M. Aldea and M. Vicente, *J. Bacteriol.* (1986) **166** 985–992.
95. E. Bi, K. Dai, S. Subbarao, B. Beall and J. Lutkenhaus, *Res. Microbiol.* (1991) **42** 249–252.
96. M. Doi, M. Wachi, F. Ishino, S. Tomioka, M. Ito, Y. Sakagami, A. Suzuki and M. Matsuhashi, *J. Bacteriol.* (1988) **170** 4619–4624.
97. A.C. Robinson, J.F. Collins and W.D. Donachie, *Nature* (1987) **328** 766.
98. M. Ikeda, T. Sato, M. Wachi, H.K. Jung, F. Ishino, Y. Kobayashi and M. Matsuhashi, *J. Bacteriol.* (1989) **171** 6375–6378.
99. B. Joris. G. Dive, A. Henriques, P.J. Piggot and J.M. Ghuysen, *Mol. Microbiol.* (1990) **4** 513–517.
100. W.D. Donachie and K.J. Begg, *Res. Microbiol.* (1990) **141** 75–89.
101. T. Ogura, T. Tomoyasu, T. Yuki, S. Morimura, K.J. Begg, W.D. Donachie, H. Mori, H. Niki and S. Hiraga, *Res. Microbiol.* (1991) **142** 279–282.
102. R. D'Ari, A. Jaffé, P. Bouloc and A. Robin, *J. Bacteriol.* (1988) **170** 65–70.
103. T. Komano, R. Utsumi and M. Kawamukai, *Res. Microbiol.* (1991) **142** 269–277.
104. P. Bouloc, A. Jaffé and R. D'Ari, *EMBO J.* (1989) **8** 317–323.
105. P. Palomeque-Messia, S. Englebert, M. Leyh-Bouille, M. Nguyen-Distèche, C. Duez, S. Houba, O. Dideberg, J. Van Beeumen and J.M. Ghuysen, *Biochem. J.* (1991) **279** 223–230.
106. M. Nguyen-Distèche, M. Leyh-Bouille, S. Pirlot, J.M. Frère and J.M. Ghuysen, *Biochem. J.* (1986) **235** 167–176.
107. P.E. Reynolds, in *Antibiotic Inhibition of Bacterial Cell Surface Assembly and Function* (eds P. Actor, L. Daneo-Moore, M.L. Higgins, M.R.J. Salton and G.D. Shockman), American Society for Microbiology, Washington, D.C. (1988), pp. 343–351.
108. A.W. Wyke, J.B. Ward, M.V. Hayes and N.A.C. Curtis, *Eur. J. Biochem.* (1981) **119** 389–393.
109. N.A.C. Curtis, M.V. Hayes, A.W. Wyke and J.B. Ward, *FEMS Microbiol. Lett.* (1980) **9** 263–266.
110. W.P. Hammes, *Eur. J. Biochem.* (1976) **70** 107–113.
111. P.W. Wrezel, L.F. Ellis and F.C. Neuhaus, *Antimicrob. Agents Chemother.* (1986) **29** 432–439.
112. S.T. Shepherd, H.A. Chase and P.E. Reynolds, *Eur. J. Biochem.* (1977) **78** 521–532.
113. A. Rousset, M. Nguyen-Disteche, R. Minck and J.M. Ghuysen, *J. Bacteriol.* (1982) **152** 1042–1048.
114. F. Beise, H. Labischinski and P. Giesbrecht, *FEMS Microbiol. Lett.* (1980) **55** 195–202.
115. M. Shohayeb and I. Chopra, *J. Gen. Microbiol.* (1987) **133** 1733–1742.
116. W. Park and M. Matsuhashi, *J. Bacteriol.* (1984) **157** 538–544.
117. G.E.D. Jackson and J.L. Strominger, *J. Biol. Chem.* (1984) **259** 1483–1490.
118. T. Kobayashi, Y.F. Zhu, N. Nicholls and J.O. Lampen, *J. Bacteriol.* (1987) **169** 3873–3878.
119. J.O. Lampen, N.J. Nicholls, A.J. Salerno, M.J. Grossman, Y.F. Zhu and T. Kobayashi, in *Antibiotic Inhibition of Bacterial Cell Surface Assembly and Function* (Eds P. Actor, L. Daneo-Moore, M.L. Higgins, M.R.J. Salton and G.D. Shockman), American Society for Microbiology, Washington, D.C. (1988), pp. 481–487.
120. B. Joris, P. Ledent, T. Kobayashi, J.O. Lampen and J.M. Ghuysen, *FEMS Microbiol. Lett.* (1990) **70** 107–113.
121. S.J. Rowland and K.G.H. Dyke, *EMBO J.* (1989) **8** 2761–2773.
122. W. Tesch, C. Ryffel, A. Strässle, F.H. Kayser and B. Berger-Bächi, *Antimicrob. Agents Chemother.* (1990) **34** 1703–1706.
123. D.F.J. Brown and P.E. Reynolds, *FEBS Lett.* (1980) **122** 275–278.
124. C.E. Buchanan and J.L. Strominger, *Proc. Natl. Acad. Sci. USA* (1973) **72** 1816–1820.
125. T.J. Dougherty, A.E. Koller and A. Tomasz, *Antimicrob. Agents Chemother.* (1980) **18** 730–737.
126. A.F. Giles and P.E. Reynolds, *Nature* (1979) **280** 167–168.
127. R. Hakenbeck, M. Tarpay and A. Tomasz, *Antimicrob. Agents Chemother.* (1980) **17** 364–371.
128. P.B. Percheson and L.E. Bryan, *Antimicrob. Agents Chemother.* (1980) **18** 390–396.
129. B.G. Spratt, *Nature* (1978) **274** 713–715.

130. S. Zighelboim and A. Tomasz, *Antimicrob. Agents Chemother.* (1980) **17** 434–442.
131. P.J. Hedge and B.G. Spratt, *Eur. J. Biochem.* (1985) **151** 111–121.
132. P.J. Hedge and B.G. Spratt, *Nature* (1985) **318** 478–480.
133. P. Taschner, N. Ypenburg, B.G. Spratt and C.L. Woldringh, *J. Bacteriol.* (1988) **170** 4828–4837.
134. F. Malouin and L.E. Bryan, *Antimicrob. Agents Chemother.* (1986) **30** 1–5.
135. J.G. Cannon and P.F. Sparling, *Ann. Rev. Genet.* (1984) **38** 111–113.
136. H. Faruki and P.F. Sparling, *Antimicrob. Agents Chemother.* (1986) **30** 856–860.
137. T.J. Dougherty, *Antimicrob. Agents Chemother.* (1986) **30** 649–652.
138. C.G. Dowson, A.E. Jephcott, K.R. Gough and B.G. Spratt, *Mol. Microbiol.* (1989) **3** 35–41.
139. J.A. Brannigan, I.A. Tirodimos, Q.Y. Zhang, C.G. Dowson and B.G. Spratt, *Mol. Microbiol.* (1990) **4** 913–919.
140. B.G. Spratt, *Nature* (1988) **332** 173–176.
141. B.G. Spratt, Q.Y. Zhang, D.M. Jones, A. Hutchison, J.A. Brannigan and C.G. Dowson, *Proc. Natl. Acad. Sci. USA* (1989) **86** 8988–8992.
142. R. Lujan, Q.Y. Zhang, J.A.S. Nieto, D.M. Jones and B.G. Spratt, *Antimicrob. Agents Chemother.* (1991) **35** 300–304.
143. C.G. Dowson, A. Hutchison and B.G. Spratt, *Mol. Microbiol.* (1989) **3** 95–102.
144. C.G. Dowson, A. Hutchison, J.A. Brannigan, R.C. George, D. Hansman, J. Linares, A. Tomasz, J.M. Smith and B.G. Spratt, *Proc. Natl. Acad. Sci. USA* (1989) **86** 8842–8846.
145. C.G. Dowson, A. Hutchison, N. Woodford, A.P. Johnson, R.C. George and B.G. Spratt, *Proc. Natl. Acad. Sci. USA* (1990) **87** 5858–5862.
146. T.J. Coffey, C.G. Dowson, M. Daniels, J. Zhou, C. Martin, B.G. Spratt and J.M. Musser, *Mol. Microbiol.* (1991) **5** 2255–2260.
147. G. Laible, R. Hakenbeck, M.A. Sicard, B. Joris and J.M. Ghuysen, *Mol. Microbiol.* (1989) **3** 1337–1348.
148. J. Garcia Bustos and A. Tomasz, *Proc. Natl. Acad. Sci. USA* (1990) **87** 5415–5419.
149. M.D. Song, M. Wachi, M. Doi, F. Ishino and M. Matsuhashi, *FEBS Lett.* (1987) **221** 167–171.
150. D.M. O'Hara, C.R. Harrington and P.E. Reynolds, *FEMS Microbiol. Lett.* (1989) **57** 97–104.
151. J. Pierre, R. Williamson, M. Bornet and L. Gutmann, *Antimicrob. Agents Chemother.* (1990) **34** 1691–1694.
152. B.R. Lyon and R. Skurray, *Microbiol. Rev.* (1987) **51** 88–134.
153. P.R. Matthews and P.R. Stewart, *J. Gen. Microbiol.* (1988) **134** 1455–1464.
154. K. Hiramatsu, E. Suzuki, H. Takayama, Y. Katayama and T. Yokota, *Antimicrob. Agents Chemother.* (1990) **34** 600–604.
155. K. Ubukata, R. Nonoguchi, M. Matsuhashi and M. Konno, *J. Bacteriol.* (1989) **171** 2882–2885.
156. K. Ubukata, R. Nonoguchi, M.D. Song, M. Matsuhashi and M. Konno, *Antimicrob. Agents Chemother.* (1990) **34** 170–172.
157. B. Berger-Bächi, L. Barberis-Maino, A. Strässle and F.H. Kayser, *Mol. Gen. Genet.* (1989) **219** 1263–1269.
158. B.J. Hartman and A. Tomasz, *Antimicrob. Agents Chemother.* (1986) **29** 85–92.
159. K. Murakami and A. Tomasz, *J. Bacteriol.* (1989) **171** 874–879.
160. B.L. de Jonge, H. de Lencastre and A. Tomasz, *J. Bacteriol.* (1991) **173** 1105–1110.
161. H. Maidhof, B. Reinicke, P. Blümel, B. Berger-Bächi and H. Labischinski, *J. Bacteriol.* (1991) **173** 3507–3513.
162. N.H. Georgopapadakou, S.A. Smith and D.P. Bonner, *Antimicrob. Agents Chemother.* (1982) **22** 172–175.
163. A. Tomasz, H.B. Drugeon, H.M. de Lencastre, D. Jabes, L. McDougall and J. Bille, *Antimicrob. Agents Chemother.* (1989) **33** 1869–1874.
164. S. Al Obeid, L. Gutmann and R. Williamson, *J. Antimicrob. Chemother.* (1990) **26** 613–618.
165. R. Fontana, R. Cerini, P. Longoni, A. Grossato and P. Canepari, *J. Bacteriol.* (1983) **155** 1343–1350.
166. G. Piras, A. El Kharroubi, J. Van Beeumen, E. Coeme, J. Coyette and J.M. Ghuysen, *J. Bacteriol.* (1990) **172** 6856–6862.

167. R. Williamson, C. le Bouguenec, L. Gutmann and T. Horaud, *J. Gen. Microbiol.* (1985) **131** 1933–1940.
168. A. El Kharroubi, P. Jacques, G. Piras, J. Van Beeumen, J. Coyette and J.M. Ghuysen, *Biochem. J.* (1991) **280** 463–469.
169. R. Fontana, *J. Antimicrob. Chemother.* (1985) **16** 412–416.
170. A.M. Hadonou, J.-M. Wilkin, L. Varetto, B. Joris, J. Lamotte-Brasseur, D. Klein, J.-M. Ghuysen and J.-M. Frère, *Eur. J. Biochem.* (1992) (In Press).
171. A.M. Hadonou, M. Jamin, M. Adam, B. Joris, J. Dusart, J.-M. Ghuysen and J.-M. Frère, *Biochem J.* (1992) **282** 495–500.
172. J. Brannigan, A. Matagne, F. Jacob, C. Damblon, B. Joris, D. Klein, B.G. Spratt and J.-M. Frère, *Biochem. J.* (1991) **278** 673–678.
173. C. Bourguignon-Bellefroid, J.-M. Wilkin, B. Joris, R.T. Aplin, C. Houssier, F.G. Prendergast, J. Van Beeumen, J.-M. Ghuysen and J.-M. Frère, *Biochem. J.* (1992) **282** 361–367.
174. H. Adachi, M. Ishiguro, S. Imajo, T. Ohta and H. Matsuzawa, *Biochemistry* (1992) **31** 430–437.
175. J.K. Broome-Smith, P.J. Hedge and B.G. Spratt, *EMBO* (1985) **4** 231–235.
176. N. Houba-Hérin, H. Hara, M. Inouye and Y. Hirota, *Mol. Gen. Genet.* (1985) **201** 499–504.

6 β-Lactamase: mechanism of action

S.G. WALEY

6.1 Introduction

Previous chapters have described how β-lactam antibiotics act, and how the β-lactam ring is built up. This chapter describes how the four-membered ring is broken down. This is a topic of chemical and clinical interest; β-lactamases are enzymes that bacteria produce to defend themselves against β-lactam antibiotics.[1] Since β-lactamases represent a main cause of antibiotic resistance, they can limit the effectiveness of antibiotics and contribute to the ravages of infectious disease. The mechanism of β-lactamase action represents a challenge with twin goals: (i) to understand how β-lactamases act, and how their action differs from that of the transpeptidases described in chapter 5; and (ii) to exploit this understanding. The exploitation will take the form of improved antimicrobially active β-lactams that are inert to β-lactamases, or improved β-lactamase inhibitors.

6.1.1 Methods used to study β-lactamase mechanisms

β-Lactamase mechanisms have been studied for some years[2,3] but the more powerful methods were not available until relatively recently. The two main approaches for characterizing mechanisms are structural and kinetic. The overall reaction is hydrolysis of the β-lactam ring (Figure 6.1). There are few substrates that are not β-lactams, but certain esters and especially thiol esters, have been shown to be substrates[4] (see also section 6.3.2). The rate of hydrolysis of β-lactams is usually measured either by the evolution of protons (at pH >6), or by any change in the absorbance. Defining the mechanism, then, consists of establishing the structures of intermediates, together with proving by kinetics that they do indeed lie on the reaction pathway. Moreover, the part played by the enzyme must be, as far as possible, defined. In particular, the identification of active site groups and a satisfying explanation of the roles that they play lie at the heart of enzyme mechanisms. When we understand β-lactamase mechanisms we shall be able to see why the acyl-enzymes from transpeptidases (chapter 5) are so much

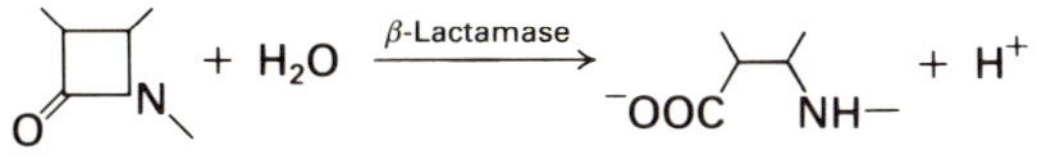

$$\text{(β-lactam ring)} + H_2O \xrightarrow{\beta\text{-Lactamase}} {}^{-}OOC \ldots NH- + H^+$$

Figure 6.1 The reaction catalysed by β-lactamases.

more stable than those from β-lactamases; this will have practical implications for the design of β-lactamase inhibitors (chapter 7).

The main high-resolution method used for studying β-lactamase structures has been crystallography. The impressive advances made in the last five years are described later. The application to studying mechanisms is, however, only just starting. The main difficulty may be stated simply: intermediates do not (normally) persist. Good substrates generate intermediates with lifetimes of milliseconds. One hope here lies in carrying out crystallography at low temperatures. In this way, a 'frozen snapshot' may be achieved.

At the present time the most accessible procedure is to study very slow substrates or inhibitors; in fact both have been used (for different β-lactamases) and these results are described later.

Kinetics are used to decide whether a postulated intermediate lies on the reaction pathway. There is also the important question of which intermediates, if any, will accumulate during the reaction. The answer to this question is governed by kinetic criteria. Structural methods will provide data hard to interpret if the high molecular weight component is a mixture of free enzyme and two intermediates. This is a real possibility with efficient enzymes and good substrates.

Kinetics also have an importance in their own right. Information on the energetics of β-lactamase action derives from kinetic data. Mechanistic information is provided by the 'effects' aspect of kinetics, such as pH-dependence, kinetic isotope effects, and the effects of viscosogens.

Over the last five years the knowledge of β-lactamase structures on the one hand, and the applications of site-directed mutagenesis on the other, have provided investigators with the wherewithal to *start* thinking sensibly about mechanisms. The, mostly unfulfilled, need for detailed and accurate knowledge about the kinetics of the mutants has now become very evident.

6.1.2 Lessons from amino acid sequences

A penetrating comparative analysis of β-lactamase (and peptidase) sequences has provided a general framework for comparing penicillin-recognizing proteins.[5,6] The alignment of class A β-lactamases forms the basis for the numbering system used here;[7] moreover it should not be forgotten that correct decisions about which residues are conserved depends on a correct alignment.

6.1.3 *Mechanistic classes of β-lactamases*

The main division of β-lactamases is into serine enzymes and zinc enzymes (Figure 6.2); the former have an active-site serine and function by a covalent acyl-enzyme mechanism, whereas the latter are metalloenzymes and appear to involve only non-covalent intermediates. Comparison of amino acid sequences has led to separating the serine enzymes into two classes, called class A and class C. All zinc β-lactamases are often lumped together as class B enzymes, although only some of them are known to have closely related amino acid sequences. The properties and primary structures of class A β-lactamases differ considerably.[8] Class A is a diverse class. On the other hand, the members of class C resemble each other closely — they are members of a close-knit family.[9,10] The OXA β-lactamases[11] are allocated to class D; their sequences differ much from those of other β-lactamases, but they are serine enzymes; they are not further discussed because little has been reported about their mechanisms.

6.1.4 *Arrangement of chapter*

The general approach taken is to describe, separately for classes A, B and C, the work on mechanisms, and then to follow this by an account of the structures of the active sites. On the whole, kinetics provides descriptions of mechanisms, but crystallography is required for explanations. The β-lactamases that are discussed in most detail are listed in Table 6.1.

6.2 Acyl-enzyme mechanism of β-lactamase action: class A β-lactamases

6.2.1 *Discovery of 'essential serine'*

There were many years of frustrating endeavours in the search for active site residues in β-lactamases. In hindsight, the enhanced reactivity towards

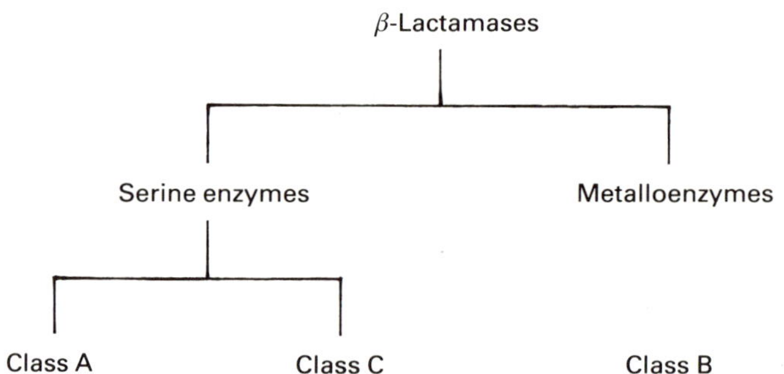

Figure 6.2 Mechanistic classification of β-lactamases. From Knott-Hunziker *et al. Biochem. J.* (1982) **207** 315–322.

Table 6.1 Main β-lactamases discussed in chapter 6.

Name used	Description	Type	Class
RTEM[a] β-lactamase	Plasmid-encoded periplasmic enzyme from *Escherichia coli*	Serine	A
PC1 β-lactamase	Extracellular enzyme from *Staphylococcus aureus* PC1	Serine	A
β-Lactamase I	Extracellular enzyme from *Bacillus cereus* 569/H	Serine	A
β-Lactamase II	Extracellular enzyme from *Bacillus cereus* 569/H	Zinc	B
P99 β-lactamase	Periplasmic enzyme from *Enterobacter cloacae*	Serine	C
Citrobacter β-lactamase	Periplasmic enzyme from *Citrobacter freundii*	Serine	C
Licheniformis β-lactamase	Extracellular enzyme from *Bacillus licheniformis* 749/C	Serine	A
Albus G β-lactamase	Extracellular enzyme from *Streptomyces albus* G	Serine	A

[a] RTEM-1 and RTEM-2 β-lactamases differ structurally by one residue, and there are no reported kinetic differences.

several acylating agents so conveniently displayed by serine proteases proved a misleading guide. The true lead came from the reaction of β-lactamase I with a particular β-lactam, 6-β-bromopenicillanic acid,[12,13] a reaction discussed in chapter 7. The residue labelled is a serine residue, now known as Ser-70. The essential nature of this serine was soon confirmed by findings from site-directed mutagenesis (section 6.2.6).

6.2.2 *Demonstration of acyl-enzyme intermediate*

The first clear evidence for an acyl-enzyme intermediate in β-lactamase action came from the interaction of the RTEM enzyme with the sluggish substrate cefoxitin.[14] The use of Fourier transform infrared measurements during the course of reaction showed that an ester was formed as an intermediate. Thus the acyl-enzyme mechanism (Figure 6.3) became established as the key concept in understanding β-lactamase action. In the acyl-enzyme mechanism there are three steps. In the first, enzyme and substrate combine reversibly to give a non-covalent enzyme–substrate complex. In the second the covalent acyl-enzyme is formed, and in the third it is hydrolysed. The

$$E + S \underset{k_{-1}}{\overset{k_1}{\rightleftharpoons}} ES \xrightarrow{k_2} E\text{-acyl} \xrightarrow{k_3} E + P$$

Figure 6.3 Acyl-enzyme mechanism.

rate constants for acylation and deacylation are k_2 and k_3, and if one is considerably smaller than the other then the corresponding step is regarded as the rate-determining chemical step (see Appendix). For the non-enzymic counterpart, namely the alcoholysis of penicillins, the rate-limiting step is the breakdown of the tetrahedral intermediate (chapter 4), and so we may surmise that, for such a step, the enzyme interacts particularly well with the transition state.

6.2.3 Substrate-induced inactivation: altered acyl-enzyme?

A characteristic and fascinating property of β-lactamases is their tendency to be inactivated during the course of the hydrolysis of certain substrates.[15–19] This property does in fact throw some light on mechanism, because it facilitates the isolation of an acyl-enzyme, albeit an altered one. The interaction of β-lactamase I and cloxacillin (**1**) leads to a covalent intermediate,[20,21] shown by electrospray mass spectrometry to have the expected molecular weight for an acyl-enzyme (S.J. Thornewell and R.T. Aplin, unpublished experiments). Moreover, the acyl-enzyme can be isolated by rapid centrifugal minigel chromatography;[21] it regenerates fully active enzyme, but with a rate constant for regain of activity considerably smaller than k_{cat}, showing that there is a branched mechanism (Figure 6.4). The main point of interest is how the acyl-enzyme is altered.

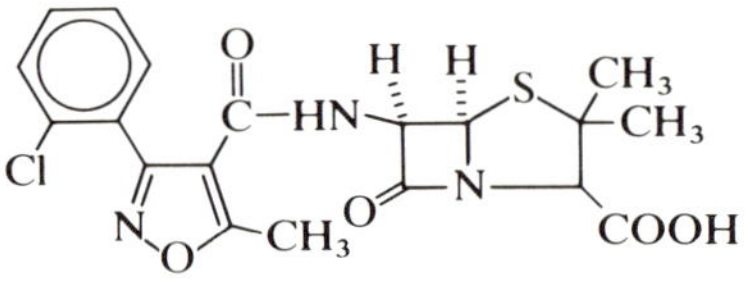

(**1**)

Tritium-hydrogen exchange experiments suggested that the conformation was more labile,[22] and this has been confirmed by the enhanced susceptibility to acid, heat and proteolysis. However, the secondary structure appeared unchanged,[21] and the same applies to the other system examined in detail, namely the PC1 β-lactamase and quinacillin.[23] The properties of the isolated acyl-enzyme appear to resemble those of the last intermediates

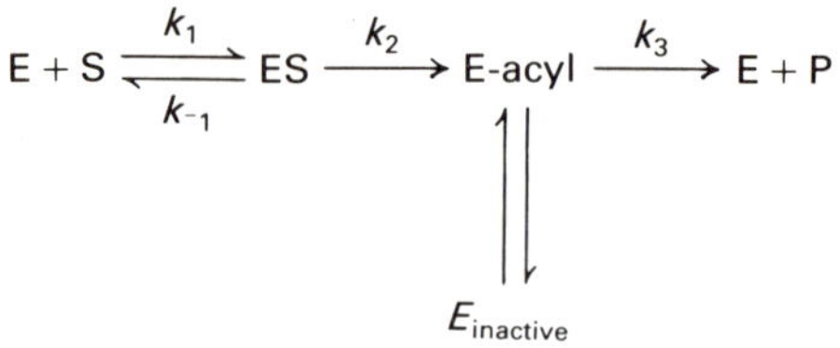

$$E + S \underset{k_{-1}}{\overset{k_1}{\rightleftharpoons}} ES \xrightarrow{k_2} E\text{-acyl} \xrightarrow{k_3} E + P$$
$$E\text{-acyl} \rightleftharpoons E_{inactive}$$

Figure 6.4 Branched acyl-enzyme mechanism.

in protein folding, such as those that lack a few final intramolecular interactions.[24]

Progress curves for the hydrolysis of methicillin by β-lactamase I show a burst, but instead a lag is observed with the *Streptomyces albus* G β-lactamase.[8] Even less is known about the molecular basis underlying this phenomenon.

6.2.4 *Substrate specificity*

A detailed comparison of four enzymes (the *licheniformis* β-lactamase, the *albus* G β-lactamase, and the β-lactamases from *Actinomadura* R39 and from *Streptomyces cacoi*) showed that although nearly all the penicillins tested were good substrates, the cephalosporins varied greatly, with k_{cat}/K_m varying from around $1\,mM^{-1}\,s^{-1}$ to $1\,\mu M^{-1}\,s^{-1}$.[8] When information for the other class A β-lactamases listed in Table 6.1 is included, the picture becomes even more complicated.[25]

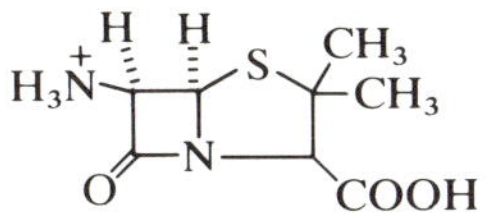

(2)

The specificity of one enzyme, β-lactamase I, has been studied systematically. Comparison of the rates of enzymic hydrolysis of penicillins with the base catalysed reaction gave an acceleration of about 10^8 fold: this corresponds to a hastening from decades to seconds. 6-Aminopenicillanic acid (6-APA) (**2**) is a relatively poor substrate, but the rate-enhancement brought about by the enzyme is still about 10^6 fold. Thus the side chains do not contribute greatly to binding. The variation of k_{cat}/K_m with length of side chain for the hydrolysis of alkyl penicillins increases to a maximum with the hexyl derivative. The free energy of transfer of the methylene group from water to the enzyme in the transition state was $1.46\,kJ\,mol^{-1}$, a relatively small value.[26] An even smaller value ($0.88\,kJ\,mol^{-1}$) was found for alkyl cephalosporins.[27] There is little information about the effects of side chain structure on individual rate constants, but for some penicillins with non-polar side chains the acylation rate constant varied less with structure than did the deacylation rate constant, as if the approach of water to the acyl-enzyme were being hindered.[28]

6.2.5 *Kinetics of acyl-enzyme mechanism*

6.2.5.1 Determination of rate constants from progress curves. The combination of single-turnover and steady-state measurements has enabled

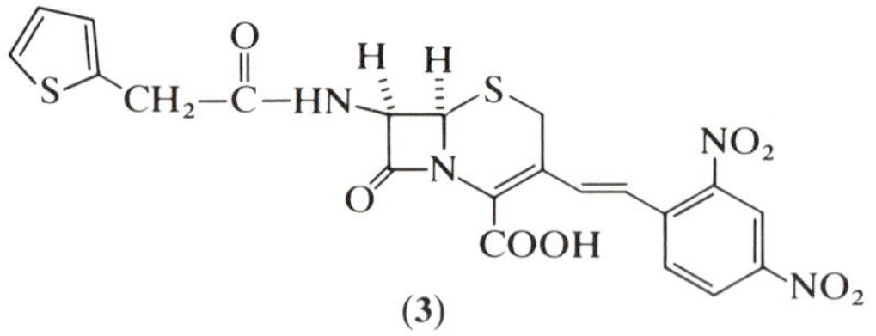

(3)

the rate constants for acylation and deacylation in the acyl-enzyme mechanism to be determined. For β-lactamase I and nitrocefin (**3**) k_2/k_3 was thus found to be 0.42.[28] This is probably the best way of determining k_2/k_3, because there is a check on the consistency of the values; here the K_m calculated from the single turnover experiments was 57 μM, and the K_m from steady-state experiments was found to be 60 μM. This is satisfactory agreement. However, nitrocefin and either β-lactamase I or PC1 gave no spectroscopic evidence for an intermediate in cryosolvents.[29,30] It is not clear why the acyl-enzyme was not detected spectroscopically — conversion of nitrocefin into the α-methyl ester alters the absorption spectrum; non-covalent intermediates were detected during the hydrolysis of nitrocefin by β-lactamase II[31] (see section 6.4.3.1). Probably there was enough change in the rate constants with medium and temperature to lead to the acyl-enzyme being less than 10–20% of the total enzyme (instead of 30%[28]), under which conditions it might well escape detection.

Spectroscopic evidence for intermediates has, however, been provided by the use of fluorescent substrates with the PC1 β-lactamase; quenching the reaction with acid showed that the intermediate was covalently bound.[32,33] The calculated values of the rate constants for acylation and deacylation were comparable for this fluorescent cephalosporin, but for a fluorescent penicillin the rate constant for acylation was about 20 times that for deacylation.[34] However, for a different β-lactamase, namely β-lactamase I, and for different cephalosporins, the rate constants for deacylation were considerably greater than those for acylation.[35]

6.2.5.2 Determination of rate constants from amount of acyl-enzyme. Penicilloyl esters react with mercuric chloride (the penamaldate reaction); the enhanced absorption at 282 nm has been used to measure the steady-state concentration of acyl-enzyme after the reaction had been quenched with acid.[36] The reaction was carried out for about 0.1 s. Four, class A β-lactamases have been examined by the penamaldate procedure. The fraction of enzyme that is present as acyl-enzyme was 0.6 ± 0.1 for the hydrolysis of benzylpenicillin catalysed by β-lactamase I, or the *licheniformis* β-lactamase, or the PC1 β-lactamase, or the RTEM β-lactamase[28,36] (also A. Matagne, unpublished experiments). Thus the rate constants for acylation and deacylation are comparable for these four, class A β-lactamases. This, together with the high values for the rate constants (a few thousand per second) was taken as evidence that these β-lactamases are fully efficient enzymes.[28]

With β-lactamase I and FAP — 6β-[(furylacryloyl)amino]-penicillanic acid — or dansylpenicillin, the extent of labelling was 0.8, but only if reaction was carried out in cryosolvent at −40°C rather than at ordinary temperatures.[37] With PC1 β-lactamase, however, the extent of labelling of the enzyme by FAP was 0.8, both at −40°C and at 20°C.[30]

For benzylpenicillin and the PC1 β-lactamase there is some variation in the reported extents of labelling, with values of 0.64 by the penamaldate method, and 0.86[28] and 1.01–1.20[34] by the use of tritiated substrate being given. The conditions differed: reaction for about one-tenth of a second at pH 7 and 20°C[28] or for three seconds at pH 9 and 0°C[34], and the methods of quenching the reactions also differed.

6.2.5.3 Role of diffusion in catalysis by class A β-lactamases. Efficient enzymes may catalyse reactions so well that the overall rate at low substrate concentrations is affected by, or even limited by, the frequency with which enzyme and substrate collide. This can be shown by the dependence of k_{cat}/K_m on the viscosity of the medium. When this procedure was applied to hydrolyses catalysed by β-lactamase I, by the PC1 β-lactamase, and by the RTEM β-lactamase, good substrates were indeed found to display such a dependence.[28,38] A valuable control was provided by the lack of variation of k_{cat}/K_m with viscosity for a poor substrate. This work provided values for k_1 and k_{-1}/k_2 which are otherwise hard to obtain.

When rapid quench, or single-turnover, methods to find k_2/k_3 are combined with methods to find k_{-1}/k_2 and k_1 then there is enough information to find all the four rate constants of the acyl-enzyme mechanism. Either acylation or deacylation can be rate-determining, but often neither is completely so.[28]

6.2.5.4 pH-dependence of kinetic parameters. The pH-dependence of k_{cat} and k_{cat}/K_m has been reported for several substrates and inhibitors.[37,39–42] The pH-dependence of k_{cat}/K_m gives pK_a values of about 5 and 8.5 which, on the simplest hypothesis, are assigned to groups in the free enzyme that govern activity; this constitutes one of the less reliable ways of identifying active-site groups. The simplest hypothesis, however, is unlikely to apply to the acid limb of the pH-dependence of k_{cat}/K_m. The significance of k_{cat}/K_m is likely to vary with pH; although k_{-1} is approximately equal to k_2 at neutral pH, it may well be much greater at lower pH values. This situation has been shown to give anomalous pK_a values.[43] Moreover, the structures (sections 6.2.6 and 6.3.4) show that the charged groups in the active site participate in such a complex network of hydrogen bonds and are so shielded from bulk solvent that perturbed pK_a values are to be expected.

The ionizations characterized by pK_a values of about 5 and 8.5 have been ascribed to a carboxyl group and an amino group, respectively;[42,44] their nature is discussed further in section 6.3.4.

Nearly all the information available relates to kinetic parameters rather than individual rate constants but, where there is information, the pH-dependence of k_2 and k_3 is similar, suggesting that the same ionisable groups are involved in acylation and deacylation.[33,36]

6.2.5.5 Solvent kinetic isotope effects. Solvent kinetic isotope effects have been measured for the hydrolysis of penicillins by class A β-lactamases.[28,40] The solvent deuterium isotope effects on k_{cat} and k_{cat}/K_m differ, indicating two steps in the reaction mechanism. Indeed, there was no significant isotope effect on k_{cat}/K_m in any of the reactions studied. The value of the isotope effect on k_{cat} was in the range 1.5 to 2.2; moreover, individual rate constants for acylation and deacylation gave similar values. These results show that a proton transfer is important for both processes. Whether one or two protons were 'in flight' in the transition states could not be determined.[28]

6.2.5.6 Inhibitors. Inhibitors of β-lactamases have been hard to find; since they are discussed in chapter 7, only a few aspects relevant to catalysis are mentioned here. Borates and boronic acids are competitive inhibitors,[45,46] and the pH-dependence, and kinetics, of their interaction with β-lactamases have been investigated.[41] The pH-dependence of the inhibition constant for β-lactamase I and $C_6H_5CH_2CONHCH_2B(OH)_2$ — phenylacetamidomethaneboronic acid — can be accounted for in terms of pK_a values of 4.7 and 8.2, and a mechanism in which the unionized boronic acid combines only with a form EH of the enzyme, but not with EH_2 or E. The curve has effectively zero slope at pH 6.5; this is equivalent to saying that no protons are given off (or taken up) on binding, unless there is compulsory proton uptake by some group on the protein. Similar results have been obtained for the RTEM β-lactamase and boric acid.[42] Since this is an equilibrium process, the assignment of the pK_a values to groups on the enzyme is more firmly based than when the pH-dependence of k_{cat}/K_m is similarly used. The force of this argument is somewhat weakened, however, by the fact that the kinetics revealed a two-step binding mechanism.

The crystallographic results (section 6.2.6.5) show that the boron in the complex with boric acid is covalently bound to the side chain of the active-site Ser-70. The arrangement around the boron appears trigonal, and so it may be that the explanation of the shape of the curve is that the boron is neutral rather than negatively charged as might have been expected. Aspects of boronic acids as inhibitors are discussed further in sections 6.2.6.5 and 6.3.3.4. Another inhibitor is α-methyl benzylpenicilloate: the K_i for β-lactamase I is 0.7 mM;[47] the ester is a (poor) substrate for class C β-lactamases (section 6.3.2).

6.2.6 *Structural studies on class A β-lactamases*

6.2.6.1 Comparison of tertiary structures. An important advance was reported when the arrangement of the secondary structure elements of a D-alanyl–D-alanine peptidase from *Streptomyces* R61 was compared with those of the *licheniformis* β-lactamase and β-lactamase I.[48,49] The discovery of an extensive region of common tertiary structure strongly suggested that the two groups of enzymes had evolved by divergence from a common ancestor. Subsequent studies on the R61 enzyme,[50] and on the β-lactamases which will be described (Table 6.2), show that the active sites are also related (see also chapter 5).

Table 6.2 Crystallographic studies on β-lactamases.

β-Lactamase	Resolution (Å)	R-factor	Reference
PC1	2.5	0.284	51
PC1	2.0 (refined)	0.163	52
Licheniformis	2.0 (refined)	0.15	54, 55
S. albus G	3.0		56
	1.7 (refined)		Unpublished
β-Lactamase I	1.7 (refined)	0.22	59
Citrobacter	2.0 (refined)	0.182	99
β-Lactamase II (Cd)	3.5		120

6.2.6.2 PC1 β-lactamase. The structure of the PC1 β-lactamase has been solved at 2.0 Å resolution;[51,52] the present account is confined to aspects relating to the enzymic mechanism. It may be noted at the outset that the electron density in the region of the active site is well defined, and indeed the active site residues have some of the lowest temperature factors.

The structure consists of two closely-associated domains, with the active site serine (Ser-70) lying in a crevice between the domains. Moreover, Ser-70 is at the N-terminus of a helix, a location where there is a formal positive charge associated with the helix. The next question is: what groups co-operate with this serine?

Two other active-site groups will now be introduced: Lys-73 and Glu-166. These interact with each other by a 2.8 Å salt bridge, and (especially Lys-73) are close to Ser-70. Also nearby are two more groups, Ser-130 and Lys-234. The role of these groups has been probed by model-building, with ampicillin (**4**) as substrate. The formation of an acyl-enzyme with Ser-70 meant that the substrate was positioned so that the thiazolidine carboxylate was removed from solvent and its charge stabilized by a salt bridge with Lys-234. Finally, the most satisfactory position for the substrate resulted in interactions

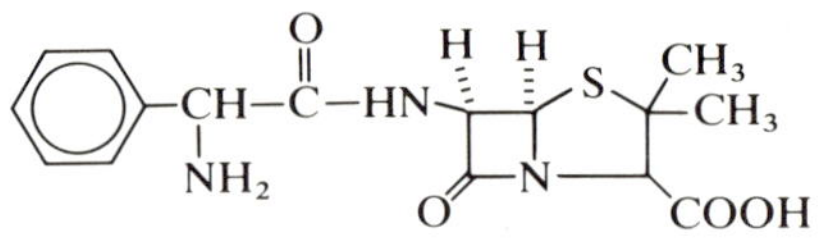

(4)

between the oxygen of the β-lactam carbonyl and the main chain NH groups of residues 70 and 237. This at once suggested an important factor in catalysis: the stabilization of the (full or partial) negative charge on the oxygen atom of the tetrahedral intermediate.

The next stage in constructing an hypothesis about catalysis entailed a detailed geometrical comparison with serine proteases. Here, in marked contrast to β-lactamases, a histidine activates the active-site serine, essentially by general base catalysis; the positional counterpart of this histidine in the β-lactamase is between the ammonium group of Lys-73 and the side chain oxygen atom of Ser-130. This does not make it immediately obvious just how Ser-70 is activated — we need a base for general base catalysis. It is not only the OH of Ser-70 that needs to be activated: the attacking water must be activated during the second phase of reaction, namely deacylation. The carboxylate of Glu-166 was suggested for this latter role. This suggestion has been widely accepted. There has been less agreement about the mode of activation of Ser-70; this will be further discussed later. The OH group of Ser-130 interacts with Ser-70, Lys-73 and Lys-234 but it is perhaps unexpectedly dispensable, as judged by the activity of the mutant in which Ser-130 is replaced by Ala (section 6.2.8).

There are several unusual structural features associated with the active site. One is a *cis* peptide bond between residues 166 and 167; this is regarded as important in defining the precise position of residues in the active site gully. Another feature is the unusual ϕ, ψ angles of residues 69 and 220, implying strain. When such features occur they are commonly connected with active sites.[53] Finally, there is an internal cavity beneath the active site that is filled with six solvent molecules; this, and a nearby cavity containing three solvent molecules, partially isolate the active site loop. This loop might be readily deformed; its instability may be lessened by the salt bridge between Arg-164 and Asp-179.

The analysis of penicillin-recognizing enzymes referred to earlier[6] noted four structural elements characterizing the active site. These were, for the PC1 β-lactamase: S(70)TSK, S(130)DN, E(166)IELN and D(233)KSG. Some of the interactions of components of the four structural elements have been noted above.

6.2.6.3 Licheniformis *β-lactamase.* The structure of the β-lactamase of *Bacillus licheniformis* 749/C has been solved, and refined, at 2.0 Å[54,55] (Table 6.2); the results bearing on the mechanism are now outlined.

Although the *licheniformis* β-lactamase was crystallized at pH 5.5, and the PC1 β-lactamase at (nominally) pH 8, there are no obvious active-site differences that might be attributable to this difference. The active-site groups Lys-73 and Glu-166 are close to Ser-70; the distances are shown in Figure 6.5, which also features the postulated position of a penicillin. The positioning of the penicillin was partly based on analogy with the R61 peptidase; here the details of substrate-binding have been elucidated,[50] an achievement facilitated by the relative stability of the acyl-enzyme. Most of the interactions depicted in Figure 6.5 show residues that are now familiar from the account of the PC1 β-lactamase. The direction of attack of the OH group of Ser-70 on the α-face of the β-lactam ring is clear. The penicillin side chain is hydrogen-bonded to the main chain carbonyl of residue 237, and is relatively exposed; the range of side chains in penicillin substrates is thus accounted for.

An analysis of the water molecules associated with the protein has yielded much interesting information; many of the water molecules are similarly

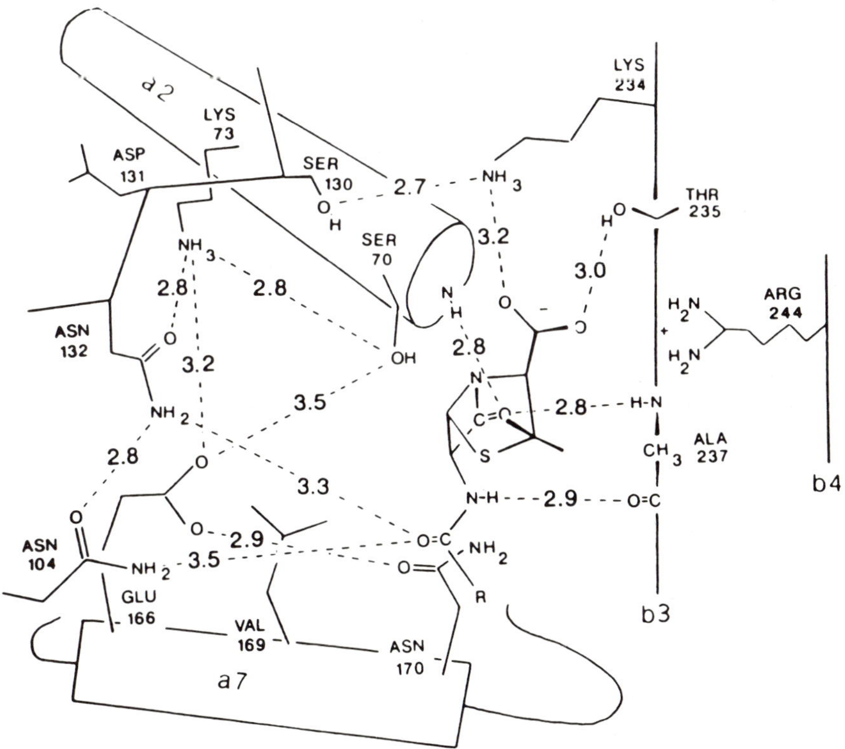

Figure 6.5 Schematic diagram of a penicillin in the active site of the *licheniformis β-lactamase*. From Moews *et al. Proteins: structure, function, and genetics* (1990) **7** 156–171.

situated in the two protein molecules of the asymmetric unit, and the active site contains at least seven water molecules. One that is thought to be retained when substrate is bound is a candidate for the water that effects deacylation.

The calculated solvent exposures of Lys-73, Glu-166 and Lys-234 were 0, 1.5 and 0 Å^2, respectively: future work will have to be directed towards the expected properties of (presumably) charged groups in such an eclectic environment.

6.2.6.4 Albus *G β-lactamase.* The refinement of the earlier[56] structure has yet to be published, but has been used in modelling to show, *inter alia*, the locations and interactions of ordered water molecules in the active site,[57] and the mechanism of acyl transfer (J. Lamotte-Brasseur, G. Dive, O. Dideberg, P. Charlier, J.M. Frère and J.M. Ghuysen, personal communication).

6.2.6.5 β-lactamase I. The crystallographic studies on β-lactamase I[49,58,59] described here concern the structure of the complex (section 6.2.5.6) observed after boric acid has diffused into the crystal. The most striking features are that (i) the O atom of the side chain of the active-site Ser-70 forms a covalent bond with the B atom; and (ii) the B has only two other

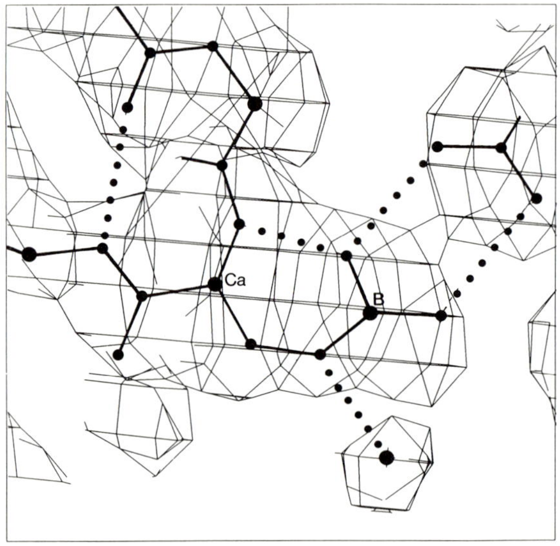

Figure 6.6 Electron-density map of the active site of β-lactamase I with boric acid covalently bound (C. Baguley, D.I. Stuart and D.C. Phillips, unpublished experiments). The Ca is the Cα of Ser-70, the group towards the north-east corner is the side chain of Asn-170, and the species towards the bottom is a water molecule.

groups bound covalently, and the three O atoms and the B atom are in the same plane (Figure 6.6). Thus, the arrangement around the boron atom is trigonal rather than tetrahedral. Here the observed structure mimics that of an acyl-enzyme. As a result, it is an important confirmation of the model-building described previously for other class A β-lactamases, in that there is a hydrogen bond (moreover an unusually short, 2.2 Å, one) between the NH of Ser-70 and the O atom of one of the B–OH groups; here we have direct evidence for the postulated 'oxy-anion hole'.

The trigonal boron explains the findings described in section 6.2.5.6 that, with a boronic acid, there is no proton evolution (at pH 6.5).

6.2.7 *Sketch for β-lactam hydrolysis by class A β-lactamases*

The kinetic and structural studies described in the preceding sections lead to a working hypothesis depicted in Figure 6.7. In this hypothesis, the role of Glu-166 is to deprotonate the hydroxyl group of Ser-70 as it attacks the β-lactam carbonyl group, and to protonate the departing nitrogen atom. Correspondingly, Glu-166 deprotonates the attacking water molecule in the hydrolysis of the acyl-enzyme. The role of Lys-73, although not so readily defined, is no less important, and it engages in a charge–charge interaction with Glu-166, both in the unliganded β-lactamases and in the boric acid complex of β-lactamase I. The foregoing is a minimalist picture, and important stereochemical aspects, and the suggested parts played by Lys-234 and Ser-130, can be gleaned from Figure 6.5.

6.2.8 *Altered enzymes: mutants and mutagenesis*

Site-directed mutagenesis (sometimes optimistically called protein engineering) has become an ardently exploited gambit of enzymologists. There are, however, some pitfalls that should be borne in mind;[60] one such

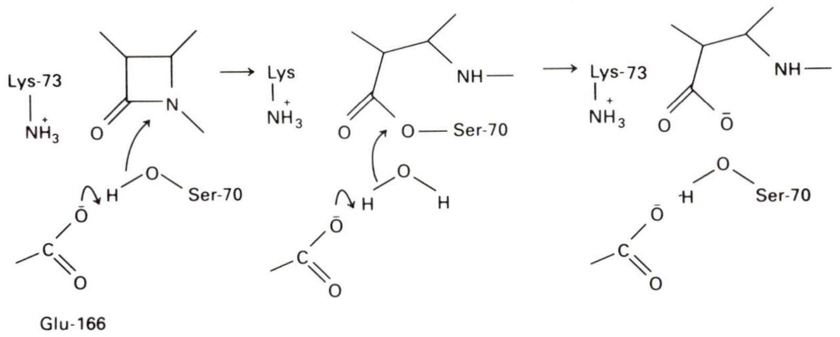

Figure 6.7 Schematic picture of the possible co-operation between Glu-166 and Ser-70 in the action of class A β-lactamases. From Gibson *et al. Biochem. J.* (1990) **272** 613–619.

was, in fact, encountered in work on a β-lactamase.[61] In the widely adopted nomenclature the single-letter code for amino acids is used, and E116D, for example, refers to a mutant in which glutamate 166 is replaced by aspartate.

The first use of this technique to probe protein function led to a variant of the RTEM β-lactamase in which the active-site serine (Ser-70) and the next residue (Thr-71) were interchanged.[62] This double mutant was inactive. There has been much further work on mutants of Ser-70 and Thr-71,[63,64] and in particular on the S70C mutant, which retains appreciable activity,[65,66] but has interestingly different properties.[42,66,67] The pH-dependence of the inactivation of the thiol β-lactamase (the S70C mutant) by alkylation and acylation has been used to construct hypotheses about the pK_a values, and indeed the nature, of active-site groups.[42,67] The mutant S70C of the *albus* G β-lactamase had greatly reduced activity (e.g. k_{cat}/K_m 0.015% of wild-type with ampicillin as substrate) and an unusual pH-dependence of k_{cat}/K_m; even the mutant S70A unexpectedly retained some activity.[68]

Some elegant experiments with the mutant K234H of the *albus* G β-lactamase[69] will now be described. Lys-234 was the site proposed for binding the carboxylate of the substrate (section 6.2.6). Both k_{cat} and k_{cat}/K_m had an 'alkaline' pK_a of about 6.4 for the hydrolysis of benzylpenicillin, some 3.5 units lower than the value for the wild-type enzyme. Moreover, the His residue in the mutant had a pK_a of 6.4, as found by nmr spectroscopy. So here the pK_a of the residue matches the pK_a in k_{cat}/K_m. Moreover, at pH 5.5 the k_{cat} for the mutant was only reduced two-fold. These results suggest that the enzyme is virtually fully functional as long as there is a positive charge at this position. Experiments on the mutants K234A and K234E of the *licheniformis* β-lactamase further stress the importance of Lys-234.[44]

The methyl ester of benzylpenicillin, a poor substrate for the wild-type *albus* G β-lactamase, was a better substrate for the R220L mutant.[57] Thus, when the negative charge in the substrate was absent, the lack of a positive charge in the enzyme was no longer deleterious. Model-building showed that Arg-220 interacted with Thr-216, which itself interacted with Thr-235. The carboxylate of the substrate is hydrogen-bonded to the side chain of Thr-235.[57]

The specificity of β-lactamases may be altered in mutants; thus, when Ala-237 in RTEM β-lactamase was replaced by threonine or asparagine the normal preference for penicillins over cephalosporins was reduced.[70,71] Indeed, the latter substitution gave enzyme with enhanced activity towards cephalosporins and diminished activity towards penicillins. Recall that the backbone NH of residue 237 forms part of the postulated oxyanion hole. Other replacements in this region show the importance of the 'SDN' loop, residues 130 to 132. Replacement of Asn-132 by serine in the *albus* G enzyme left activity towards penicillins relatively little changed, but greatly decreased activity towards cephalosporins.[72] For example, k_{cat}/K_m for benzylpenicillin and cephalothin were reduced two-fold and 2000-fold, respectively. Model-building showed that penicillins docked into the active site could form hydrogen bonds between the amide side chain of the substrate and the side chain of either Asn-132 or (in the mutant) Ser-132. Serine-130 turns out to be dispensable; thus S130A and S130G had k_{cat}/K_m 2.5% and 17% of wild-type;[73] these results seem to preclude a central role for Ser-130.

So far, there have been few studies of mutant β-lactamases in which individual rate constants have been measured (see Appendix). However, for two single mutations, and the corresponding double mutant, this has been done with β-lactamase I. The groups are Glu-166 and Lys-73.[74,75] The mutant E166D had greatly reduced activity. The rate constants for both acylation and deacylation were reduced about 2000-fold; it is noteworthy that both processes were affected similarly by this mutation. On the other hand, in the K73R mutant, and in the double mutant (E166D + K73R), the rate constants for acylation were decreased about 100-fold and 10000-fold respectively. Enzyme–substrate combination was also affected. Thus k_1 was decreased about 20- to 60-fold, and k_{-1} was decreased by about 500-fold. This was not expected — indeed it is sometimes *assumed* that these rate constants will not be altered. The mutants may well have a slightly altered conformation in which the active site is less accessible. The kinetic barrier diagram (Figure 6.8) for the E166D mutant shows that it is the transition states that are affected, or, in other words, that the partitioning of the intermediates is little altered.

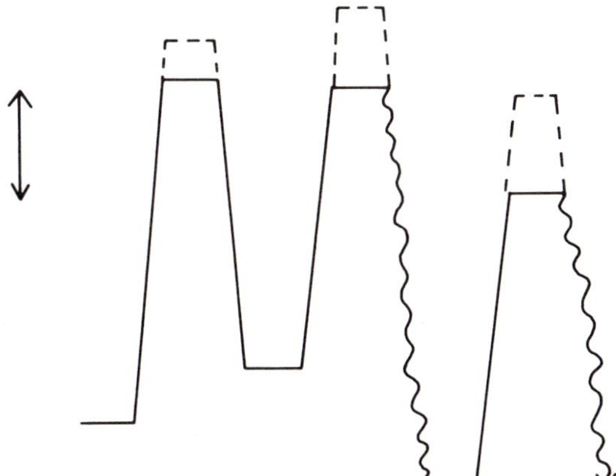

Figure 6.8 Kinetic barrier diagram for the hydrolysis of benzylpenicillin by wild-type β-lactamase I and mutant E166D. Unbroken lines denote wild-type β-lactamase I and broken lines the mutant E166D in this kinetic barrier diagram (Burbaum, Raines, Albery and Knowles *Biochemistry* (1989) **28** 9293–9305) for 1 μM substrate. The wavy lines indicate uncertainty about the level of the second minimum, the acyl-enzyme, and the factr that the reaction is in practice irreversible. The arrow denotes 20 kJ mol^{-1}. From Gibson *et al. Biochem. J.* (1990) **272** 613–619.

Mutants of the RTEM β-lactamase, in which Glu-166 is replaced by alanine, glutamine or asparagine, can be treated with radioactive benzylpenicillin; the putative acyl-enzymes were relatively long-lived.[76] The authors of the article averred that Glu-166 was inessential for acylation. The present author does not concur — the mutations could have reduced the rate constant for acylation by a large factor; acylation would still have been seen.

A number of unwelcome mutants of the RTEM β-lactamase have been encountered; these have arisen by chance and have been selected for, as they confer resistance to some of the newer β-lactam antibiotics. This often appears to have been achieved at the expense of activity against the best substrates.[77,78] The mutations lie in regions bordering the active site;[79,80] the β-lactamases have mostly not yet been studied in great detail. Site-directed mutagenesis of the RTEM β-lactamase has supplemented this work. The mutants E164K, R164S and E240K, and some double mutants, were examined and some striking increases in activity were observed, e.g. k_{cat} for ceftazidime increased from $0.0016\,s^{-1}$ to $3.4\,s^{-1}$ in R164S.[81] If Arg-164 is playing the same role in the RTEM β-lactamase as it does in the PC1 β-lactamase, where it forms a charge–charge interaction with a buried Asp, it is an important residue — but it is by no means obvious why the mutant has altered activity. A variety of other replacements has also been carried out, by novel methods, with the RTEM β-lactamase.[82–84]

Such are some of the kinetic findings that structural studies will have to explain; many questions have been raised, but few so far answered, by the results of site-directed mutagenesis.

6.3 The acyl-enzyme mechanism of β-lactamase action: class C β-lactamases

6.3.1 Identification of active-site serine

The labelling of the active site in class C β-lactamases was effected by the use of sluggish substrates as well as inhibitors;[85,86] the site labelled, again a serine, is Ser-64.

6.3.2 Substrate specificity of class C β-lactamases

The k_{cat} for cephalosporins is commonly larger than that for penicillins with class C β-lactamases. In the past, this difference from the class A enzymes led to class C β-lactamases being called cephalosporinases. This is misleading. Specificity is measured by k_{cat}/K_m not k_{cat}. With this criterion there is no such marked preference of class C β-lactamases.[10,87] Certain substrates, such as cloxacillin, have such low values of k_{cat} that they behave as, and are often regarded as, inhibitors; the acyl-enzyme accumulates and K_m is correspondingly low.[9,87] A non-β-lactam substrate is α-methyl benzylpenicilloate, which is hydrolysed by the *Pseudomonas* β-lactamase at about 1% of the rate of benzylpenicillin.[87] Other non-β-lactam substrates are discussed in section 6.3.3.2.

6.3.3 *Study of the acyl-enzyme intermediate*

6.3.3.1 Mass spectroscopy. The direct observation of an acyl-enzyme intermediate has been achieved by electrospray mass spectroscopy.[88] The P99 β-lactamase was observed to have a molecular mass of 39 203 ± 7 Da (calculated 39 206 Da). During the hydrolysis of carbenicillin by the class C P99 β-lactamase, a new set of peaks was observed. These had mass shift 378.7 Da (calculated for acyl-enzyme 378.4 Da), and their abundance was approximately 70% after 5 min and 20% after 10 min. These values are consistent with the kinetic data;[9] under somewhat different conditions, $1/k_{cat}$ was 10 min. Some other β-lactams gave more stable acyl-enzymes — aztreonam, oxacillin and flucloxacillin; on the other hand, acyl-enzymes were not detected when benzylpenicillin or ampicillin were used. These results suggest that the procedure used requires relatively long-lived acyl-enzymes.

6.3.3.2 Transfer reactions

(a) *Penicillins.* Certain differences have made studying the mechanism of class C β-lactamases easier; in particular, the intermediate acyl-enzyme was, surprisingly, found to react with methanol to give the penicilloyl α-methyl ester,[87] a reaction that could not be detected with the class A enzymes. This difference has yet to be explained. The advantage of a reaction in which methanolysis as well as hydrolysis takes place is that an intermediate can be demonstrated from measurements of how the rates of formation of the different products, acid and ester, vary with the concentration of methanol.

(b) *Depsipeptides.* The hydrolysis of depsipeptides (compounds in which the NH of a peptide is replaced by an oxygen atom) is discussed here because most of the work has been done with the class C P99 β-lactamase, although some activity was reported with class A and B enzymes.[4,89–93] Moreover, the aminolysis of depsipeptides is another example of a transfer reaction essentially confined to class C enzymes. The steady-state kinetics were complicated by parallel enzyme-catalysed hydrolysis and aminolysis. The anticipated enzyme–D-phenylalanine complex, however, was thought not to be formed. The kinetics of product inhibition demanded the existence of two separate sites on the enzyme at which the peptide product could bind. Only one of these sites apparently overlaps with the substrate binding site. The nature of the second site is less clear; it can be occupied by the peptide product of the reaction without inhibition of hydrolysis of the substrate. A similar feature had in fact already been suggested from measurements of the hydrolysis of Dnp-APA — 2,4-dinitrophenyl penicillanic acid — by β-lactamase P99 at −20.8°C.[29] The product bound to the enzyme, as shown by a change in A_{359}, but did not inhibit the hydrolysis. There must therefore be a site on the enzyme for the product, which is distinct from the site at which substrate binds leading to reaction.

(c) *An acyclic amide.* One acyclic amide is now known to be a β-lactamase substrate, namely N-(phenylacetyl)glycyl-D-aziridine-2-carboxylate (**5**).

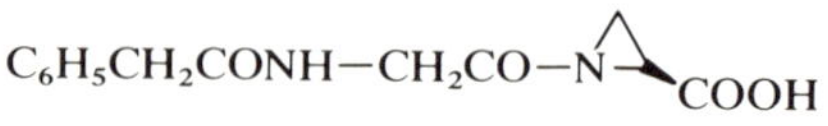

$C_6H_5CH_2CONH{-}CH_2CO{-}N$ (aziridine) COOH

(**5**)

This compound contains an amide bond in which the nitrogen atom is part of a three-membered ring (thus the amide is acyclic only on the 'carbonyl side'). The aziridine is a substrate for β-lactamases, and for the R61 DD-peptidase (chapter 1); e.g. k_{cat} for the P99 β-lactamase is $130\,s^{-1}$. It is a considerably better substrate for class A β-lactamases than are depsipeptides. Like the depsipeptides, the aziridine takes part in transfer reactions, and a similar complex machanism appears to hold.

6.3.3.3 pH-dependence. The pH-dependence of k_{cat}/K_m for the *Pseudomonas* β-lactamase on cephalosporin C (**6**) gave pK_a values of 5.2 and 9.7. The pH-dependence of k_{cat} was more complex, and there were pK_a values of 4.7, 7.4 and 10.1.[95] For deacetylcephalosporin C (**7**) and its lactone

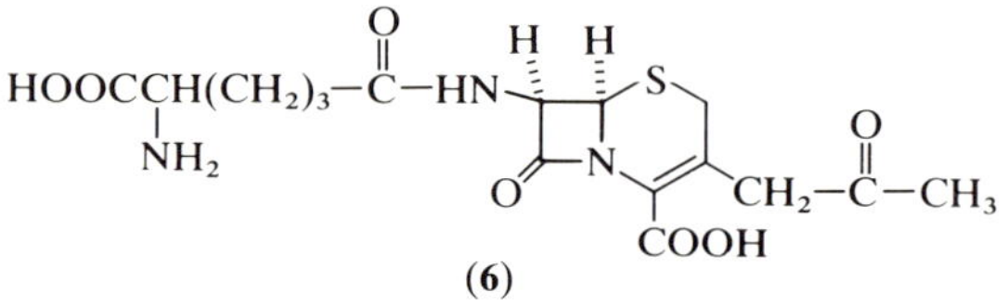

(**6**)

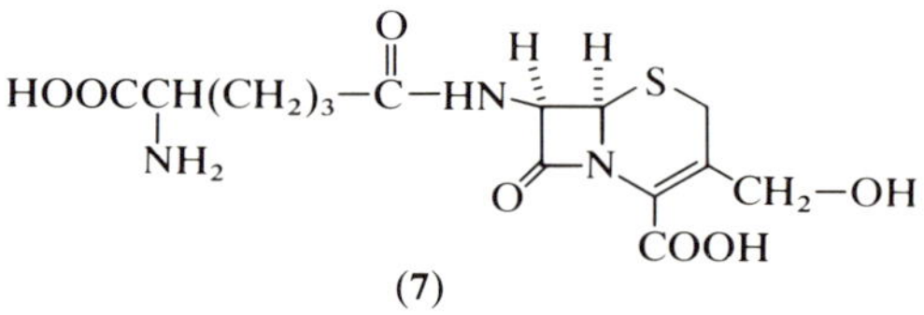

(**7**)

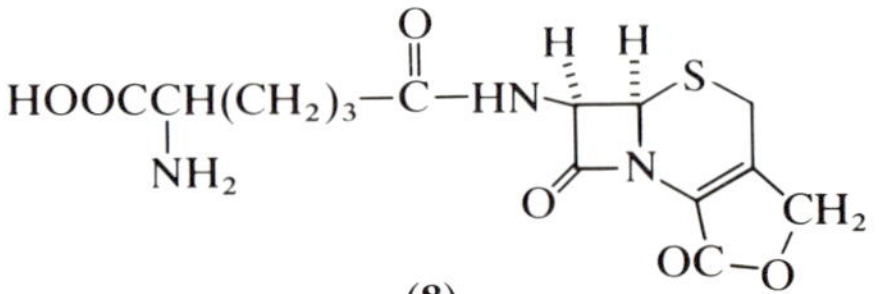

(**8**)

(**8**), the pH-dependence of k_{cat}/K_m for hydrolysis by the β-lactamase from *Enterobacter cloacae* 908R (similar to the P99 enzyme) differed. The importance of the carboxylate group in penicillins and cephalosporins is emphasized by the differences in pH-dependence when this group is lacking.

6.3.3.4 Inhibitors. Aromatic boronic acids are good competitive inhibitors of class C β-lactamases,[93,96] and react by a two-step binding process.[41] There was a resonance in the ^{11}B nmr spectrum of dansylamidophenylboronic acid at 8.2 ppm, which, in the presence of the P99 β-lactamase, changed to −17.4 ppm for the bound species; this chemical shift is similar to that (−16 ppm) of borate at pH 12. This suggests a tetrahedral (negatively charged) boron, or conceivably a trigonal boron interacting with a negative charge.[97] A phosphonate monoester inhibits the P99 β-lactamase, probably by phosphonylation of the active site serine;[98] this interesting finding opens the way for ^{31}P nmr investigations.

6.3.4 Structural studies on class C β-lactamases

There is only one high-resolution electron-density map for a class C β-lactamase, namely that of the enzyme from *Citrobacter freundii*.[99] There are two domains, and the active-site serine (identified as described previously) lies at the bottom of the gully between the domains. The larger domain contains a nine-stranded antiparallel β-sheet; the first five resemble those in other β-lactamases (and the R61 transpeptidase, see below). However, two of the other strands of this domain lack a counterpart in class A β-lactamase structures. The smaller domain contains seven α-helices; the active-site serine (residue 64) is at the N-terminal end of the α-helix, and the conserved Lys-67 is on the same side of this helix. The structure therefore resembles those of class A β-lactamases, although the amino acid sequences are very different.

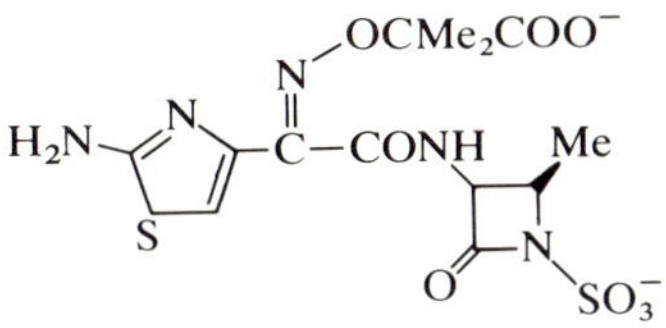

$OCMe_2COO^-$
N
H_2N N C–CONH Me
S N
O SO_3^-

(**9**)

Fortunately, the monocyclic β-lactam (monobactam) aztreonam (**9**), forms a relatively long-lived acyl-enzyme with class C β-lactamases,[9] which enabled the structure of the acyl-enzyme to be solved (at 2.5 Å).[99] The carbonyl group derived from the β-lactam interacted with two main chain NH groups (of Ser-64 and Ser-318), and was bound to the oxygen of Ser-64.

The side chain NH of aztreonam formed a hydrogen bond with the main chain carbonyl of Ser-318. There was an inferred rotation of about 70° about the C-3–C-4 axis when the β-lactam ring was cleaved.

Tyr-150, as the phenolate, is believed to act as a general base. The O^- of Tyr-150 deprotonates the OH of Ser-64 as this attacks the β-lactam carbonyl; subsequently the proton is transferred to the newly-formed NH group (here a sulphamate). In the (slower) second stage of deacylation, the O^- of Tyr-150 deprotonates the OH of the incoming water molecule. Again there has to be rotation about the C-3–C-4 bond to permit the water molecule to take the position previously occupied by the newly formed NH group.

The key feature of this hypothesis is that the pK_a of Tyr-150 is lowered from the usual value of about 9.5 to about 5.5. The presence of two, nearby, positively charged amino groups is held to be adequate to account for this decrease of pK_a. It would be useful to have independent confirmation that the Tyr group is indeed playing this role.

The currently favoured ideas thus ascribe comparable roles to Glu-166 in class A β-lactamases and Tyr-150 in class C β-lactamases. Although their Ca positions are different, when the Tyr-150 of class C swings down its OH is in the same general region as the carboxylate of Glu-166 in class A.

One implication of this comparison is that the pK_a of Glu-166 in class A β-lactamases, which forms a charge–charge interaction with Lys-73, would be expected to be <3, displaced downwards from an unperturbed value of 4.5; similarly, the pK_a of the lysine amino group would be expected to be perturbed upwards from about 9.5 to >11. Although this argument leaves out solvation, which is in fact the main driving force in ionization, solvation might be comparable in class A and class C enzymes, and so the argument is probably valid.

6.3.5 *Site-directed mutagenesis*

There have been relatively few reported examples of site-directed mutagenesis of class C β-lactamases, and they concern the enzyme from *Citrobacter freundii*. The replacement of the conserved Lys-67 (the active-site serine being residue 64) to obtain K67E, K67T and K67R showed that only the last of these retained appreciable activity.[100] The difference in the pH-profiles between wild-type enzyme and the K67R mutant suggests some perturbation in the active-site, and cannot reasonably be ascribed to a difference in pK_a values of the lysine and arginine residues. Another series of mutants of the same enzyme comprised D217E, D217T and D217K; these were all fully active, and indeed had enhanced relative activity towards oxyiminocephalosporins.[101]

6.4 Metalloenzyme mechanism of β-lactamase action

6.4.1 Discovery of a zinc β-lactamase: β-lactamase II

Bacillus cereus was found to contain a second β-lactamase, distinguished from β-lactamase I by its ability to hydrolyse cephalosporin C;[102] the new enzyme was called β-lactamase II. This enzyme was found to require zinc ions for activity;[103] some other metal ions, such as cobalt and cadmium, could replace zinc.[104] These properties distinguished β-lactamase II from other β-lactamases known at the time; there are still relatively few zinc β-lactamases known. Except where otherwise stated in the following account, the β-lactamase II is from *B. cereus* 569/H/9 (the same strain that is used for preparing β-lactamase I). There is more than one metal-binding site in β-lactamase II, but only the active site will be discussed here.

6.4.2 Substrate specificity

A main characteristic of β-lactamase II is that it catalyses the hydrolysis of nearly all β-lactams, and that the k_{cat}/K_m is not very different for many penicillins and cephalosporins.[27,105–107] Moreover, the usual β-lactamase inactivators such as clavulanic acid or penicillanic acid sulphone do not inactivate β-lactamase II but instead are substrates. On the other hand, chelating agents inactivate β-lactamase II but not serine β-lactamases.

6.4.3 Role of metal

6.4.3.1 Kinetics. The action of β-lactamase II at low (sub-zero) temperatures has been investigated with two systems: (i) cobalt β-lactamase II and benzylpenicillin; and (ii) zinc β-lactamase II and nitrocefin.[31] Moreover, the former system was also studied by rapid-scanning stopped-flow spectroscopy; this furnished a welcome and seldom-provided test of the validity of measurements in mixed aqueous solvents; all the main features of the low-temperature work were displayed at ordinary temperatures.[108] The kinetics showed that there was a branched mechanism, with two intermediates, ES^1 and ES^2.

The nature of the intermediates was probed as follows. When the reaction (with [enzyme] > [substrate]) was stopped soon after mixing the low molecular weight component of ES^1 was substrate, not product. Thus ES^1 was a non-covalent complex. Similarly, after low-temperature chromatography, ES^2 could be isolated, and its stability to acid suggested that it too was a non-covalent complex.

The properties of cobalt lend themselves to study by circular dichroism and magnetic circular dichroism, as well as by electron paramagnetic resonance,[108,109] and cobalt β-lactamase II and the complexes referred to above were examined. The conclusion was that the properties of unliganded enzyme and ES^2 suggested that there were five ligands to the metal, whereas in ES^1 there were four.

6.4.3.2 Spectroscopic studies. It turned out, rather remarkably, that it was possible to determine not only the nature of the groups, but also their position in the sequence as determined by protein sequencing[110] and by nucleic acid sequencing.[111] This fortunate circumstance depended on three of the ligands being identified by ^{1}H nmr as histidine residues; these were free to titrate in the apoenzyme but did not titrate in the zinc enzyme.[112] The histidine residue acting as zinc ligands were identified by differential tritium exchange.[113] The same ligands interacted with cobalt, when this replaced zinc in the active-site.[114]

Further spectroscopic studies[115] reinforced the idea[104] that a thiol group (the protein contains only one residue of cysteine) was one of the groups interacting with the metal.

6.4.3.3 Chemical modification and site-directed mutagenesis. Group modification of β-lactamase II with a carbodi-imide plus a nucleophile led to inactivation; the group modified was identified as Glu-37.[116] However, when Glu-37 in β-lactamase II from *B. cereus* 5/B/6[117,118] is replaced with glutamine, the mutant enzyme confers resistance of *E. coli* to ampicillin or cephalosporin C to an undiminished extent.[119] Similarly, activity was retained in E212Q and, to a lesser extent, in H28N.

6.4.4 Structural studies on a metallo-β-lactamase

The crystallographic studies on β-lactamase II have utilized the enzyme in which cadmium replaced zinc; the crystals were grown from poly(ethylene glycol) and cross-linked with glutaraldehyde.[120] The electron-density map was interpreted at 3.5 Å; the histidine residues 86, 88 and 210 interacted with the cadmium, as did the thiol group of cysteine 168 (Figure 6.9). This interaction appeared relatively weak, with a Cd–S distance of approximately 4.5 Å; this fits in with other evidence that the metal is not bound tightly.[31,108] It is easy to see that a bulky group on Glu-37 (see previously) would preclude substrate binding; the postulated role of this glutamate as a general base is, however, excluded — it is too distant from the metal. The same also applies to Glu-212. By analogy with other zinc metalloenzymes, hydrolysis might be effected by a Zn/OH species. However, the details of catalysis by β-lactamase II are still comparatively obscure.

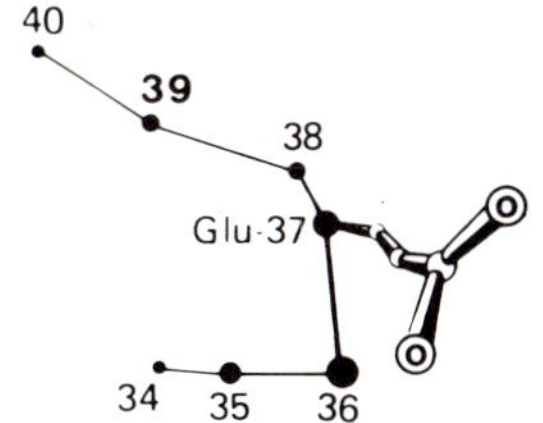

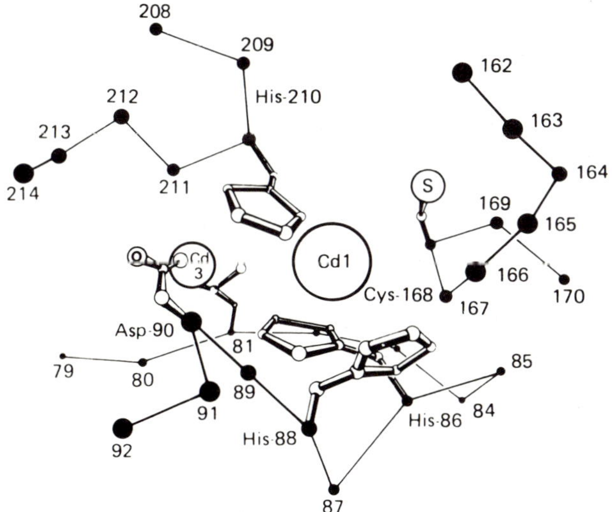

Figure 6.9 Structure of β-lactamase II in the region of the principal metal-ion site, Cd1. One of the minor sites, Cd3, is also shown. The course of segments of the polypeptide chain is represented by the positions of the α-carbon atoms, and side chains of residues involved in metal-binding — and of Glu-37 — are shown. From Sutton *et al. Biochem. J.* (1987) **248** 181–188.

6.4.5 Other metallo-β-lactamases

For many years β-lactamase II from *B. cereus* was the only β-lactamase known to require metal ions for activity. Now several such are recognized. Some are clearly related to β-lactamase II, but others appear not to be. There has not yet been much work reported on the mechanisms of these other metallo-β-lactamases.

Two strains of the important anaerobic pathogen *Bacteroides fragilis* produce a zinc-dependent β-lactamase; remarkably, the amino acid sequence is about 30% identical with that of β-lactamase II, and the key cysteine and histidine residues are present.[121,122]

The β-lactamase L1 from *Pseudomonas maltophilia*,[123] which is a tetramer, is a zinc enzyme.[124] There are several differences from β-lactamase II — the thiol group is inessential, and nickel can replace zinc. This β-lactamase resembles β-lactamase II in hydrolysing monocyclic β-lactams.

6.4.6 *Mechanism and medicine*

The most obvious clinical aspect of mechanistic work on β-lactamases concerns their inhibition, which is discussed in chapter 7. Perhaps the most directly relevant feature concerns the close similarity between serine β-lactamases and the penicillin-binding enzymes that are the targets for the antibiotics, and are discussed in chapter 5. The unexpected finding of a zinc β-lactamase in a common pathogen (see section 6.4.5) is a reminder of the medical relevance of fundamental studies.

6.4.7 *Mechanism and evolution*

We may regard mechanism as a clue to evolution — or evolution as a clue to mechanism. The dangers of trying to correlate two things we do not know very much about is perhaps outweighed by the fascination of both topics. Central to this discussion is the somewhat vexed question of the function of β-lactamases. Pollock,[125] who concluded that β-lactamases are detoxifying agents, also put forward a tentative scheme for the evolution of these enzymes. There is now, of course, much more information, and detailed phylogenetic trees for class A β-lactamases have been put forward.[126,127] These build on the important concept that there is a family of penicillin-recognizing enzymes, and that the members of the family contain an active-site serine.[5,128] Our present structural knowledge strongly suggests the evolutionary relationship between serine β-lactamases and the serine DD-peptidases described in chapter 5. This has been supported by the recruitment of peptidase activity into a β-lactamase.[29] Depsipeptides and the aziridine (section 6.3.3.2) form a connecting link between serine peptidases and serine β-lactamases. On the other hand, there are no signs of a relationship between zinc β-lactamases and zinc DD-peptidases. Both serine and zinc β-lactamases must have been selected for an inability to hydrolyse ordinary (planar) peptides, as well as an ability to hydrolyse the non-planar bond in β-lactams.[94]

Acknowledgements

I thank Professor L.N. Johnson for her hospitality, Professor J.-M. Frère for hospitality, advice and unpublished manuscripts, and Dr Fink, Dr Herzberg, Dr Knox and Dr Pratt for unpublished manuscripts. I also thank Dr Stuart for discussions and permission to cite unpublished work.

Appendix

The interpretation of the results of site-directed mutagenesis of serine β-lactamases is not as straightforward as at first appears. The difficulties arise because, in general, the four rate constants of acyl-enzyme mechanism are not measured. Instead k_{cat}, and K_m (or k_{cat}/K_m), are determined. It is worthwhile to consider the significance of the Michaelis–Menten parameters in more detail.

Comparison of Michaelis constants of wild-type and mutant

The acyl-enzyme mechanism (Figure 6.3) leads to the following expression for K_m:

$$K_m = \frac{k_{-1} + k_2}{k_1} \; \frac{k_3}{k_2 + k_3}$$

We usually want to infer something about the binding of substrate to enzyme from the value of K_m, therefore we require the affinity, or dissociation constant, K_s, defined by $K_s = k_{-1}/k_1$. It is easy to see that $K_m = K_s$ when (i) $k_{-1} \gg k_2$, or equivalently, $k_1 \gg k_{cat}/K_m$, and, as well (ii) $k_2 \ll k_3$. It is less well known that (iii) $K_m = K_s$ when $k_{-1} = k_3$.[28] When a mutant is being compared with wild-type β-lactamase then it is sufficient that conditions (i) and (ii), or condition (iii), should hold for *both* mutant and wild-type for the comparison of K_m values to provide information about the relative affinities. Condition (i) does not hold for β-lactamases and good substrates when it has been tested,[38,28] and condition (ii) is not generally fulfilled for class A β-lactamases and penicillins as substrates. Condition (iii) is unlikely to hold accurately for both wild-type and mutant, but it could be approximately fulfilled.

If a mutation is known *only* to affect binding, as shown by k_{cat} being unaltered, and if condition (i) holds, then the ratio (K_m of mutant/K_m of wild-type) does indeed give the corresponding K_s ratio. Now, condition (i) is more likely to hold for a poor substrate, and so this is an argument for basing mechanistic inferences on measurements with poor substrates.

The specificity constant

The specificity constant, k_{cat}/K_m, is often thought to be useful as a measure of transition state affinity. The difficulty is that the limits of k_{cat}/K_m are:

$$\frac{k_2}{K_s} < \frac{k_1 k_2}{k_{-1} + k_2} < k_1$$

so that k_{cat}/K_m for the wild-type enzyme might be nearly k_1, but if mutation causes a larger decrease in k_2 than in k_{-1} (quite a likely possibility) then k_{cat}/K_m for the mutant will have a different significance. If, however, condition (i) held for both wild-type and mutant, then the significance of k_{cat}/K_m would be unaltered. Here again then, the use of a poor substrate could simplify things.

Another aspect of the significance of k_{cat}/K_m in the acyl-enzyme mechanism is that it is given by:

$$\frac{k_{cat}}{K_m} = \frac{k_1 k_2}{k_{-1} + k_2}$$

and so k_3 is not represented. Thus a mutation that only affects k_3 does not alter k_{cat}/K_m. Clearly, k_{cat}/K_m does not measure the affinity of a β-lactamase for a transition state for deacylation.

The frequent finding that k_{cat} is greatly decreased, together with K_m being little changed, is incompatible with a mutation in which only k_3 is changed; if *only* k_3 is changed, K_m has to decrease as much as k_{cat} does.

These conclusions are based on the usual form of the acyl-enzyme mechanism, as given in Figure 6.3. More complicated forms, such as those containing two successive reversible steps, would of course give different expressions.

When double mutants are being studied, some general inferences about whether the mutations affect the same or different steps can be made. If both mutations affect the same step then the double mutant may show synergistic effects, whereas if the two mutations affect two different steps then the effects may be less than additive.[130]

Numerical examples

Some of the points made above are exemplified by the values in Table 6.3. Both mutants have the same Michaelis–Menten parameters, but the affinities differ by 100-fold. In mutant B the affinity is the same as that of wild-type enzyme, but the K_m has decreased 50-fold. In mutant B, k_{cat}/K_m is unchanged from the wild-type value because k_3 is the only altered rate constant and does not enter into the expression for k_{cat}/K_m. In mutant A, k_{cat}/K_m is again unaltered, but this time because the changes in k_{-1} and k_2 lead to cancellation.

Table 6.3 Rate constants and kinetic parameters.

	Wild type	Mutant A	Mutant B
k_1	40	40	40
k_{-1}	4000	40	4000
k_2	4000	40	4000
k_3	4000	4000	40
k_{cat}	2000	39.6	39.6
k_{cat}/K_m	20	20	20
K_m	100	1.98	1.98
K_s	100	1	100

The units of k_1 and k_{cat}/K_m are micromolar^{-1} s^{-1}, of the rate constants s^{-1}, and of K_m and K_s micromolar.

Rate-determining step

There are several possible definitions for a rate-determining step in an enzyme catalysed reaction, but here we consider the acyl-enzyme mechanism in the steady-state. When the concentration of substrate is much larger than K_m the rate at a given enzyme concentration is governed by k_{cat}; k_{cat} is given by $k_2k_3/(k_2 + k_3)$. A change in k_3 makes a proportional change in k_{cat} measured by

$$\partial \ln k_{cat}/\partial \ln k_3 = 1/(1 + k_3/k_2)$$

This is also the fraction (F) of total enzyme that is present as acyl-enzyme under these conditions. Also, the quotient k_3/k_2 is given by $(1 - F)/F$.

References

1. S.G. Waley, *Sci. Prog. (Oxford)* (1988) **72** 579–597.
2. E.P. Abraham, in *The Enzymes*, Vol. 1, (Eds J.B. Sumner and K. Myrback), Academic Press, New York (1951), pp. 1170–1185.
3. M.R. Pollock, in *The Enzymes*, 2nd ed. Vol. 4, (Eds P.D. Boyer, H. Lardy and K. Myrback), Academic Press, New York (1960), pp. 269–278.
4. M. Adam, C. Damblon, B. Plaitin, L. Christiaens and J.-M. Frère, *Biochem. J.* (1990) **270** 525–529.
5. B. Joris *et al.*, *Biochem. J.* (1988) **250** 313–324.
6. B. Joris, P. Ledent, O. Dideberg, E. Fonze, J. Lamotte-Brasseur, J.A. Kelly, J.-M. Ghuysen and J.-M. Frère, *Biochem. J.* (1991) In Press.
7. R.P. Ambler, A.F.W. Coulson, J.-M. Frère, J.-M. Ghuysen, B. Joris, M. Forsman, R.C. Levesque, G. Tiraby and S.G. Waley, *Biochem. J.* (1991) **276** 269–272.
8. A. Matagne, A.-M. Misselyn-Bauduin, B. Joris, T. Erpicum, B. Granier and J.-M. Frère, *Biochem. J.* (1990) **265** 131–146.
9. M. Galleni and J.-M. Frère, *Biochem. J.* (1988) **255** 119–122.
10. M. Galleni, G. Amicosante and J.-M. Frère, *Biochem. J.* (1988) **255** 123–129.
11. D. Mossakowska, N.A. Ali and J.W. Dale, *Eur. J. Biochem.* (1989) **180** 309–318.
12. R.F. Pratt and M.J. Loosemore, *Proc. Natl. Acad. Sci. USA* (1978) **75** 4145–4149.
13. V. Knott-Hunziker, B.S. Orlek, P.G. Sammes and S.G. Waley, *Biochem. J.* (1979) **177** 365–367.

14. J. Fisher, J.G. Belasco, S. Khosla and J.R. Knowles, *Biochemistry* (1980) **19** 2895–2901.
15. N. Citri, A. Samuni and N. Zyk, *Proc. Natl. Acad. Sci. USA* (1976) **73** 1048–1052.
16. Y. Klemes and N. Citri, *Biochim. Biophys. Acta* (1979) **567** 401–409.
17. R.H. Pain and R. Virden, in *Beta Lacatmases* (Eds J.M.T. Hamilton-Miller and J.T. Smith), Academic Press, London (1979), pp. 141–180.
18. J.-M. Frère, *Biochem. Pharmacol.* (1981) **30** 549–552.
19. S.G. Waley, *Biochem. J.* (1991) **279** 87–94.
20. P.A. Kiener, V. Knott-Hunziker, S. Petursson and S.G. Waley, *Eur. J. Biochem.* (1980) **109** 575–580.
21. A.L. Fink, K.M. Behner and A.K. Tan, *Biochemistry* (1987) **26** 4248–4258.
22. P.A. Kiener and S.G. Waley, *Biochem. J.* (1977) **165** 279–285.
23. K.C. Persaud, R.H. Pain and R. Virden, *Biochem. J.* (1986) **237** 727–730.
24. H. Christensen and R.H. Pain, *Eur. J. Biophys.* (1991) **19** 221–229.
25. R. Labia, M. Barthelemy, C. Farbre, M. Guionie and J. Peduzzi, in *Beta-lactamases* (Eds J.M.T. Hamilton-Miller and J.T. Smith), Academic Press, London (1979) pp. 429–442.
26. S.C. Buckwell, M.I. Page and J.L. Longridge, *J. Chem. Soc., Perkin Trans. 2* (1988) 1809–1813.
27. S.C. Buckwell, M.I. Page, J.L. Longridge and S.G. Waley, *J. Chem. Soc., Perkin Trans. 2* (1988) 1823–1827.
28. H. Christensen, M.T. Martin and S.G. Waley, *Biochem. J.* (1990) **266** 853–861.
29. S.J. Cartwright and S.G. Waley, *Biochemistry* (1987) **26** 5329–5337.
30. R. Virden, A.K. Tan and A.L. Fink, *Biochemistry* (1990) **29** 145–153.
31. R. Bicknell and S.G. Waley, *Biochemistry* (1985) **24** 6876–6887.
32. E.G. Anderson and R.F. Pratt, *J. Biol. Chem.* (1981) **256** 11401–11404.
33. E.G. Anderson and R.F. Pratt, *J. Biol. Chem.* (1983) **258** 13120–13126.
34. R.F. Pratt, T.S. McConnell and S.J. Murphy, *Biochem. J.* (1988) **254** 919–922.
35. R. Bicknell and S.G. Waley, *Biochem. J.* (1985) **231** 83–88.
36. M.T. Martin and S.G. Waley, *Biochem. J.* (1988) **254** 923–925.
37. S.J. Cartwright, A.K. Tan and A.L. Fink, *Biochem. J.* (1989) **263** 905–912.
38. L.W. Hardy and J.F. Kirsch, *Biochemistry* (1984) **23** 1275–1282.
39. S.G. Waley, *Biochem. J.* (1975) **149** 547–551.
40. L.W. Hardy and J.F. Kirsch, *Biochemistry* (1984) **23** 1282–1287.
41. I.E. Crompton, B.K. Cuthbert, G. Lowe and S.G. Waley, *Biochem. J.* (1988) **251** 453–459.
42. A.K. Knap and R.F. Pratt, *Biochem. J.* (1991) **273** 85–91.
43. M. Renard and A.R. Fersht, *Biochemistry* (1973) **12** 4713–4718.
44. L.M. Ellerby, W.A. Escobar, A.L. Fink, C. Mitchinson and J.A. Wells, *Biochemistry* (1990) **29** 5797–5806.
45. O. Dobozy, I. Mile, I. Ferencz and V. Csanyi, *Acta Biochim. Biophys. Acad. Sci. Hung.* (1971) **6** 97–105.
46. P.A. Kiener and S.G. Waley, *Biochem. J.* (1978) **169** 197–204.
47. M. Jones, S.C. Buckwell, M.I. Page and R. Wrigglesworth, *J. Chem. Soc., Chem. Commun.* (1989) 70–71.
48. J.A. Kelly *et al.*, *Science* (1986) **231** 1429–1431.
49. B. Samraoui, B.J. Sutton, R.J. Todd, P.J. Artymiuk, S.G. Waley and D.C. Phillips, *Nature* (1986) **320** 378–380.
50. J.A. Kelly, J.R. Knox, H. Zhao, J.-M. Frère and J.-M. Ghuysen, *J. Molec. Biol.* (1989) **209** 281–295.
51. O. Herzberg and J. Moult, *Science* (1987) **236** 694–701.
52. O. Herzberg, *J. Molec. Biol.* (1991) **217** 701–719.
53. O. Herzberg and J. Moult, *Proteins: Struct. Funct. Genet.* (1991) **11** 223–229.
54. P.C. Moews, J.R. Knox, O. Dideberg, P. Charlier and J.-M. Frère, *Proteins: Struct. Funct. Genet.* (1990) **7** 156–171.
55. J.R. Knox and P.C. Moews, *J. Molec. Biol.* (1991) **220** 435–455.
56. O. Dideberg, P. Charlier, J. Wèry, P. Dehottay, J. Dusart, T. Erpicum, J.-M. Frère and J.-M. Ghuysen, *Biochem. J.* (1987) **245** 911–913.
57. F. Jacob-Dubuisson, J. Lamotte-Brasseur, O. Dideberg, B. Joris and J.-M. Frère, *Protein Eng.* (1991) **4** 811–819.

58. R. Aschaffenburg, D.C. Phillips, B.J. Sutton, G. Baldwin, P.A. Kiener and S.G. Waley, *J. Molec. Biol.* (1978) **120** 447–449.
59. C. Baguley, *D. Phil. Thesis*, University of Oxford (1990).
60. P. Schimmel, *Acc. Chem. Res.* (1989) **22** 232–233.
61. M.J. Toth, E.J. Murgola and P. Schimmel, *J. Molec. Biol.* (1988) **201** 451–454.
62. G. Dalbadie-McFarland, L.W. Cohen, A.D. Riggs, C. Morin, K. Itakura and J.H. Richards, *Proc. Natl. Acad. Sci. USA* (1982) **79** 6409–6413.
63. G. Dalbadie-McFarland, J.J. Neitzel and J.H. Richards, *Biochemistry* (1986) **25** 332–338.
64. S.C. Schultz and J.H. Richards, *Proc. Natl. Acad. Sci. USA* (1986) **83** 1588–1592.
65. I.S. Sigal, B.G. Harwood and R. Arentzen, *Proc. Natl. Acad. Sci. USA* (1982) **79** 7157–7160.
66. I.S. Sigal, W.F. DeGrado, B.J. Thomas and S.R. Petteway, *J. Biol. Chem.* (1984) **259** 5327–5332.
67. A.K. Knap and R.F. Pratt, *Proteins: Struct. Funct. Genet.* (1989) **6** 316–323.
68. F. Jacob, B. Joris and J.-M. Frère, *Biochem. J.* (1991) **277** 647–652.
69. J. Brannigan, A. Matagne, F. Jacob, C. Damblon, B. Joris, D. Klein, B.G. Spratt and J.-M. Frère, *Biochem. J.* (1991) **278** 673–678.
70. A. Hall and J.R. Knowles, *Nature* (1976) **264** 803–804.
71. W.J. Healey, M.R. Labgold and J.H. Richards, *Proteins: Struct. Funct. Genet.* (1989) **6** 275–283.
72. F. Jacob, B. Joris, O. Dideberg, J. Dusart, J.-M. Ghuysen and J.-M. Frère, *Protein Eng.* (1990) **4** 79–86.
73. F. Jacob, B. Joris, S. Lepage, J. Dusart and J.-M. Frère, *Biochem. J.* (1990) **271** 399–406.
74. P.J. Madgwick and S.G. Waley, *Biochem. J.* (1987) **248** 657–662.
75. R.M. Gibson, H. Christensen and S.G. Waley, *Biochem. J.* (1990) **272** 613–619.
76. H. Adachi, T. Ohta and H. Matsuzawa, *J. Biol. Chem.* (1991) **266** 3186–3191.
77. K. Bush, *Antimicrob. Agents Chemother.* (1989) **33** 259–263.
78. K. Bush and S.B. Singer, *Infection* (1989) **17** 429–433
79. E. Collatz, R. Labia and L. Gutman, *Molec. Microbiol.* (1990) **4** 1615–1620.
80. M. Boissinot and R.C. Levesque, *J. Biol. Chem.* (1990) **265** 1225–1230.
81. J.A. Sowek, S.B. Singer, S. Ohringer, M.F. Malley, T.J. Dougherty, J.Z. Gougoutas and K. Bush, *Biochemistry* (1991) **30** 3179–3188.
82. D.K. Dube and L.A. Loeb, *Biochemistry* (1989) **28** 5703–5707.
83. A.R. Oliphant and K. Struhl, *Proc. Natl. Acad. Sci. USA* (1989) **86** 9094–9098.
84. C.J. Noren, S.J. Anthony-Cahill, M.C. Griffith and P.G. Schultz, *Science* (1989) **244** 182–188.
85. V. Knott-Hunziker, S. Petursson, G.S. Jayatilake, S.G. Waley, B. Jaurin and T. Grundström, *Biochem. J.* (1982) **201** 621–627.
86. B. Joris, J. Dusart, J.-M. Frère, J. Van Beeumen, E.L. Emanuel, S. Petursson, J. Gagnon and S.G. Waley, *Biochem. J.* (1984) **223** 271–274.
87. V. Knott-Hunziker, S. Petursson, S.G. Waley, B. Jaurin and T. Grundström, *Biochem. J.* (1982) **207** 315–322.
88. R.T. Aplin, J.E. Baldwin, C.J. Schofield and S.G. Waley, *FEBS Lett.* (1990) **277** 212–214.
89. R.F. Pratt and C.P. Govardhan, *Proc. Natl. Acad. Sci. USA* (1984) **81** 1302–1306.
90. C.P. Govardhan and R.F. Pratt, *Biochemistry* (1987) **26** 3385–3395.
91. S. Pazhanisamy, P. Govardhan and R.F. Pratt, *Biochemistry* (1989) **28** 6863–6870.
92. S. Pazhanisamy and R.F. Pratt, *Biochemistry* (1989) **28** 6870–6875.
93. S. Pazhanisamy and R.F. Pratt, *Biochemistry* (1989) **28** 6875–6882.
94. B.P. Murphy and R.F. Pratt, *Biochemistry* (1991) **30** 3640–3649.
95. R. Bicknell, V. Knott-Hunziker and S.G. Waley, *Biochem. J.* (1983) **213** 61–66.
96. T. Beesley, N. Gascoyne, V. Knott-Hunziker, S. Petursson, S.G. Waley, B. Jaurin and T. Grundström, *Biochem. J.* (1983) **209** 229–233.
97. J.E. Baldwin, T.D.W. Claridge, A.E. Derome, B.D. Smith, M. Twyman and S.G. Waley, *J. Chem. Soc., Chem. Commun.* (1991) 573–574.
98. R.F. Pratt, *Science* (1989) **246** 917–919.
99. C. Oefner, A. D'Arcy, J.J. Daly, K. Gubernator, R.L. Charnas, I. Heinze, C. Hubschwerlen and F.K. Winkler, *Nature* (1990) **343** 284–288.

100. K. Tsukamoto, K. Tachibana, N. Yamazaki, Y. Ishii, K. Ujiie, N. Nishida and T. Sawai, *Eur. J. Biochem.* (1990) **188** 15–22.
101. K. Tsukamoto, R. Kikura, R. Ohno and T. Sawai, *FEBS Lett.* (1990) **264** 211–214.
102. G.G.F. Newton and E.P. Abraham, *Biochem. J.* (1956) **62** 651–658.
103. L.D. Sabath and E.P. Abraham, *Biochem. J.* (1966) **98** 11c–13c.
104. R.B. Davies and E.P. Abraham, *Biochem. J.* (1974) **143** 129–135.
105. R.B. Davies, E.P. Abraham and J. Melling, *Biochem. J.* (1974) **143** 115–127.
106. E.P. Abraham and S.G. Waley, in *Beta-Lactamases* (Eds J.M.T. Hamilton-Miller and J.T. Smith), Academic Press, London (1979), pp. 311–338.
107. S.C. Buckwell, M.I. Page, S.G. Waley and J.L. Longridge, *J. Chem. Soc., Perkin Trans. 2* (1988) 1815–1821.
108. R. Bicknell, A. Schäffer, S.G. Waley and D.S. Auld, *Biochemistry* (1986) **25** 7208–7215.
109. J.L. Myers and R.W. Shaw, *Biochim. Biophys. Acta* (1989) **995** 264–272.
110. R.P. Ambler, M. Daniel, J. Fleming, J.-M. Hermoso, C. Pang and S.G. Waley, *FEBS Lett.* (1985) **189** 207–211.
111. M. Husain, A. Carlino, M. Madonna and J.O. Lampen, *J. Bact.* (1985) **164** 223–229.
112. G.S. Baldwin, A. Galdes, H.A.O. Hill, B.E. Smith, S.G. Waley and E.P. Abraham, *Biochem. J.* (1978) **175** 441–447.
113. G.S. Baldwin, S.G. Waley and E.P. Abraham, *Biochem. J.* (1979) **179** 459–463.
114. A. Galdes, H.A.O. Hill, G.S. Baldwin, S.G. Waley and E.P. Abraham, *Biochem. J.* (1980) **187** 789–795.
115. G.S. Baldwin, A. Galdes, H.A.O. Hill, S.G. Waley and E.P. Abraham, *J. Inorg. Biochem.* (1980) **13** 189–204.
116. C. Little, E.L. Emanuel, J. Gagnon and S.G. Waley, *Biochem. J.* (1986) **233** 465–469.
117. R.B. Davies, E.P. Abraham, J. Fleming and M.R. Pollock, *Biochem. J.* (1975) **145** 409–411.
118. H.M. Lim, J.J. Pène and R.W. Shaw, *J. Bact.* (1988) **170** 2873–2878.
119. H.M. Lim and J.J. Pène, *J. Biol. Chem.* (1989) **264** 11682–11687.
120. B.J. Sutton, P.J. Artymiuk, A. Cordero-Borboa, C. Little, D.C. Phillips and S.G. Waley, *Biochem. J.* (1987) **248** 181–188.
121. J.S. Thompson and M.H. Malamy, *J. Bact.* (1990) **172** 2584–2593.
122. B.A. Rasmussen, Y. Gluzman and F.P. Tally, *Antimicrob. Agents Chemother.* (1990) **34** 1590–1592.
123. Y. Saino, F. Kobayashi, M. Inoue and S. Mitsuhashi, *Antimicrob. Agents Chemother.* (1982) **22** 564–570.
124. R. Bicknell, E.L. Emanuel, J. Gagnon and S.G. Waley, *Biochem. J.* (1985) **229** 791–797.
125. M.R. Pollock, *Proc. Roy. Soc. (B)* (1971) **179** 385–401.
126. N. Pastor, D. Piñero, A.M. Valdés and X. Xoberón, *Molec. Microbiol.* (1990) **4** 1957–1965.
127. A. Huletsky, F. Couture and R.C. Levesque, *Antimicrob. Agents Chemother.* (1990) **34** 1725–1732.
128. J.-M. Ghuysen, *Ann. Rev. Microbiol.* (1991) **45** 37–67.
129. Y.-H. Chang, M.R. Labgold and J.H. Richards, *Proc. Natl. Acad. Sci. USA* (1990) **87** 2823–2827.
130. A. Kuliopulos, P. Talalay and A.S. Mildvan, *Biochemistry* (1990) **29** 10271–10280.

7 β-Lactamase: inhibition

R.F. PRATT

7.1 Introduction

The search for β-lactamase inhibitors began immediately after it was realized that β-lactamases existed, since it was recognized that such enzymes threatened the clinical application of the amazingly effective, then new, β-lactam antibiotics.[1–3] Early experiments did not, however, reveal any strikingly effective inhibitors, nor any general class of inhibitory molecules.[4,5] Although semisynthetic penicillins, resistant to β-lactamases, were subsequently developed and found, in some circumstances, to potentiate the effects of β-lactamase-susceptible penicillins,[6,7] these molecules, either alone or in combination, were unable to overcome the general problem of the β-lactamases.[8,9]

By 1970, it was very clear that the β-lactamases did present a real threat to the continued efficacy of β-lactam antibiotics.[10] Thus, during the period between 1970 and 1980, an intensive search for potent β-lactamase inhibitors was begun in many parts of the world. This involved, on one hand, the screening of natural sources and, on the other, molecular studies of the β-lactamases themselves. These efforts were rewarded by the discovery of a number of very effective mechanism-based inhibitors of β-lactamases, for example cefoxitin, clavulanic acid, thienamycin, penicillanic acid sulfone and 6-β-bromopenicillanic acid. These molecules and their analogs have, by and large, dominated both the design and the clinical standing of β-lactamase inhibitors since 1980.

There have been many reviews of the literature on β-lactamases and β-lactamase inhibitors, from many different points of view. The present one will deal primarily with mechanistic aspects of these inhibitors, i.e. the chemical mechanisms by which they inactivate β-lactamases. The author firmly believes that a close examination of these mechanisms of inhibition, taking into account the essential features of the structures of both the inhibitor and the enzyme active site, gives the most direct information on which to base future design of novel inhibitors. Since it is also now clear that the serine β-lactamase active site has much in common with the D-alanyl–D-alanine transpeptidase-carboxypeptidase (DD-peptidase) active site, mechanisms of inhibition of the former enzymes may well be relevant to those of the latter, and thus also to antibiotic design.

This review will commence with a brief overview of the features of the active sites of the β-lactam-recognizing enzymes that are relevant to β-lactamase inhibitor mechanisms. An analysis of the literature on β-lactamase inhibitors will then be presented, including major themes from prior to 1985, and more exhaustive coverage of developments since then. A short appraisal of the clinical situation and general conclusions follow. Mechanistic aspects of β-lactamase inhibitors have been previously reviewed,[11–14] but not, it seems, for some time. The present review is designed to complement and update an earlier one by the same author[15], and covers the literature until mid-1991.

7.2 β-Lactamase and DD-peptidase active sites: structure and mechanism

As described in chapter 6, β-lactamases have been most usefully classified as molecules into three[15] or, perhaps[16] four groups, on the basis of amino acid sequence homology. Of these, three classes, A, C and D, employ a double displacement mechanism of catalysis (Scheme 7.1), with an acyl-enzyme intermediate (**1**) and where the primary active site nucleophile is a serine hydroxyl group.

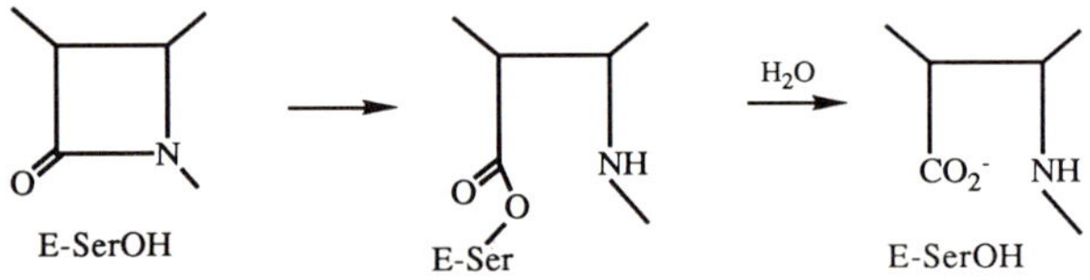

E-SerOH → (**1**) (E-Ser ester, NH) —H_2O→ CO_2^- NH, E-SerOH

Scheme 7.1

Crystal structures,[17,18] amino-acid sequence comparisons,[16,19] and functional studies[20] suggest that the class A, C and D or serine β-lactamases, together with bacterial cell wall DD-peptidases, comprise a superfamily[19] of β-lactam-recognizing enzymes of very similar active site structure and function. It now seems very likely that the serine β-lactamases are evolutionary descendants of the DD-peptidases that catalyse the cross-linking of the peptidoglycan of bacterial cell walls and are the primary sites of interaction of β-lactam antibiotics with bacteria.[17–21]

Apart from the essential active site serine residue (Ser-70 of the class A enzymes, according to the numbering system of Ambler[22,23]), one might expect to find other functional groups at these active sites that play an essential role in substrate binding and/or catalysis, and which therefore have been conserved. The class A β-lactamases have been studied most intensely in this regard, and a combination of evidence from crystal structures,[24–26] site-specific mutagenesis,[27–30] and chemical modification studies[31–33] suggests that the functional components of the active site are as represented in Figure 7.1. On the basis of model-building by the crystallographers,[24,25] and other evidence,[30,34,35] it has been suggested that the Lys-234 ammonium ion interacts with the carboxylate of the substrate, the carboxylate of Glu-166 plays a general base role — assisting nucleophilic attack of the Ser-70 hydroxyl group on the substrate — and the Lys-73 ammonium ion acts as an electrostatic catalyst. There is, however, no certainty here. It has been proposed, for example,[21,36] that Glu-166 only catalyses the hydrolysis of the acyl-enzyme but not the acylation reaction; the evidence for this too, however, is not completely unambiguous (compare the reported effect of the Glu-166-Asp mutation on acylation and deacylation rates in the *Bacillus cereus* I[28] and TEM[36] enzymes). Acylation and deacylation of Ser-70 is also probably assisted, as in serine proteinases,[37] by an oxyanion hole[24,25,38] consisting of two amide NH hydrogen bond donors. Other structural features that are believed to be involved in substrate binding have been identified, e.g. the β3 strand of β-sheet[24,25] and the 'SDN loop'.[24,25,39]

It seems likely, from the structural evidence now available, that residues homologous to Ser-70, Lys-73, Lys-234 (the latter is a histidine in the DD-peptidase of *Streptomyces* R61[19]) and the oxyanion hole play similar roles in class C β-lactamases,[40] and in the *Streptomyces* R61 DD-peptidase.[41,42] The situation with respect to Glu-166 is again much less clear. The general base role of Glu-166 may be taken over by the phenolate of Tyr-150 (in the amino acid sequence of the β-lactamase of *Citrobacter freundii*) in class C enzymes[40] and may be absent entirely in DD-peptidases[41,42] where facile hydrolysis of penicilloyl-enzymes does not occur (although that of peptides and depsipeptides does).

The evolution of a β-lactamase from a DD-peptidase has been proposed to involve the following.

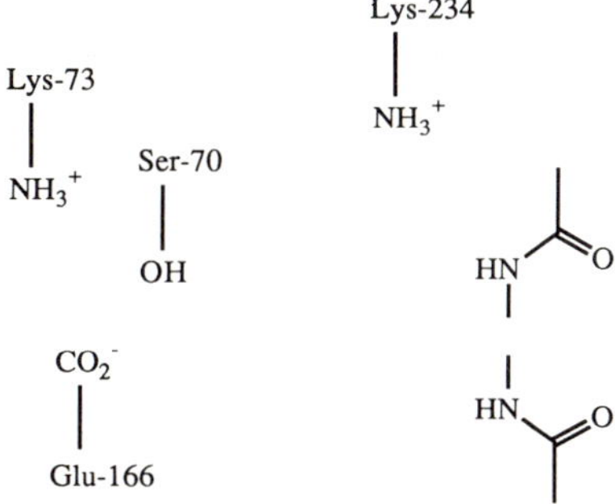

Figure 7.1 Functional components of the active sites of class A β-lactamases.

1. A change in the overall shape of the substrate binding pocket such that a molecule close to tetrahedral at the heteroatom of the scissile bond can be productively accommodated, but a planar peptide no longer can.[43]
2. Introduction of the capacity to catalyse the hydrolysis of a covalent penicilloyl-enzyme, perhaps by replacement of a hydrophobic residue in the binding site by a catalytic group, e.g. Glu-166, and an attendant water molecule.[42]

It is interesting that the reverse conversion, from a β-lactamase to a DD-peptidase, has been achieved, partly at least, by replacement of a 28 amino acid segment of the TEM β-lactamase with the corresponding region of *Escherichia coli* PBP5, a DD-peptidase.[44] It may be, however, that preventing such proteins from having some peptidase activity is more difficult than producing it.

The major conclusion from the above discussion is that very similar functional groups and structural features appear to be present as targets for the inhibitors of all β-lactam-recognizing enzymes of the serine class.

The class B β-lactamases, as zinc metalloenzymes, have little that is structurally or mechanistically in common with the serine enzymes. It might be supposed that their mechanism of action would resemble that of carboxypeptidase A, involving general base-assisted attack of a metal-coordinated water molecule on a metal-bound carbonyl group of the substrate,[45] but there is little conclusive evidence at this point. Although intermediates have been observed to accumulate during catalysis,[46,47] they do not appear to have substrate coordinated to the metal ion. There is evidence for a non-accumulating intermediate, with metal ion coordination to the carbonyl heteroatom of a β-lactam, from studies of the interactions between an 8-thionocephalosporin and *B. cereus* β-lactamase II containing metal ions of differing thiophilicity.[48] Although chemical modification studies[49] suggested that a particular carboxyl group (Glu-37) is essential for catalysis by *B. cereus* β-lactamase II, the crystal structure, at the resolution published to date,[50] does not seem to place this group close enough to the active site metal ion for it to be a catalytic functional group; Glu-212 was suggested as an alternative. Site-specific mutagenesis of Glu-37 and Glu-212 did not affect activity,[51] although conversion of Asp-90, one of the next closest acidic residues to the metal ion, to Asn or Glu gave rise to an inactive enzyme.[52] The mechanism therefore remains problematic.

7.3 Inhibitors of the serine β-lactamases

7.3.1 Mechanism-based inhibitors

Most specific and effective β-lactamase inhibitors to date are themselves β-lactams, i.e. substrate analogs. Although they can be described as poor substrates, this term is of course ambiguous, and one must clearly distinguish between poor substrates that are not recognized by an enzyme and thus are able to avoid it, and those that interact with it strongly and thus can inhibit its action against other β-lactams in solution. In principle, both are of importance in meeting the challenge of β-lactamases, but the present review will deal only with the latter group.

In general, the inert complexes formed on interaction of serine β-lactamases with the β-lactam inhibitors discovered to date are acyl-enzymes rather than non-covalent Michaelis-type complexes. Since the inhibitors have thus passed along a significant part of the normal substrate reaction coordinate, they are best thought of as mechanism-based inhibitors.[53,54] Since they are β-lactams and thus, in principle, complete substrates, they are not, in general, 'reaction coordinate inhibitors',[55] a class of incomplete substrates that possess only part of the substrate structure and which can progress to a certain point along the reaction coordinate but no further.

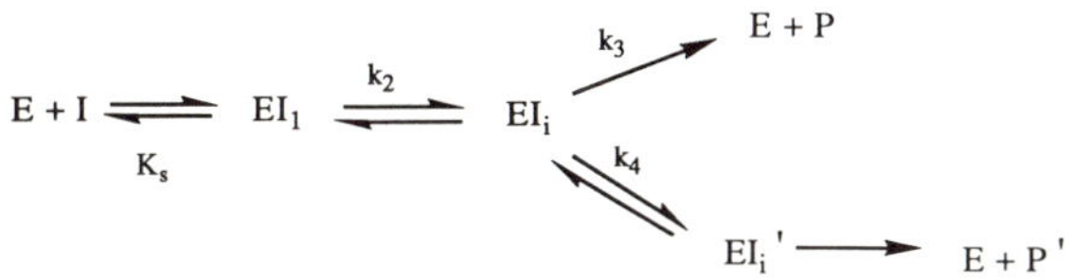

$$E + I \underset{K_s}{\rightleftharpoons} EI_1 \overset{k_2}{\rightleftharpoons} EI_i \xrightarrow{k_3} E + P$$

$$EI_i \overset{k_4}{\rightleftharpoons} EI_i' \longrightarrow E + P'$$

Scheme 7.2

The reaction pathway and reaction coordinate diagram for a mechanism-based inhibitor, I, are shown in Scheme 7.2 and Figure 7.2 respectively. A normal intermediate on the reaction coordinate of a substrate, EI_i, partitions between normal turnover, yielding E and P, and a reaction path not available to normal substrates and leading to a non-productive, or more slowly productive, complex EI_i'. If, in the steady state, a significant proportion of the total enzyme is present in the only slowly (or not at all) turning-over complex EI_i', the enzyme will be partly or essentially completely inhibited. In Scheme 7.2, P is the product of normal turnover of I, and P′ is the product, if any, of slow turnover of EI_i'; in many cases,

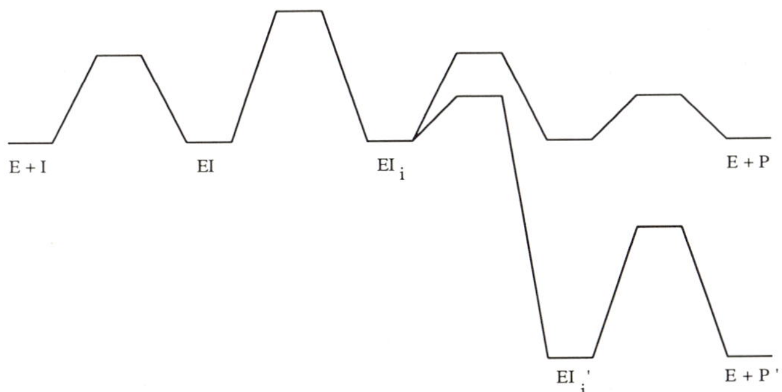

Figure 7.2 Reaction coordinate diagram for a mechanism-based inhibitor, I.

although not in the diagram shown, the most facile route from EI'_i to free enzyme will involve (slow) reversion to EI_i. The kinetics of the enzyme inactivation and reactivation are complicated.[56–58] Bush and Sykes[59] have made practical suggestions for approaching these situations.

In the mechanism-based inhibitors of β-lactamases, EI_i is the acyl-enzyme intermediate (**1**, Scheme 7.1) and EI'_i is a rearranged acyl-enzyme. There is, in general, and in the case of β-lactamases in particular, a wide variety of chemical reactions whereby an inert EI'_i can be derived from EI_i — some involve covalent chemistry, some non-covalent, some both; some involve crucial interactions between inhibitor and enzyme, and some do not. A classification of mechanism-based inhibitors, based on these differences, has been introduced,[15] which appears, to the present author at least, to have considerable merit, particularly with respect to β-lactamases where a wide range of mechanistically different mechanism-based inhibitors are known. Subsequent discussion of the mechanism-based inhibitors of β-lactamases will therefore follow this scheme of classification.

7.3.1.1 Active covalent (class 1 (a)) In this class, the formation of EI'_i involves creation of a reactive functional group in EI_i, which covalently, and usually irreversibly, modifies the enzyme. Although this class, comprising the most overt of the mechanism-based inhibitors, was the first identified and exploited, and although there are many examples based on the diversion of the acyl-enzymes of serine proteinases,[60] it has not yet provided any important β-lactamase inhibitors. The compound (**2**)[61] designed to follow the example of the *o*-halomethylphenyl ester and 6-halomethylcoumarin inhibitors of chymotrypsin,[62] did not, unfortunately, inhibit a variety of β-lactamases. More elaborate analogs were also unsuccessful,[63] although the strategy has apparently succeeded in an elastase inhibitor.[64]

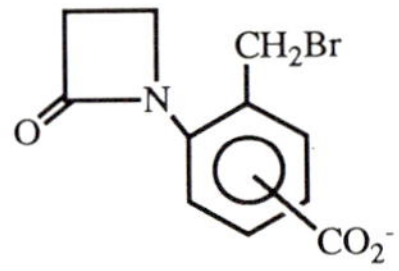

(2)

More recently, the N-nitroso-β-lactams (**3**) have been prepared[65] by analogy to the N-nitrosamide inhibitors of chymotrypsin first devised by White *et al.*[66] Although the parent β-lactams were not inhibitors, (**3a**) and (**3b**) showed inhibitory activity against several β-lactamases. Two mechanisms of inhibition can be readily envisaged (Scheme 7.3), one of which (A) would fall under this class of inhibition, while the other (B) would fall under class 2(a), described in section 7.3.1.3. Compound (**3b**) exhibited only transient inhibition, typical of a class 2(a) inhibitor (Scheme 7.3, B), while (**3a**) gave rise to a more irreversible inhibition and may thus represent an example of class 1(a) inhibition (Scheme 7.3, A). The compounds (**3**) appeared too unstable to spontaneous degradation in solution to have promise of practical application.

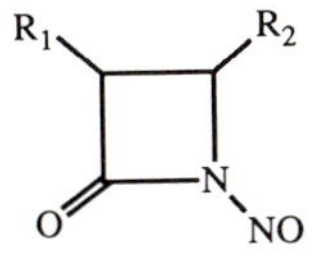

(**3a**) $R_1 = H, R_2 = Ph$
(**3b**) $R_1 = Ph, R_2 = H$

Thus, very little application of this type of inhibitor has been made to β-lactamases (or DD-peptidases). Despite this, in view of the considerable precedent with serine proteinases[60] and taking into account the extension of β-lactamase specificity to acyclic substrates,[20] it seems likely that effective inhibitors of this class could be devised.

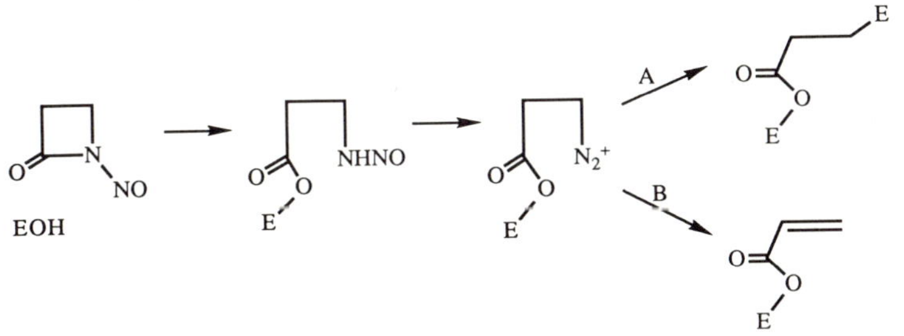

Scheme 7.3

7.3.1.2 Active non-covalent (class 1(b)) Here, rearrangement of the enzyme-bound inhibitor occurs to form a complex that is inert, not because of the change in structure of the inhibitor itself, but rather because of changes induced by rearrangement of the inhibitor in the conformation of the active site of the enzyme or in the position of the bound inhibitor. A number of well-known and important inhibitors belong to this class, in particular, and previously identified,[15] the penems, carbapenems and cephalosporins.

With respect to the penems and carbapenems, Knowles and co-workers[67,68] demonstrated that olivanic acids, which are carbapenems of general structure (**4**), progressively inhibited the TEM β-lactamase by diversion, through covalent rearrangement, of the normal Δ^2-pyrroline intermediate (**5**) into the tautomeric and thermodynamically more stable Δ^1-pyrroline (**6**) (Scheme 7.4). Since there is no obvious reason why (**6**) should deacylate by ester hydrolysis much more slowly than (**5**), the components of the enzyme active site must be somehow differently placed with respect to the inhibitor in (**6**) leading to less effective catalysis of deacylation.

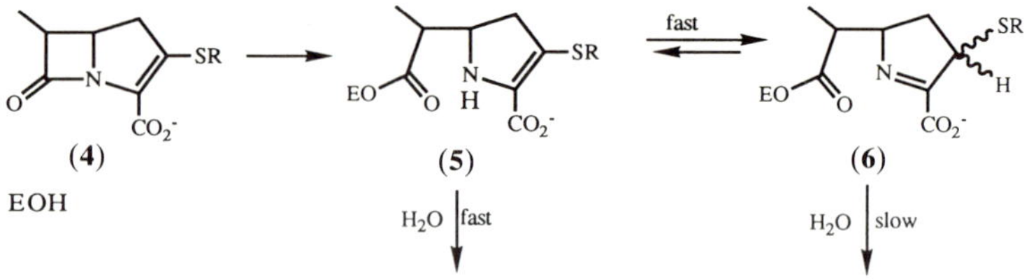

Scheme 7.4

It is clear from the early literature (reviewed for example by Fisher[12]) that penems and carbapenems are, in general, not only poor substrates of most serine β-lactamases, but also inhibitors, and often very potent ones. The kinetics of the inhibition, showing the progressive and often transient characteristics typical of mechanism-based inhibitors, has been described for several β-lactamases from classes A, C and D.[69–71] The important carbapenems, thienamycin and imipenem, have been studied kinetically in some detail.[72–74] The mechanistic basis of the inhibition does not seem to have been further studied although the mechanism of Scheme 7.4 seems likely; it is reported that the product of imipenem hydrolysis is the rearranged Δ^1-pyrroline (**6**).[74] There was also an indication of hysteresis in these reactions,[73,74] i.e. that the free enzyme released after interaction with these compounds (including passage through EI'_i) was in an unnatural and more fragile conformation with unnatural substrate specificity. This phenomenon, which will be dealt with in more detail below (section 7.3.1.4), is suggestive of an altered enzyme conformation in EI'_i.

A very similar situation applies with cephalosporins ((**7**), $R' = H$) and cephamycins ((**7**), $R' = OMe$), which interact with β-lactamases and DD-peptidases according to Scheme 7.5. If X is a sufficiently good leaving group such that $k'_3 > k_3$, a second acyl-enzyme (**9**) is produced from the initially formed species (**8**) prior to deacylation. In cases where this elimination of X does occur at the acyl-enzyme stage, the cephem often acts as an inhibitor since (**9**) is much more stable to enzyme catalysed hydrolysis than the normal intermediate (**8**) and can accumulate as a relatively inert enzyme–substrate complex.

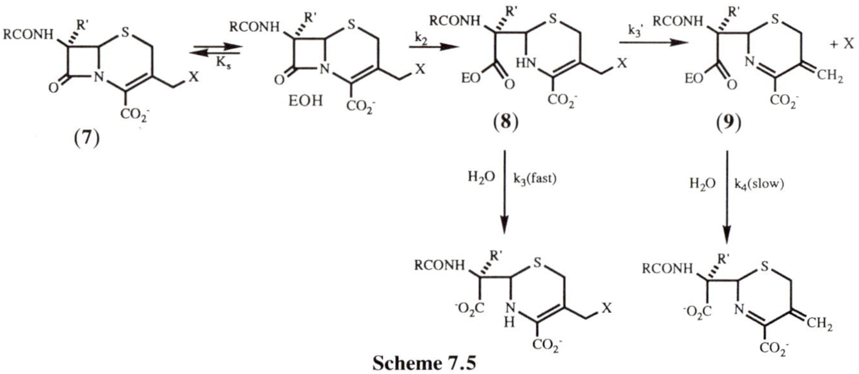

Scheme 7.5

The existence of (**9**) as an accumulating species has been established in the inhibition of the class A *Staphylococcus aureus* PC1 β-lactamase[75] by first-generation cephalosporins, of the class A TEM β-lactamase by cephamycins,[76] of various class C enzymes by cephamycins and third-generation cephalosporins,[77,78] and of the *Streptomyces* R61 DD-peptidase by first-generation cephalosporins.[79] Comparable cephems lacking a 3′-leaving group are generally found to hydrolyse more rapidly. In cases such as that of the class C β-lactamase of *Enterobacter cloacae* P99 and first-generation cephalosporins, a 3′-leaving group effect of this sort is not observed,[78] probably because the enzyme catalyses the hydrolysis of (**8**) so efficiently, i.e. because $k_3 \gg k_3'$ (Scheme 7.5). Through these effects on β-lactamases and/or DD-peptidases, a cephem with a good 3′-leaving group may, all other things being equal, be a more effective antibiotic than one without such a leaving group.[80,81]

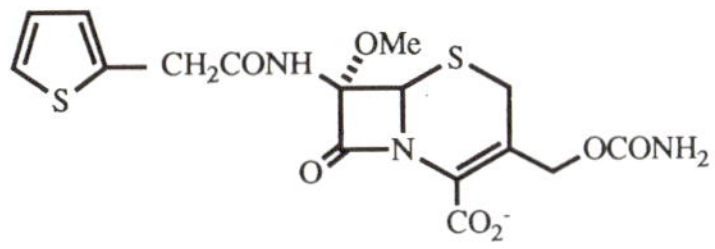

(**10**)

In cephamycins ((**7**), R′ = OMe), the 3′-leaving group effect and that of the 7-α-OMe group can reinforce each other. For example, on interaction of cefoxitin (**10**) with the class A TEM β-lactamase, the 7-α-OMe group is seen to weaken non-covalent binding and slow down both the acylation and deacylation steps. In some cases, for example the class A *B. cereus* β-lactamase I, this seems sufficient to preclude any attack of the enzyme on cefoxitin. In others, such as the TEM β-lactamase, the deacylation rate is diminished enough to permit departure of the 3′-leaving group, i.e. $k_3' > k_3$, such that (**9**) accumulates and cefoxitin is an effective inhibitor. The important third-generation cephalosporin moxalactam (**11**) also inhibits β-lactamases through formation of inert complexes of structure analogous to (**9**).[78] The 7-α-formamido group is apparently also able to potentiate the 3′-leaving group effect and give rise to inert acyl-enzymes.[82]

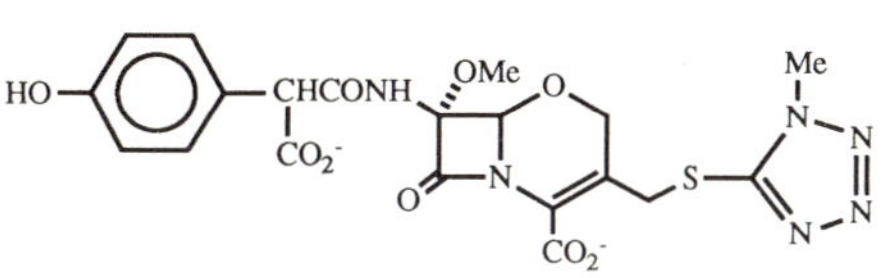

(**11**)

A variety of other recently employed cephem side chains, also providing steric bulk directly adjacent to the 7-position, may have a similar effect, leading to β-lactamase inhibition. Examples include the side chains found in cefotetan (**12**)[83] and in the alkoxime side chain of cefotaxime (**13**) and many other third-generation cephalosporins.[78]

Finally, with regard to leaving group effects, it is interesting to note that an

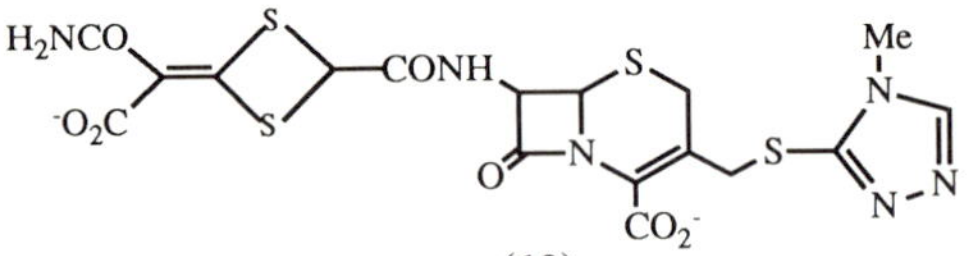

(12)

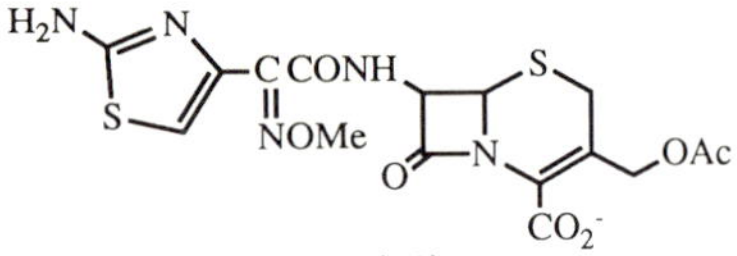

(13)

elimination reaction analogous to that observed in cephems is possible in 2′-substituted carbapenems and penems, and has in fact been observed to occur in the latter compounds by Franceschi and coworkers[84,85] and by Iwata *et al.*[86] Alkaline and renal dehydropeptidase I catalysed hydrolysis of these compounds generate a product of structure (**14**),[84] closely similar to (**6**) and (**9**). Such products are also found in metabolites from rats.[85] Perrone and coworkers have suggested that penems with 2′-leaving groups do have strikingly high antibiotic activity because of this elimination reaction, but direct tests of the proposition do not seem to have been made.[87,88]

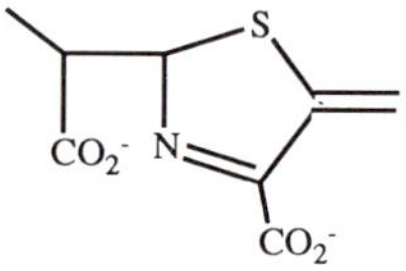

(14)

The phenomenon of antibiotic 'trapping' by β-lactamases as a mechanism of bacterial resistance to cephems and (carba)penems,[89] to the extent that it exists, must largely consist of the formation of inert acyl-enzymes of the type already described.[90]

7.3.1.3 Passive covalent (class 2(a)) These inhibitors are distinguished by the rearrangement of EI_i into an inert EI_i', not by covalent or non-covalent modification of the protein, but through formation, by internal rearrangement of the bound inhibitor — a chemically inert inhibitor structure whose further reaction the enzyme is unable to catalyse. In the case of β-lactamases this process involves formation of a hydrolytically stable acyl-enzyme from the normal labile form.

This category includes the now-classic β-lactamase inhibitors clavulanic acid, penicillanic acid sulfone and 6-β-bromopenicillanic acid. These and their close analogs undergo, as is described in more detail below, rearrangement of the normal acyl-enzyme into a chemically much more inert vinylogous carbamate (**15**) (Scheme 7.6), but differ in their mode of achieving this structure.

The mechanism of action of clavulanic acid has been investigated in considerable detail, particularly by Knowles and coworkers.[91–93] The major events appear to be as depicted in Scheme 7.7, where, competitively with normal turnover, the oxazolidine ring of the initial acyl-enzyme (**16**) opens, and is trapped in the open position by ketonization of the enolate formed, to

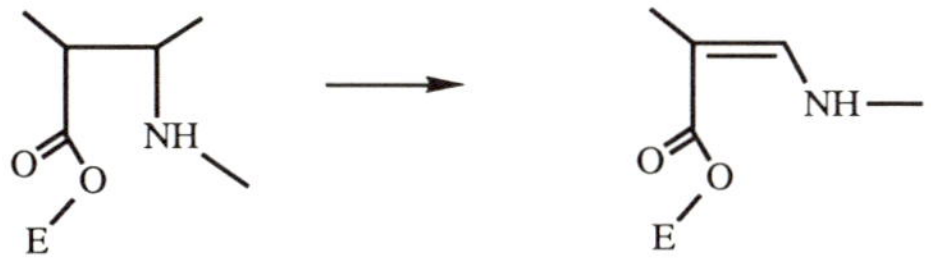

Scheme 7.6

give the imine (**17**). Loss of a now quite acidic proton from the 6-position yields the enamine or vinylogous carbamate structure (**18**), whose formation, transiently at least, inhibits the enzyme. In the case of the *Staphylococcus aureus* PC1 β-lactamase, there is evidence of two inhibitory complexes.[94,95] The initially formed (**18**) had an absorption maximum at around 295 nm, which slowly changed with time into one at 277 nm. Cartwright and Coulson[94] suggested isomerization of (**18**) to the more stable *trans* form (**19**) to explain this observation. Only a chromophore of λ_{max} 281 nm, which may correspond to (**19**), was observed with the TEM β-lactamase.[92] Since (**19**) would most reasonably be formed via C-5–C-6 single bond rotation in (**17**), this conformational event may be more facile at the active site of the TEM enzyme.

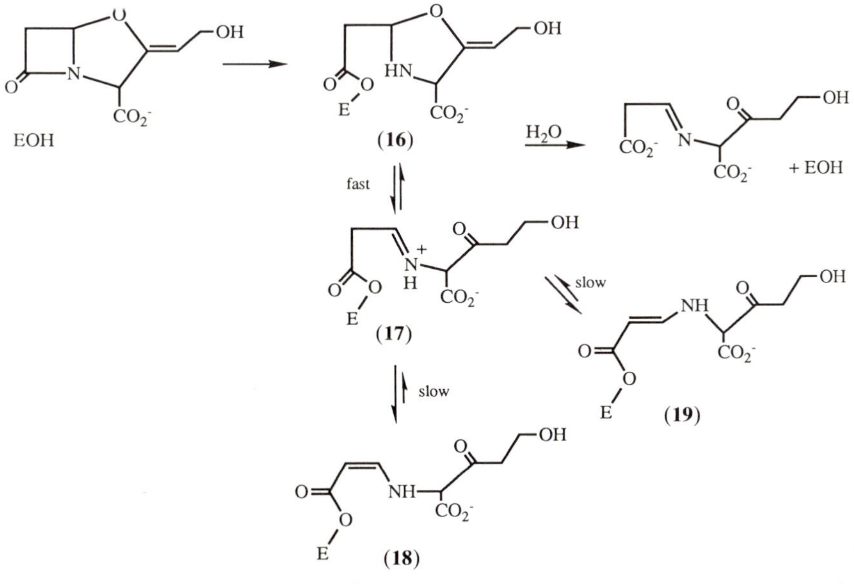

Scheme 7.7

Generally slow — depending on the enzyme concerned — reversion to free enzyme, and thus reactivation, must occur either through reformation of (**17**) from (**18**) or (**19**), particularly at low pH or, also possible, slow hydrolysis of (**18**) and (**19**). By making measurements at lower temperatures, Rizwi *et al.*[95] showed that the loss of *S. aureus* PC1 β-lactamase activity preceded enamine formation, i.e. that either (**16**) or (**17**) (or both) was also slowly hydrolysed. It may be that the imine (**17**) is inert to hydrolysis for the same reason that obtains with the penem and cephalosporin complexes, (**6**) and (**9**) respectively.

Finally, under some circumstances, with the TEM enzyme for example,[93] it appears that complexes even more resistant to hydrolysis can be achieved, involving cross-linking of clavulanate to the enzyme. Fragmentation of the clavulanate can apparently also occur, such as in the proposed transimination reaction of (**17**) yielding (**20**).[13] These complexes have, however, not been fully characterized structurally.[93]

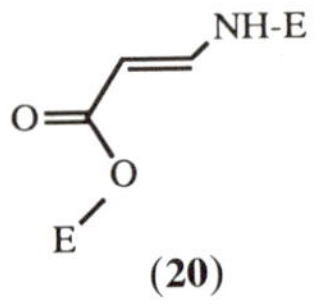

(**20**)

In general, class C β-lactamases are not as susceptible to inhibition by clavulanic acid as are the class A enzymes.[96] Mechanistic studies[97] suggested that much the same chemistry as described above did occur but that these enzymes are more effective at direct hydrolysis of clavulanic acid than are the class A enzymes. No irreversible modification of the *Ent. cloacae* P99 β-lactamase was detected.[97]

As has been much discussed elsewhere,[11–15] the mechanism of inhibition

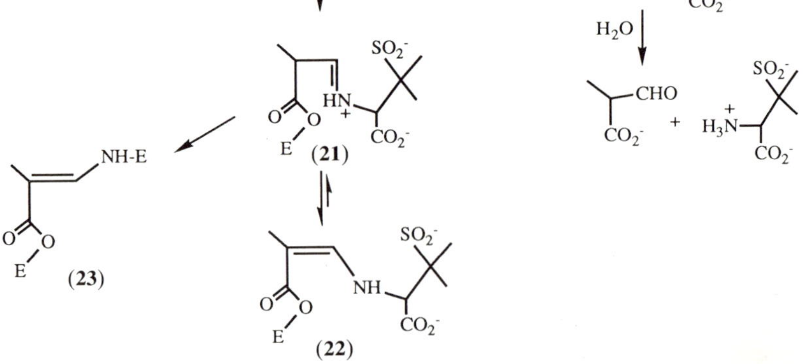

(**21**) (**22**) (**23**)

Scheme 7.8

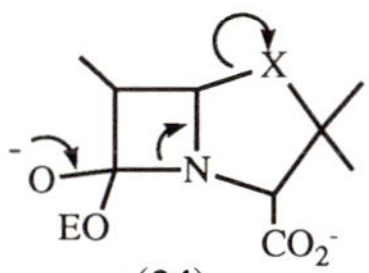

(24)

of β-lactamases by penicillin sulfones (Scheme 7.8) has much in common with that of clavulanic acid. Both transient inactivation through formation of (**22**) and irreversible inactivation, probably through formation of (**23**), is observed. The identity of (**23**) from inactivation of the TEM β-lactamase by penicillanic acid sulfone, and (**20**) from clavulanic acid is a strong point in favor of these mechanisms.[98] Two mechanistic points suggested in the earlier literature but now abandoned are: (i) a concerted breakdown (**24**) of tetrahedral intermediates and opening of the fused five-membered ring;[99] and (ii) opening of the five-membered ring by elimination across the C-5–C-6 bond (**25**).[100] There is no evidence for the presence of, or necessity for a specific base to remove the (highly acidic) C-6 proton of (**17**) or (**21**).

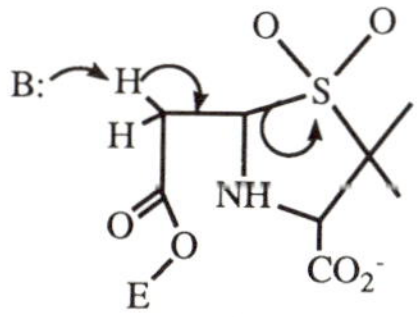

(25)

Penicillanic acid sulfone itself (sulbactam) is generally believed to be a less effective inhibitor of class A β-lactamases than clavulanic acid, but more effective against class C enzymes.[101] The effectiveness of the sulfone can be modulated through the C-6 substituent[94,102–107] or through the C-2 substituents.[108–111] The clinically important sulfone of the latter type, tazobactam (**26**), also appears to be somewhat more effective than clavulanic acid against class C enzymes.[110,112]

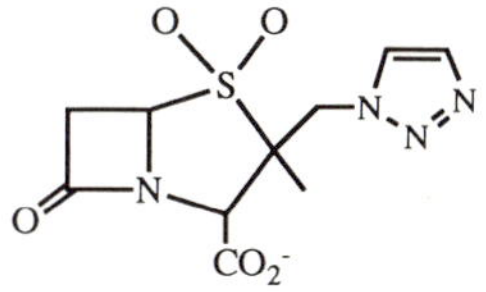

(26)

Various products of reactions of penicillanic acid sulfone in solution, enzyme catalysed and otherwise, may also be inhibitory. For example, Brenner and Knowles[98] showed that malonsemialdehyde, which is rapidly formed in solution from the product of penicillanic acid sulfone hydrolysis (Scheme 7.8), extensively modified the TEM β-lactamase, although in that case the enzyme was not inhibited. Yamaguchi, Sawai and coworkers have described a more intriguing phenomenon. On incubation of cloxacillin sulfone in phosphate buffer, an inhibitor of the class C β-lactamase of *Citrobacter freundii* is slowly produced.[113,114] The identity of the inhibitor does not seem to have been subsequently revealed, but it could conceivably be, from the information given, a penicilloyl phosphate or the oxazolinone derived from it. At any event, the results would seem to presage a new type of inhibitor.

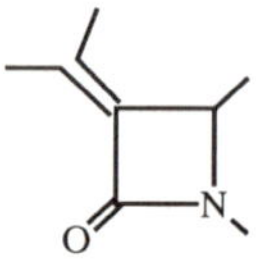

(27)

Another source of an inert acryloyl-enzyme intermediate is the 6(7)-exo-methylene penam (cephem) structure (**27**). Here the β-lactam provides sufficient reactivity for an acyl-enzyme to be formed, which is then immediately an acryloyl derivative. The first of these structures to be described as β-lactamase inhibitors were the naturally occurring asparenomycins (**28**),[115,116] which are 6-exomethylene-carbapenems. Synthetic variants, acetylmethylene penicillanic acid (**29**)[117] and 6-[(*Z*)-methoxymethylene] penicillanic acid (**30**)[118] then appeared in the literature essentially simultaneously. The *E*-isomer of (**29**) is also a β-lactamase inhibitor but is generally less effective than (**29**).[119]

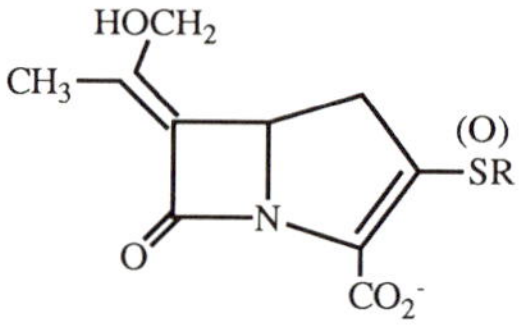

(28) 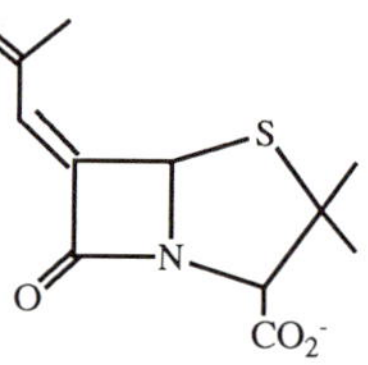 **(29)** 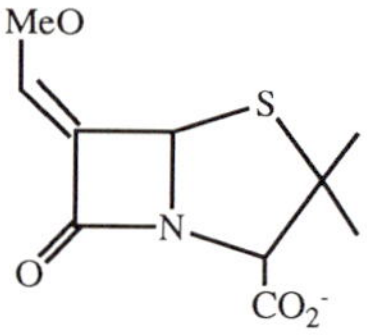 **(30)**

A more recent series has the general structure (**31**) where R is a five-membered ring heterocycle[120,121] and, in particular, a 1,2,3-triazole derivative.[122] Again, the *Z*-isomers are the more effective inhibitors.[120] Simple 6-ethylidene derivatives have also been described.[123] All are potent inhibitors of both class A and C β-lactamases.

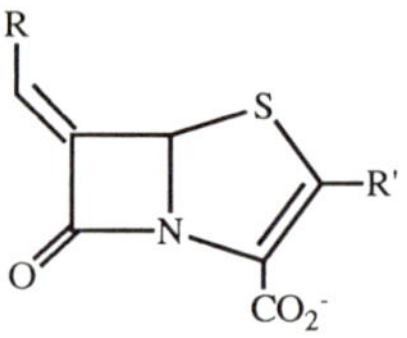

(31)

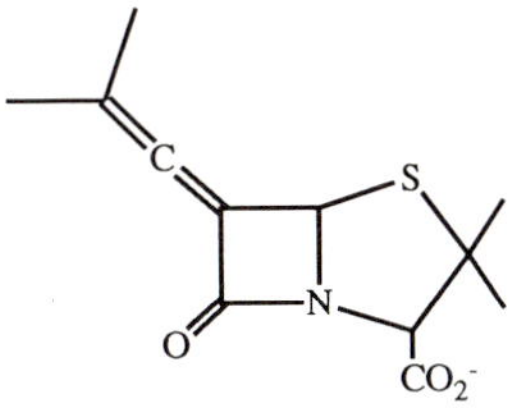

(**32**)

In a further elaboration of this theme, 6-vinylidenepenams (**23**) have been prepared as β-lactamase inhibitors.[124]

These molecules, as previously discussed,[15] can give rise to a number of stabilized structures, both alone and by further interaction with the protein; acetylmethylene penicillanic acid, (**29**), for example, is thought to form a quite stable complex of structure (**33**) (or perhaps (**34**) or (**35**)) on inhibition

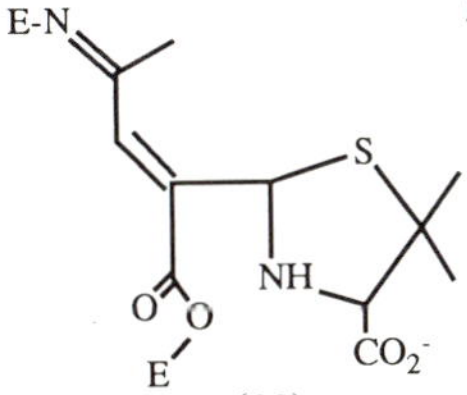

(**33**) 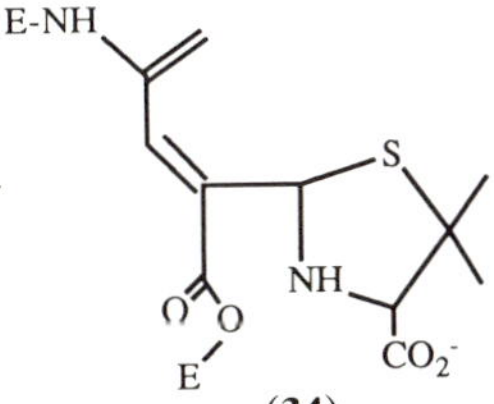 (**34**) 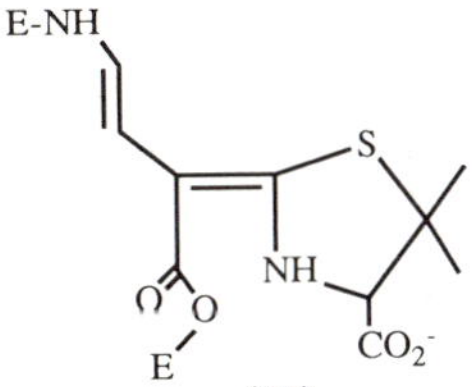 (**35**)

of the TEM β-lactamase.[125] On further incubation of the inactivated enzyme, this complex slowly rearranges to (**36**), which is even more stable to hydrolysis; the very slow rate of formation of (**36**), however, probably makes it irrelevent to any practical application of (**29**) as an inhibitor, of this enzyme at least. As might be expected if formation of (**33**)–(**35**) were important to the observed inhibition, (**37**) ($R = SO_2R'$, SO_2Ar, SOR', CO_2R' and CO_2^-) appeared to be much less effective than (**29**).[126]

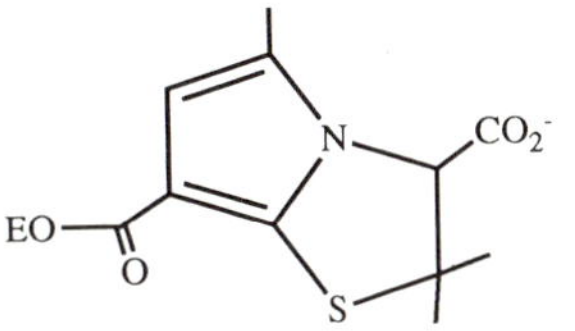

(**36**) 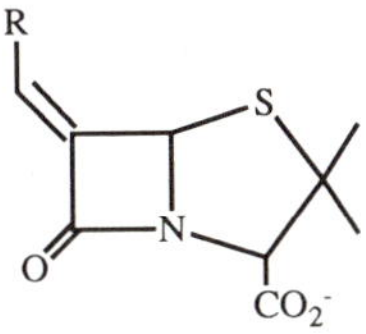 (**37**)

A combination of an exomethylene penam and a sulfone is seen in the compound (**38**). Model studies suggest that on nucleophilic cleavage of the β-lactam ring, such as by a serine β-lactamase, the rearrangement depicted in Scheme 7.9 could occur, giving rise to a hydrolytically inert bicyclic structure (**40**).[127] Compound (**38**) is an impressive inhibitor of both class A

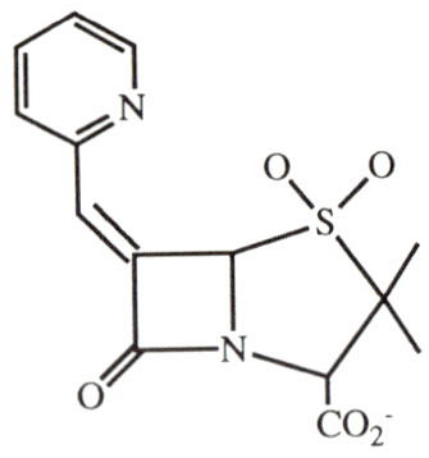

(38)

and class C β-lactamases.[128] The rearrangement of Scheme 7.9 does not seem to have been demonstrated to occur at the enzyme active site, but compounds without the nucleophilic nitrogen of the pyridine ring in the *ortho* position do not seem to be as generally effective as inhibitors. The non-rearranged species **(39)**, possibly supplemented by addition of an enzymic nucleophile to form **(41)**,[129] could also be sufficiently inert to confer significant inhibitory properties — inhibition by the phenyl analog has been reported elsewhere.[104,129] Detailed kinetic and structural studies would be needed to decide between the relative contributions of these alternatives.

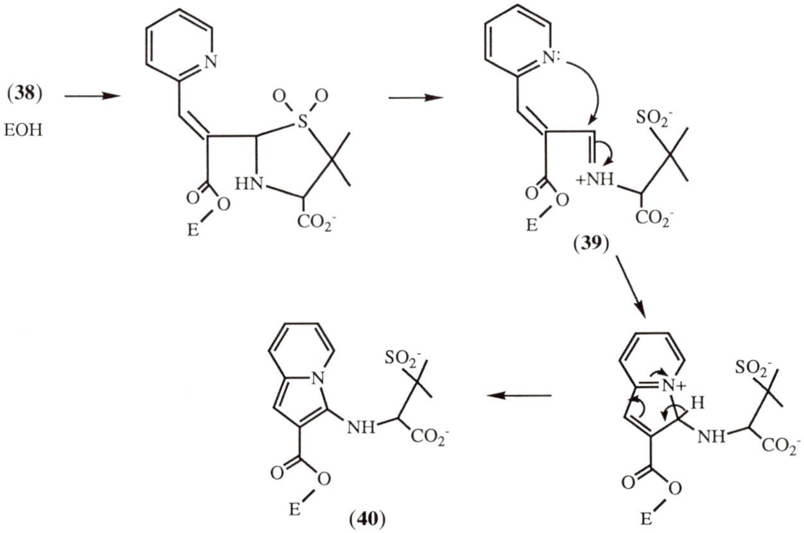

(39)

(40)

Scheme 7.9

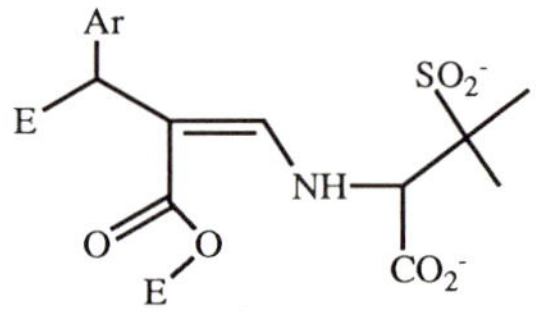

(41)

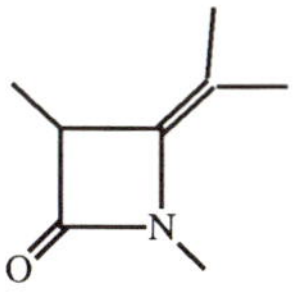

(42) 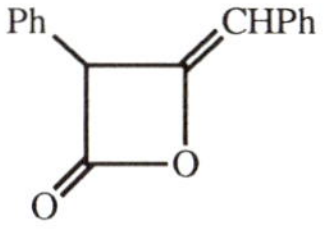 (43)

A possible recent addition to this class of inhibitor may indicate that structures such as (**42**) may also give rise to inhibitors. Tanizawa and co-workers[130] have reported that the diketene (**43**) is an irreversible inhibitor of *B. cereus* β-lactamase I. It is possible that Scheme 7.10 obtains. Alternatively, the ketone of (**44**) may form an imine and thence an enamine with an enzymic lysine residue. The azetidine-2,4-dione (**45**), however, was reported to be inactive as a β-lactamase inhibitor.[131]

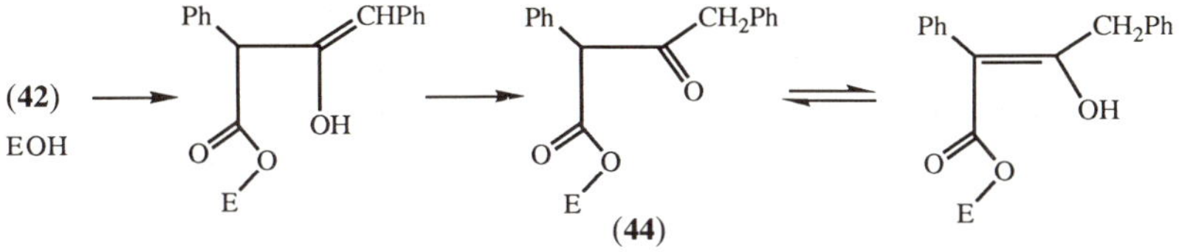

(44)

Scheme 7.10

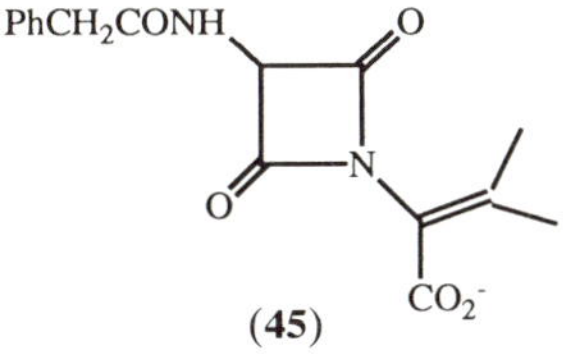

(45)

Finally, the 6-β-halopenicillanic acids should be mentioned. Of these, 6-β-bromopenicillanic acid (**46**) was the first discovered,[132,133] and was shown to be a good inhibitor of class A β-lactamases. The iodo analog is, in general, comparably active, but 6-β-chloropenicillanic acid is much less so.[134,135] The effectiveness of these compounds is significantly less against class C β-lactamases,[135] and indeed this difference has been proposed as diagnostic of the two groups of enzymes.[136] In passing it might be noted that Bush[137] has proposed the relative effectiveness of clavulanic acid and aztreonam as inhibitors to make the same distinction. The early mechanisms proposed for the inactivation reaction,[138–140] although correct in outline, were probably incorrect in detail. In particular, a mechanism (e.g. Scheme

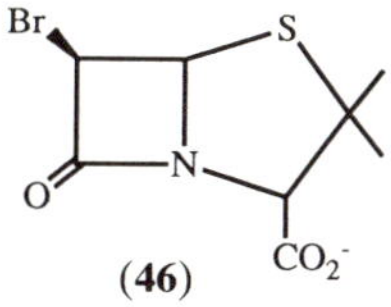

(46)

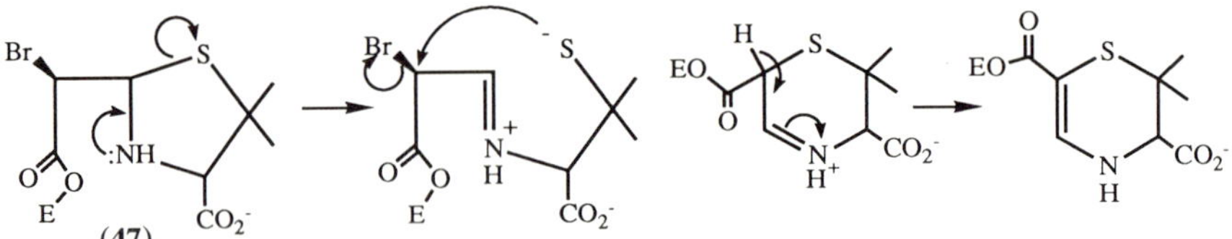

Scheme 7.11

7.11) requiring opening of the thiazolidine ring of the initially formed acyl-enzyme (**47**)[140] is most likely wrong. Careful studies of model systems[141] indicate that the mechanism of Scheme 7.12 is more likely to be correct, and there is in fact evidence that the reaction at a β-lactamase active site does follow just this pathway.[142] An important point here is that it is unlikely that the thiazolidine ring of (**47**) can open quickly enough to allow the observed rates of rearrangement and inhibition (see chapter 4). Thus, Scheme 7.12 includes displacement of the bromide prior to thiazolidine ring opening (**49**) and the formation of a bicyclic sulfonium ion intermediate (**50**). This kinetic barrier should be noted as a general feature in inhibitor design. It explains why penicillins are not themselves β-lactamase inhibitors through thiazolidine ring opening and rearrangement to (**15**) at the acyl-enzyme

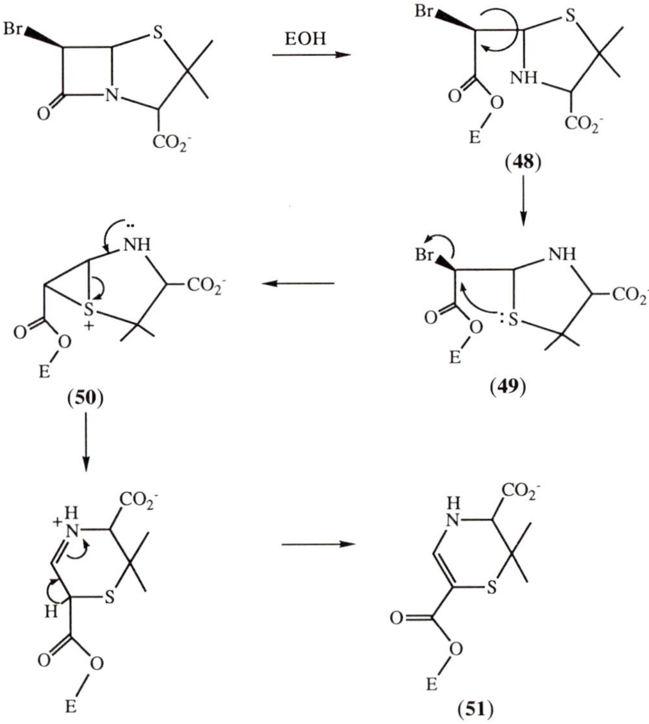

Scheme 7.12

stage, and a kinetic feature of sulfones as inhibitors. A requirement of the mechanism of Scheme 7.12 is rotation about the C-5–C-6 bond in the initial acyl-enzyme (**48**) to allow *trans* displacement of the bromide (**49**). This rotation would be promoted by unfavorable steric interaction between the 6-β substituent and the thiazolidine sulfur atom; it is noticeable, for example, that such a conformational motion has been observed in the crystal structure of β-lactam antibiotics bound in acyl-enzymes of the *Streptomyces* R61 DD-peptidase,[41] and of the class C β-lactamase of *Citrobacter freundii*.[40] It is possible that the rate of this rotation permitted by the active site structure is a significant determinant of the effectiveness of 6-β-bromo- and 6-β-iodo-penicillanic acids as β-lactamase inhibitors. Another significant feature of the 6-β-halo-penicillanic acid inhibition pathway is that it is clean and irreversible; the final inert vinylogous amide (**51**), unlike (**18**)/(**19**) from clavulanic acid or (**22**) from penicillanic acid sulfone, cannot revert to a more readily hydrolysed β-imino ester. β-Lactamases do not catalyse the hydrolysis of (**51**) at any significant rate.

Scheme 7.12 probably also describes the mechanism of inhibition of *B. cereus* β-lactamase I by diazotized 6-β-aminopenicillanic acid (**52**)[143] where the leaving group from (**53**) is dinitrogen (Scheme 7.13).

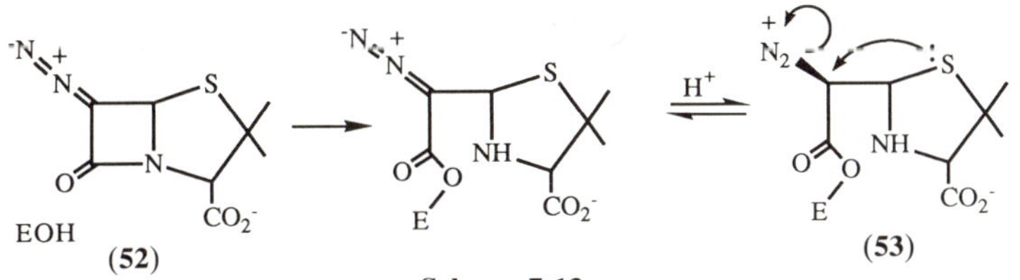

(**52**) (**53**)

Scheme 7.13

More recent analogs of the 6-β-halopenicillanic acids are the 6-spiro-epoxides (**54**) and (**55**)[144] prepared by Bycroft and coworkers. The stereochemistry was established by X-ray crystallographic investigation of a sulfoxide of the (**54**) series. Typically, X and Y were halogen, OMe, NHR and NHAr groups. Many of them had β-lactamase-inhibitory and antibacterial activity; in general the (**55**) series was more active in both respects than the (**54**).

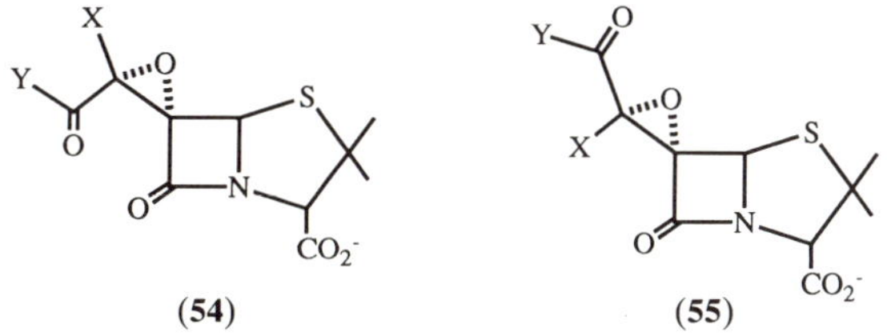

(**54**) (**55**)

Some details of the inhibition of *B. cereus* β-lactamase I by two of these compounds (**54**) and (**55**) (Y = PhNH) have been reported.[145,146] Both

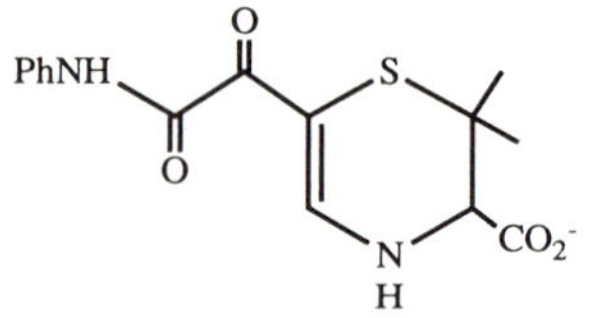

(56)

compounds are substrates and mechanism-based inhibitors, with values for the ratio of turnover to inactivation of 10 800 : 1 and 2080 : 1 respectively. The turnover product from the enzyme was identified as **(56)**, which strongly suggests a 6-β-halopenicillanate-like rearrangement. Consequently, the inactivation mechanism is likely to be that of Scheme 7.14.

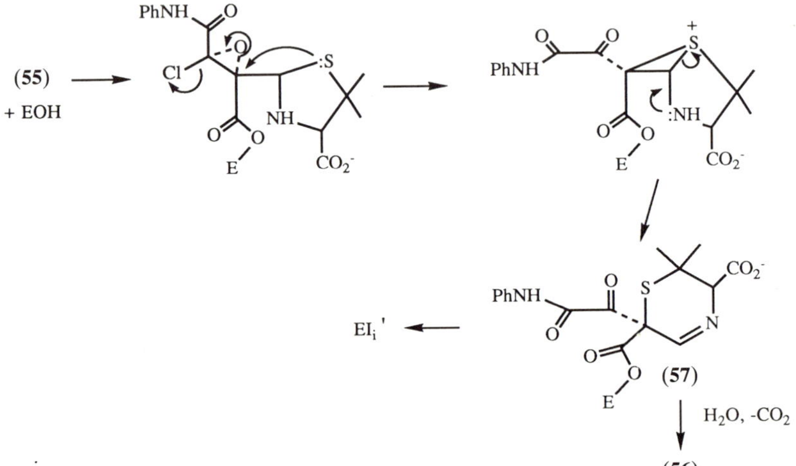

Scheme 7.14

An interesting difference between **(55)** and **(46)** is that the leaving group from C-6 during the rearrangement has an α-stereochemistry in the original β-lactam ring in **(55)** rather than β as in **(46)**. 6α-Halopenicillanic acids have little β-lactamase-inhibitory ability.[132,134] The latter is thought to arise from differences in steric interactions in the alternative sulfonium ion intermediates **(58)** and **(59)** and associated transition states.[141] In the 6-halopenicillanic acids, where R = H, **(59)** is more stable because of the smaller steric interaction between R and the β-methyl group in **(59)** than in **(58)**. The situation is less clear with the spiroepoxides where R=–CCl(OH)CONHPh,

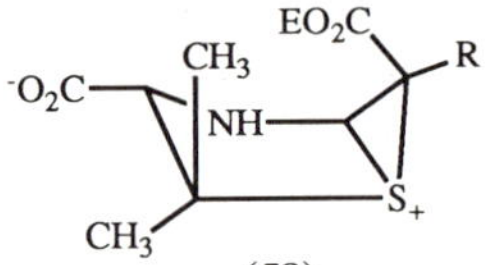

(58) 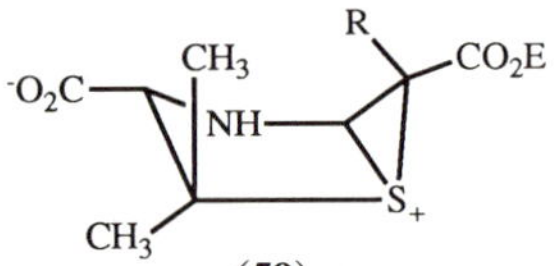 **(59)**

and in fact (**58**) may be thermodynamically preferred. The 6-β-spiro-epoxides do not seem to be available for comparison. Certainly (**54**) and (**55**) are much poorer inhibitors of *B. cereus* β-lactamase I than is 6-β-bromo-penicillanic acid.[132] The rate constants reported[145,146] do seem large enough, however, to preclude a mechanism requiring thiazolidine ring opening.[141] The nature of EI'_i, the irreversibly inhibited enzyme, is not yet known. Imine formation between (**57**) and an enzymic amine group is possible, and conformational changes also seem likely.[146]

The side chains of (**54**) and (**55**), unlike those of most good β-lactamase substrates and antibiotics, are largely rigid. Since (**55**) but not (**54**) has antibacterial properties, Shute *et al.*[147] have proposed that the antibiotically active conformation of benzylpenicillin, i.e. that conformation which binds to cell wall DD-peptidases, is best represented by (**60**), which is the stable conformation of the penicillin that best overlaps with the (**55**).

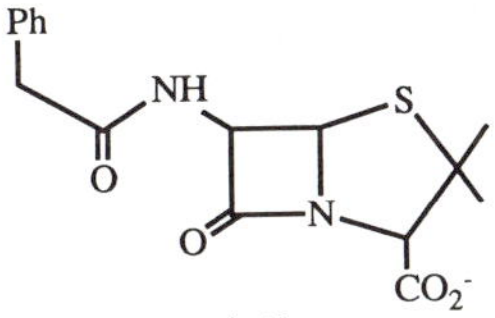

(**60**)

An interesting, but to date apparently unsuccessful attempt to produce novel β-lactamase inhibitors of this class was that of Marchand-Brynaert *et al.*[148] who synthesized a series of monocyclic β-lactams of general structure (**61**) where L is a leaving group, –SCO(O,S,NH)R or $-SO_2R$. It was hoped that these molecules might be mechanism-based inhibitors according to Scheme 7.15. Disappointingly, no inhibition of class A, B, or C β-lactamases was observed, nor significant antibiotic activity. It was not reported whether any of the compounds were β-lactamase substrates. The mechanism of Scheme 7.15 may apply, however, for some monocyclic β-lactam inhibitors of elastase.[149]

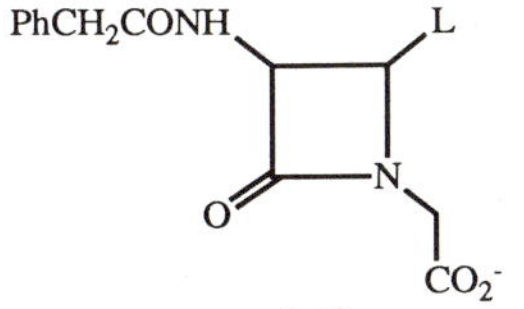

(**61**)

A question that has often been raised with respect to the β-lactamase inhibitors of this class, and certainly not an idle one in view of the existence of a wide range of inhibitors of class 2(b), as discussed in section 7.3.1.4, is to what extent the inhibition produced by the molecules of the present class resides in the chemical stability of the rearranged inhibitors as opposed to conformational changes in the active site induced by the molecular

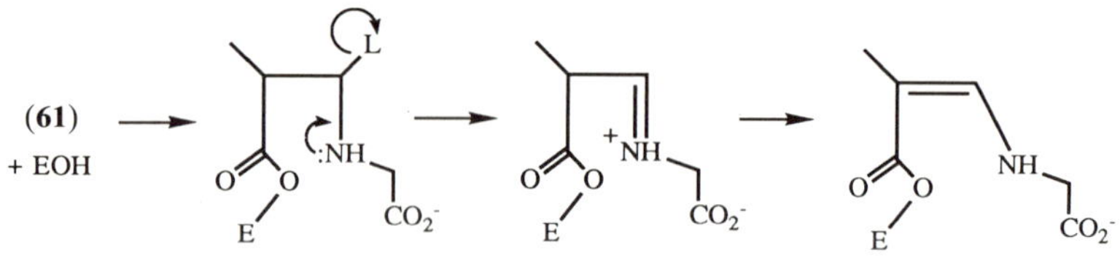

Scheme 7.15

rearrangements. On one hand, the chemical stability of vinylogous carbamates seems sufficient to explain much of the inhibition observed but, on the other, it seems likely that the active site must suffer considerable conformational discomfort during the rearrangements described above, where the inhibitors, in general, undergo a significant change in shape and distribution of functional groups. Irrespective of the relative contributions of the two factors, it is certainly possible to search by physical means for the presence of protein conformational changes in the inactivated proteins. Several such investigations have in fact been carried out.

Dmitrienko, Viswanathan and coworkers,[150] arguing that an electron-withdrawing group at the 6-position of a penicillin sulfone would stabilize the vinylogous carbamate (**22**) and thus produce a more effective inhibitor, prepared the sulfone (**62**) and showed that it was indeed more effective than

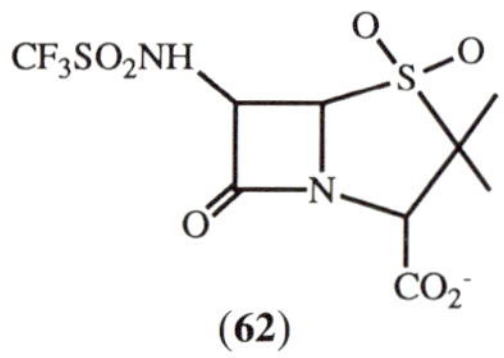

(**62**)

penicillanic and sulfone itself against *B. cereus* β-lactamase I. The inhibition appeared to be completely irreversible. The irreversibility of the inhibition may relate to the acidity of the 6-trifluorosulfonamide group, which would be anionic at neutral pH, since the analogous N-methylsulfonamide measurably partitioned between transiently and irreversibly inhibited species.[151] Absorption spectra and experiments with a radioactively labelled inhibitor suggested that the inert complex had the structure (**22**). Differential scanning calorimetry measurements indicated that the irreversibly inactivated enzyme denatured at a temperature some 9°C lower than did the native enzyme. Further, optical rotatory dispersion measurements suggested substantial changes in secondary structure, including the likely loss of considerable α-helix.

Fink, Ellerby and Bassett[152] have extended these results. They found that although the transiently inhibited enzyme, again *B. cereus* β-lactamase I, produced by penicillanic acid sulfone (**22**) had, by the criterion of far UV circular dichroism, the same structure as the native enzyme, this was not so with sulfones containing bulky 6-arylamido side chains. The latter compounds at hgh pH inhibited the enzyme essentially irreversibly (although the inhibition could be reversed on lowering the pH). The circular dichroic spectra of the high pH complexes differed from that of the native enzyme in a way that could be interpreted in terms of the loss of 8% of α-helix.

Thus, it seems clear that one form of enzyme derived from irreversible inactivation by certain sulfones contains a protein conformation significantly different from the native, and it is likely that a major source of the irreversibility lies with the conformational change. Acting in this mode the sulfones would be better categorized in class 1(b). It is also clear, however, that significant (although transient) inhibition can be achieved by sulfones with little or no conformational change, presumably through the chemical stability of vinylogous carbamates. This also seems to be the case in the stablest complex achieved on interaction of clavulanate with the *S. aureus* β-lactamase.[95] It should be recalled, however, that the transient nature of these complexes may more reflect their ability to revert to (**21**) than their susceptibility to direct deacylation; that derived from 6-β-bromopenicillanic acid (**51**) and *B. cereus* β-lactamase I is firmly irreversible, and has a far UV circular dichroism spectrum essentially identical to that of the native enzyme (B. George and R.F. Pratt, unpublished results). This mode of inhibition therefore seems more closely that of class 2(a). Finally, the achievement of irreversibility by cross-linking to the enzyme, as in (**23**), may best be considered to reflect a class 1(a) mode of action.

In general, it seems likely that some conformational accommodation will occur to all these inhibitors, but only in extreme cases will it be unambiguously clear whether the covalent rearrangement of the inhibitor (class 2(a)) or non-covalent distortion of the enzyme (class 1(b)) is mainly responsible for the inhibition; in many cases both may be present and each equally sufficient. Clavulanic acid and the sulfones appear unusual in their ability to inactivate β-lactamases in a number of different ways. This versatility may be important to their general clinical effectiveness.

7.3.1.4 Passive non-covalent (class 2(b)) In this mechanistic variant, the structure of the inhibitor induces a non-covalent rearrangement, i.e. a conformational change, of a normal enzyme/substrate intermediate into an inert, conformationally abnormal form. This class of inhibitors could superficially be included in a group labelled 'poor substrates.' There are, however, many reasons why a substrate may be poor, apart from that just mentioned. Many substrates are poor for very clear chemical reasons — because of the addition of electronically deactivating substituents slowing down the first step of catalysis, for example, or because for steric reasons they do not bind well to the active site. Such compounds would not generally be classified as inhibitors, although the fact that they are not may well be instructive. Only substrates that are poor because a refractory and unnatural intermediate accumulates should be included in the present class of mechanism-based inhibitors. Distinction should also be made between the class of inhibitors of interest here, and molecules that possess some features of a substrate — enabling them to commence the catalytic path — but not other features; the enzyme is therefore trapped as a complex of normal structure, but which cannot complete the reaction. These could be termed partial or incomplete substrates, although the previously mentioned term 'reaction coordinate inhibitors' has been coined for them.[55]

β-Lactamases appear to be extremely susceptible to inhibitors of the passive non-covalent type. In general, it is the acyl-enzyme intermediate which is conformationally diverted. It appears likely that, on turnover of normal substrates, there is a conformational change on formation of the acyl-enzyme, enforced, perhaps, by rotation about the C-5–C-6 bond (in the case of a penicillin), and necessary, perhaps, to allow access of water to the acyl-enzyme ester carbonyl group.[34,40] The protein conformation of the acyl-enzyme is apparently less stable, or more mobile and thus more easily diverted, than that of the native enzyme.

This curious property of the acyl-enzyme has long been recognized and remarked upon, particularly with respect to class A β-lactamases.[153,154] It is most clearly observed through the consequences of turnover of certain substrates (A-type substrates[155]) that are characterized by possession of bulky and/or rigid[156,157] substituents closely adjacent to the β-lactam ring. Turnover of these substrates includes partitioning of the acyl-enzyme between a normal path and one leading to another acyl-enzyme — more resistant to hydrolysis than the 'normal' one — through a protein conformational rearrangement of some sort. Thus, these types of substrates can be recognized as inhibitors of the passive non-covalent class. They can be detected through the branched pathway kinetics of a mechanism-based inhibitor or, in cases where $k_4 \gg k_3$ (Scheme 7.2), through formation of an unusually stable acyl-enzyme intermediate — 'unusually' with respect to comparable substrates lacking the particular structural feature of the type described above and 'stable' kinetically rather than thermodynamically, although the two are correlated in these cases.

The first-discovered examples of this type of inhibitor were the early β-lactamase-resistant penicillins such as methicillin and cloxacillin. Interaction of these compounds with class A β-lactamases was observed to lead to time-dependent transitions to less-active enzyme forms.[155] More recently, the 7-α-methoxy group of cephamycins, for example cefoxitin, has been shown to give rise to conformationally distorted acyl-enzymes with certain β-lactamases,[158,159] through a combination of the effects of the methoxy group and the 3′-leaving group;[76] 7-α-substituents of cephems, in general, may have this effect.[82] Thus the cephamycins contain elements of class 1(b) and class 2(b) inhibitors. It should be noted, however, that these effects are not general to all β-lactamases — in some cases, the 7-α-substituent precludes any covalent interaction with the enzyme, and thus the compound would be, if anything, a reversible non-covalent inhibitor.

Many β-lactams, apart from those mentioned above, give inert acyl-enzymes with β-lactamases, and may therefore be inhibitors belonging to this class. The oximino-containing side chains characteristic of third-generation cephalosporins, for example, appear to give rise to inert acyl-enzymes with class C β-lactamases when incorporated into cephem,[77,78,160] monocyclic β-lactam,[161–163] and perhaps penam[164,165] structures. Small substituents having the α-stereochemistry and placed adjacent to the β-lactam carbonyl group often appear to produce this kind of inhibitor. The α-methoxy and α-formamido substituents in cephems have already been noted above, but they are also effective in penicillins[166,167] and monocyclic systems.[168] The 7-α-(1-hydroxyethyl) substituent, such as found in imipenem, interestingly, also seems effective and, in recent times, has been successfully incorporated into other penem,[86] oxapenem,[169] cephem[170,171] and oxacephem systems;[172] the *R*-sulfoxides and sulfones of these cephems are also β-lactamase inhibitors, particularly and intriguingly, as allyl esters.[173]

The stereochemistry at the 4-position of a β-lactam ring may also be an important variable, at least in monocyclic systems. Matsuda *et al.*[174] have recently shown that the return of activity of a β-lactamase which had been inhibited by monobactams was faster when the substituents (one of which was a β-oximinoacyl side chain) were *cis* to each other across the C-3–C-4 (**63**) bond (as in natural penicillins) than when they were *trans* (**64**), indicating that the latter compounds may be the better β-lactamase inhibitors. A variety of *cis*-3-amido-4-arylazetidinones have, however, been reported as β-lactamase inhibitors[175–181] but their mode of action is unclear and their activity probably limited.[175,177]

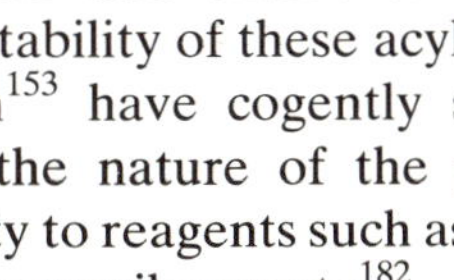

(**63**) 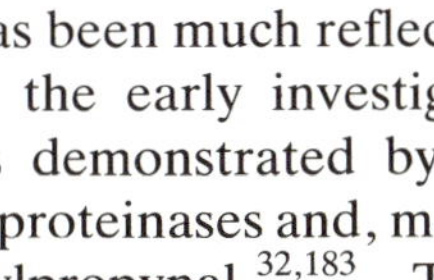 (**64**)

The nature and extent of the conformational event leading to the hydrolytic stability of these acyl-enzymes has been much reflected on; Pain and Virden[153] have cogently summarized the early investigations. The change in the nature of the protein was demonstrated by changes in susceptibility to reagents such as iodine and proteinases and, more recently, *p*-hydroxymercuribenzoate[182] and phenylpropynal.[32,183] The conformational transitions can be slowed down or eliminated by antibodies to β-lactamases[153,184,185] or by intramolecular chemical cross-linking of the enzymes,[186] and the induced conformation preserved by cross-linking.[187,188] Kiener and Waley[189] showed that the rates of labile proton exchange from *B. cereus* β-lactamase I increased markedly on formation of acyl-enzymes with methicillin and cloxacillin, indicating a more open or more mobile protein conformation. Persaud *et al.*[190] investigated the physical nature of the inert acyl-enzyme formed between the *S. aureus* PC1 β-lactamase and quinacillin. Viscosity measurements and far UV circular dichroism spectra showed that the protein had not undergone gross unfolding, but enhanced proton exchange rates and an enhanced susceptibility to denaturation by urea showed that the protein was different in some more subtle way from the native form. Fink *et al.*[191] have investigated the physical properties of *B. cereus* β-lactamase I after interaction with methicillin, cloxacillin and naficillin. Their results and conclusions were similar to those of Persaud *et al.* The inactivated enzyme had far UV circular dichroism and fluorescence spectra essentially the same as the native enzyme, but was more sensitive to thermal and acid-induced denaturation, and to trypsinolysis than was the native enzyme. Finally, Hardy and Kirsch[192] have isolated the inactive complex between the *S. aureus* enzyme and dicloxacillin and examined its reactivation in detail. They suggest, but do not provide evidence for, an alternative explanation for the phenomena described above, whereby these substrates acylate an amino acid residue other than Ser-70. There seems to be no independent evidence for this proposition, although the site of attachment of these type-A substrates to class A β-lactamases has not been demonstrated. It might be noted, however, that cloxacillin, which forms an inert complex with typical class C β-lactamases,[193] has been shown to acylate the active site serine hydroxyl group in one such case.[194] Transient inactivations of class C enzymes have been observed in the presence of both bulky penicillins[195] and cephems,[196] although in the latter case this may be due, in part at least, to the effects of the C-3′-leaving group, as described above.

It is still not possible to directly correlate these observations with specific changes in protein structure. Pain and coworkers[153,186,197] proposed some time ago that the conformational inactivation resulted from interdomain motion. Subsequently, crystal structures showed the β-lactamases to be two-domain proteins with the active site in a cleft between the domains.[24,25,30] The inspection of the crystal structures indicates loose packing, and hence the possibility of conformational flexibility between the Ω loop (comprised of residues 163–178 of a class A β-lactamase) and the α-2 helix which bears the active site Ser-70 and Lys-73 residues.[24] Such flexibility, which may include a transition between two discrete conformations of the loop, dictated by a *cis* peptide between residues 166 (the conserved and probably catalytic glutamate of class A β-lactamases) and 167,[198] may be responsible for the transient phenomena.[24,25,198]

A final matter of some interest with respect to these inhibitors is the question of whether the protein conformational changes that are apparently present in the inert acyl-enzymes persist beyond deacylation, i.e. whether there is hysteresis. The most direct evidence on this point lies in the observations of Citri and coworkers,[155,199] and subsequently noted by others,[73,74] that the kinetics of turnover of one substrate, for example a normal (type-S) one, were influenced by another, for example a type-A substrate, in ways other than rapid competitive inhibition of one by the other. The data can certainly be interpreted in terms of hysteresis, but it is not clear whether it could not also be explained by various combinations of the complex kinetics of Scheme 7.2, inhibition by unstable products, and perhaps, by the existence of a second substrate or product analog binding site.[200] Hysteretic phenomena do not seem to have been directly observed by physical means with a single substrate or substrate analog. The situation is complicated, intriguing, and seemingly unresolved.

7.3.2 Transition-state analog inhibitors

Although the effectiveness of this type of inhibitor, both in general[201] and for serine proteinases in particular,[37] has been clearly demonstrated, few have yet been devised for β-lactamases. As with acyl-transfer enzymes in general,[37,202] the aim would be to produce a stable substrate analog with a tetrahedral and preferably anionic moiety in place of the carbonyl group of the scissile bond; classic examples would be tetrahedral aldehyde or ketone adducts, boronate adducts, or phosphonates. These inhibitors would, in principle, bind tightly through interaction with the oxyanion hole.[24,25,38] Synthetic difficulties, stemming from the likely instability of some of the most obvious β-lactam targets, for example the phosphonolactam system (**65**),[203] have probably limited development in this area. Nonetheless, there have been some successes as described below.

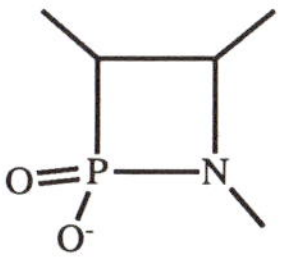

O=P—N
O⁻

(**65**)

Several groups, for example, have prepared deazapenicillin derivatives. Gordon *et al.*[204] prepared (**66**)–(**68**). These molecules did not, however, inhibit the TEM β-lactamase nor the *Streptomyces* R61 DD-peptidase.

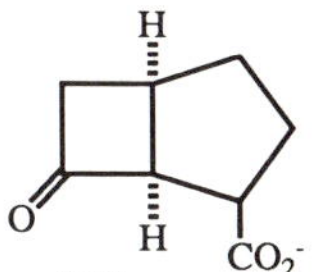

H H H
O H CO_2^- O H CO_2^- O H CO_2^-

(**66**) 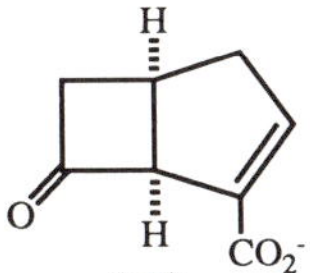 (**67**) 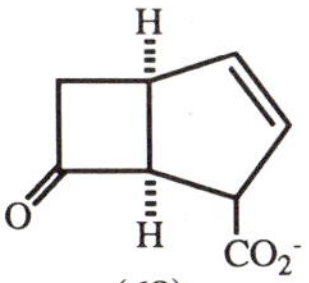 (**68**)

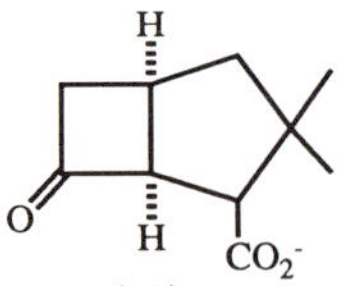

(69) 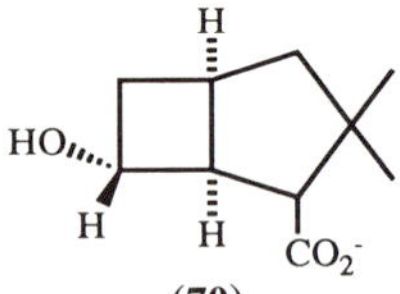 (70) 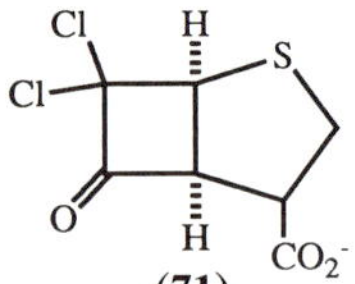 (71) 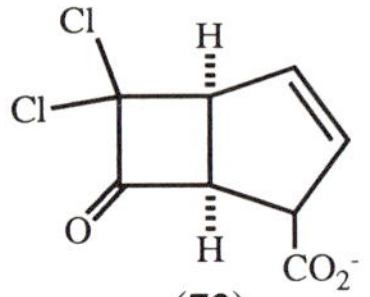 (72)

Compounds (**69**) and (**70**) have also been prepared[205] although no inhibition was reported. In view of the likely attack of β-lactamases on the α-face of β-lactams,[24,25,206] the β-epimer of the alcohol (**70**) might still be of interest. Chlorinated analogs (**71**) and (**72**), which might be anticipated to form more stable tetrahedral adducts with the active site serine hydroxyl group, have also been reported[204,207,208] but again with no report of inhibitory or antibiotic activity. Crystallographic investigation showed, however, that (**71**) did bind to the active site of the *Streptomyces* R61 DD-peptidase.[209] N-Acetyldeazathienamycin (**73**) and some other deaza analogs have also been prepared and reported to act in synergy with benzylpenicillin against β-lactamase-producing strains of *S. aureus*.[210] The azetidine (**74**), which is also tetrahedral at the penicillin 7-position, was reported to be a very weak β-lactamase inhibitor.[211]

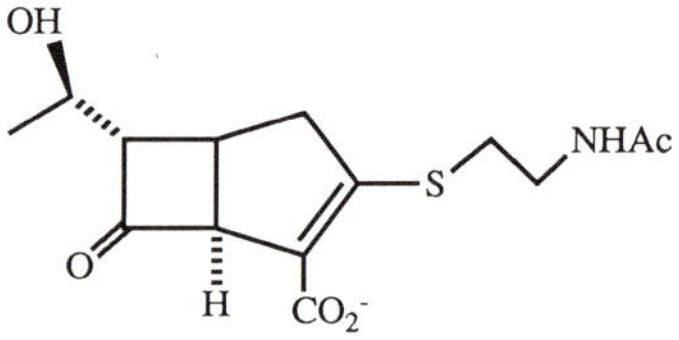

(73) 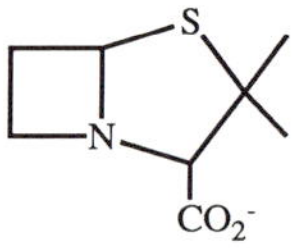 (74)

More promising perhaps were the oxapenams (**75**) and (**76**) prepared by Lowe and Swain.[212] Although they did not show antibiotic activity, (**75**) inhibited the R61 DD-peptidase ($I_{50} \sim 1\,mM$), and both (**75**) and (**76**) gave rise to time-dependent inhibition of the TEM β-lactamase and of *B. cereus* β-lactamase I. Unfortunately, the preliminary report does not seem to have been followed up. Hints of acyclic aldehyde and fluoroketone inhibitors have appeared in the patent literature[213,214] but no data seem to have been published.

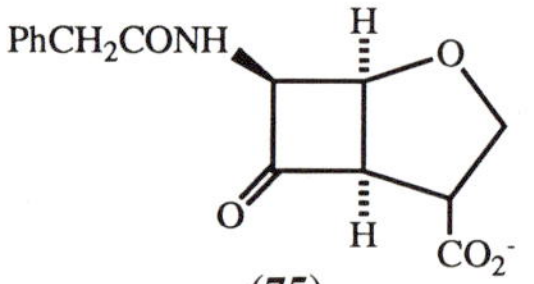

(75) 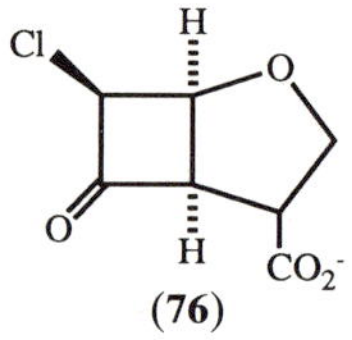 (76)

The sultams (**77**)[215] and (**78**),[216] although uncharged at the tetrahedral sulfur, have also been prepared, and are possible inhibitors of this type.

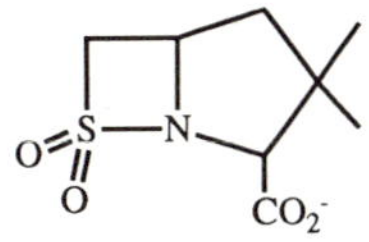

(77) 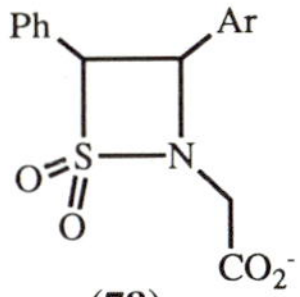 (78)

Only the absence of antibiotic activity has been reported with respect to (**77**) but (**78**) was found not to inhibit *B. cereus* β-lactamase I.

One might have hoped for more detailed testing of all of these derivatives, although it is certainly possible that such investigations were carried out in many cases but proved negative.

Borate and boronates inhibit serine proteinases by formation of a negatively-charged, tetrahedral adduct with the active site serine hydroxyl group (**79**). Serine β-lactamases are also inhibited by these compounds[34,217–219] and may be purified by means of boronate affinity

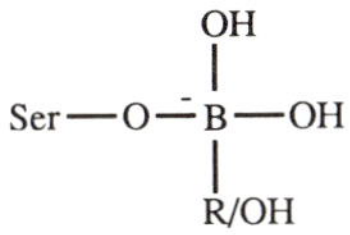

Ser—O—$\bar{B}$(OH)(OH)—R/OH

(79)

columns.[220] Class C β-lactamases, in general, appear more susceptible to inhibition by these reagents.[220] More recently, Crompton *et al.*[221] have synthesized boronates of structure more analogous to those of substrates (**80**)–(**82**). These compounds were effective inhibitors of both class A and class C β-lactamases at micromolar concentrations. Rapid kinetic methods revealed a two-step binding process, as also seen with certain boronates and serine proteinases[222] and, frequently, with transition-state analog inhibitors in general;[223] the slower step may well be a protein conformational change. Recent ^{11}B nmr measurements by Baldwin *et al.*[224] have revealed that the complex between the class C β-lactamase of *Enterobacter cloacae* P99 and 3-dansylamidophenylboronic acid contains the expected tetrahedral boronate.

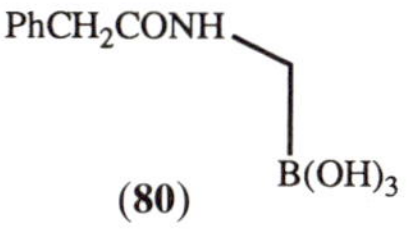

$PhCH_2CONH$—CH_2CH_2—$B(OH)_3$ (80)

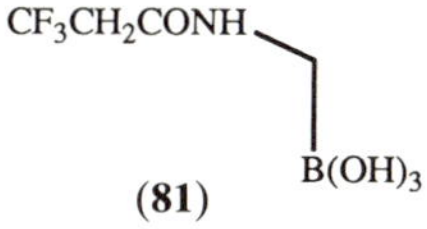

CF_3CH_2CONH—CH_2CH_2—$B(OH)_3$ (81)

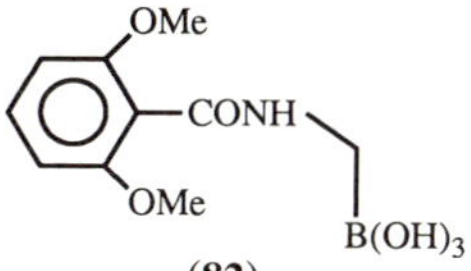

2,6-(OMe)$_2$C$_6$H$_3$CONH—CH_2CH_2—$B(OH)_3$ (82)

Phosphonate monoester inhibitors of β-lactamases have also been described. These were inspired by the discovery that certain acyclic depsipeptides (**83**) were β-lactamase substrates[225] and the fact that the β-lactamase active site contained considerable positive charge.[24,25,34,35] Phosphonate

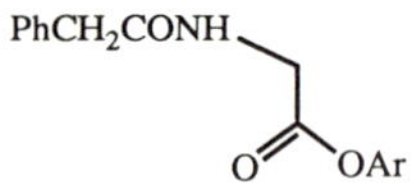

$PhCH_2CONH$–CH_2–C(=O)OAr

(83)

$PhCH_2CONH$–CH_2–P(=O)(O⁻)OAr

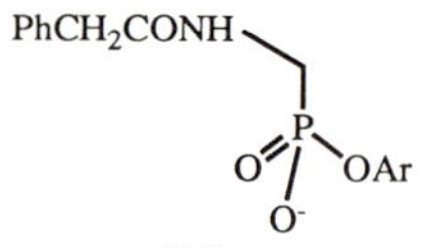

(84)

monoesters (**84**) were therefore prepared and shown to inhibit both class A and class C β-lactamases.[226,227] Weak inhibition of the *Streptomyces* R61 DD-peptidase was also reported,[227] and therefore phosphonate antibiotics may also be possible. Although it was anticipated on the basis of the general stability of phosphonate monoesters to nucleophilic cleavage that compounds (**84**) would act as stable transition-state analogs, as they do of carboxypeptidase A,[228] inactivation of β-lactamases by (**84**) was found to be accompanied by a concerted and stoichiometric release of ArOH. The proposed inactivation reaction is represented in Scheme 7.16 and involves phosphonylation of an active site residue, which is most likely to be the serine hydroxyl group. A ^{31}P nmr spectrum of a phosphonate-inactivated sample of *Enterobacter cloacae* P99 β-lactamase, denatured with guanidine hydrochloride, displays a single resonance of chemical shift appropriate for an alkyl phosphonate monoester monoanion.[229] The chemical stability of the adduct also supports the structure (**85**). As shown in Scheme 7.16, there is also a slow dephosphonylation reaction involving release of phenylacetamidomethylphosphonic acid and reactivation of the enzyme.

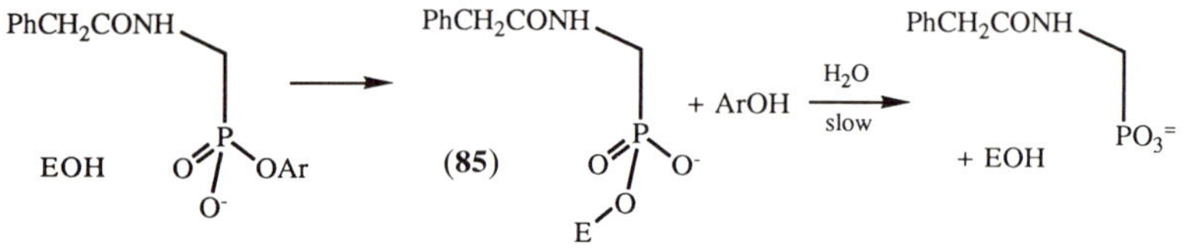

EOH + $PhCH_2CONH$–CH_2–P(=O)(O⁻)OAr ⟶ $PhCH_2CONH$–CH_2–P(=O)(O⁻)–O–E **(85)** + ArOH $\xrightarrow[\text{slow}]{H_2O}$ $PhCH_2CONH$–CH_2–$PO_3^{=}$ + EOH

Scheme 7.16

The rate of the inactivation reaction is strongly dependent on the leaving group ability of the aryloxy group; the alkyl phosphonate (**86**) is not an inhibitor. The P99 enzyme was much more susceptible to these inhibitors than were several class A enzymes. The phosphonate (**87**) was a good inhibitor of the class C enzyme but does not inhibit class A enzymes; better leaving groups are needed to inhibit the latter.[227] The greater susceptibility of class C than class A enzymes to these inhibitors is also observed with the

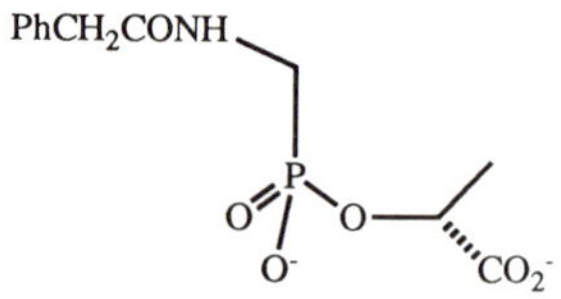

$PhCH_2CONH$–CH_2–P(=O)(O⁻)–O–CH(CH_3)CO_2^-

(86)

$PhCH_2CONH$–CH_2–P(=O)(O⁻)–O–C_6H_4–CO_2^-

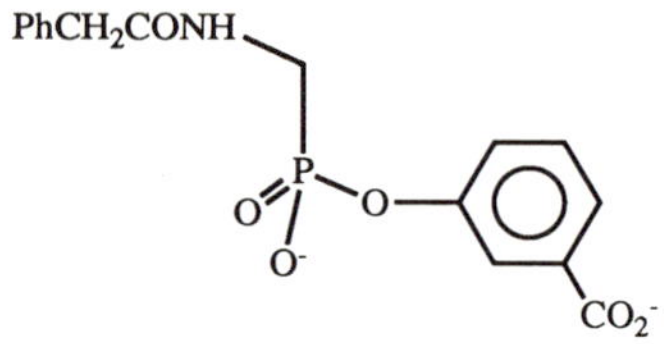

(87)

boronates, and presumably reflects the same differences in active site structure.

The unexpected phosphonylation reaction (Scheme 7.16) has been proposed to reflect the marked ability of the β-lactamase active site to stabilize tetrahedral anions, such as the tetrahedral intermediate of the β-lactamase reaction, by electrostatic means. The conserved Lys-73, whose ammonium ion is directly adjacent to the active site serine hydroxyl group[24,25,33] may be involved in such electrostatic stabilization,[34] in support of the oxyanion hole[24,25,38] and the macrodipole of the α-2-helix.[24,25] The lysine ammonium ion may also provide specific stabilization to the transition state for phosphonylation of the serine hydroxyl group by the phosphonate inhibitors in a reaction closely analogous to the normal acylation reaction (Scheme 7.17). The adduct (**85**) would therefore be an analog of the tetrahedral intermediate involved in deacylation of the enzyme.[226,227]

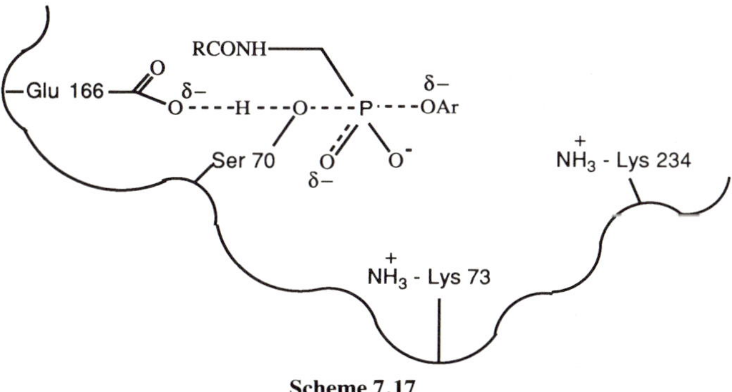

Scheme 7.17

7.3.3 Other inhibitors

There are few new additions to this category. No functional group specific reagents have been usefully employed recently, although phenylpropynal has been shown to be an interesting cross-linker of lysine residues, but not specific to β-lactamases.[32,183] Page and coworkers[230] have shown that α-methyl benzylpenicilloate (**88**) is a rather better (0.67 mM vs. ≥ 50 mM) inhibitor, apparently fast, reversible and non-covalent, of *B. cereus* β-lactamase I than is benzylpenicilloate itself. They interpret this in terms of an unfavorable electrostatic interaction between the Glu-166 carboxylate of the active site and the α-carboxylate of benzylpenicilloate, which, of course,

$PhCH_2CONH$ S

MeO_2C NH

CO_2^-

(**88**)

is not present when (**88**) is bound. β-Lactamase I, unlike class C enzymes,[231] does not catalyse the hydrolysis of (**88**). Another analog of izumenolide[232] and dotriacolide[233] has been described.[234] These detergent-like molecules inhibit not only β-lactamases but a wide range of other enzymes. Other curiosities from natural sources include a β-lactamase-binding protein from *Streptomyces clavuligerus* that is also an inhibitor[235] and β-lactamase-inhibitory materials from higher plants.[236]

7.4 Inhibitors of class B (metallo) β-lactamases

There has been little advance on this front either, and no effective inhibitors other than metal sequestering reagents have yet been found or devised. Some notable failures should be mentioned, however. On the basis of the success of phosphonates and phosphonamidates with carboxypeptidase A,[228,237] Bartlett and coworkers prepared the cyclic phosphonate (**89**) as a potential inhibitor of *B. cereus* β-lactamase II. It apparently had little or no effect on the enzyme.[238] In a similar vein, although depsipeptides of structure (**83**) are substrates of *B. cereus* β-lactamase II,[20,43] the phosphonates (**86**) and (**87**) have little inhibitory activity[226] with this enzyme. These observations are rather difficult to understand in terms of the evidence for productive metal–ion substrate contact[48] and proposed mechanisms of catalysis.[48,49]

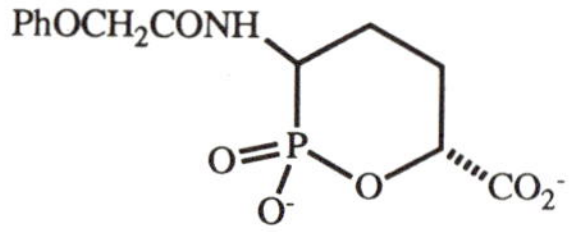

$PhOCH_2CONH$; O=P ; O^- ; O ; CO_2^-

(**89**)

The recent demonstration that a β-lactamase capable of catalysing the hydrolysis of penems, and thus providing *Bacterioides fragilis* with resistance to these antibiotics, is a class B β-lactamase[239,240] is likely to stimulate activity in this area.

7.5 Clinical indications

β-Lactamase inhibitors have been shown to potentiate the activity of β-lactamase-susceptible β-lactam antibiotics in circumstances where bacterial resistance derives from β-lactamase production.[241,242] Until recently, however, the preferred strategy employed by pharmaceutical companies to overcome resistance of this type has been to develop either β-lactamase-resistant β-lactams[243] or molecules that are effective both as antibiotics, and as β-lactamase inhibitors; the latter were generally first recognized as molecules that were resistant to the action of β-lactamases. This strategy avoids the 'polypharmacy' problem presented by β-lactamase inhibitors as separate molecules.

Since 1985, the clinical application of β-lactamase inhibitors has apparently become more attractive.[244–246] This is due largely, no doubt, to the impressive commercial success of 'augmentin' — the mixture of clavulanic acid and amoxicillin — and, more recently, 'unasyn' — the sulbactam (penicillanic acid sulfone) and ampicillin mixture. 'Timentin' (clavulanic acid and ticarcillin) has also been successful and the soon-emerging combination of tazobactam (**26**) and piperacillin appears promising. Many comparative studies of these combinations, particularly of augmentin vs. unasyn, under a wide variety of *in vitro* and *in vivo* circumstances, have now been reported. References 247–255 provide typical examples. In general, the sulfone combinations seem more effective than augmentin against organisms producing chromasomal cephalosporinases while the reverse holds for plasmid-borne penicillinases.

In view of these successes, and the number of effective β-lactamase inhibitors and β-lactam antibiotics now available, it is not surprising to find a variety of new combinations in various stages of development. A mixture of brobactam (6-β-bromopenicillanic acid) and ampicillin has recently been compared favorably with augmentin.[256] Other variants that have been examined in at least a preliminary way include clavulanic acid or sulbactam with third-generation cephalosporins,[249,257–260] imipenem with ampicillin,[261] various penems with amoxicillin and cephalosporins[262,263] and a 6-methylenepenam with amoxicillin.[264] It is also interesting to note in passing that some indication of synergy between β-lactam antibiotics and boronates has been detected.[220,265]

The mechanism-based β-lactamase inhibitors (as β-lactams themselves) face essentially the same obstacles in reaching their site of action and achieving their goal as do their antibiotic partners. First, there are problems of stability — intrinsic, and under therapeutically relevant conditions.[266] Sulbactam, for example, seems more stable, generally, than clavulanic acid.[267] Stability has always been a problem with penems[268] and clavems.[269] 6-Acetylmethylenepenicillanic acid (**29**) appeared to fail *in vivo* through its low stability/high reactivity,[270] a finding which initiated the search for stabler pro-drugs.[271] Penems and carbapenems are unique in that they are also susceptible to degradation *in vivo* by renal dehydropeptidase I. The realization of this problem led to the development of cilastatin, a dehydropeptidase I inhibitor, as an important complement to imipenem.[272] All new penems are now routinely examined for stability to dehydropeptidase I,[273] and the search for specific inhibitors of this enzyme continues.[274,275] The therapeutic effectiveness of a β-lactamase inhibitor will also depend on the level of β-lactamase production by the target organism. If the inhibitor is also a β-lactamase inducer, as it may well be if it is a β-lactam, the benefit of its presence could well be nullified. It seems that clavulanic acid is only a weak inducer of β-lactamases, although somewhat stronger than the sulfones.[276–279] Pharmacokinetics, as always, are important,[280] as is the ability — where gram-negative targets are involved — to penetrate the outer membrane.[256,281,282] Although there are significant problems, therefore, in the application of β-lactamase inhibitors to clinical practice, there seems no reason — in principle at least — why these cannot be overcome, just as they have been with the β-lactam antibiotics themselves.

7.6 Retrospects and prospects

The β-lactamase inhibitors have proven themselves to be no less fascinating chemically than the β-lactamase substrates. A wide range of mechanistic approaches to β-lactamase inhibitors is now available to guide further design. Their standing in pharmaceutical circles has waxed, waned, and waxed again with the perceived importance of β-lactamases in bacterial resistance to β-lactam antibiotics. They seem at present to be poised, basking in the glow of the success of augmentin and unasyn, perhaps gathering momentum for a major expansion throughout the β-lactam antibiotic market. The inhibitors now available, and logical analogs of them, appear capable of successfully fine-tuning the specificity of established β-lactams against most foreseeable challenges from β-lactamases.

Future research in this area should include the following.

1. The marshalling of β-lactams and β-lactamase inhibitors into combinations capable of dealing with newly arising β-lactamases. The current case in point would be the TEM and SHV mutants that can catalyse hydrolysis of third-generation cephalosporins more successfully than the parent enzymes.[283–286] The recently recognized threat of a class B β-lactamase[239,240] might also be met with β-lactamase inhibitors although, to date, no specific inhibitors of this class of enzyme have been discovered.
2. The development of β-lactamase inhibitors from newly discovered substrate and transition state analogs of β-lactamases and DD-peptidases. These would include γ-lactams,[287,288] bicyclic pyrazolidinones,[289,290] lactivicins,[291–293] and phosphonate monoesters.[226,227]
3. The employment of the emerging β-lactamase crystal structures in both β-lactam antibiotic, and β-lactamase inhibitor design. At present, there are at least four class A,[18,24,25,26,198] one class B,[50] and two class C[40,294] β-lactamase crystal structures at various stages of refinement. These structures, along with improved molecular modelling methods,[295–298] could well represent the major source of progress over the next decade. In principle, one should be able to design both

β-lactamase-resistant antibiotics, and more effective β-lactamase inhibitors, through optimization of presently known structures, and by devising new ones. The latter could well derive from features of β-lactamase active sites that overlap with those of DD-peptidases[43,200] and thus merge with antibiotic design.

4. In a related area, specifically designed β-lactams have been recently found to be impressive serine proteinase inhibitors, for example of elastase[64,149,299] and enkephalinase.[300] The variety of motifs employed in the β-lactamase inhibitors may well also find application in this area.

Thus, further exploration of β-lactamase inhibitors is clearly warranted and seems likely to occur into the foreseeable future. Discoveries in this area may also be relevant to the other mechanisms of bacterial resistance to β-lactam antibiotics that involve alterations in penicillin-binding proteins.[301,302]

References

1. E.P. Abraham and E. Chain, *Nature* (1940) **146** 837.
2. W.M.M. Kirby, *Science* (1944) **99** 452–453.
3. A. Biondi, Jr. and C.C. Dietz, *Proc. Soc. Exptl. Biol. Med.* (1945) **60** 55–58.
4. R.D. Reid, L.C. Felton and M.A. Pitroff, *Proc. Soc. Exp. Biol. Med.* (1946) **63** 438–443.
5. O.K. Behrens and L. Garrison, *Arch. Biochem. Biophys.* (1950) **27** 94–98.
6. J.M.T. Hamilton-Miller, J.T. Smith and R. Knox, *Nature (London)* (1964) **201** 867–868.
7. R. Sutherland and F.R. Batchelor, *Nature (London)* (1964) **201** 868–869.
8. M. Cole, S. Elson and P.D. Fullbrook, *Biochem. J.* (1972) **127** 295–308.
9. C.H. O'Callaghan and A. Morris, *Antimicrob. Agents Chemother.* (1972) **2** 442–448.
10. G.N. Rolinson, *Proc. R. Soc. Lond. (B)* (1971) **179** 403–410.
11. S.J. Cartwright and S.G. Waley, *Med. Res. Rev.* (1983) **3** 341–382.
12. J. Fisher, in *Antimicrobial Drug Resistance* (Ed. L.E. Bryan), Academic Press, Orlando, Florida (1984) pp. 33–79.
13. J.R. Knowles, *Acc. Chem. Res.* (1985) **18** 97–104.
14. A.L. Fink, *Pharm. Res.* (1985) 55–61.
15. R.F. Pratt, in *Design of Enzyme Inhibitors as Drugs* (Eds M. Sandler and H.J. Smith), Oxford University Press, Oxford (1989) pp. 178–205.
16. D. Mossakowska, N.A. Ali and J.W. Dale, *Eur. J. Biochem.* (1989) **180** 309–318.
17. J.A. Kelly, O. Dideberg, P. Charlier, J.P. Wery, M. Libert, P.C. Moews, J.R. Knox, C. Duez, C. Fraipont, B. Joris, J. Dusart, J.-M. Frère and J.-M. Ghuysen, *Science* (1986) **231** 1429–1431.
18. B. Samraoui, B.J. Sutton, R.J. Todd, R.P. Artymiuk, S.G. Waley and D.C. Phillips, *Nature (London)* (1986) **320** 378–380.
19. B. Joris, J.-M. Ghuysen, G. Dive, A. Renard, O. Dideberg, P. Charlier, J.-M. Frère, J.A. Kelly, J.C. Boyington, P.C. Moews and J.R. Knox, *Biochem. J.* (1988) **250** 313–324.
20. R.F. Pratt and C.P. Govardhan, *Proc. Natl. Acad. Sci. USA* (1984) **84** 1302–1306.
21. D.J. Tipper and J.L. Strominger, *Proc. Natl. Acad. Sci. USA* (1965) **54** 1133–1141.
22. R.P. Ambler, *Phil. Trans. R. Soc. London (B)* (1980) **289** 321–331.
23. R.P. Ambler, A.F.W. Coulson, J.-M. Frère, J.-M. Ghuysen, B. Joris, M. Forsman, R.C. Levesque, G. Tiraby and S.G. Waley, *Biochem. J.* (1991) **276** 269–272.
24. O. Herzberg and J. Moult, *Science* (1987) **236** 694–701.
25. P.C. Moews, J.R. Knox, O. Dideberg, P. Charlier and J.-M. Frère, *Proteins: Structure, Function and Genetics* (1990) **7** 156–171.

26. O. Dideberg, P. Charlier, J.-P. Wery, P. Dehottay, J. Dusart, T. Erpicum, J.-M. Frère and J.-M. Ghuysen, *Biochem. J.* (1987) **245** 911–913.
27. P.J. Madgwick and S.G. Waley, *Biochem. J.* (1987) **248** 657–662.
28. R.M. Gibson, H. Christensen and S.G. Waley, *Biochem. J.* (1990) **272** 613–619.
29. S.C. Schultz, S.S. Carroll and J.H. Richards, in *Proteins: Structure and Function* (Ed. J.J. L'Italien), Plenum Press, New York (1987), pp. 521–528.
30. L.M. Ellerby, W.A. Escobar, A.L. Fink, C. Mitchinson and J.A. Wells, *Biochemistry* (1990) **29** 5797–5806.
31. C. Little, E.L. Emanuel, J. Gagnon and S.G. Waley, *Biochem. J.* (1986) **240** 215–219.
32. M.E. Grace and R.F. Pratt, *J. Biol. Chem.* (1987) **262** 16778–16785.
33. A.K. Knap and R.F. Pratt, *Proteins: Structure, Function and Genetics* (1989) **6** 316–323.
34. A.K. Knap and R.F. Pratt, *Biochem. J.* (1991) **273** 85–91.
35. A.P. Laws and M.I. Page, *J. Chem. Soc., Perkin Trans 2* (1989) 1577–1581.
36. H. Adachi, T. Ohta and H. Matsuzawa, *J. Biol. Chem.* (1991) **266** 3186–3191.
37. J. Kraut, *Ann. Rev. Biochem.* (1977) **46** 331–358.
38. B.P. Murphy and R.F. Pratt, *Biochem. J.* (1988) **256** 669–672.
39. F. Jacob, B. Joris, S. Lepage, J. Dusart and J.-M. Frère, *Biochem. J.* (1990) **271** 399–406.
40. C. Oefner, A. D'Arcy, J.J. Daly, K. Gubernator, B. Charnas, I. Heinz, C. Hubschwerlen and F.K. Winkler, *Nature (London)* (1990) **343** 284–288.
41. J.A. Kelly, J.R. Knox, H. Zhao, J.-M. Frère and J.-M. Ghuysen, *J. Mol. Biol.* (1989) **209** 281–295.
42. J.R. Knox and J.A. Kelly, in *Chemical and Biochemical Problems in Molecular Recognition* (Ed. S.M. Roberts), Royal Society of Chemistry, London (1989), pp. 46–55.
43. B.P. Murphy and R.F. Pratt, *Biochemistry* (1991) **30** 3640–3649.
44. Y.-H. Chang, M.R. Labgold and J.H. Richards, *Proc. Natl. Acad. Sci. USA* (1990) **87** 2823–2827.
45. D.W. Christianson, P.R. David and W.N. Lipscomb, *Proc. Natl. Acad. Sci. USA* (1987) **84** 1512–1517.
46. R. Bicknell and S.G. Waley, *Biochemistry* (1985) **24** 6876–6887.
47. R. Bicknell, A. Schaffer, S.G. Waley and D.S. Auld, *Biochemistry* (1986) **25** 7208–7215.
48. B.P. Murphy and R.F. Pratt, *Biochem. J.* (1989) **258** 765–768.
49. C. Little, E.L. Emanuel, J. Gagnon and S.G. Waley, *Biochem. J.* (1986) **233** 465–469.
50. B.J. Sutton, P.S. Artymiuk, A.E. Cordero-Borboa, C. Little, D.C. Phillips and S.G. Waley, *Biochem. J.* (1987) **248** 181–188.
51. H.M. Lim and J.J. Pène, *J. Biol. Chem.* (1989) **264** 11682–11687.
52. H.M. Lim, R.K. Iyer and J.J. Pène, *Biochem. J.* (1991) **276** 401–404.
53. A. Maycock and R. Abeles, *Acc. Chem. Res.* (1976) **9** 313–319.
54. C. Walsh, *Tetrahedron* (1982) **7** 871–909.
55. D.W. Christianson and W.N. Lipscomb, *Acc. Chem. Res.* (1989) **22** 62–69.
56. S. Tatsunami, N. Yago and M. Hosoe, *Biochim. Biophys. Acta* (1981) **662** 226–235.
57. R.B. Silverman, in *Mechanism-Based Enzyme Inactivation: Chemistry and Enzymology*, Vol. I, CRC Press, Boca Raton, Florida (1988), Ch. 1.
58. T. Funaki, Y. Takanohashi, H. Fukazawa and I. Kuruma, *Biochim. Biophys. Acta* (1991) **1078** 43–46.
59. K. Bush and R.B. Sykes, *Antimicrob. Agents Chemother.* (1987) **30** 6–10.
60. R.B. Silverman, in *Mechanism-Based Enzyme Inactivation: Chemistry and Enzymology*, Vol. I, CRC Press, Boca Raton, Florida (1988), Ch. 5.
61. M. Zrihen, R. Labia and M. Wakselman, *Eur. J. Med. Chem. Chim. Ther.* (1983) **18** 307–314.
62. M. Vilkas, in *Enzyme-Activated Irreversible Inhibitors* (Eds N. Seiler, M.J. Jung and J. Koch-Weser), Elsevier-North Holland, Amsterdam (1978), pp. 323–335.
63. R. Joyeau, H. Molines, R. Labia and M. Wakselman, *J. Med. Chem.* (1988) **31** 370–374.
64. M. Wakselman, R. Joyeau, R. Kobaiter, N. Bogetto, I. Vergely, J. Maillard, V. Okochi, J.-J. Montague and M. Rebaud-Ravaud, *FEBS Lett.* (1991) **282** 377–381.
65. K. Tanizawa, K. Sanroh and Y. Kanaoka, *FEBS Lett.* (1989) **250** 218–220.
66. E.H. White, D.F. Roswell, I.R. Pollitzer and B.R. Branchini, *J. Amer. Chem. Soc.* (1975) **97** 2290–2291.
67. R.L. Charnas and J.R. Knowles, *Biochemistry* (1981) **20** 2732–2737.

68. C.J. Easton and J.R. Knowles, *Biochemistry* (1982) **21** 2857–2862.
69. K. Okonogi, Y. Nozaki, A. Imada and M. Kuno, *J. Antibiot.* (1981) **34** 212–217.
70. K. Okonogi, S. Harada, S. Shinagawa and A. Imada, *J. Antibiot.* (1982) **35** 963–971.
71. A. Yamaguchi, T. Hirata and T. Sawai, *Antimicrob. Agents Chemother.* (1984) **25** 348–353.
72. T. Hashizume, A. Yamaguchi, T. Hirata and T. Sawai, *Antimicrob. Agents Chemother.* (1984) **25** 149–151.
73. T. Hashizume, A. Yamaguchi and T. Sawai, *Chem. Pharm. Bull.* (1988) **36** 676–684.
74. J. Monks and S.G. Waley, *Biochem. J.* (1988) **253** 323–328.
75. W.S. Faraci and R.F. Pratt, *Biochemistry* (1985) **24** 903–910.
76. W.S. Faraci and R.F. Pratt, *Biochemistry* (1986) **25** 2934–2941.
77. R.L. Charnas and R.L. Then, *Rev. Infect. Dis.* (1988) **10** 752–760.
78. L.J. Mazzella and R.F. Pratt, *Biochem. J.* (1989) **259** 255–260.
79. W.S. Faraci and R.F. Pratt, *Biochem. J.* (1986) **238** 309–312.
80. D.B. Boyd, *J. Med. Chem.* (1984) **27** 63–66.
81. M. Narisada, J. Nishikawa, F. Watanabe and Y. Terui, *J. Med. Chem.* (1987) **30** 514–522.
82. K. Murakami, M. Doi and T. Yoshida, *Antimicrob. Agents Chemother.* (1986) **30** 447–452.
83. R. Labia, A. Morand and J. Peduzzi, *J. Antimicrob. Chemother.* (1983) **11**(Suppl. A) 153–157.
84. G. Cassinelli, R. Corigli, P. Orezzi, G. Ventrella, A. Bedeschi, E. Perrone, D. Borghi and G. Franceschi, *J. Antibiot.* (1988) **41** 984–987.
85. G. Franceschi, E. Perrone, M. Alpegiani, A. Bedeschi, C. Della Bruna and F. Zarini, in *Recent Advances in the Chemistry of β-Lactam Antibiotics* (Eds P.H. Bentley and R. Southgate) Royal Society of Chemistry, London (1989), pp. 222–246.
86. H. Iwata, R. Tanaka and M. Ishiguro, *J. Antibiot.* (1990) **43** 901–903.
87. M. Alpegiani, A. Bedeschi, E. Perrone, F. Zarini and G. Franceschi, *Heterocycles* (1985) **23** 2255–2270.
88. E. Perrone, M. Alpegiani, A. Bedeschi, F. Guidici, F. Zarini, G. Franceschi, C. Della Bruna, D. Jabes and G. Meinardi, *J. Antibiot.* (1986) **39** 1351–1355.
89. R.L. Then and P. Angehrn, *Antimicrob. Agents Chemother.* (1982) **21** 711–717.
90. D.M. Livermore, *J. Antimicrob. Chemother.* (1987) **20** 7–13.
91. J. Fisher, R.L. Charnas and J.R. Knowles, *Biochemistry* (1978) **17** 2180–2184.
92. R.L. Charnas, J. Fisher and J.R. Knowles, *Biochemistry* (1978) **17** 2185–2189.
93. R.L. Charnas and J.R. Knowles, *Biochemistry* (1981) **20** 3214–3219.
94. S.J. Cartwright and A.F.W. Coulson, *Nature (London)* (1979) **278** 360–361.
95. I. Rizwi, A.K. Tan, A.L. Fink and R. Virden, *Biochem. J.* (1989) **258** 205–209.
96. H.C. Neu and K.P. Fu, *Antimicrob. Agents Chemother.* (1978) **14** 650–655.
97. C. Reading and T. Farmer, *Biochem. J.* (1981) **199** 779–787.
98. D.G. Brenner and J.R. Knowles, *Biochemistry* (1984) **23** 5833–5839.
99. P.H. Crackett and R.J. Stoodley, *Tetrahedron Lett.* (1984) **25** 1295–1298.
100. D.G. Brenner and J.R. Knowles, *Biochemistry* (1981) **20** 3680–3687.
101. M. Arisawa and R.L. Then, *J. Antibiot.* (1982) **35** 1578–1583.
102. J. Fisher, R.L. Charnas, S.M. Bradley and J.R. Knowles, *Biochemistry* (1981) **20** 2726–2731.
103. P. Mezes, A. Clarke, G. Dmitrienko and T. Viswanatha, *FEBS Lett.* (1982) **143** 265–267.
104. C.D. Foulds, M. Kasmirak and P.G. Sammes, *J. Chem. Soc., Perkin Trans. 2* (1985) 963–968.
105. T.W. Hall, S.N. Maiti, R.G. Micetich, P. Spevak, S. Yamabe, N. Ishida, M. Kajitani, M. Tanaka and T. Yamasaki, in *Recent Advances in the Chemistry of β-Lactam Antibiotics* (Eds A.G. Brown and S.M. Roberts) The Royal Society of Chemistry, London (1985), pp. 242–254.
106. A. Yamaguchi, A. Adachi, T. Hirata, H. Adachi and T. Sawai, *J. Antibiot.* (1985) **35** 83–93.
107. S. Hanessian and M. Alpegiani, *Tetrahedron* (1989) **45** 941–950.
108. W.J. Gottstein, L.B. Crast, Jr., R.G. Graham, V.J. Haynes and D.N. McGregor, *J. Med. Chem.* (1981) **24** 1531–1534.
109. W.J. Gottstein, U.J. Haynes and D.N. McGregor, *J. Med. Chem.* (1985) **28** 518–522.

110. R.C. Micetich, S.N. Maiti, P. Spevak, T.W. Hall, S. Yamabe, N. Ishida, M. Tanaka, T. Yamazaki, A. Nakai and K. Ogawa, *J. Med. Chem.* (1987) **30** 1469–1474.
111. H. Tanaka, M. Tanaka, A. Nakai, S. Yamada, N. Ishida, T. Otani and S. Torii, *J. Antibiot.* (1988) **41** 579–582.
112. M. Akova, Y. Yang and D.M. Livermore, *J. Antimicrob. Chemother.* (1990) **25** 199–208.
113. A. Yamaguchi, M. Nemoto, A. Adachi, T. Inaba and T. Sawai, *J. Antibiot.* (1986) **39** 1744–1753.
114. A. Yamaguchi, T. Inaba and T. Sawai, *J. Antibiot.* (1988) **41** 690–693.
115. K. Tanaka, J. Shoji, Y. Terui, N. Tsuji, E. Kondo, M. Mayama, Y. Kawamura, T. Hattori, K. Matsumoto and T. Yoshida, *J. Antibiot.* (1981) **34** 909–911.
116. K. Murakami, M. Doi and T. Yoshida, *J. Antibiot.* (1982) **35** 39–45.
117. M. Arisawa and R.L. Then, *J. Antibiot.* (1982) **35** 1578–1583.
118. D.G. Brenner and J.R. Knowles, *Biochemistry* (1984) **23** 5839–5846.
119. S. Adam, R.L. Then and P. Angehrn, *J. Antibiot.* (1987) **40** 108–109.
120. N.J.P. Broom, K. Coleman, P.A. Hunter and N.F. Osborne, *J. Antibiot.* (1990) **43** 76–82.
121. I. Bennett, N.J.P. Broom, G. Bruton, S. Calvert, B.P. Clarke, K. Coleman, R. Edmondson, P. Edwards, D. Jones, N.F. Osborne and G. Walker, *J. Antibiot.* (1991) **44** 331–337.
122. K. Coleman, D.R.J. Griffin, J.W.J. Page and P.A. Upshon, *Antimicrob. Agents Chemother.* (1989) **33** 1580–1582.
123. M.J. Basker and N.F. Osborne, *J. Antibiot.* (1990) **43** 70–75.
124. J.D. Buynak, H.B. Borate, C. Husting, T. Hurd, J. Vallabh, J. Matthew, J. Lambert and U. Siriwardane, *Tetrahedron Lett.* (1988) **29** 5053–5056.
125. M. Arisawa and S. Adam, *Biochem. J.* (1983) **211** 447–454.
126. D. Häbich and K. Metzger, *Heterocycles* (1986) **24** 289–296.
127. Y.L. Chen, C.-W. Chang and K. Hedberg, *Tetrahedron Lett.* (1986) **27** 3449–3452.
128. Y.L. Chen, C.-W. Chang, K. Hedberg, K. Guarino, W.M. Welch, L. Kiessling, J.A. Retsema, S.L. Haskell, M. Anderson, M. Manousos and J.F. Barrett, *J. Antibiot.* (1987) **40** 803–822.
129. C.D. Foulds, M. Kosmirak, A.C. O'Sullivan and P.G. Sammes, in *Recent Advances in the Chemistry of β-Lactam Antibiotics* (Eds A.G. Brown and S.M. Roberts), Royal Society of Chemistry, London (1985), pp. 255–265.
130. K. Tanizawa, K. Santoh and Y. Kanaoka, *Chem. Pharm. Bull.* (1989) **37** 824–825.
131. A.C. Kaura and R.J. Stoodley, *J. Chem. Soc., Perkin Trans 1* (1988) 2813–2820.
132. R.F. Pratt and M.J. Loosemore, *Proc. Natl. Acad. Sci. USA* (1978) **75** 4145–4149.
133. V. Knott-Hunziker, B.S. Orlek, P.G. Sammes and S.G. Waley, *Biochem. J.* (1979) **177** 365–367.
134. W. von Daehne, *J. Antibiot.* (1980) **33** 451–452.
135. R. Wise, J.M. Andrews and N. Patel, *J. Antimicrob. Chemother.* (1981) **7** 531–536.
136. F. DeMeester, J.-M. Frère, S.G. Waley, S.J. Cartwright, R. Virden and F. Lindberg, *Biochem. J.* (1986) **239** 575–580.
137. K. Bush, *Rev. Infect. Dis.* (1988) **10** 681–690.
138. J. Fisher, J.G. Belasco, R.L. Charnas, S. Khosla and J.R. Knowles, *Phil. Trans. R. Soc. London (B)* (1980) **289** 309–319.
139. H.A.O. Hill, P.G. Sammes and S.G. Waley, *Phil. Trans. R. Soc. London (B)* (1980) **289** 333–344.
140. S.A. Cohen and R.F. Pratt, *Biochemistry* (1980) **19** 3996–4003.
141. R.F. Pratt and D.J. Cahn, *J. Amer. Chem. Soc.* (1988) **110** 5096–5104.
142. F. DeMeester, A. Matagne, G. Dive and J.-M. Frère, *Biochem. J.* (1989) **257** 245–249.
143. T.G. Heckler and R.A. Day, *Biochim. Biophys. Acta* (1983) **745** 292–300.
144. B.W. Bycroft, R.E. Shute and M.J. Begley, *J. Chem. Soc., Chem. Commun.* (1988) 274–276.
145. B.W. Bycroft, L. Gledhill, R.E. Shute and P. Williams, *J. Chem. Soc., Chem. Commun.* (1988) 1610–1612.
146. L. Gledhill, P. Williams and B.W. Bycroft, *Biochem. J.* (1991) **276** 801–807.
147. R.E. Shute, D.E. Jackson and B.W. Bycroft, *J. Computer-Aided Molec. Design* (1989) **3** 149–164.
148. J. Marchand-Brynaert, R. Laub, F. DeMeester and J.-M. Frère, *Eur. J. Med. Chem.* (1988) **23** 561–571.

149. R.A. Firestone, P.L. Barker, J.M. Pisano, B.M. Ashe and M.E. Dahlgren, *Tetrahedron* (1990) **46** 2255–2262.
150. G.I. Dmitrienko, C.R. Copeland, L. Arnold, M.E. Savard, A.J. Clarke and T. Viswanatha, *Bioorg. Chem.* (1985) **13** 34–46.
151. A.J. Clarke, P.S. Mezes, S.F. Vice, G.I. Dmitrienko and T. Viswanatha, *Biochim. Biophys. Acta* (1983) **748** 389–397.
152. A.L. Fink, L.M. Ellerby and P.M. Bassett, *J. Amer. Chem. Soc.* (1989) **111** 6871–6873.
153. R.H. Pain and R. Virden, in *Beta-Lactamases* (Eds J.M. Hamilton-Miller and J.T. Smith), Academic Press, London (1979), pp. 141–180.
154. N. Citri, in *Beta-Lactam Antibiotics* (Ed. S. Mitsutashi), Japan Sci. Soc. Press, Tokyo (1981), pp. 225–232.
155. N. Citri, A. Samuni and N. Zyk, *Proc. Natl. Acad. Sci. USA* (1976) **73** 1048–1052.
156. P.C. Blanpain, J.B. Nagy, G.H. Laurent and F.V. Durant, *J. Med. Chem.* (1980) **23** 1283–1292.
157. A. Sammuni and A.Y. Meyer, *Mol. Pharmacol.* (1978) **14** 704–709.
158. N. Citri and N. Zyk, *Biochem. J.* (1982) **201** 425–427.
159. N. Citri, A. Kalkstein, A. Samuni and N. Zyk, *Eur. J. Biochem.* (1984) **144** 333–338.
160. R. Labia, A. Morand, C. Verchere-Beaur and A. Bryskier, *Am. J. Med.* (1984) **77** (Suppl. 6A) 25–27.
161. K. Bush, J.S. Freudenberger and R.B. Sykes, *Antimicrob. Agents Chemother.* (1982) **22** 414–420.
162. R.L. Then, *Chemotherapy (Basel)* (1984) **30** 398–407.
163. K. Bush, S.A. Smith, S.K. Tanaka and D.P. Bonner, *J. Antimicrob. Chemother.* (1988) **22** 801–809.
164. C. Yoshida, K. Tanaka, J. Nakano, Y. Yodo, T. Yamafuji, R. Hattori, Y. Fukuoka and I. Saikawa, *J. Antibiot.* (1986) **39** 76–89.
165. P. Brown, S.H. Calvert, P.A.C. Chapman, S.C. Cosham, A.J. Eglington, R.L. Elliot, M.A. Harris, J.D. Hinks, J. Lowther, D.J. Merrikin, M.J. Pearson, R.J. Ponsford and J.V. Syms, *J. Chem. Soc., Perkin Trans 1* (1991) 881–891.
166. K. Jules and H. Neu, *Antimicrob. Agents Chemother.* (1982) **22** 453–460.
167. W. Cullman and W. Dick, *Zbl. Bakt. Hyg. A* (1983) **256** 109–118.
168. K. Okonogi, A. Sugiura, M. Kuno, H. Ono, S. Harada and E. Higashide, *J. Antibiot.* (1985) **38** 1555–1563.
169. M. Murakami, T. Aoki, M. Matsuura and W. Nagata, *J. Antibiot.* (1990) **43** 1441–1449.
170. S. Nishimura, N. Yasuda, H. Sasaki, Y. Matsumoto, T. Kamimura, K. Sakane and T. Takaya, *J. Antibiot.* (1989) **42** 159–162.
171. S. Nishimura, N. Yasuda, H. Sasaki, Y. Matsumoto, T. Kamimura, K. Sakane and T. Takaya, *Bull. Chem. Soc. Jpn.* (1990) **63** 412–416.
172. S. Nishimura, H. Sasaki, N. Yasuda, Y. Matsumoto, T. Kamimura, K. Sakane and T. Takaya, *J. Antibiot.* (1989) **42** 1124–1132.
173. S. Nishimura, N. Yasuda, H. Sasaki, Y. Matsumoto, T. Kamimura, K. Sakane and T. Takaya, *J. Antibiot.* (1990) **43** 114–117.
174. K. Matsuda, M. Sanada, S. Nakagawa, M. Inoue and S. Mitsuhashi, *Antimicrob. Agents Chemother.* (1991) **35** 458–461.
175. N.S. Isaacs, G. Sunman and C. Reading, *J. Antibiot.* (1982) **35** 589–593.
176. Q. Gu and Z. Li, *Yaoxue Xuebao* (1985) **20** 896–901; *Chem. Abs.* (1986) **105** 190727.
177. Y. Jin, S. Zhang, Z. Ou, D. Gou and W. Mao, *Kangshengsu* (1987) **12** 211–214; *Chem. Abs.* (1987) **107** 151064.
178. D.M. Gou and W.R. Mao, *Yaoxue Xuebao* (1988) **23** 174–179; *Chem. Abs.* (1988) **110** 23594.
179. X. Tong and Z. Li, *Huaxue Xuebao* (1988) **46** 719–723; *Chem Abs.* (1989) **111** 7105.
180. S.F. Yin and W.R. Mao, *Yaoxue Xuebao* (1990) **25** 340–344; *Chem. Abs.* (1991) **114** 23600.
181. S. Yin and W. Mao, *Huaxue Xuebao* (1990) **48** 1212–1215; *Chem. Abs.* (1991) **114** 142930.
182. G. Amicosante, A. Oratore, M.C. Marinucci, M.I. Tizzani, F. Scazzocchio, C. Crifo and R. Strom, *Biochem. Int.* (1987) **14** 15–26.
183. D.P. Schenkein and R.F. Pratt, *J. Biol. Chem.* (1980) **255** 45–48.
184. N. Zyk and N. Citri, *Biochim. Biophys. Acta* (1968) **159** 327–339.

185. K.G.H. Dyke, *Biochem. J.* (1967) **103** 641–646.
186. E.A. Carrey, R. Virden and R.H. Pain, *Biochim. Biophys. Acta* (1984) **785** 104–110.
187. Y. Klemes and N. Citri, *Biochim. Biophys. Acta* (1979) **567** 401–409.
188. Y. Klemes and N. Citri, *Biochem. J.* (1980) **187** 529–532.
189. P.A. Kiener and S.G. Waley, *Biochem. J.* (1977) **165** 279–285.
190. K.C. Persaud, R.H. Pain and R. Virden, *Biochem. J.* (1986) **237** 723–730.
191. A.L. Fink, K.M. Behner and A.K. Tan, *Biochemistry* (1987) **26** 4248–4258.
192. L.W. Hardy and J.F. Kirsch, *Arch. Biochem. Biophys.* (1989) **268** 338–348.
193. V. Knott-Hunziker, S. Petursson, G.S. Jayatilake, S.G. Waley, B. Jaurin and T. Grundstrom, *Biochem. J.* (1982) **201** 621–627.
194. B. Joris, J. Dusart, J.-M. Frère, J. van Beeumen, E.L. Emanuel, S. Petursson, J. Gagnon and S.G. Waley, *Biochem. J.* (1984) **223** 271–274.
195. M. Galleni and J.-M. Frère, *Biochem. J.* (1988) **255** 119–122.
196. M. Galleni, G. Amicosante and J.-M. Frère, *Biochem. J.* (1988) *255* 123–129.
197. E.A. Carrey and R.H. Pain, *Biochim. Biophys. Acta* (1978) **533** 12–22.
198. O. Herzberg, *J. Mol. Biol.* (1991) **217** 701–719.
199. N. Zyk and N. Citri, *Biochim. Biophys. Acta* (1968) **151** 306–308.
200. S. Pazhanisamy and R.F. Pratt, *Biochemistry* (1989) **28** 6875–6882.
201. L. Frick and R. Wolfenden, in *Design of Enzyme Inhibitors as Drugs* (Eds M. Sandler and H.J. Smith) Oxford University Press, Oxford (1989), pp. 19–48.
202. P.A. Bartlett, *Stud. Org. Chem. (Amsterdam)* (1985) **20** 439–450.
203. E.S. Gubnitskaya, Z.T. Semashko, V.S. Parkhomenko and A.V. Kirsanov, *J. Gen. Chem. USSR* (1980) **50** 1746–1757.
204. E.M. Gordon, J. Pluscec and M.A. Ondetti, *Tetrahedron Lett.* (1981) **22** 1871–1874.
205. O. Meth-Cohn, A.J. Reason and S.M. Roberts, *J. Chem. Soc., Chem. Commun.* (1982) 90–92.
206. M.I. Page, *Adv. Phys. Org. Chem.* (1987) **23** 165–270.
207. B.E. Tamczuk, *Ph.D. Thesis*, University of Connecticut; *Diss. Abst.* (1980) **41** 576.
208. G. Lange, M.E. Savard, T. Viswanathan and G.I. Dmitrienko, *Tetrahedron Lett.* (1985) **26** 1791–1795.
209. J.A. Kelly, J.R. Knox, P.C. Moews, G.J. Hite, J.B. Bartolone, H. Zhao, B. Joris, J.-M. Frère and J.-M. Ghuysen, *J. Biol. Chem.* (1985) **260** 6449–6458.
210. A.J. Cocuzza and G.A. Boswell, *Tetrahedron Lett.* (1985) **26** 5363–5366.
211. P.G. Sammes, S. Smith and G.T. Woolley, *J. Chem. Soc., Perkin Trans. 1* (1984) 2603–2610.
212. G. Lowe and S. Swain, *J. Chem. Soc., Chem. Commun.* (1983) 1279–1281.
213. D. Schirlin and M. Jung, *Eur. Pat. Appl.* (1989); *Chem. Abst.* (1989) **110** 173757.
214. P. Bey, M. Angelastro and S. Mehdi, *Eur. Pat. Appl.* (1990); *Chem. Abst.* (1991) **114** 123084.
215. F. Cavagna, W. Koller, A. Linkies, H. Rehling and D. Reuschling, *Angew, Chem. Int. Ed.* (1982) **21** 548–549.
216. P. Loiseau, M. Bonnafous and Y. Adam, *Eur. J. Med. Chem.-Chim. Ther.* (1984) **19** 569–570.
217. O. Dobozy, I. Mile, I. Ferencz and V. Csanyi, *Acta Biochim. Biophys. Acad. Sci. Hung.* (1971) **6** 97–105.
218. P.A. Kiener and S.G. Waley, *Biochem. J.* (1978) **169** 197–204.
219. T. Beesley, N. Gascoyne, V. Knott-Hunziker, S. Petursson, S.G. Waley, B. Jaurin and T. Grundstrom, *Biochem. J.* (1983) **209** 229–233.
220. S.J. Cartwright and S.G. Waley, *Biochem. J.* (1984) **221** 505–512.
221. I.E. Crompton, B.K. Cuthbert, G. Lowe and S.G. Waley, *Biochem. J.* (1988) **251** 453–459.
222. C.A. Kettner and A.B. Shenvi, *J. Biol. Chem.* (1984) **259** 15106–15112.
223. J.F. Morrison and C.T. Walsh, *Adv. Enzymol. Relat. Areas Mol. Biol.* (1988) **61** 201–301.
224. J.E. Baldwin, T.D.W. Claridge, A.E. Derome, B.D. Smith, M. Twyman and S.G. Waley, *J. Chem. Soc., Chem. Commun.* (1991) 573–574.
225. C.P. Govardhan and R.F. Pratt, *Biochemistry* (1987) **26** 3385–3395.
226. R.F. Pratt, *Science* (1989) **246** 917–919.
227. J. Rahil and R.F. Pratt, *Biochem. J.* (1991) **275** 793–795.

228. J.E. Hanson, A.P. Kaplan and P.A. Bartlett, *Biochemistry* (1989) **28** 6294–6305.
229. J. Rahil and R.F. Pratt, *Biochemistry* (In Press).
230. M. Jones, S.C. Buckwell, M.I. Page and R. Wrigglesworth, *J. Chem. Soc., Chem. Commun.* (1989) 70–71.
231. V. Knott-Hunziker, S. Petursson, S.G. Waley, B. Jaurin and T. Grundstrom, *Biochem. J.* (1982) **207** 315–322.
232. W.C. Liu, G. Astle, J.S. Wells, Jr., W.H. Trejo, P.A. Principe, M.L. Rathnum, W.L. Parker, O.R. Kocy and R.B. Sykes, *J. Antibiot.* (1980) **33** 1256–1261.
233. Y. Ikeda, S. Kondo, T. Sawa, M. Tsuchiya, D. Ikeda, M. Hamada, T. Takeuchi and H. Umezawa, *J. Antibiot.* (1981) **34** 1628–1630.
234. M. Uyeda, H. Oshima, K. Nishi, K. Suzuki, S. Yahara, T. Nohara and M. Shibata, *J. Enzyme Inhib.* (1989) **2** 279–294.
235. J.L. Doran, B.K. Leskiw, S. Aippersbach and S.E. Jensen, *J. Bacteriol.* (1990) **172** 4909–4918.
236. M. Jiminez-Valera, A. Ruiz-Bravo and A. Ramos-Cormenzana, *J. Antibiot. Chemother.* (1987) **19** 31–37.
237. N.E. Jacobsen and P.A. Bartlett, *J. Amer. Chem. Soc.* (1981) **103** 654–657.
238. P.A. Bartlett, L.J. Vanmaele and W.B. Kezer, *Bull. Soc. Chim. Fr.* (1986) 776–780.
239. J.S. Thompson and M.H. Malamy, *J. Bact.* (1990) **172** 2584–2593.
240. B.A. Rasmussen, Y. Gluzman and F.P. Tally, *Antimicrob. Agents Chemother.* (1990) **34** 1590–1592.
241. C.J. Easton and J.R. Knowles, *Antimicrob. Agents Chemother.* (1984) **26** 358–363.
242. H.C. Neu, *Pharmac. Ther.* (1985) **30** 1–18.
243. M.H. Richmond, *Rev. Infect. Dis.* (1979) **1** 30–38.
244. R.N. Jones, *J. Reprod. Med.* (1988) **33** (Suppl.) 574–578.
245. G.R. Donowitz and G.R. Mandell, *New Engl. J. Med.* (1988) **318** 419–426.
246. P.C. Fuchs and A.L. Barry, *J. Reprod. Med.* (1990) **35** (Suppl.) 317–321.
247. M.R. Jacobs, S.C. Aronoff, S. Johenning and S. Yamabe, *J. Antimicrob. Chemother.* (1986) **18** 177–184.
248. M.R. Jacobs, S.C. Aronoff, S. Johenning, D.M. Shlaes and S. Yamabe, *Antimicrob. Agents Chemother.* (1986) **29** 980–985.
249. S. Kobayashi, S. Arai, S. Hayashi and T. Sakaguchi, *Antimicrob. Agents Chemother.* (1989) **33** 331–335.
250. P.J. Masters, C.E. Cooper, K.E. Griffin, B. Slocum and R. Sutherland, *J. Drug Dev.* (1989) Suppl. 2 51–54.
251. N.A. Kuck, N.V. Jacobus, P.J. Petersen, W.J. Weiss and R.T. Testa, *Antimicrob. Agents Chemother.* (1989) **33** 1964–1969.
252. F. Higashitani, A. Hyodo, N. Ishida, M. Inoue and S. Mitsuhashi, *J. Antimicrob. Chemother.* (1990) **25** 567–574.
253. K.E. Aldridge, A. Henderberg and C.V. Sanders, *J. Antimicrob. Chemother.* (1990) **25** 873–875.
254. J. Kempers and D.M. MacLaren, *J. Antimicrob. Chemother.* (1990) **26** 598–599.
255. K.S. Thomson, D.A. Weber, C.C. Sanders and W.E. Sanders, Jr., *Antimicrob. Agents Chemother.* (1990) **34** 622–627.
256. N.H. Melchior and J. Keiding, *J. Antimicrob. Chemother.* (1991) **27** 29–40.
257. P.C. Fuchs, R.N. Jones and A.L. Barry, *Diagn. Microbiol. Infect. Dis.* (1989) **8** 61–65.
258. M.M. Martin, A.M. Castillo, J. Liebana, P. Mesa, G. Piedrola and J. Gutierrez, *Med. Lab. Sci.* (1990) **47** 163–167.
259. A. Bauernfeind, *Infection (Munich)* (1990) **18** 56–60.
260. C.C. Knapp, J. Sierra-Madero and J.A. Washington, *Diagn. Microbiol. Infect. Dis.* (1990) **13** 45–49.
261. K. Saionji, M. Shitara, N. Miyake, K. Inoue and I. Kabayashi, *Jpn. J. Antibiot.* (1990) **43** 1685–1697.
262. T. Niwa, T. Yoshida, A. Tamura, Y. Kazuno, S. Inouye, T. Ito and M. Kojima, *J. Antibiot.* (1986) **39** 943–955.
263. I. Bennett, N.J.P. Broom, G. Bruton, S. Calvert, B.P. Clarke, K. Coleman, R. Edmondson, P. Edwards, D. Jones, N.F. Osborne and G. Walker, *J. Antibiot.* (1991) **44** 331–337.

264. K. Coleman, D.R.J. Griffin, J.W.J. Page and P.A. Upshon, *Antimicrob. Ag. Chemother.* (1989) **33** 1580–1587.
265. G. Amicosante, A. Felici, B. Segatore, L. DiMarzio, N. Franceschini and M. Di Girolamo, *J. Chemother. (Florence)* (1989) **1** 394–398.
266. H. Erichsen, *Laboratoriumsmedizin.* (1990) **14** 137–138.
267. A. Wildfeuer and K. Raeder, *Arznein. Forsch.* (1991) **41** 70–73.
268. R.B. Woodward, *Proc. R. Soc. London (B)* (1980) **289** 239–250.
269. C.E. Newall, in *Recent Advances in the Chemistry of β-Lactam Antibiotics* (Ed. G.I. Gregory), Royal Society of Chemistry, London (1981), pp. 151–169.
270. M. Arisawa and R.L. Then, *J. Antibiot.* (1983) **36** 1372–1379.
271. S. Adam, R. Then and P. Angehrn, *J. Antibiot.* (1986) **39** 833–838.
272. F.M. Kahan, H. Kropp, J.G. Sundelof and J. Birnbaum, *J. Antimicrob. Chemother.* (1983) **12** (Suppl. D) 1–35.
273. I.S. Bennett, N.J.P. Broom, K. Coleman, S. Coulton, P.D. Edwards, I. Francois, D.R.J. Griffin, N.F. Osborne and P.M. Woodall, *J. Antibiot.* (1991) **44** 338–343.
274. S. Hashimoro, H. Murai, K. Nitta, A. Fujie, M. Okuhara, M. Kohsaka and H. Imanaka. *J. Antibiot.* (1990) **43** 281–285.
275. Y.-Q. Wu and S. Mobashery, *J. Med. Chem.* (1991) **34** 1914–1916.
276. F. Moosdeen, J. Keeble and J.D. Williams, *Rev. Infect. Dis.* (1986) **8** (Suppl. S) 562–568.
277. T.H. Farmer and C. Reading, *J. Drug Dev.* (1989) Suppl. 2 47–50.
278. D.A. Weber and C.C. Sanders, *Antimicrob. Agents Chemother.* (1990) **34** 156–158.
279. A. Kazmierak, X. Cordin, J.M. Duez, E. Siebor, A. Pechinot and J. Sirot, *J. Int. Med. Res.* (1990) **18** (Suppl. 4) 67–77.
280. B. Hampel, H. Lode, G. Bruckner and P. Koeppe, *Drugs* (1988) **35** (Suppl. 7) 29–33.
281. H. Nikaido, in *Handbook of Experimental Pharmacology*, Vol. 91 (Ed. L.E. Bryan) Springer-Verlag, Berlin (1989), pp. 1–34.
282. R. Wise, *J. Antimicrob. Chemother.* (1982) **9** (Suppl. B) 31–40.
283. K. Bush, *J. Antimicrob. Chemother.* (1989) **24** 831–840.
284. A. Philippon, R. Labia and G. Jacoby, *Antimicrob. Agents Chemother.* (1989) **33** 1131–1136.
285. L. Gutman, M.D. Kitzis and J.F. Acar, *J. Int. Med. Res.* (1990) **18** (Suppl. 4) 37–47.
286. D.J. Payne and S.G.B. Amyes, *J. Antimicrob. Chemother.* (1991) **27** 255–261.
287. D.B. Boyd, B.J. Foster, L.D. Hatfield, W.J. Hornback, N.D. Jones, J.E. Munroe and J.K. Schwartzendruber, *Tetrahedron Lett.* (1986) **27** 3457–3460.
288. J.E. Baldwin, C. Lowe, C.J. Schofield and E. Lee, *Tetrahedron Lett.* (1986) **27** 3461–3464.
289. L.N. Jungheim, S.K. Sigmund and J.W. Fisher, *Tetrahedron Lett.* (1987) **28** 285–288.
290. L.N. Jungheim, S.K. Sigmund, N.D. Jones and J.K. Schwartzendruber, *Tetrahedron Lett.* (1987) **28** 289–292.
291. S. Harada, S. Tsubotani, T. Hida, H. Ono and H. Okasaki, *Tetrahedron Lett.* (1986) **27** 6229–6232.
292. S. Harada, S. Tsubotani, T. Hida, K. Koyama, M. Kondo and H. Ono, *Tetrahedron* (1988) **44** 6589–6606.
293. N. Tamura, Y. Matsushita, K. Yoshioka and M. Ochiai, *Tetrahedron* (1988) **44** 3231–3240.
294. P. Charlier, O. Dideberg, J.-M. Frère, P.C. Moews and J.R. Knox, *J. Mol. Biol.* (1983) **171** 237–238.
295. S. Wolfe, M. Khalil and D.F. Weaver, *Canad. J. Chem.* (1988) **66** 2715–2732.
296. K.A. Durkin, M.J. Sherrod and D. Liotta, *J. Org. Chem.* (1989) **54** 5839–5841.
297. S.K. Chung and D.F. Chodosh, *Bull. Korean Chem. Soc.* (1989) **10** 185–190.
298. L.N. Jungheim, D.B. Boyd, J.M. Indelicato, C.E. Pasini, D.E. Preston and W.E. Alborn, Jr., *J. Med. Chem.* (1991) **34** 1732–1739.
299. J.B. Doherty, B.M. Ashe, L.W. Argenbright, P.L. Barker, R.J. Bonney, G.O. Chandler, M.E. Dahlgren, C.P. Dorn, Jr., P.E. Finke, R.A. Firestone, D. Fletcher, W.K. Hagmann, R. Mumford, L. O'Grady, A.L. Maycock, J.M. Pisano, S.K. Shah, K.R. Thompson and M. Zimmerman, *Nature (London)* (1986) **322** 192–194.
300. P.S. Williams, R.D.E. Sewell, H.J. Smith and J.P. Gonzalez, *J. Enzyme Inhib.* (1989) **3** 91–101.

301. B.G. Spratt, in *Handbook of Experimental Pharmacology*, Vol. 91 (Ed. L.E. Bryan), Springer-Verlag, Berlin (1989), pp. 77–100.
302. G.A. Jacoby and G.L. Archer, *New Engl. J. Med.* (1991) **324** 601–612.

8 Novel β-lactam structures — the carbacephems

R.D.G. COOPER

8.1 Introduction

In 1991, Oxford University celebrated the 50th anniversary of the first clinical use of penicillin. From the initial discovery to the present day, the improvement in the clinical management of infectious diseases has represented a significant achievement on the part of modern medicine. However, just because bacterial infections have generally become manageable, this does not mean that defenses can be relaxed. Bacteria are continually adapting to conditions and evolving mechanisms to overcome selective pressures. Over the last 50 years we have seen the development of several other families of β-lactam antibiotics — the cephalosporins, the carbapenems, the monobactams and, most recently, the carbacephems — each class having some unique property that has increased the value of the β-lactam antibiotics. This latter class, the carbacephems, are only now being introduced into clinical practice and have shown some interesting and unexpected properties that could establish them as an important class.

8.2 Nomenclature

It is correct to refer to the carbacephems as 4,2,0-bicyclooctane derivatives; however, for simplicity, the same nomenclature will be used that has been used for the penicillins and cephalosporins for many decades. Thus the ring numbering system used will be as shown in Figure 8.1.

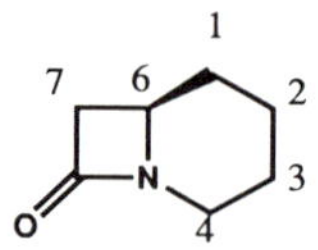

Figure 8.1 Nomenclature of the carbacephems.

8.3 Synthesis

The synthetic strategies that have evolved to construct this molecule have attempted to solve three major problems:

(i) the synthesis must result in a chiral molecule of 6*R*,7*S* stereochemistry;
(ii) the *cis* stereochemistry of the substitutions at C-6 and C-7 is required; and
(iii) methodology is required to construct the 4,2,0-bicyclooctane ring system.

Most of the syntheses have not utilized the naturally occurring β-lactams, i.e. penicillins and cephalosporins, as starting points primarily because of the challenges associated with the construction of a C-6–C-1 bond with the less thermodynamically stable stereochemistry. Recently though, this problem may well have been solved in an innovative manner (see section 8.3.8).

8.3.1 *Nuclear synthesis*

The first synthesis of the basic 4,6-membered ring system, a 3-phenyl substituted carbacepham (**1**), was reported in 1965.[1] The synthetic strategy utilized a diazo-insertion reaction of the α-diazo-amide (**2**). This same principle was later used to synthesize the substituted carbacepham (**3**) when the α-diazo ester (**4**) on photolysis furnished the β-lactam (**5**) as a mixture of *cis* and *trans*-isomers in the ratio of 1 : 2. The *cis*-isomer was converted into the acylhydrazide (**6**), via condensation of the acid obtained from (**5**) with butyl carbazate and acid removal of the t-butyl protecting group. Curtius rearrangement of (**6**) in the presence of butyl alcohol yielded the carbamate

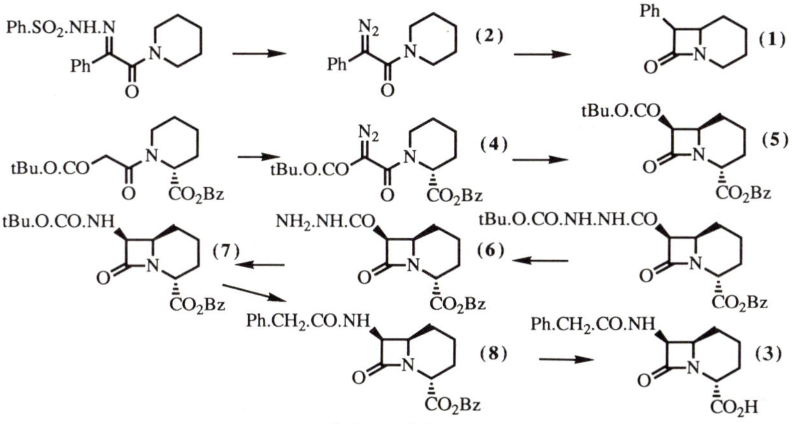

Scheme 8.1

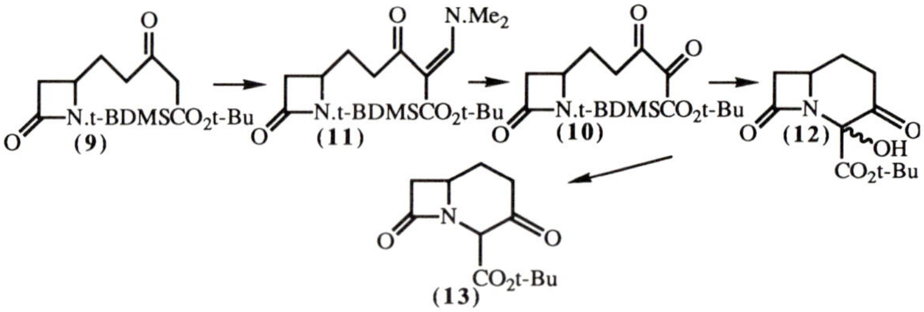

Scheme 8.2

(7), which was then converted into the phenylacetyl derivative (**8**) (see Scheme 8.1).

An alternative strategy for the synthesis of the basic 4,2,0 ring system utilized the β-keto ester (**9**).[2] The α-diketone (**10**) was prepared by reaction of the enaminoketone (**11**) with singlet oxygen. After desilylation, dehydration furnished the bicyclic α-hydroxy ketone (**12**). Reduction to the carbacephem (**13**) was accomplished using trimethylsilyl iodide (see Scheme 8.2).

8.3.2 *Merck synthesis*

Although there were several earlier syntheses of the basic 4,2,0-1-azabicyclooctane ring system (**1**), it was not until 1974 that a 1-carba analogue of a microbiologically active cephalosporin was first synthesized[3,4] (Scheme 8.3) by the Merck group. They prepared the racemic carbacephems (**14**), (**15**) and (**16**) analogous to the cephalosporins cephalothin, cefoxitin, and cefamandole (Scheme 8.4). The strategy followed the initial synthesis of the β-lactam ring using the Staudinger reaction followed by final closure of the six-membered ring via a Horner–Emmons procedure. The protected ketone (**17**) was oxidized to the aldehyde (**18**), and the Schiff base (**19**) formed from condensation of (**18**) with the α-aminophosphonoacetate was reacted with the ketene formed from azidoacetyl chloride and triethylamine at −78°C. A stereospecific reaction occurred resulting in the racemic *cis*-β-lactam (**20**) in 30% yield.

Deketalization also hydrolyzed the acetoxy group; thus, following reacetylation the acetoxy ketone (**21**) was obtained. Cyclization was achieved using sodium hydride in glyme and yielded the carbacephem (**22**) in 62%. Simultaneous hydrogenation of the azide group and benzyl ester, followed by acylation of the nucleus (**23**) with thienyl acetyl chloride resulted in the carbacephem, (±)carbacephalothin (**14**).

When reduction of (**22**) was carried out using hydrogen sulfide/triethylamine, a selective reduction of the azido group resulted in the amino-ester (**24**). Thienylation and 7-methoxylation with lithium methoxide/t-butyl hypochlorite afforded the 7-α-methoxy derivative (**25**). When attempts

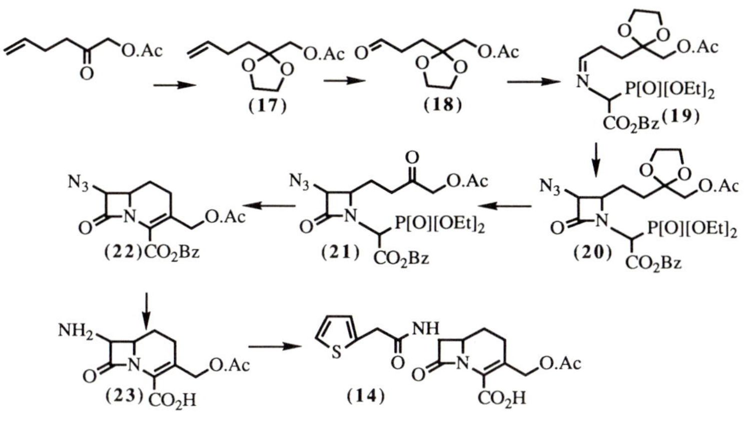

Scheme 8.3

were made to remove the benzyl ester from (**25**) the reaction was found to be surprisingly sluggish, and a competitive hydrogenolytic cleavage of the acetoxy group occurred resulting in (**26**) as the major product. This problem was circumvented by converting the nucleus (**23**) to the benzhydryl ester (**27**) and repeating the above sequence to obtain the carbacephem (**28**). Conversion of (**28**) to the carbacephem, (±)carbadethiacefoxitin (**15**) was accomplished by hydrolysis to the hydroxy derivative (**29**) using citric acetyl esterase followed by carbamoylation with chlorosulfonyl isocyanate.

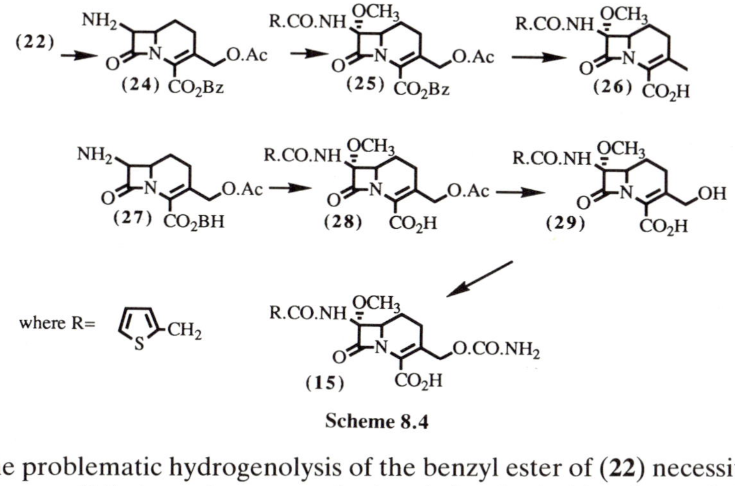

where R= (thienyl)CH_2

Scheme 8.4

The problematic hydrogenolysis of the benzyl ester of (**22**) necessitated a retreat to (**21**) in order to synthesize (±)carbadethiacefamandole (**16**) (Scheme 8.5). Utilization of the acid-cleavable *p*-methoxybenzylester of the aminophosphonate, following the same sequence of reactions, furnished the hydroxy acid (**30**), which was then reprotected as the benzhydryl ester. Mesylation and displacement with the sodium salt of N-methyl tetrazole gave the tetrazole derivative (**31**). Following the same sequence the amine (**32**) was obtained; on acylation and removal of the ester group, this gave the carbacephem, ($\pm$)carbadethiacephamandole (**16**).

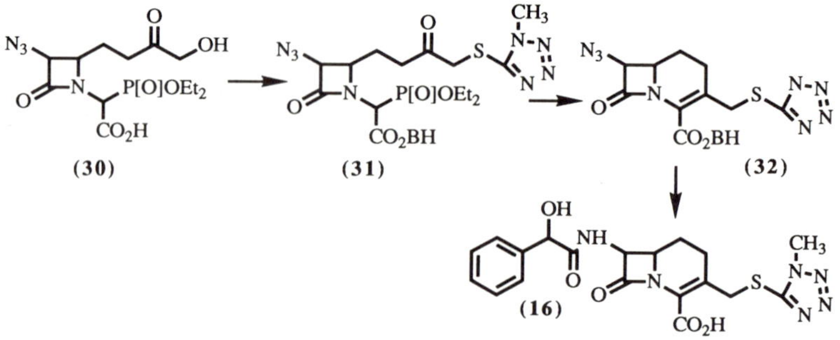

Scheme 8.5

8.3.3 *Shionogi synthesis*

In an attempt to develop a chiral carbacephem synthesis, the imine (**33**) was used in the Staudinger reaction with azidoacetyl chloride (Scheme 8.6).[5] The chirality of the β–γ unsaturated amino-ester originated in penicillin from which the amine (**34**) was prepared by a previously known route. The imine (**33**) formed from pentenal and (**34**) was condensed with azidoacetyl chloride to give (**35**) and (**36**). Unfortunately, no diastereoselectivity was obtained in this reaction, and a 1 : 1 ratio of products was obtained. After double bond isomerization, reduction of the azide (**37**) followed by acylation gave (**38**). Oxidation of the olefin to the epoxide (**38**) and subsequent opening of the epoxide with N-methyl tetrazole thiol, followed by oxidation of the secondary alcohol afforded the thio ketone (**40**). The α–β double bond was elaborated into the phosphorane (**41**) through the procedure of ozonolysis, reduction to the α-hydroxy ester, formation of the α-chloro compound, and treatment with triphenyl phosphine. Wittig ring closure gave the carbacephem (**42**), from which the amide side chain was removed via iminochloride formation. This nucleus (**43**) was used for a selected structure activity relationship (SAR) study of 3′ substituted carbacephems.

Reduction of (**42**) with magnesium gave the exomethylene derivative (**44**) (Scheme 8.7), treatment of which with triethylamine resulted in only the Δ^3 compound (**45**), a departure from cephalosporin chemistry in which Δ^2/Δ^3 mixtures were obtained. Ozonolysis of (**44**) gave the 3-hydroxy derivative (**46**), which was reported to be very unstable. Whilst (**46**) could be trapped with diazomethane to furnish the 3-methoxy compound (**47**) attempts to convert it into the 3-mesylate (**48**) or the 3-chloro derivative were unsuccessful. Alternatively, treatment of the epoxide (**39**) with acid gave the diol, which was protected as its isopropylidene derivative (**49**) while the α–β

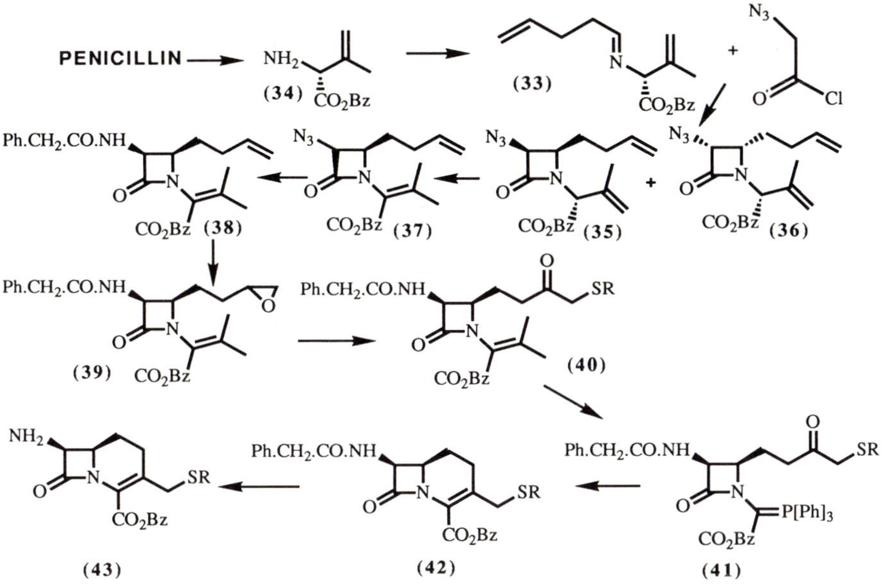

Scheme 8.6

unsaturated ester moiety was converted into the phosphorane (**50**) using established procedures. Oxidative cleavage of the diol gave the aldehyde (**51**) which, on ring closure, resulted in the 3-hydrocarbacephem derivative (**52**). Cleavage of the side chain gave the nucleus (**53**), which then allowed an SAR study of the 3-hydrocarbacephem nucleus.

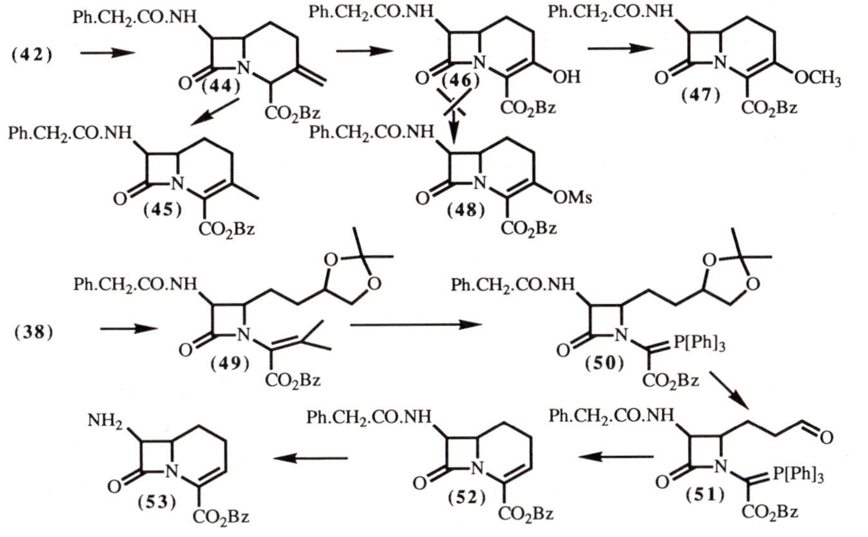

Scheme 8.7

A variation of this synthesis afforded the 1,2 dehydrocarbacephems (Scheme 8.8). Use of the imine from cinnamaldehyde and (**34**) resulted in the β-lactam (**54**). Ozonolysis, Wittig 3-carbon homologation, followed by phosphorane formation resulted in the ketone (**55**). This could not be closed directly, presumably because of the steric requirement of the *trans* stereochemistry of the double bond. This was alleviated by the addition of thiophenol to the double bond when ring closure readily occurred to give the 1-thiophenyl derivative (**56**). Elimination to the dehydrocarbacephem (**57**) could be achieved by oxidation of (**56**) to the sulphoxide and the thermal elimination of phenyl sulphenic acid. Reduction of the 1,2 double bond of (**57**) gave the previously prepared carbacephem (**45**). Manipulation of (**57**) to prepare potentially biologically active derivatives (e.g. (**58**)) was achieved by standard procedures.

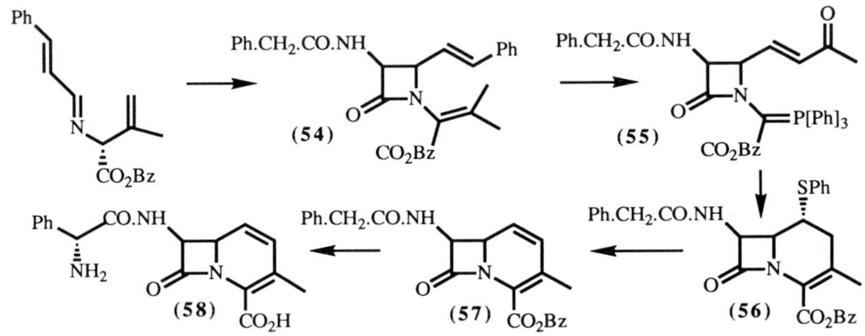

Scheme 8.8

8.3.4 Bristol synthesis

The chiral azetidinone (**59**) had been used previously in the synthesis of 2-oxacephems.[7–9] This synthesis was by the condensation of azido acid chloride with the imine (**60**). Conversion of the isopropylidene derivative (**59**) to the enoltriflate (**61**) was accomplished using standard procedures. Treatment of (**61**) with malonate anion realized a cyclization to the carbacephem (**62**) (Scheme 8.9). When R = CH_3, reduction followed by acylation furnished the 2,2′-dicarbomethoxy derivative ((**63**), R = CH_3). Use of dibenzyl malonate allowed simultaneous removal of all three benzyl groups from the 2,2′ disubstituted derivative ((**64**), R = $PhCH_2$–) to give the 2-carboxy carbacephem (**65**).

From the t-butyl malonate, the diester ((**64**), R = t-Bu) was obtained, which, on treatment with trifluoroacetic acid, furnished the diacid (**65**). Refluxing in toluene gave a 3 : 1 mixture of the Δ^3 and Δ^2 isomers ((**66**) and (**67**)). From the 2-carboxy-Δ^3 derivative (**65**), the 2-carbomethoxycarbacephem (**68**) was obtained. Decarboxylation of (**66**) could be achieved by treatment with triethylamine in pentanol; DBN then smoothly isomerized the Δ^2 isomer (**69**) to the Δ^3 isomer (**70**) from which the 3-methylcarbacephem (**71**) was derived.

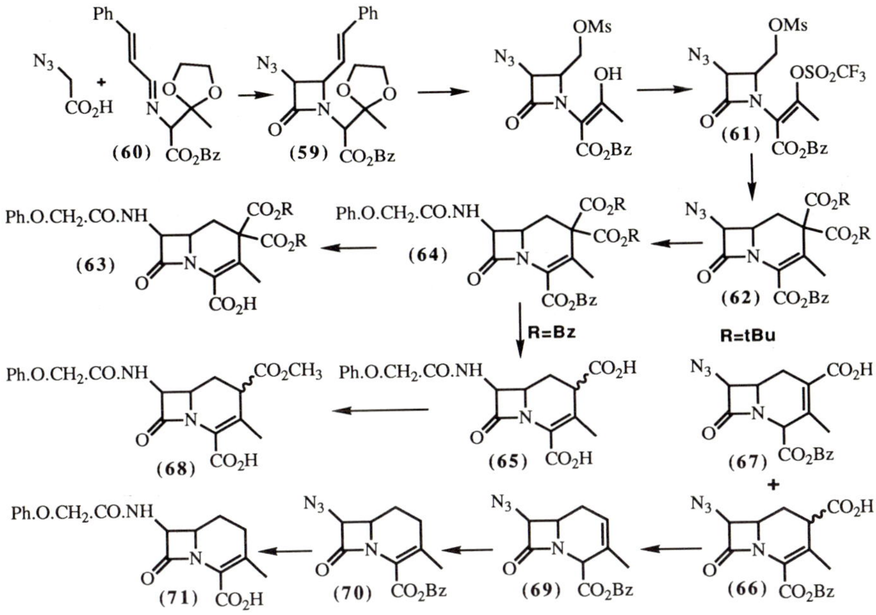

Scheme 8.9

From the Δ^2 isomer (**69**), treatment with NBS yielded the 1-bromo compound (**72**), which, on treatment with base, eliminated the elements of hydrogen bromide resulting in the 1,2 dehydro compound (**73**) (Scheme 8.10). Addition of bromine to (**69**) gave the dibromo compound (**74**) as only one isomer; elimination of HBr from (**74**) yielded the 2-α-bromo-Δ^3

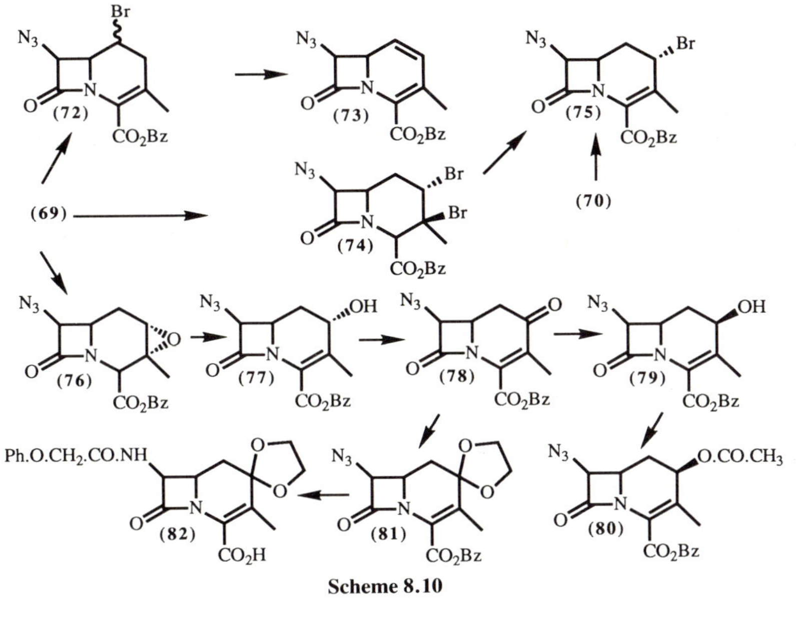

Scheme 8.10

derivative (**75**), a compound which was also obtained by treatment of (**70**) with *N*-bromosuccinimide. The Δ^2 derivative (**69**) also yielded the α-epoxide (**76**) on treatment with *m*-chloroperbenzoic acid; on subsequent treatment with DBN this was converted to the allylic alcohol (**77**), which could then be oxidized to the ketone (**78**) with manganese dioxide. Reduction with sodium borohydride gave the β-alcohol (**79**), which was then converted to its acetate (**80**). The ketal (**81**) of the ketone (**78**) was also prepared using ethylene glycol/*p*-toluene sulfonic acid. These intermediates were then converted to the 7-amido carbacephems, e.g. (**82**).

8.3.5 Kyowa Hakko synthesis

This synthesis followed the same basic strategy previously elaborated in the Merck synthesis, with the added important refinement that a resolution was incorporated, resulting in the preparation of chiral carbacephem derivatives. Furthermore, the objective was to develop a synthesis to carbacephems having a 3-H and then to elaborate this intermediate into other 3-substituted carbacephems (Scheme 8.11).[10]

The pentenal (**83**) was condensed with t-butyl phosphonoglycine (**84**), and the resulting imine condensed with azidoacetyl chloride to afford the *cis*-β-lactam (**85**) in good yield. Cleavage of the double bond with osmium tetroxide, and ring closure using a Horner–Emmons procedure gave the carbacephem (**86**). The azido group was reduced catalytically to the achiral amino-ester (**87**). Acylation of the acid (**88**) with D-phenylglycine, and HPLC separation of the resulting diastereomers afforded the chiral carbacephem (**89**). Alternatively, (**89**) was prepared by incubation of the racemic amino acid with D-phenylglycine methyl ester and intact cells of *Pseudomonas melanoqeuum* KY-8541, a penicillin-acylase-producing, β-lactamase-deficient mutant. A chiral acylation produced (**89**) selectively. Acylation of (**88**) with phenyl acetyl chloride, and treatment of the amide (**90**) with *Kluyvera citrophila* KY-7844 resulted in an enantioselective hydrolysis yielding the chiral nucleus (**88**) which, on acylation, resulted in a variety of chiral 7-amidocarbacephem derivatives.

This synthesis was also modified by using the methyl hexenal (**91**). Interestingly, the selectivity of the Staudinger reaction decreased with the addition of the methyl substituent, causing the formation of (**92**) as a *cis*/*trans* mixture (3 : 2). Following the previous procedure, the two amino acids (**93** and **94**) were produced in a 2-α : 2-β ratio of 4 : 1. These two nuclei were then further elaborated into the 2-methyl substituted carbacephems.

Another variation of this basic scheme began with the butenal (**95**) prepared from furan (Scheme 8.12).[11] This afforded 1-thiosubstituted carbacephems (**96**), which — via oxidation to the sulphoxide and pyrolysis — yielded the 1,2 dehydroderivative (**97**). This derivative could also be obtained by a direct cyclization using diazabicyclooctane as base. The 1,2

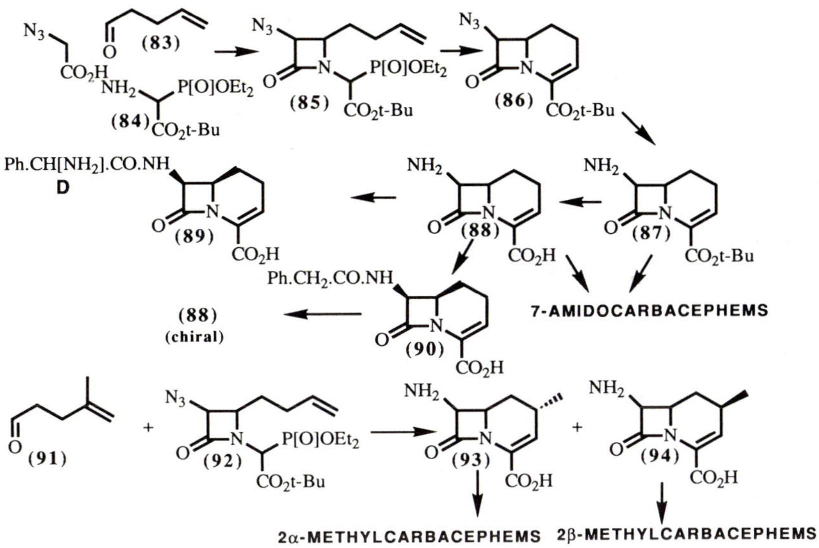

7-AMIDOCARBACEPHEMS

2α-METHYLCARBACEPHEMS 2β-METHYLCARBACEPHEMS

Scheme 8.11

unsaturation could be reduced catalytically, resulting in the previously synthesized derivative (**86**). Reduction and removal of the ester group from (**97**) with trifluoroacetate (TFA) afforded the nucleus (**98**), which was utilized in the synthesis of a series of 1,2 dehydrocarbacephems. The versatility of the 3-H derivative was demonstrated by its conversion into the 3-chloro carbacephem nucleus (**99**) (Scheme 8.13).[12] Addition of phenylthiol to (**86**) gave the thio-ether (**100**) as only one isomer. Oxidation to the sulphoxide using *m*-chloroperbenzoic acid followed by a Pummerer rearrangement using trifluoroacetic anhydride gave the Δ^3 carbacephem (**102**). However, conversion of (**102**) to the enol (**103**) was unsuccessful. Treatment of the sulphoxide (**101**) with sulphuryl chloride yielded the

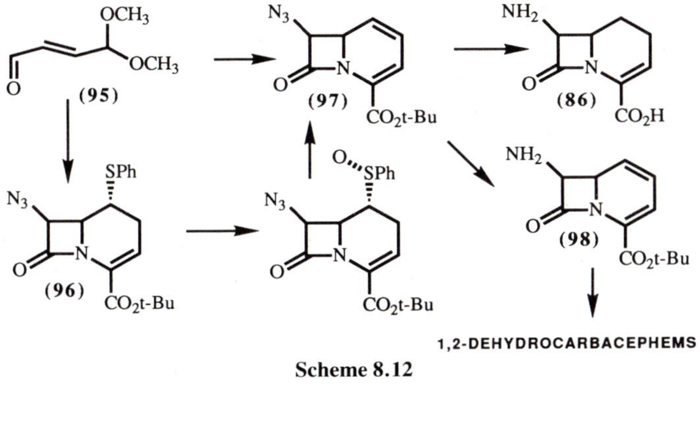

1,2-DEHYDROCARBACEPHEMS

Scheme 8.12

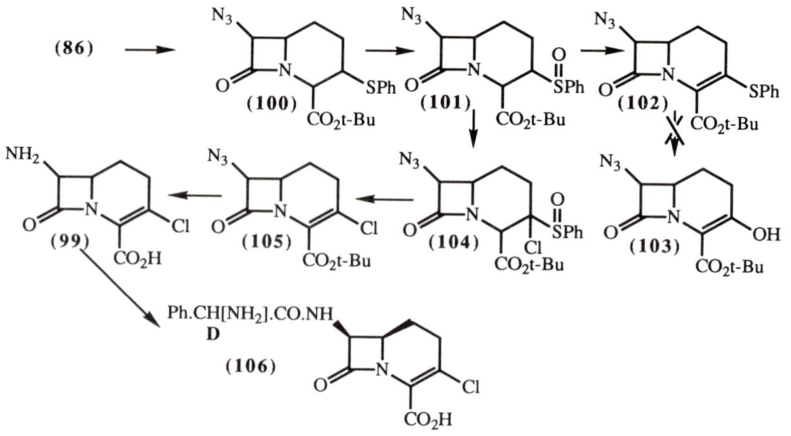

Scheme 8.13

α-chlorosulfoxide (**104**), from which smooth elimination of phenyl sulphenic acid resulted in the 3-chlorocephem (**105**).

This sequence was also successfully carried out on the 2-β-methyl derivative to yield the 2-β-methyl 3-chlorocarbacephem; however, the elimination totally failed in the 2-α-methyl case. Reduction of the azide group of (**105**) followed by acidic cleavage of the butylester gave the desired nucleus (**99**). This was converted into the chiral carbacephem derivatives using the previously described enzymatic procedures, which gave both the chiral nucleus and the chiral phenylglycine carbacephem derivative (**106**) [loracarbef].

8.3.6 Evans synthesis

This synthetic strategy evolved from the initial observations[13] that utilizing a chiral center in the ketene fragment of the Staudinger reaction resulted in a high degree of chiral control in the $2 + 2$ addition reaction (Scheme 8.14). The efficiency of the control depended on the position and nature of this chiral center; however, it was possible to obtain >95% ee in the reaction. L-phenylglycine was used as the center, inducing the chirality into the Staudinger reaction by converting the amino acid into the oxazolidinone glycine (**106**).[14] Addition to the imine (**108**) afforded a high yield, both chemically and diastereomerically, of the *cis*-β-lactam (**109**).

After reduction of the double bond the chiral auxiliary was reductively removed using lithium/ammonia to yield (**110**). This reductive process also accomplished two other reactions, removal of the N-benzyl group and reduction of the *m*-methoxyphenyl ring to the cyclohexadiene moiety. After protection of the amine as the t-butyloxycarbonyl derivative (**111**) the reason for using the *m*-methoxyphenyl group became apparent, as an

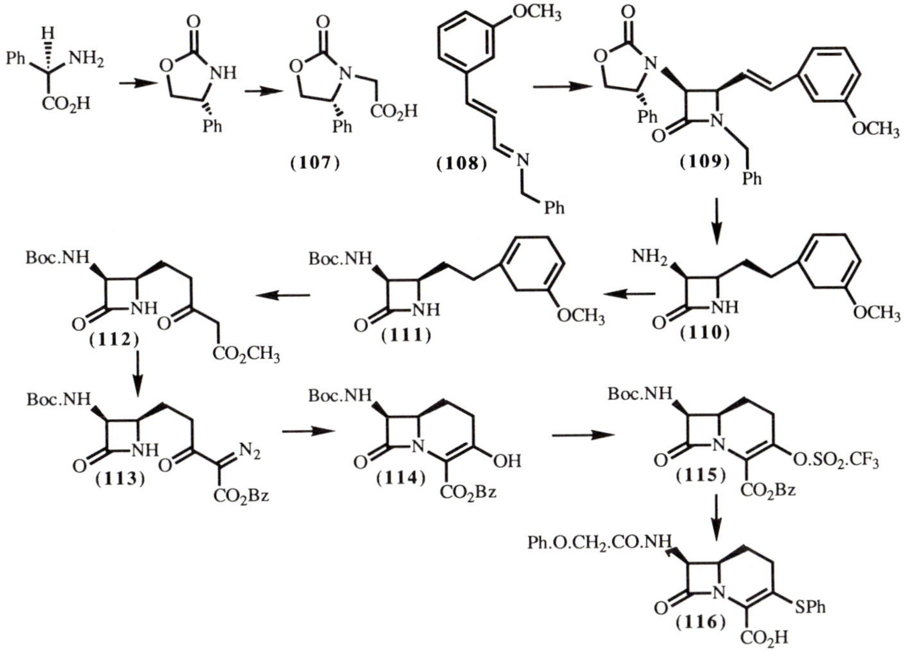

Scheme 8.14

ozonolytic cleavage of the cyclohexadiene ring afforded the β-keto ester (**112**). The ester was exchanged to the benzyl ester, which, on reaction with tosyl azide, furnished the diazo compound (**113**). Treatment with rhodium acetate yielded the carbacephem enol (**114**) via a carbene insertion into the NH bond of the azetidinone. Although this enol had been previously claimed by several groups as being too unstable to isolate, it now appeared to have reasonable stability and could be reacted with triflic anhydride to form the 3-triflyl carbacephem (**114**), from which the 3- thio derivatives (e.g. (**116**)) could be prepared.

A variation of this theme[15] involved condensation of the chiral glycine derivative (**107**) with the imine (**117**) produced from furylacrolein (Scheme 8.15). Reduction of the olefin, and removal of the chiral auxiliary followed by acylation with phenoxyacetyl chloride resulted in (**118**). Ozonolysis furnished the carboxylic acid (**119**). A chain extension to the β–keto ester (**120**) was accomplished using the magnesium salt of *p*-nitrobenzyl malonate. This synthesis successfully converted the enol (**114**) to the 3-chloronucleus (**121**) in one step using triphenyl phosphite dichloride. Acylation with the enamine-protected D-phenylglycine, followed by deprotection with zinc/acetic acid resulted in loracarbef (**106**).

Other chiral auxiliaries in the ketene portion of the Staudinger have also been investigated (Scheme 8.16). Generally, it was found that tartarimide

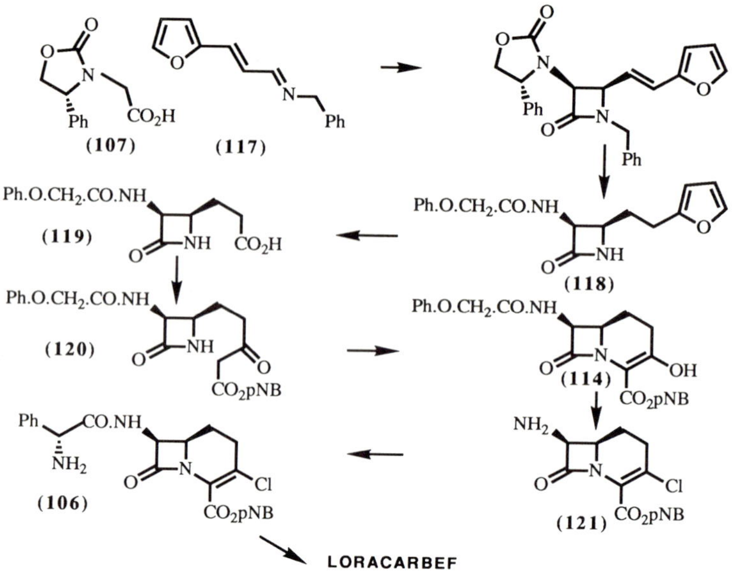

LORACARBEF

Scheme 8.15

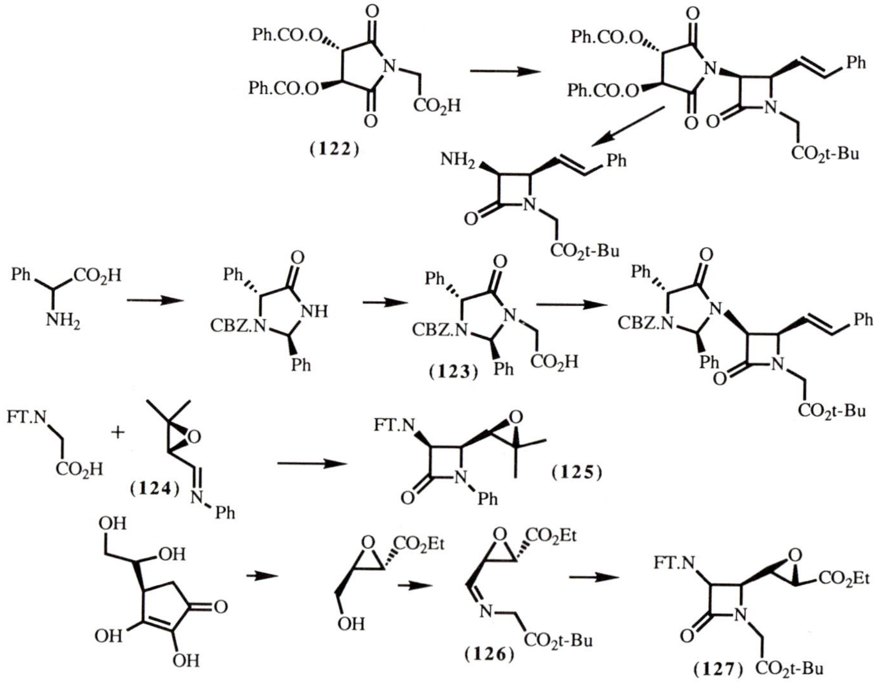

Scheme 8.16

derivatives (e.g. (**121**)) could also function as chiral auxiliaries giving ees in the range of 90–95%.[16,17] Furthermore, these auxiliaries had the advantage of being able to be removed without destruction of the chiral center. Another extension of this principle was exemplified by using the chirality of D-phenylglycine as the key chiral directing group.[16] The imidazolone (**123**) was synthesized from D-phenylglycine, its utilization afforded >95% ee in the Staudinger reaction. This could then be directly used in the synthesis of loracarbef (**106**) without having to remove the chiral auxiliary.

An alternative site for incorporation of a chiral directing group was in the aldehyde part of the imine, where a chiral epoxide functionality was found to induce a high degree of diastereofacial selectivity into the Staudinger reaction. For example, the chiral epoxide (**124**) derived from β-β′ dimethyl acrylaldehyde afforded a high yield of the β-lactam product (**125**) with ee ≥ 95%.[18] When the epoxide (**126**) derived from ascorbic acid was used, the β-lactam obtained (**127**) would then be elaborated into the carbacephem nucleus.[19]

8.3.7 Dieckman synthesis

This synthesis was directed towards the 3-hydroxy carbacephem, and the derivatives that could then be elaborated from this versatile intermediate.[20] The method utilized the Dieckman cyclization to close the six-membered ring. It also demonstrated the value of the furan ring as a carboxy group synthon (Scheme 8.17). Thus, the imine (**128**) prepared from furylacrolein and t-butyl glycinate was condensed with azido acid chloride to give the

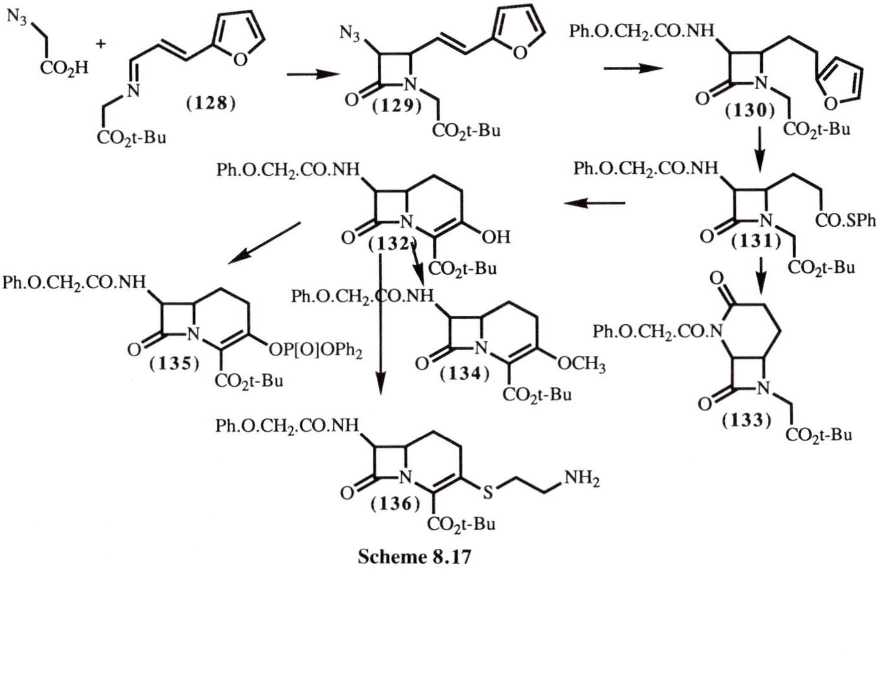

Scheme 8.17

β-lactam (**129**) in 72% yield. Reduction of the azide, and acylation with phenoxy acetyl chloride afforded (**130**). Ozonolysis of the furan ring gave the carboxylic acid, which was then converted to the thiophenyl ester (**131**). Subsequent treatment with lithium bis-trimethylsilylamide (3 equivalents) effected the desired Dieckman cyclization to the enol (**132**) obtained as a stable crystalline compound, whereas use of 2 equivalents of base furnished the alternate cyclization product (**133**).

The enol (**132**) could be converted to the methyl ether (**134**) with diazomethane or via the phosphonate (**135**) into the 3-thio-ethers e.g. (**136**).

An extensive study of this Dieckman reaction has been published[21] where the phenyl ester was found to be as effective as the original thiophenyl ester.

This synthesis has been further adapted to furnish the chiral enol (**132**) by the treatment of the amino acid (**137**) with methyl phenoxyacetate in the presence of penicillin amidase.[22] The enzymatic process furnished only the desired chiral amido-acid (**138**). This was then converted to the enol via the Dieckman cyclization of the *p*-nitrobenzyl ester (**139**). This method represented the efficient synthesis of loracarbef shown in Scheme 8.18.

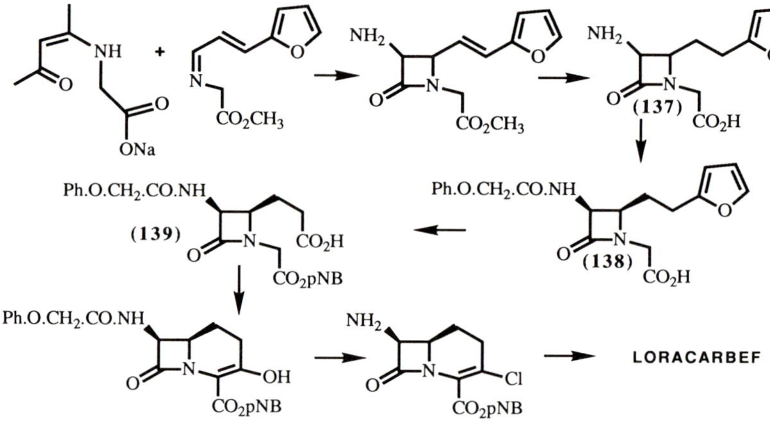

Scheme 8.18

8.3.8 *Synthesis from penicillin*

Potential advantages to using penicillin as a starting material are that the molecule is available, inexpensive, and that it contains the β-lactam ring with substituents containing the correct chirality. The problem is primarily one of constructing the C-1–C-6 bond while retaining the thermodynamically less stable stereochemistry at that position.

This was partially solved by the discovery that certain seco-penicillins could be successfully allylated at the C-4 position (seco-penicillin numbering) using a single electron process (Scheme 8.19).[23] In contrast to the ionic process, which produces only the *trans*-isomer, the radical method

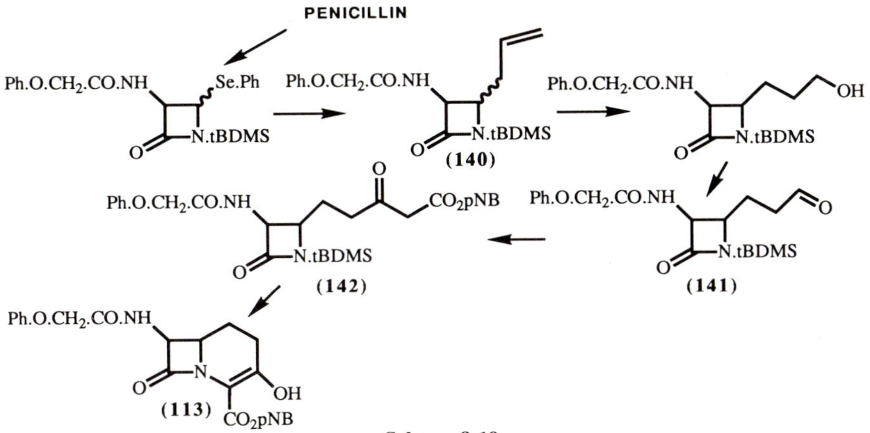

Scheme 8.19

gave a *cis* : *trans* mixture in the ratio of 1 : 2. Although not of optimum efficiency, this method did allow the synthesis of the *cis*-substituted azetidinone (**140**). Hydroboration of the terminal olefin, subsequent oxidation to the aldehyde (**141**), and chain extension with *p*-nitrobenzyl phosphonoacetate yielded the α-β-unsaturated ester,[24] which was then converted to the β-keto ester (**142**) using a palladium catalysed oxidation. Acid cleavage of the silyl protecting group gave the intermediate (**119**), which had previously been converted to the carbacephem enol (**113**).

This difficult control of stereoselectivity of C–C bond formation has been solved in an innovative manner. Penicillin was converted by established procedures via the acetoxyazetidinone (**143**) into the 4-phenylsulphone substituted azetidinone (**144**). Alkylation of (**144**) with allyl bromide gave the 4-allyl azetidinone (**145**) having the requisite *cis* stereochemistry of the newly formed C–C bond with the 3-amido substituent.[25] After desilyation, reductive cleavage of the phenyl sulphone moiety could be accomplished with a stereochemical control of 94 : 6 by using the highly hindered reducing agent, $Li(t\text{-}BuO)_3AlH$. Less sterically demanding reducing agents produced greater amounts of the *trans*-isomer. The *cis*-azetidinone (**146**) was alkylated on nitrogen with *p*-nitrobenzyl bromoacetate, and the double bond was then oxidized with potassium permanganate to the carboxylic acid (**147**). This compound has previously been converted into loracarbef using the Dieckman cyclization route. This represents a highly stereospecific route from penicillin (Scheme 8.20).

8.4 Stability of the carbacephs

The dogma amongst the β-lactam aficianados has always been that to increase microbiological activity one has to increase the reactivity of the

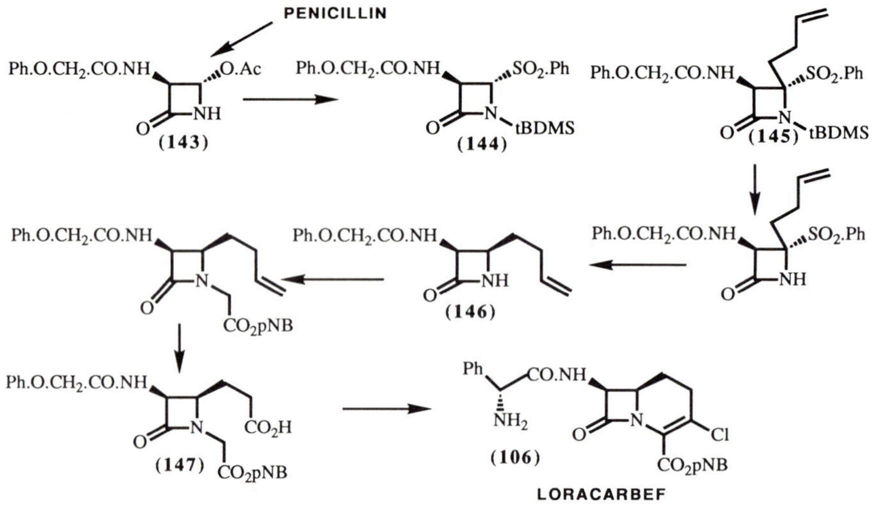

Scheme 8.20

β-lactam carbonyl and thereby increase its rate of reaction with the target bacterial penicillin binding proteins. Many attempts to corroborate this relationship have been made by seeking a correlation of some physical property of the molecule with the microbiological activity of the molecule. The infrared stretching frequency of the β-lactam carbonyl[26] and C-13 nmr[27] are just two parameters that have been reported to be a direct indication of the activity. In addition, theoretical calculations have correlated the inductive effect with activity for a series of direct 3-substituted cephems.[28] An empirically derived parallelism between general reactivity of the β-lactam ring to hydroxide ion, and the microbiological activity of the β-lactam-containing molecule has been found. This has led to the general perception that any increase in activity will result in a corresponding decrease in stability. In addition, for any given β-lactam nucleus, the degree of reactivity to hydroxide ion reaches a maximum when the 7-side chain is phenylglycine, presumably due to the availability of an internal attack of the side chain α-amino group on the β-lactam carbonyl. For this reason, attempts to prepare phenylglycyl cephems with 3-substituents having greater inductive effect than chloro have been unsuccessful, as have attempts to prepare phenylglycyl derivatives of other β-lactam nuclei, e.g. the monobactams. Thus, in the cephems the 7-phenylglycine-3-chloro-cephem (cefaclor) represents the best balance between activity and stability that could be achieved.

The one property that has distinguished the carbacephems from the cephems has been in their stability to hydroxide ion, for it was observed that the 7-phenylglycine-3-chloro-carbacephem was substantially more stable at physiological pH than was the corresponding cephem.[29] This greater stability could result in significant clinical advantages in that, without competing serum decomposition, one could obtain higher and longer serum levels for the carbacephems; in addition, it has allowed the synthesis of several new classes of carbacephems that have no parallel cephem derivative. Using the carbacephem nucleus it has been possible to increase the inductive effect of the 3-position and retain chemical stability. This advantage is shown in Table 8.1, where the stability of a series of cephems and carbacephems to conditions of pH 10 at 37°C are shown. In all cases the carbacephems were 10–50 fold more stable to attack by hydroxide ion than were the cephalosporins.[30]

Another study has examined the differences in stability of certain 3-thio-substituted carbacephems with their cephalosporin analogues.[31] As shown in Table 8.2 the cephalosporins were found to be very unstable with a $T_{1/2}$ at pH 6.6 of >0.6 hr, whereas the carbacephems were noted as having excellent stability. This difference is particularly intriguing, as the two family classes have comparable microbiological activity, and it calls into question the assumption that the level of activity corresponds to the rate of reaction towards base. It clearly shows that hydroxide ion hydrolysis is not a good model for transpeptidase activity.

Table 8.1 Comparison of carbacephems and cephalosporins to reaction with base.

R	R′	Z	K_s/K_c	$T_{1/2}$ (hr)
Phenylglycyl	Cl	S	29.1	0.734
Phenylglycyl	Cl	CH_2		21.4
Phenoxyacetyl	OMs	S	21.4	0.363
Phenoxyacetyl	OMs	CH_2		7.79
Phenoxyacetyl	Cl	S	20.4	0.774
Phenoxyacetyl	Cl	CH_2		15.8
Phenoxyacetyl	CH_2OCH_3	S	24.7	1.88
Phenoxyacetyl	CH_2OCH_3	CH_2		46.5
Phenoxyacetyl	CH_3	S	49.6	8.74
Phenoxyacetyl	CH_3	CH_2		433
Thienylacetyl	CO_2CH_3	S	7.8	0.111
Thienylacetyl	CO_2CH_3	CH_2		0.867
Thienylacetyl	CO_2H	S	32.3	2.04
Thienylacetyl	CO_2H	CH_2		66.0
Aminothiazole methoxime	H	S	20.2	6.08
Aminothiazole methoxime	H	CH_2		123
Aminothiazole methoxime	$CH_2NC_5H_5+$	S	29.5	0.375
Aminothiazole methoxime	$CH_2NC_5H_5+$	CH_2		11.0

Table 8.2 Stability of certain 3-thio-substituted carbacephems with their cephalosporin analogues.

Ph.CH[NH_2].CO.NH
X
N
O
R
CO_2H

X	R	$T_{1/2}$ (hr) at 30°C
CH_2	2-Thio-N-methyltetrazole	Stable (pH 7.4)
S	2-Thio-N-methyltetrazole	>0.5 (pH 6.6)
CH_2	4-Thiopyridine	Stable (pH 7.4)
S	4-Thiopyridine	0.6 (pH 6.6)

8.5 Structure–activity relationships in the carbacephems

Much of the impetus for research on the carbacephems is the direct result of the discovery of the activity and chemical stability of the 7-phenylglycyl-3-chloro derivative (**106**) [loracarbef]. Biological evaluation of (**106**) indicated that it possessed a similar spectrum of activity to cefaclor but was substantially superior in its chemical stability.[29,32] It was also determined to have the same mechanism of action as the cephalosporins in that it bound to the penicillin binding proteins. It exhibited two to three times the affinity to PBP 1A and PBP 4 than did cefaclor, and about two times less affinity to PBP 1B. Both loracarbef and cefaclor had similar, good affinities for PBP 3. Consequently the carbacephems possessed the desirable antibiotic properties of good activity, high chemical stability and a bactericidal mechanism of action. This has resulted in loracarbef being investigated clinically. It has also motivated the medicinal chemists to explore some new directions in structure–activity relationships. In the following SAR discussions, only derivatives that have no corresponding cephalosporin analogue will generally be analyzed.

8.5.1 1-Position

A difference in the carbacephems in relationship to the cephems is the availability of positions 1 and 2, which the medicinal chemist can exploit in order to modify or increase the microbiological activity of the compounds. The only possibility of variation of the 1-position in the cephems has been a change in the oxidation level of the sulphur atom, with corresponding general loss of broad spectrum activity. Although there has been a small amount of exploration at the 1- and 2-positions of the carbacephems, the results have been uniformly disappointing.

The 1-exomethylene-2,2-dimethyl derivative had some activity, but again

Table 8.3 Activity of 1-substituted carbacephems.

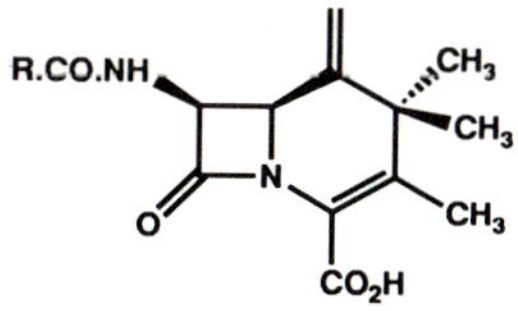

Organism	A*	B*	C*	D*
B. subtilis	1.5	1	1.5	1
S. lutea	2	1	0.12	0.03
S. aureus	4	1	8	0.3
E. coli	1.5	1	5	1

* Compounds tested are a mixture of the two enantiomers
A: R = $PhOCH_2CONH-$
B: deacetoxycephalosporin V
C: R = $PhCH(NH_2)CONH-$
D: cephalexin

substantially less than the unsubstituted compounds (Table 8.3).[33] The 1,2 dehydro derivative, however, did possess gram-negative activity comparable to cefazolin (**A** in Table 8.3), but considerably less than the corresponding saturated compound (**F** in Table 8.6).[34] In Table 8.4 can be

Table 8.4 Activity of 1,2-disubstituted carbacephems.

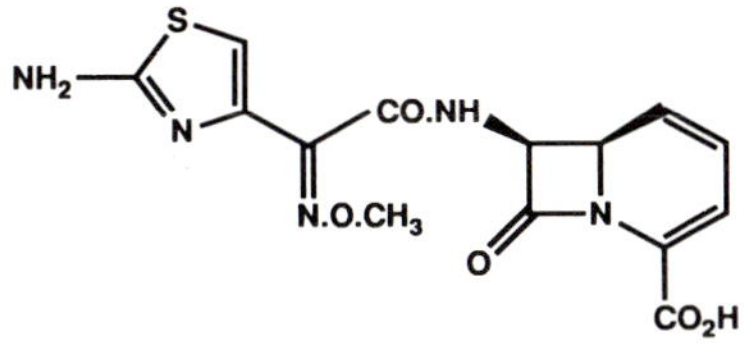

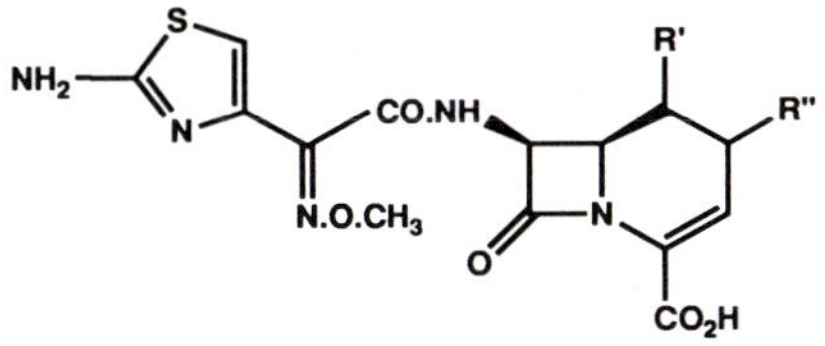

Organism	A*	B*	C*	D*	E*	F*
S. aureus 209P	>50	>100	>50	>100	12.5	50
E. coli NIHJ	25	>100	6.25	6.25	0.05	0.2
K. pneumoniae 8045	6.25	>100	1.56	1.56	0.02	0.05
S. marcescens T-26	>50	>100	50	50	0.04	3.12
P. mirabilis 1287	6.25	>100	3.12	3.12	0.1	0.2
P. vulgaris 6897	6.25	>100	–	–	–	–
P. rettgeri 4289	6.25	>100	–	–	–	–

* Compounds tested are a mixture of the two enantiomers
A: 1,2 dehydroderivative
B: R′ = α-OH, R″ = β-HO
C: R′ = α-OH, R″ = β-Cl
D: R′ = α-OH, R″ = β-Br
E: R′ = H, R″ = α-OH
F: R′ = H, R″ = β-OH

Table 8.5 Activity of 2-substituted carbacephems.

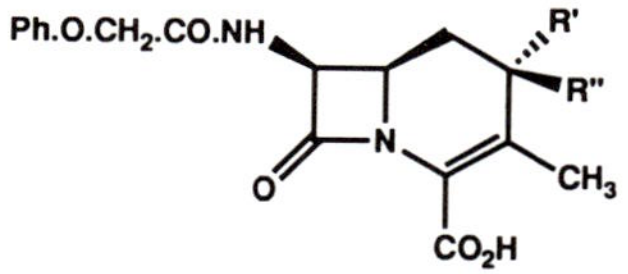

Organism	**A*** $R' = CO_2CH_3$ $R'' = CO_2CH_3$	**B*** $R' = H$ $R'' = CO_2CH_3$	**C*** $R' = H$ $R'' = CO_2H$	**D*** $R' = H$ $R'' = H$	**E*** $R' = OH$ $R'' = H$	**F*** $R' = OAc$ $R'' = H$	**G*** $R' = R = O$	**H*** $R' = R'' = OCH_2O$	**I*** $R' = H$ $R'' = OH$	**J*** $R' = H$ $R'' = OAc$	**K** Deacetoxy-ceph V
S. aureus Smith	>125	>32	>125	4	16	125	0.5	16	8	2	1
S. aureus 1633	>125	63	125	8	32	>125	125	>125	32	4	2
S. pyogenes	16	16	>125	4	8	125	0.13	16	4	0.5	2
E. coli Juhl	>125	>125	125	125	>125	>125	>125	>125	>125	>125	>125

* Compounds tested are a mixture of the two enantiomers.

seen the influence of hydroxy and halogen group substitution at the 1- and 2-positions. The conclusions that can be drawn from a limited amount of data are (i) α-substitution appears to be more advantageous than β-; and (ii) disubstitution markedly decreases the activity.[35]

8.5.2 2-Position

Examination of Table 8.5 shows that compared to the compound where R = R′ = H, the 2-α-acetoxy compound has somewhat better gram-positive activity, and in overall activity is comparable to the deacetoxycephem.[36] The 2-keto compound has the greatest activity; unfortunately with this combination of side chain and 3-substituent no gram-negative activity could be expected and no other derivatives with this substituent have been reported.

A more systematic study was done on the 2-methyl substituted compounds[37] on a nucleus having a hydrogen at the 3-position, and phenylglycine and aminothiazole methoxime side chains. These substituents will normally give compounds with a high level of activity and thus it will be possible to assess the activity contributions of the methyl substitution (see Table 8.6). In the cases of both side chains the β-methyl group is less active than the corresponding α-methyl group, especially against gram-negative

Table 8.6 Activity of 2-substituted carbacephems.

R.CO.NH
R″
R‴
N
R′
O
CO_2H

Organism	A	B	C	D	E	F	G
S. aureus	0.4	0.4	0.78	50	12.5	12.5	3.12
S. epidermidis	3.12	3.12	6.25	>100	12.5	12.5	12.5
E. coli	3.12	50	12.5	0.20	0.05	0.05	100
K. pneumoniae	0.78	6.25	3.12	0.2	0.05	0.05	>100
S. marcescens	–	>100	>100	6.25	1.56	0.78	>100
P. mirabilis	12.5	>100	50	0.78	0.05	0.05	>100
P. vulgaris	50	>100	50	0.78	0.05	0.05	>100
P. morganii	50	>100	100	0.78	0.05	0.05	>100
P. rettgeri	12.5	>100	>100	0.4	0.05	0.05	>100
P. aeruginosa	–	>100	>100	>100	>100	50	>100

A: R = phenylglycyl, R′ = H, R″ = R‴ = H
B: R = phenylglycyl, R′ = H, R″ = H, R‴ = Me
C: R = phenylglycyl, R′ = H, R″ = Me, R‴ = H
D: R = aminothiazole methoxime, R′ = H, R″ = H, R‴ = Me
E: R = aminothiazole methoxime, R′ = H, R″ = Me, R‴ = H
F: R = aminothiazole methoxime, R′ = R″ = R‴ = H
G: R = thienylacetyl, R′ = R″ = R‴ = $OCOCH_3$

organisms (**B** vs. **C** and **D** vs. **E**). With both side chains, neither methyl substitution offers any advantage over the unsubstituted compounds (**A** and **F**).

From the data shown in Table 8.7 another study has shown that the 2-α-methyl substitution has greater activity than the 2-β-substitution (**G** vs. **F**) and, in this case, the 2-α is comparable to the unsubstituted compound (**G** vs. **E**) and has the same activity as cefotaxime (**M**). Heterogroup substitution shows an increase in activity against *Pseudomonas* without, in these examples, any appreciable difference in the stereochemistry of the substitution (**J** vs. **K**). In this study the best compound appears to be the 2-α-hydroxy derivative (**H**).[38]

8.5.3 3-Position

This section will be devoted to derivatives that have unusual or unexpected properties or are of a new type of substitution pattern. The greater stability of the carbacephem nucleus has allowed the synthesis and evaluation of several new series of antibiotics.

8.5.3.1 3-Quaternary ammonium substituted carbacephems The synthesis of the 3-substituted quaternary ammonium carbacephems was achieved by the direct displacement of the enol triflate (**148**) with a nitrogen base (Scheme 8.21).[39] The quaternization proceeded successfully with aromatic heterocycles that possessed the requisite nucleophilicity. Generally, little Δ^2 isomer was produced; however, with saturated tertiary amines (e.g. N-methyl morpholine) the reaction was extremely facile but produced exclusively the Δ^2 isomer. When 2-substituted pyridines were used as the quarternary base a mixture of two products, which could be observed in the nmr spectrum and on HPLC, was obtained. These were found to be rotational isomers around the C-3–N bond. However, the energy barrier was insufficient to allow their separation and individual evaluation.

These derivatives were found to be excellent antibiotics, similar to the ATMO 3′ quaternary cephems, but with two exceptions. The first was the lack of activity against *Pseudomonas aeruginosa*. By comparing the *in vitro* susceptibility of a wild-type strain of *P. aeruginosa* to that of an isogenic mutant strain with these derivatives, it was determined that this lack of activity was due to changes in outer membrane permeability and not to differences in affinity to essential penicillin binding proteins. It was also determined that this lack of activity towards *P. aeruginosa* was due to the lack of the 3′-methylene group and not to the difference between CH_2 and S in the nucleus.

The second major difference was in the potency of the carbacephems against the stably derepressed, β-lactamase-producing strains of *Enterobacter cloacae*. These strains constitutively produce high levels of a type I

Table 8.7 Structure activity relationship of 2-substituted carbacephems.

Organism	A*	B*	C*	D*	E*	F*	G*	H*	I*	J*	K*	L*	M*
S. aureus 209P	0.4	0.4	0.78	3.12	3.12	100	1.56	3.12	3.13	6.25	3.13	6.25	0.78
S. epidermidis	3.12	3.12	6.25	12.5	12.5	>100	3.12	25	–	–	–	–	1.56
E. coli NIHJ	3.12	50	12.5	6.25	0.02	0.2	0.02	0.02	0.78	0.1	0.39	0.05	0.1
K. pneumoniae 8045	0.78	6.25	3.12	3.12	0.01	0.2	0.01	0.01	0.39	0.02	0.2	0.02	0.1
S. marcescens T-26	–	>100	100	100	0.2	6.25	0.4	0.02	6.25	6.25	25	6.25	0.78
P. mirabilis	12.5	>100	50	25	0.01	0.78	0.01	0.05	–	–	–	–	0.02
*P. vulgaris*6897	50	>100	50	25	0.01	0.78	0.01	0.01	0.02	0.01	0.01	0.01	0.01
P. morgani 4298	50	>100	100	12.5	0.05	0.78	0.02	0.01	0.78	0.01	0.2	0.02	0.05
P. rettgeri 4289	12.5	>100	>100	6.25	0.01	0.4	0.01	0.01	0.12	0.01	0.02	0.01	0.01
P. aeruginosa 145	–	>100	>100	>100	50	>100	>100	6.25	12.5	12.5	12.5	12.5	50

* Denotes that the compounds tested are a mixture of the two enantiomers.

A: R = phenylglycyl, R″ = R‴ = H
B: R = phenylglycyl, R″ = H, R‴ = Me
C: R = phenylglycyl, R″ = Me, R‴ = H
D: Phenylglycyl, R″ = OH, R‴ = H
E: R = aminothiazole methoxime, R″ = R‴ = H
F: R = aminothiazole methoxime, R″ = H, R‴ = Me
G: R = aminothiazole methoxime, R″ = Me, R‴ = H
H: R = aminothiazole methoxime, R″ = OH, R‴ = H
I: R = aminothiazole methoxime, R″ = H, R‴ = 5-thio-N-phenyltetrazole
J: R = aminothiazole methoxime, R″ = H, R‴ = 5-thio-N-methyltetrazole
K: R = aminothiazole methoxime, R″ = 4-thiopyridine, R‴ = H
L: R = aminothiazole methoxime, R″ = azido, R‴ = H
M: cefotaxime

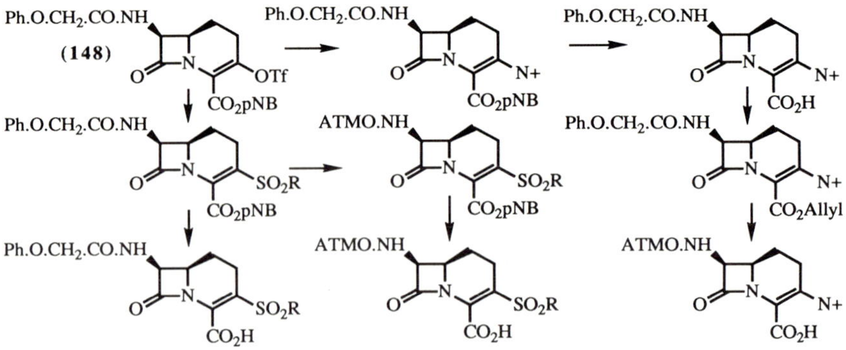

Scheme 8.21

β-lactamase and express a high level of resistance to most third-generation cephalosporins. Table 8.8 shows that these 3-quaternary carbacephems have activity against *E. cloacae* 265A at concentrations <4 $\mu g\, ml^{-1}$.

As previously discussed, these derivatives were found to be stable to basic hydrolysis whereas the 3-quaternary cephems have not been able to be synthesized because of their chemical instability.

8.5.3.2 3-Sulphone substituted carbacephems Another series of derivatives that was synthesized in an effort to take advantage of the superior carbacephem stability was the 3-sulphone derivatives.[40] The synthesis again employed displacements on the 3-triflate, in this case using the sodium salt of the aryl/alkyl sulphonate (Scheme 8.21).

The activity, as shown in Table 8.9, is good when compared to cefotaxime, with gram-positive activity improving with increase in lipophilicity of the sulphone substituent. The compounds have no activity against the organisms containing the type I β-lactamase, e.g. *E. cloacae* 265A. This is presumably due to instability to this enzyme and, possibly, poor penetration across the outer membrane of these poorly penetrable organisms.

The 3-sulphone carbacephems as shown in the section on stability are slightly less stable than ceftazidime and, from a stability aspect, are thus quite viable compounds.

8.5.3.3 3-Carboalkoxy substituted carbacephems The enol triflate (**148**) again proved to be a valuable intermediate, since a palladium catalysed carbonylation in alchoholic solvents yielded the 3-carboalkoxy derivatives (Scheme 8.22).[41]

The ATMO derivatives possessed potent broad spectrum activity very similar to that of cefotaxime;[42] although resistant to the TEM β-lactamases, they appeared to be less effective against the organisms containing the type I β-lactamase, e.g. *E. cloacae* 265A. Increasing the lipophilicity of the ester group appeared to increase the activity against *E. cloacae* 265A with the best

Table 8.8 Activity of 3-quaternary substituted carbacephems.

Organism	**A**	**B**	**C**	**D**	**E**	**F**	**G**	**H**	**I**	**J**	Ceftazidime
S. aureus	4	4	2	2	2	2	4	2	1	128	4
S. aureus(β+)	8	8	8	8	16	8	8	4	2	128	8
S. pyogenes	0.03	0.06	0.03	0.06	0.06	0.03	0.03	0.03	0.03	128	0.06
S. pneumoniae	0.03	0.06	0.03	0.06	0.015	0.03	0.06	0.015	0.015	128	0.06
H. influenzae	0.5	0.5	0.5	0.5	0.5	0.25	0.5	0.25	0.12	128	0.06
E. coli	0.06	0.12	0.12	0.25	2.0	0.06	0.25	0.06	4.0	128	0.25
E. coli (Tem)	0.03	0.5	0.5	1.0	0.12	0.12	0.12	0.06	1.0	128	0.12
K. pneumoniae	0.06	0.06	0.03	0.12	0.008	0.06	0.12	0.06	0.06	128	0.06
E. cloacae	0.12	0.12	0.12	0.25	2.0	0.12	0.25	0.06	4.0	128	0.25
E. cloacae 265A	2.0	1.0	1.0	2.0	4.0	2.0	1.0	0.25	8.0	128	32
S. typhi	0.03	0.06	0.03	0.12	0.5	0.03	0.12	0.06	1.0	128	0.12
P. aeruginosa	128	128	128	128	128	128	128	128	128	128	2.0
S. marcescens	0.12	0.25	0.06	0.25	16	0.25	0.25	0.06	8.0	128	0.25
M. morganii	0.12	0.5	0.25	1.0	8.0	0.25	0.25	0.06	2.0	128	0.5
P. rettgeri	0.06	0.25	0.50	4.0	8.0	0.5	0.25	0.06	8.0	128	0.25

A: R = OCH_3, N = pyridinium
B: R = OCH_3, N = 3,4-N-methylimidazopyridinium
C: R = OCH_3, N = pyrindinium
D: R = OCH_3, N = tetrahydroquinolinium
E: R = OCH_3, N = 5-*p*-MeO, phenyl (2,3)*b*-thienopyridinium
F: R = OCH_3, N = N-methylimidazolium
G: R = OCH_3, N = 2,5-dimethylpyridinium
H: R = OCH_3, N = 3,4-dimethylpyridinium
I: R = OCH_2Ph, N = 3,4-dimethylpyridinium
J: R = OCH_3, N = 4-dimethylaminopyridinium

Table 8.9 Activity of 3-sulphone substituted carbacephems.

Organism	A	B	C	D	E	F	G	Cefotaxime
S. aureus	16	8.0	8.0	8.0	4.0	8.0	4.0	2.0
S. pyogenes	0.06	0.03	0.03	2.0	0.015	1.0	0.015	0.015
S. pneumoniae	0.06	0.03	0.03	0.06	0.03	0.06	0.03	0.015
H. influenzae	0.25	0.25	0.25	0.5	0.25	0.25	1.0	0.015
E. coli	0.25	0.06	0.125	0.25	0.06	2.0	2.0	0.03
E. coli (Tem)	1.0	1.0	1.0	1.0	0.5	8.0	128	0.03
K. pneumoniae	0.06	0.06	0.125	0.125	0.06	1.0	2.0	0.008
E. cloacae	2.0	1.0	1.0	1.0	0.05	8.0	8.0	0.125
E. cloacae 265A	128	128	128	128	128	128	128	32
S. typhi	0.125	0.06	0.125	0.06	0.125	0.25	0.5	0.03
P. aeruginosa	128	128	128	128	128	128	128	32
P. morganii	1.0	2.0	2.0	4.0	2.0	128	64	0.25
P. rettgeri	0.5	0.125	4.0	0.25	1.0	1.0	64	0.125

A: R = CH_3
B: R = C_2H_5
C: R = C_3H_7
D: R = isoC_3H_7
E: R = cyclopropyl
F: R = phenyl
G: R = 2-thienyl

Table 8.10 Activity of 3-carboalkoxy carbacephems.

Organism	**A**	**B**	**C**	**D**	**E**	**F**	Cefotaxime
S. aureus	2.0	4.0	2.0	1.0	1.0	1.0	4.0
S. pyogenes	0.015	0.03	0.015	0.008	0.008	0.03	0.03
S. pneumoniae	0.015	0.03	0.008	0.008	0.008	0.03	0.03
H. influenzae	0.015	0.03	0.015	0.015	0.008	0.015	0.03
E. coli	0.125	0.06	0.25	0.25	0.125	0.06	0.25
E. coli (Tem)	0.125	0.25	0.06	0.125	0.06	0.06	0.5
K. pneumoniae	0.008	0.008	0.008	0.008	0.008	0.015	0.03
E. cloacae	0.125	0.125	0.25	0.05	0.25	0.125	0.5
E. cloacae 265A	32.0	64.0	8.0	8.0	32.0	16.0	32.0
S. typhi	0.015	0.008	0.03	0.06	0.03	0.03	0.125
P. aeruginosa	>128	>128	>128	>128	>128	>128	>128
M. morganii	0.25	0.5	0.06	0.03	0.03	0.06	1.0
P. rettgeri	0.5	0.25	1.0	1.0	0.25	0.5	1.0

A: R = OC_2H_5
B: R = CH_3
C: R = O*n*-Bu
D: R = OCH_2Ph
E: R = $OCH_2C{=}CH_2$
F: R = OC_6H_4*p*-OMe

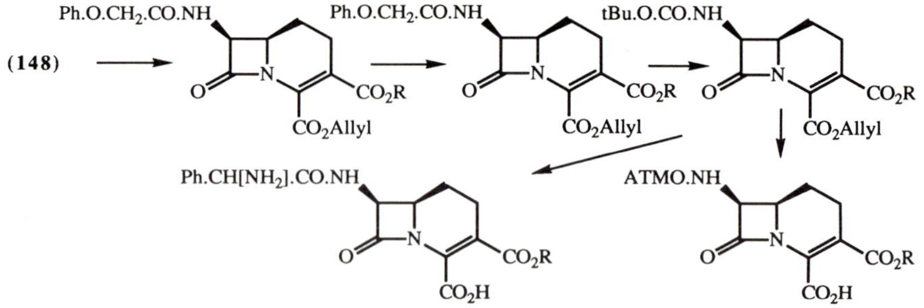

(148) → Ph.O.CH$_2$.CO.NH … CO_2R, CO_2Allyl → Ph.O.CH$_2$.CO.NH … CO_2R, CO_2Allyl → tBu.O.CO.NH … CO_2R, CO_2Allyl → ATMO.NH … CO_2R, CO_2H; Ph.CH[NH$_2$].CO.NH … CO_2R, CO_2H

Scheme 8.22

activity being 8 $\mu g\,ml^{-1}$. The most active compound reported (**D** in Table 8.10) was marginally superior to cefotaxime *in vitro*.

The phenylglycyl derivatives were also prepared and reported to be quite stable, as opposed to their cephalosporin counterparts. The activity as shown in Table 8.11 was comparable to cefaclor; however they showed poor oral absorption in the mouse. The one example of a 3-acyl derivative (**B** in Table 8.11) had poor stability, which was presumably the reason for its poor activity.

8.5.3.4 3-Heteroaryl substituted carbacephems When the enol triflate (**148**) was treated with trimethylethoxyvinylstannane in the presence of catalytic palladium, the enol ether (**149**) was formed; this could be con-

Table 8.11 Activity of 3-carboalkoxy 7-phenylglycyl carbacephems.

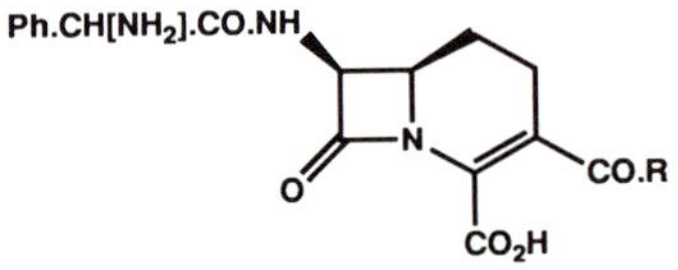

Ph.CH[NH$_2$].CO.NH … CO.R, CO_2H

Organism	Cefaclor	A	B	C
S. aureus	2.0	16.0	32.0	4.0
S. aureus (β+)	4.0	128	128	16.0
S. pyogenes	0.12	0.5	1.0	0.12
S. pneumoniae	0.25	0.5	–	0.25
H. influenzae	2.0	2.0	8.0	1.0
H. influenzae (β+)	1.0	4.0	64.0	1.0
E. coli	2.0	2.0	32.0	2.0
E. coli (Tem)	2.0	2.0	64.0	1.0
K. pneumoniae	0.5	0.5	4.0	0.5
S. typhi	0.5	2.0	16.0	0.5

A: R = OMe
B: R = Me
C: R = OC_2H_5

(148) → Ph.O.CH$_2$.CO.NH (149) CO$_2$pNB OC$_2$H$_5$ → tBu.O.CO.NH (150) CO$_2$pNB Br → ATMO.NH (151) CO$_2$H N R S

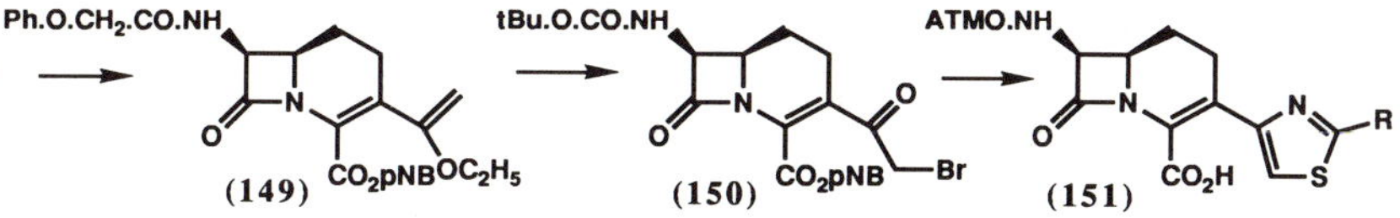

Scheme 8.23

verted, in a series of steps, to the bromo ketone (**150**).[43] This compound could be condensed with thioamides and thioureas to produce a series of 3-heteroaryl substituted carbacephems (**151**) (Scheme 8.23).

Excellent broad spectrum activity was achieved with these derivatives, together with the opportunity to vary the central design of the heteroaryl group such that activity could be directed towards specific organisms. The nitroimidazole and nitrothiazole groups (**C** and **G** in Table 8.12) imparted especially potent activity against *S. aureus* and *B. fragilis* as well as a general high level of gram-negative activity. The 3,4-dihydroxyphenyl group (**I** in Table 8.12) gave a reasonable level of activity against *P. aeruginosa* but was less active against *S. aureus*. The nitrothiazole compound also showed high potency against the type I β-lactamase-producing *E. cloacae* 265A.

8.5.3.5 3-Thiol substituted carbacephems A series of 3-thiosubstituted carbacephems was synthesized and compared to the corresponding 3-H (**A** in Table 8.13) and 3-Cl (**B** in Table 8.13) derivatives.[31] Another comparitor compound was the cephalosporin derivative (**D** in Table 8.13). The cephalosporin had no activity, which was explained by its lack of chemical stability (shown in Table 8.12). The carbacephem derivatives are again taking full advantage of the greater stability of the carbacephem nucleus. These thioderivatives were found to be more active than the 3-H or the 3-Cl compounds; certain of them were significantly more active against *S. aureus* (**F** and **I**) and one compound (**I**) was particularly active against indole-positive *Proteus*. Perhaps the most unexpected property was their potent activity against *E. faecalis*, an organism which is generally resistant to most cephalosporins. Furthermore, one of these thio derivatives (**H**) was highly active against several resistant clinical isolates of *E. faecalis*.

Acknowledgements

I would like to acknowledge my research colleagues at The Lilly Research Laboratories, in particular Larry Blaszczak, John McDonald III, John Morin Jr. and John Munroe, and Dr Hirata and his colleagues at Kyowa Hakko Kogyo Co. Research Laboratories who have been responsible for much of the investigations into this interesting new class of β-lactam antibiotics.

Table 8.12 3-Heteroaryl substituted carbacephems.

Organism	A	B	C	D	E	F	G	H	I	J
S. aureus	1.0	1.0	0.25	1.0	4.0	4.0	0.06	0.5	8.0	2.0
S. aureus	2.0	8.0	2.0	2.0	32	8.0	0.5	2.0	16	4.0
S. pyogenes	0.008	0.015	0.015	0.008	0.008	0.008	0.008	0.008	0.06	0.015
S. pneumoniae	0.008	0.015	0.008	0.008	0.015	0.015	0.015	0.008	0.125	0.03
H. influenzae	0.03	0.015	0.008	0.015	0.125	0.06	0.015	0.015	0.015	0.015
E. coli	2.0	0.03	0.25	2.0	8.0	1.0	2.0	1.0	0.125	1.0
E. coli (Tem)	1.0	0.125	0.25	–	4.0	0.5	1.0	0.5	0.06	0.5
K. pneumoniae	0.015	0.008	0.008	0.008	0.125	0.015	0.015	0.008	0.015	0.008
E. cloacae	1.0	0.125	1.0	2.0	8.0	2.0	1.0	1.0	0.25	2.0
E. cloacae	64	4.0	0.5	128	128	128	16	16	64	128
S. typhi	2.0	0.015	0.25	0.5	4.0	1.0	0.5	1.0	0.03	1.0
P. aeruginosa	128	128	32	128	128	128	128	128	8.0	128
S. marcescens	4.0	0.06	1.0	4.0	8.0	1.0	2.0	4.0	0.25	2.0
M. morganii	2.0	0.06	0.25	2.0	32	4.0	4.0	0.5	4.0	2.0
P. rettgeri	0.25	0.03	0.06	0.125	4.0	0.25	0.06	0.25	0.125	0.5

A: R = phenyl
B: R = 3-(N-methylpyridinium)
C: R = 2-(5-nitro-1,3-thiazole)
D: R = *p*-fluorophenyl
E: R = 1-[1-phenyl-1′-(3-pyridyl)]methane
F: R = amino
G: R = 2-(5-nitro-1-N-methylimidazole)
H: R = *p*-nitrophenyl
I: R = 3,4-dihydroxyphenyl
J: R = 2-furyl

Table 8.13 Activity of 3-thio substituted 7-phenylglycyl carbacephems.

Organism	A	B	C	D	E	F	G	H	I
S. aureus 219P	1.56	0.2	0.2	0.1	50	0.05	0.1	0.1	0.05
S. aureus SMITH	3.13	0.78	0.39	0.2	50	0.39	0.39	0.2	0.2
S. epidermidis	6.25	3.13	0.78	0.39	100	0.39	0.39	0.39	0.2
E. coli NIHJ	3.13	1.56	1.56	0.39	100	1.56	0.78	0.39	0.39
K. pneumoniae	1.56	0.39	0.78	0.1	100	0.78	0.39	0.1	0.2
P. mirabilis	6.25	1.56	3.13	3.13	100	0.1	1.56	0.78	0.2
P. vulgaris	100	100	100	50	100	3.13	50	12.5	3.13
P. rettgeri	12.5	50	100	100	100	100	100	100	1.56
E. cloacae	6.25	3.13	25	25	100	12.5	25	12.5	50
C. freundii	1.56	1.56	6.25	0.78	100	0.78	3.13	1.56	1.56
E. faecalis	100	25	3.13	1.56	–	6.25	1.56	0.39	–
S. pyogenes	0.78	0.2	0.05	0.01	–	0.01	0.01	0.01	–
S. pneumoniae	0.78	0.39	0.2	0.1	–	0.1	0.02	0.05	–

A: R = H
B: R = Cl
C: R = 2-thio-N-methyltetrazole
D: R = 4-thiopyridine
E: R = 4-thiopyridine
F: R = 4-thio-N-methylpyridinium
G: R = 2-thio-5-methyl-1,3,5-thiadiazole
H: R = 5-thio-1,2,3-thiadiazole
I: R = 2-thio-5-amino-1,3,5-thiadiazole

References

1. D.M. Brunwin, G. Lowe and J. Parker, *J. Chem. Soc. (C)* (1971) 3756–3762.
2. H.H. Wasserman and W.T. Han, *Tet. Lett.* (1984) 3743–3746.
3. R.N. Guthikonda, L.D. Cama and B.G. Christensen, *J. Amer. Chem. Soc.* (1974) **96** 7584–7585.
4. R.A. Firestone, J.H. Fahey, N.S. Maciejewicz, G.S. Patel and B.G. Christensen, *J. Med. Chem.* (1977) **20** 551–556.
5. S. Uyeo and H. Ona, *Chem. Pharm. Bull.* (1980) **28** 1563–1577.
6. S. Uyeo and H. Ona, *Chem. Pharm. Bull.* (1980) **28** 1578–1583.
7. T.W. Doyle, T.T. Conway, M. Casey and G. Lim, *Can. J. Chem.* (1979) **57** 222–226.
8. T.W. Doyle, T.T. Conway, G. Lim and B-Y. Luh, *Can. J. Chem.* (1979) **57** 227–232.
9. A. Martel, T.W. Doyle and B-Y. Luh, *Can. J. Chem.* (1979) **57** 614–625.
10. T. Ogasa, H. Saito, Y. Hashimoto, K. Sato and T. Hirata, *Chem. Pharm. Bull.* (1989) **37** 315–321.
11. H. Saito, H. Matsushima, C. Shiraki and T. Hirata, *Chem. Pharm. Bull.* (1989) **37** 275–279.
12. I. Matsukuma, S. Yoshiye, K. Mochida, Y. Hashimoto, K. Sato, R. Okachi and T. Hirata, *Chem. Pharm. Bull.* (1989) **37** 1239–1244.
13. D.A. Evans and E.B. Sjogren, *Tetrahedron Lett.* (1985) 3783–3787.
14. D.A. Evans and E.B. Sjogren, *Tetrahedron Lett.* (1985) 3787–3790.
15. C.C. Bodurow, B.D. Boyer, J. Brennan, C.A. Bunnell, J.E. Burks, M.A. Carr, C.W. Doecke, T.M. Eckrich, J.W. Fisher, J.P. Gardner, B.J. Graves, P. Hines, R.C. Hoying, B.G. Jackson, M.D. Kinnick, C.D. Kochert, J.S. Lewis, W.D. Luke, L.L. Moore, J.M. Morin, Jr., R.L. Nist, D.E. Prather, D.L. Sparks and W.C. Vladuchick, *Tetrahedron Lett.* (1989) 2321–2324.
16. R.D.G. Cooper, B.W. Daugherty and D.B. Boyd, *Pure and Applied Chemistry* (1987) **59** 485–492.
17. N. Ikota and A. Hanaki, *Heterocycles* (1987) **28** 418.
18. D.A. Evans and J.M. Williams, *Tetrahedron Lett.* (1988) 5065–5068.
19. J. Dunnigen and L.O. Weigel, *J. Org. Chem.* (1991) **56** 6225–6227.
20. H. Hatanaka and T. Ishimaru, *Tetrahedron Lett.* (1983) 4837–4841.
21. B.J. Jackson, J.P. Gardner and P.C. Heath, *Tetrahedron Lett.* (1990) 6317–6320.
22. M.J. Zmijewski, Jr., B.S. Briggs, A.R. Thompson and I.G. Wright, *Tetrahedron Lett.* (1991) 1621–1622.
23. L.C. Blaszczak, H.K. Armour and N.G. Halligan, *Tetrahedron Lett.* (1990) 5693–5696.
24. N.G. Halligan and L.C. Blaszczak, *Tetrahedron Lett.* (1992) (In Press).
25. C.L. Jordan, M.D. Kinnick, J.M. Morin, Jr. and R. Ternansky, Abstract 29, *201st. ACS National Meeting*, April 14–19, 1991.
26. M.I. Page, *Adv. Phys. Org. Chem.* (1987) **23** 165 and references therein.
27. D.B. Boyd, *J. Med. Chem.* (1983) **26** 1010–1013.
28. D.B. Boyd, *J. Med. Chem.* (1984) **27** 63–66.
29. I. Matsukuma, S. Yoshiiye, K. Mochida, Y. Hashimoto, K. Sato, R. Okachi and T. Hirata, *Chem. Pharm. Bull.* (1989) **37** 1239–1244.
30. L.C. Blaszczak, R.F. Brown, G.K. Cook, W.J. Hornback, R.C. Hoying, J.M. Indelicato, C.L. Jordan, A.S. Katner, M.D. Kinnick, J.H. McDonald, III, J.M. Morin, Jr., J.E. Munroe and C.E. Pasini, *J. Med. Chem.* (1990) **33** 1656–1662.
31. K. Mochida, T. Ogasa, J. Shimada, T. Hirata, K. Sato and R. Okachi, *J. Antibiot.* (1989) **42** 283–290.
32. K. Sato, R. Okachi, I. Matsukuma, K. Mochida and T. Hirata, *J. Antibiot.* (1989) **42** 1844–1853.
33. P. Herdewijn, P.J. Claes and H. Vanderhaeghe, *J. Med. Chem.* (1986) **29** 661–664.
34. US Patent 4,343,943.
35. H. Saito, F. Susuki and T. Hirata, *Chem. Pharm. Bull.* (1989) **37** 2298–2302.
36. T.W. Doyle, J.L. Douglas, B. Belleau, T.W. Conway, C.F. Ferrari, D.H. Horning, G. Lim, B-Y. Luh, A. Martel, M. Menard, L.R. Morris and M. Misiek, *Can. J. Chem.* (1980) **58** 2508–2523.

37. T. Ogasa, H. Saito, Y. Hashimoto, K. Sato and T. Hirata, *Chem. Pharm. Bull.* (1989) **37** 315–321.
38. European Patent 82,501.
39. G.K. Cook, J.H. McDonald III, W. Alborn, Jr., D.B. Boyd, J.A. Eudaly, J.M. Indelicato, R. Johnson, J.S. Kasher, C.E. Pasini, D.A. Preston and E.C-Y. Wu, *J. Med. Chem.* (1989) **32** 2442–2450.
40. T.A. Crowell, B.D. Halliday, J.H. McDonald III, J.M. Indelicato, C.E. Pasini and E.C-Y. Wu, *J. Med. Chem.* (1989) **32** 2436–2442.
41. G.K. Cook, W.J. Hornback, C.L. Jordan, J.H. McDonald III and J.E. Munroe, *J. Org. Chem.* (1989) **54** 5828–5830.
42. J.A. Eudaly, W.J. Horndack, J.M. Indelicato, M.E. Johnson, R.J. Johnson, C.L. Jordan, J.S. Kasher, J.E. Munroe, C.E. Pasini, D.A. Preston and W.E. Wright, Abstract 237, *29th Interscience Conference on Antimicrobial Agents and Chemotherapy*, Houston, 1989.
43. W.J. Hornback, J.E. Munroe and F.T. Counter, Abstract 153, *American Chemical Society National Meeting*, Washington, 1990.

9 Non-β-lactam mimics of β-lactam antibiotics

L.N. JUNGHEIM and R.J. TERNANSKY

9.1 Introduction

The β-lactam antibiotics are represented by a diverse array of chemical structural types. These include cephalosporins, cephamycins, oxa- and carbacephems, penicillins, penems, carbapenems and monobactams. In spite of this chemical diversity the β-lactam moiety present in each is capable of acylating the penicillin binding proteins (PBPs) thus interfering with bacterial cell-wall biosynthesis. In addition to being selective acylating agents, a clinically useful β-lactam must be able to penetrate the bacterial cell wall, and resist hydrolysis by β-lactamases in order to inactivate the target PBPs. This is the subject matter of the previous chapters in this book. The very low incidence of side effects, coupled with the potent antibacterial activity traditionally displayed by the β-lactams, makes them attractive targets to emulate in the search for new antibacterial agents.

Attempting to understand the structure–activity relationships (SAR) of the β-lactam family of antibiotics has recently been described as 'an impossible dream'.[1] The structural features believed necessary for antibacterial activity have changed dramatically since the structure of penicillin was first elucidated. By the early 1980s the acknowledged pharmacophore of active β-lactam antibiotics had been reduced to a sufficiently reactive azetidinone (**1**), which possesses the correct molecular shape for binding to the target PBPs. The acylating ability of the β-lactam moiety needs to be in the same range as biologically active cephalosporins or penicillins, i.e. lactam hydrolysis rates between 0.04 and 6.0 h^{-1} (pH 10, 35°C),[2–6] i.e. a second order rate constant for alkaline hydrolysis of 0.1–20 $M^{-1}s^{-1}$. The 'correct molecular shape' requires the presence of an acidic group with a separation of about 3.0–3.6 Å between the lactam carbonyl carbon and the center of the acidic group.[2,5]

In the 1970s several groups began to question the conventional wisdom that the β-lactam ring itself was essential for antibacterial activity (see chapter 2). Previous attempts, dating back to the 1940s, to prepare biologically active γ- or δ-lactam mimics of penicillin met with failure, e.g. (**2**),[7] (**3**)[8] and (**4**).[9] It was reasoned either that these compounds possessed

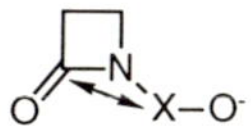

(1) 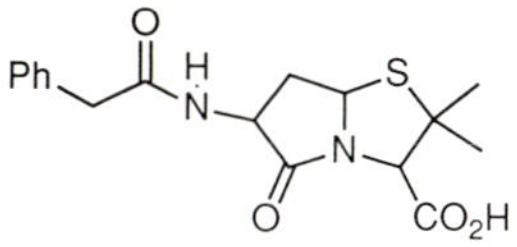 (2) 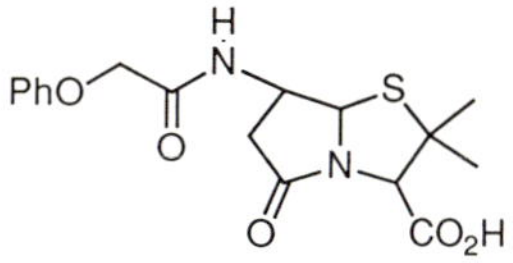 (3)

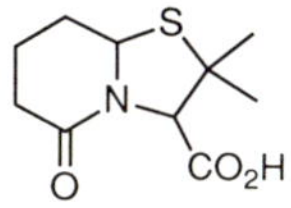

(4) 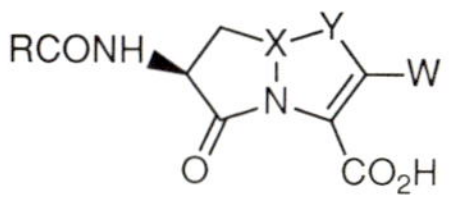 (5) 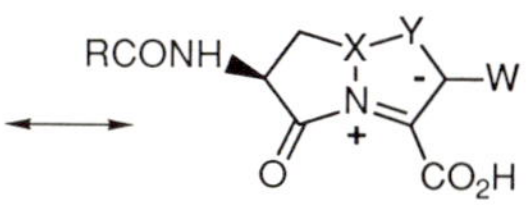 (6)

insufficient chemical reactivity to acylate the PBPs and/or that their molecular shape precluded good 'fit' into the PBPs. Chemical intuition suggested that electron-withdrawing substituents (W) at C-3 might increase the acylating ability of a γ-lactam (tautomer (**5**) vs. (**6**)) by delocalization of the lone pair of electrons on the lactam nitrogen through the olefinic bond. Furthermore, molecular modeling studies supported the hypothesis that a γ-lactam could be designed to mimic a β-lactam, with regard to chemical reactivity and molecular shape.[10] What could not be predicted *a priori* was the ability of a γ-lactam to penetrate the bacterial cell wall, or the kinetics of its interaction with a PBP. Both of these factors are critical to the ultimate success or failure of the potential antibacterial agent.

In addition to work by the Lilly group, there has been considerable effort by a number of other groups to design or discover bioisosteres of β-lactams. Two excellent reviews,[11,12] which discuss in detail efforts to prepare non-β-lactam mimics of the β-lactam antibiotics, have recently appeared. In this chapter consideration will be given mainly to the compounds or series of compounds that have demonstrated biological activity.

9.2 Pyrazolidinones

In the global search for biological surrogates of the β-lactam ring, significant attention has been given to the design of γ-lactam analogues that could potentially serve in this role. The next four sections of this chapter will document the major advances that have taken place in this endeavor over the past few years. The successful design and synthesis of γ-lactam antibacterials, as well as the isolation of naturally occurring γ-lactam-containing antibiotics (see section 9.3) have clearly challenged the long-held notion that the presence of a β-lactam ring is a prerequisite for antibacterial activity.

In designing potential β-lactam surrogates, consideration must be given both to molecular shape (for recognition by target PBPs), and to acylation ability (see section 9.1). These criteria were the key elements used by the

Lilly group in designing a series of pyrazolidinone-containing compounds as potential antibacterials. The prototype compounds synthesized in this series were designated LY173013, (**7**), and LY186826, (**8**).[13] These compounds combined key structural features taken from the cephalosporin area ('ATMO' — aminothiazoylmethoxime — side chain) and the penicillin area (carbapenem 'B' ring) with the pyrazolidinone heterocycle serving in place of the azetidinone ring. The compounds, originally prepared in racemic form, exhibited *in vitro* antibacterial activity against both gram-positive and gram-negative organisms (Table 9.1).[14] In studies with hyperpermeable mutants and their wild-type parents, the ability of (**7**) and (**8**) to permeate the outer membrane of gram-negative bacteria was found to be similar to that of the cephalosporins and penicillins.[15] In addition, the mechanism of action of this new class of compounds was shown to be the same as that of the well-known β-lactam antibiotics. Thus, the fused pyrazolidinones (**7**) and (**8**) exhibited significant binding to bacterial PBPs, especially to PBP 3.[16] Significantly, *in vivo* studies with (**8**) demonstrated its ability to protect mice infected with *Escherichia coli* and *Proteus vulgaris* when administered subcutaneously.[15,17] The documentation of these biological attributes of the two prototype [3.3.0] fused pyrazolidinones firmly established that a viable biological surrogate for the β-lactam ring has been realized!

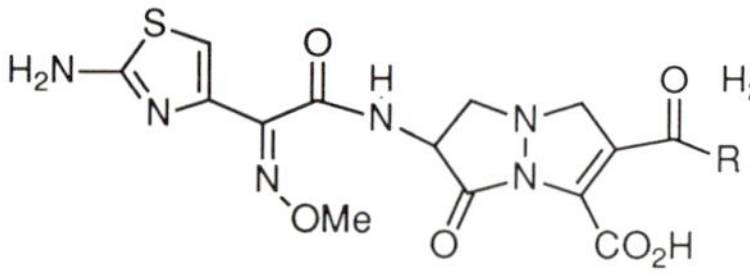

(**7**) R = OMe (LY173013)

(**8**) R = Me (LY186826)

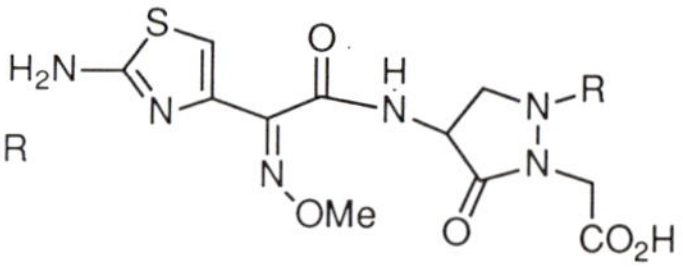

(**9**) R = $CH_2(CH_3)_2$

(**10**) R = $COCH_3$

The exciting discovery of the pyrazolidinone antibacterials prompted the Lilly group to mount a major research effort aimed at defining the scope of activity within this new class of β-lactam mimics. The systematic approach taken by the Lilly scientists involved the preparation of analogues of known, biologically-active β-lactam-containing antibacterials wherein the azetidinone ring was substituted with the pyrazolidinone moiety. Thus, γ-lactam analogues of the monobactams such as (**9**) and (**10**) were constructed, but were found to be devoid of antibacterial activity.[18] Furthermore, pyrazolidinone analogues of the 1-carba-1-dethiacephalosporins were prepared. These compounds, (**11**) and (**12**), which represent simple homologues of the original lead compounds (**7**) and (**8**) were surprisingly devoid of antibacterial activity.[19] On further study these [4.3.0] ring systems were found to be much less reactive towards hydrolysis than their [3.3.0] counterparts (**7**) and (**8**). This factor, along with the fact that the compounds did not exhibit affinity for the PBPs, may help to explain this disparity in antibacterial activity.[20] Based on these results, the Lilly group chose to focus their attention on the development of the [3.3.0] fused bicyclic pyrazolidinone series within which the original antibacterial activity was discovered.

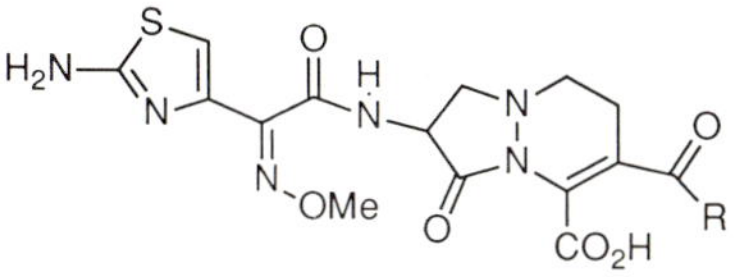

(**11**) R = OMe

(**12**) R = Me

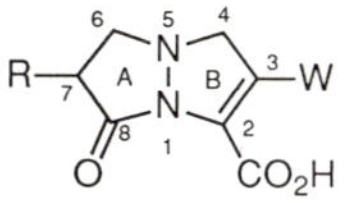

(**13**)

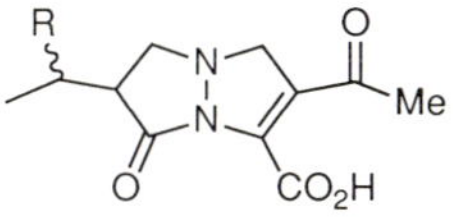

(**14**) R = OH

(**15**) R = H

The basic molecular framework of the [3.3.0] fused pyrazolidinones about which the SAR was conducted is represented by (**13**). Substitution of the hydrogens on C-4 and C-6 with methyl groups led to a loss of all antibacterial activity.[21] In addition, replacement of C-3 with a sulfur atom led to a complete loss of activity.[22] The most productive modifications in terms of improving *in vitro* antibacterial activity were those involving changes of the substituents at C-3 (W) and C-7 (R).

As with cephalosporins, good antibacterial activity was found to require the appendage of an acylamino side chain to C-7. Thus, the carbapenem-type side chains present in (**14**) and (**15**) did not support significant levels of antibacterial activity.[23] Of the acylamino side chains explored, the greatest activity was realized with those of the ATMO type. These substituents at C-7 consistently provided compounds with maximum antibacterial activities. The absolute stereochemistry of C-7 was also found to be biologically significant. Thus, independent synthesis of the *R* and *S* isomers of LY173013, (**7**), demonstrated that the antimicrobial activity of the [3.3.0] pyrazolidinones resides with the 7-*S*-isomer.[21,24] Further structure–activity studies in this series concentrated exclusively on this isomer.

The most significant improvements in biological activity were realized with variation of the substituents at C-3 (W). Thus, it was found that for W = cyano (LY255262, (**16**) and W = methylsulfanato (LY193239, (**17**) the *in vitro* activities were significantly more potent than the original prototypes (Table 9.1).[25] These compounds, bearing stronger electron-withdrawing groups in the 3-position, were also determined to be more reactive toward chemical hydrolysis when compared to pyrazolidinones (**7**) and (**8**).[6]

Table 9.1 Agar dilution MICs ($\mu g\ ml^{-1}$) of pyrazolidinone analogues.

	Compound					
Organism	**(7)**	**(8)**	**(16)**	**(17)**	**(18)**	**(19)**
Staphylococcus aureus (X1.1)	>128	32	64	32	>128	8
Streptococcus pyogenes (C203)	4	0.5	0.015	0.125	1	0.06
Haemophilis influenzae (76)	16	8	0.5	0.5	0.03	0.125
Escherichia coli (EC14)	8	2	0.125	0.06	0.03	0.03
Klebsiella pneumoniae (X26)	4	2	0.15	0.125	0.125	0.03
Enterobacter cloacae (EB5)	32	16	0.25	0.25	0.125	0.06

Although a correlation exists in this series between chemical reactivity and antibacterial activity, the observed increase in biological activity is not likely to be due simply to the increase in acylating ability. Indeed, it has been determined that compounds (**16**) and (**17**) exhibit higher affinities for the bacterial PBP enzymes than do (**7**) or (**8**).[26]

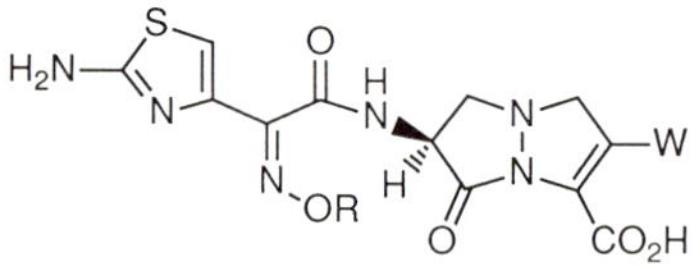

(**16**) W = CN R = Me (LY255262)

(**17**) W = SO_2Me R = Me (LY193239)

(**18**) W = SO_2Me R = $C(CH_3)_2CO_2H$

(**19**) W = SO_2Me R = CH_2CH_2F

Modification of the oxime portion of the ATMO side chain present in (**17**) was found to provide compounds with slightly altered *in vitro* activities. Thus, compound (**18**) exhibited a loss of activity against gram-positive organisms, and an increase in activity against the gram-negatives (except *Pseudomonas*). Compound (**19**) exhibited improved MICs across the spectrum of gram-positive and gram-negative organisms (Table 9.1).[27] These modifications have provided compounds exhibiting impressive antibacterial activity comparable in some cases to that observed for third-generation cephalosporins.

The discovery and development of these γ-lactam mimics of the β-lactam antibacterials represents a significant milestone in the history of the development of antimicrobial agents. The contributions of these investigations to the understanding of the mechanism of action, and the structure–activity relationships of these agents are significant.

9.3 Lactivicin

In 1986, the Takeda group reported the isolation and identification of a new antibiotic from two strains of bacteria, *Empedobacter lactamgenus* (YK-258) and *Lysobacter albus* (YK-422), which, in turn, were isolated from soil samples collected in Japan.[28] They named the new antibiotic 'lactivicin' and determined its structure to be composed of the epimeric mixture depicted by (**20**).[29,30] This novel dipeptide was discovered through a screening methodology designed to detect β-lactam-like activity.[31] The lack of a β-lactam ring in an antibiotic substance selected via this protocol was clearly very intriguing. The substance identified as lactivicin appeared to behave biologically as though it contained a β-lactam ring, even though chemical analysis excluded the presence of such a functionality. In this case, the γ-lactam ring of lactivicin was found to act as a biological mimic of the β-lactam ring.

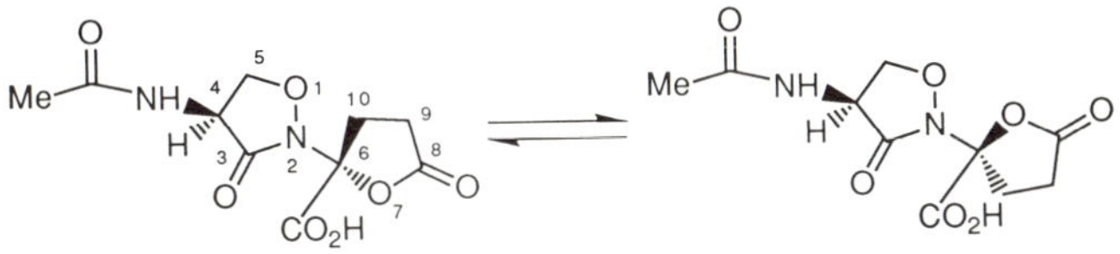

(**20**)

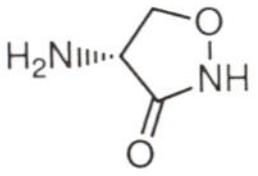

(**21**) 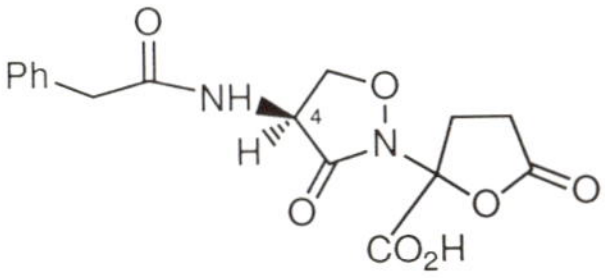 (**22**)

Lactivicin exhibits a high affinity for bacterial PBPs, especially PBP 1.[32] It displays greater activity against β-lactam-hypersensitive mutants when compared to parent strains. Lactivicin was also demonstrated to have various interactions with β-lactamases, including sensitivity to these enzymes, as well as an ability to inhibit or induce the activity of the β-lactamases. These behaviors are all consistent with a β-lactam mode of action. In spite of the structural similarity to D-cycloserine (**21**), lactivicin's mode of action has been shown to be clearly distinct. The bacterial lethality of lactivicin is considered to be primarily due to inhibition of the PBPs although activity has also been demonstrated with respect to inhibition by lactivicin of membrane sulfhydryl proteins.[32]

The *in vitro* activity of lactivicin (Table 9.2) provided much opportunity for improvement through chemical modification and SAR studies. Through these efforts, the key structural parameters required for optimal antibacterial activity of the lactivicin analogues were defined. Initial studies

Table 9.2 Agar dilution MICs ($\mu g\ ml^{-1}$) of lactivicin analogues.

	Compound		
Organism	**(20)**	**(23)**	**(24)**
Staphylococcus aureus (FDA209p)	3.13	12.5	6.25
Proteus vulgaris (IFO 3988)	100	0.78	0.39
Escherichia coli (O-111)	100	0.39	0.2
Klebsiella pneumoniae (DT)	100	0.78	0.39

were directed toward a synthesis of the parent molecule. This work led to the preparation of the 4-*S*- and 4-*R*-isomers starting with L- and D-cycloserine, respectively. The 4-*S*-phenylacetyl derivative (**22**) was shown to be significantly more active than the corresponding 4-*R*-derivative. This configuration about the carbon bearing the acylamino side chain (C-4) is the same as that required for maximum biological activity with the pyrazolidinone (section 9.2) and β-lactam classes of antibacterials.[33]

Structure–activity studies initially concentrated on modification of the side chain substituents attached to C-4. Modification of these groups showed a trend similar to that observed in the cephalosporin and pyrazolidinone series. Thus, the ATMO-type side chains, as present in (**23**) and (**24**), provided derivatives exhibiting potent *in vitro* antibacterial properties (Table 9.2).[34] Other side chains lowered the activity of these compounds but were in general more active than the parent lactivicin. The D-phenylglycyl side chain analogue was found to be biologically inactive due to its instability in aqueous solutions.[30] Further structural modifications involved nuclear changes to the basic lactivicin framework. Thus, substitution of the ring oxygen with nitrogen provided pyrazolidinone derivatives of lactivicin.

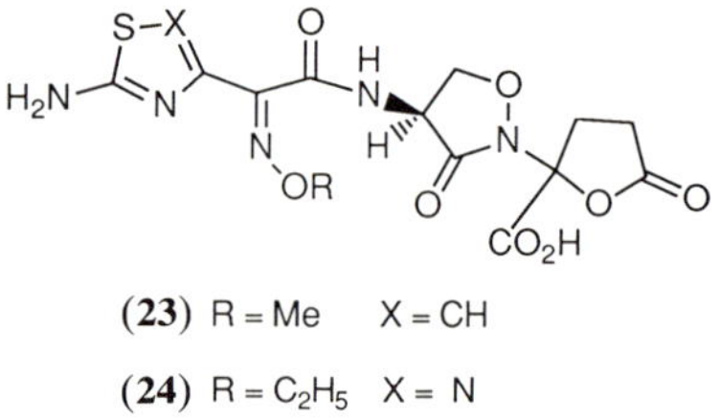

(**23**) R = Me X = CH

(**24**) R = C_2H_5 X = N

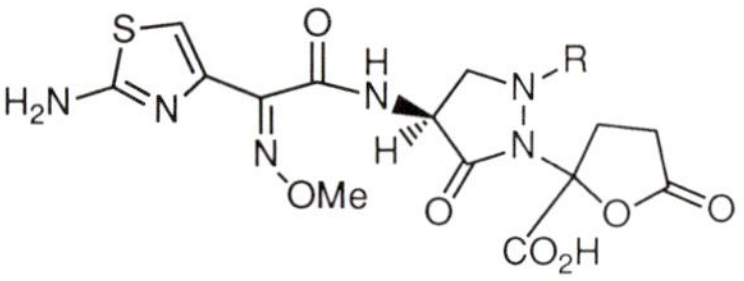

(**25**) R = $COCH_3$

(**26**) R = H

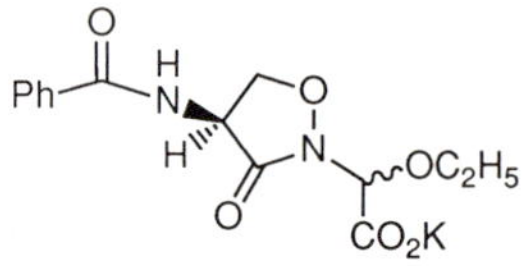

(**27**)

These derivatives, such as (**25**) and (**26**) exhibited drastically reduced antibacterial activity when compared to their oxo-analogues.[35] This activity supports the observation made by the Lilly group (section 9.2) that monocyclic pyrazolidinones were not supportive of good antibacterial activity. Finally, modification of the lactone ring of lactivicin has been investigated. Thus, preparations of compounds such as (**27**) by the Baldwin group were shown to be devoid of significant antibacterial activity.[36]

The discovery of a naturally occurring γ-lactam antibiotic that mimics the biological activity and mode of action of β-lactam antibiotics represents a major contribution to the continuing evolution of antibacterial agents. Further modifications of the lactivicin nucleus, as well as modifications to the previously discussed [3.3.0] pyrazolidinones, may one day lead to the development of a γ-lactam antibacterial for use in clinical medicine.

9.4 γ-Lactams

The Lilly group provided some of the first data that showed that γ-lactam analogues of the β-lactams could exhibit antibacterial activity. Morin and coworkers[37] reported on the synthesis of γ-lactam analogues of penems (**28**), and Munroe and coworkers[38] prepared γ-lactam analogues of carbapenems (**29**). A concurrent publication by Baldwin *et al.*[39] described the synthesis and antibacterial activity of penem analogue (**38**). Within the Lilly group, the design of these antibacterial agents was guided both by chemical intuition, and by molecular modeling. Chemical intuition suggested that electron-withdrawing substituents (W) at the C-3 position might increase the acylating ability of a γ-lactam (tautomer (**5**) vs. (**6**)) by delocalization of the lone pair of electrons on the lactam nitrogen through the olefinic bond. Molecular modeling studies supported the hypothesis that a γ-lactam could be designed to mimic a β-lactam, with regard to chemical reactivity and molecular shape.[10]

The 7-unsubstituted γ-lactam derivative (**30**) was prepared by Morin and coworkers[37] for direct comparison with the known penem (**31**), which has been shown to be a potent antibiotic.[40] In contrast to penem (**31**), γ-lactam (**30**) exhibited neither antibacterial activity against a variety of organisms, nor activity as a β-lactamase inhibitor. In order to more closely mimic the cephalosporins and penicillins a C-7 acylamino side chain was incorporated. These derivatives were of particular interest because the corresponding 6-acylamino-penems are chemically too reactive to exhibit useful antibacterial activity.[41] The 7-substituted analogues[10,37] (**32a,b**) and (**33a, b**) displayed low, yet demonstrable levels of activity against a variety of gram-positive and gram-negative organisms as shown in Table 9.3. It is interesting to note that the 7-α isomer (**32b**) was somewhat more active than the corresponding 7-β side chain isomer (**32a**), as the β-configuration is found on

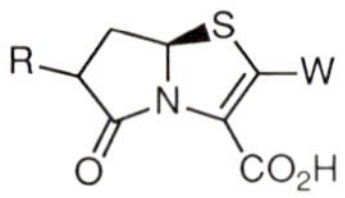

(28) 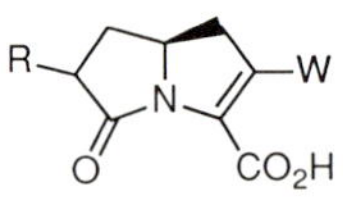 **(29)**

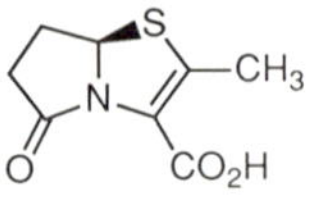

(30) 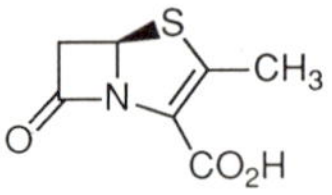 **(31)**

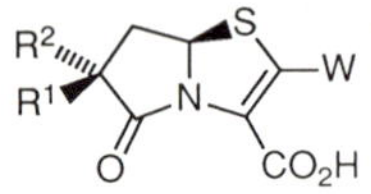

(32a) R^1 =ATMO, R^2 =H, W = CH_3
(32b) R^1 =H, R^2 = ATMO, W = CH_3
(33a) R^1 =ATMO, R^2 = W = H
(33b) R^1 =H, R^2 = ATMO, W = H
(34) R^1 =ATMO, R^2 = H, W = CO_2CH_3

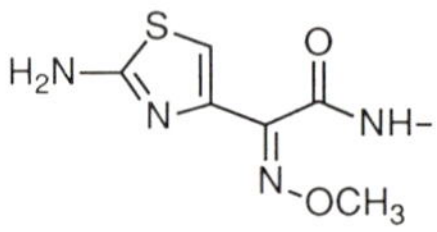

ATMO

active β-lactams bearing acylamino side chains. Preliminary studies on the mechanism of action of **(32a, b)** were consistent with inhibition of bacterial cell-wall biosynthesis. The C-3 carbomethoxy-substituted compound **(34)** was devoid of activity, in contrast to the analogous bicyclic pyrazolidinone **(7)**, which is a potent, broad spectrum antibacterial agent (see section 9.2).

Munroe and coworkers synthesized a series of γ-lactam analogues of the penems, both with[10,38] **(35a–e)** and without[10] **(36a–e)** pendant 7-acylamino side chains. It should be noted that the corresponding acylamino-substituted carbapenems are known to be chemically unstable.[42] These γ-lactam analogues of the carbapenems were generally devoid of *in vitro* antibacterial activity; however, analogues **(35a)** and **(36d)** exhibit slight activity at the highest concentrations tested. The low level of antibacterial activity exhibited by **(35a, b, e)** is noteworthy when one compares them to the closely related bicyclic pyrazolidinones **(7)**, **(16)** and **(17)**.

Table 9.3 Agar dilution MICs ($\mu g\ ml^{-1}$) of γ-lactam analogues.

	Compound				
Organism	**(32a)**	**(32b)**	**(33a)**	**(33b)**	**(7)** (S)
Staphylococcus aureus (X1.1)	>128	32	>32	>128	>128
Streptococcus pyogenes (C203)	64	4	8	64	2
Streptococcus pneumoniae (PARK)	128	8	8	64	2
Haemophilis influenzae (76)	>128	>128	32	>128	8
Escherichia coli (EC14)	>128	>128	32	>128	2
Klebsiella pneumoniae (X26)	>128	>128	8	64	2

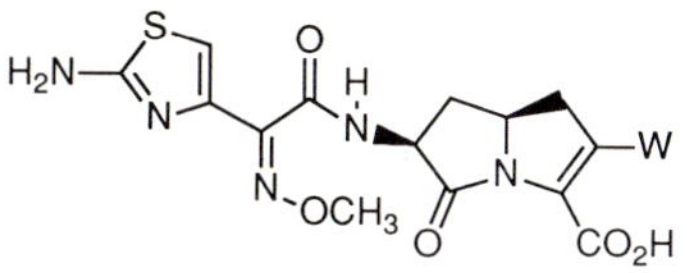

(**35a**) W = CO_2CH_3
(**35b**) W = CN
(**35c**) W = SEt
(**35d**) W = SOEt
(**36e**) W = SO_2Et

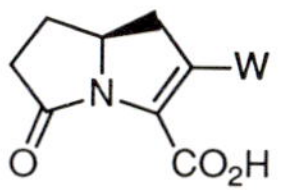

(**36e**) W = Cl
(**36a**) W = CN
(**36b**) W = SEt
(**36c**) W = SOEt
(**36d**) W = SO_2Et

The Baldwin group prepared γ-lactam (**37**) and reported that it did not possess antibacterial activity.[43] They proposed that the absence of antibacterial activity derived from the lack of reactivity of the γ-lactam relative to a β-lactam. Thus, they reasoned that a γ-lactam analogue of the 6-β-acylamino penems (**39**)[41] might show increased reactivity and hence biological activity due to delocalization of the lone pair of electrons on the lactam nitrogen, as in compound (**38**). γ-Lactam (**38**) showed 'weak but real' antibacterial activity against both *Staphylococcus aureus* and *Escherichia coli* strains.[39]

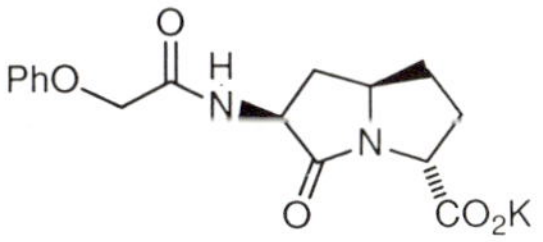

(**37**) 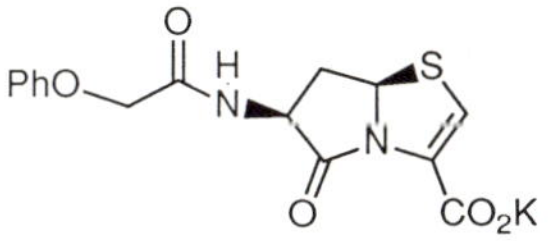 (**38**)

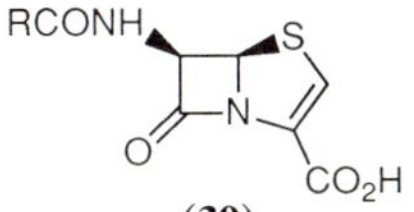

(**39**)

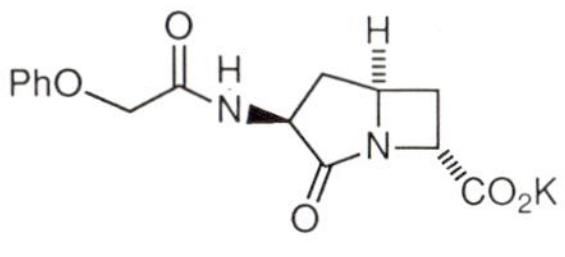

(**40**) 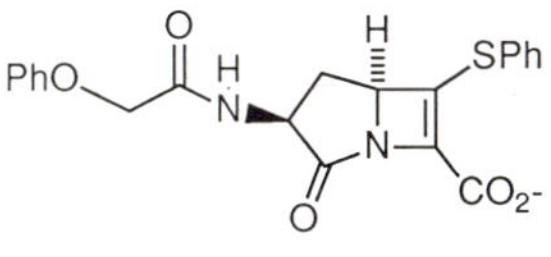 (**41**)

Another possible reason why (**37**) is inactive stems from the fact that the lactam nitrogen in this analogue is relatively planar. It has been hypothesized that the degree of pyramidal distortion of the lactam nitrogen in β-lactam antibiotics is related to antibacterial activity.[44]

Molecular modeling studies indicated that fused γ-lactam azetidine analogues, such as (**40**), would have similar pyramidal distortions to those observed for the lactam nitrogen in penicillins. Baldwin *et al.*[45] synthesized γ-lactam (**40**) and found that it was neither an antibacterial, nor an inhibitor of β-lactamases. However, Heck, of Merck Sharp & Dohme, reported in a patent that the closely related γ-lactam (**41**) exhibits antibacterial activity

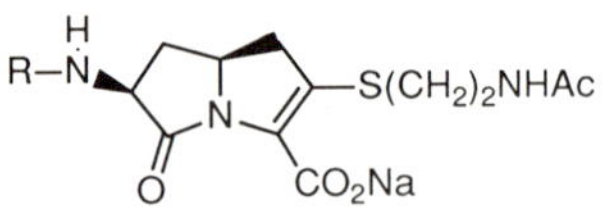

(42)

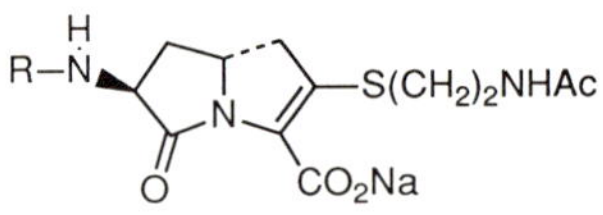

(45)

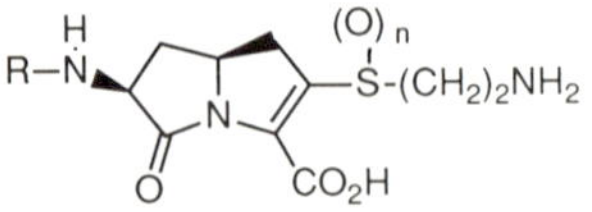

(43) n=0
(44) n=2

R =

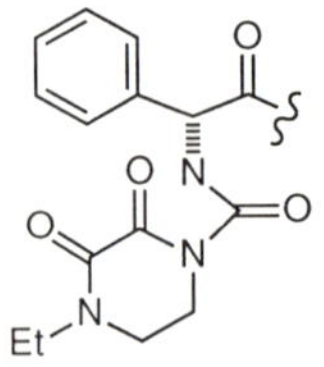

against a broad range of pathogens.[46] Unfortunately, no data were presented.

Recently Hashiguchi *et al.*, of Takeda, prepared a series of γ-lactam analogues of carbapenems.[47] Compounds **(42)**–**(44)** showed slight, but appreciable, antibacterial activity against the gram-negative organisms tested (Table 9.4). Sulfone derivative **(44)** was more potent than the corresponding sulfide **(43)**; presumably the electron-withdrawing sulfone group is activating the C–N bond as discussed previously. Interestingly, the *trans* isomer **(45)** was found to be slightly more active than the corresponding *cis* isomer **(42)**. A similar trend was reported for penem homologues **(32a, b)**.

Table 9.4 MICs ($\mu g\ ml^{-1}$) of γ-lactam analogues of carbapenems.

	Compound			
Organism	**(42)**	**(43)**	**(44)**	**(45)**
Escherichia coli (PG-8S)	–	100	50	–
Escherichia coli (PG-12)	100	–	–	25
Klebsiella pneumoniae (IFO 3317)	100	–	–	25
Proteus mirabilis (ATCC 21100)	25	>100	100	6

Several γ-lactams have been synthesized with side chains appended at the β-position relative to the lactam carbonyl, e.g. **(46)**[48], **(47)**[49] and **(48)**[43]; all of them are devoid of antibacterial activity. Likewise, the oxa-penem homologue **(49)**[50] and 1-hydroxy-carbapenem homologue **(50)**[51] were reported to be inactive. None of the monobactams **(51)**[52], **(52)**[11] or **(53)**[53] possesses antibacterial activity; however **(52)** was reported to inhibit the β-lactamase isolated from *Enterobacter cloacae* P99 at very high concentration (1 mM).

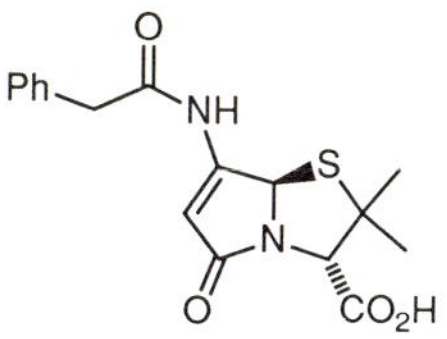

(46) 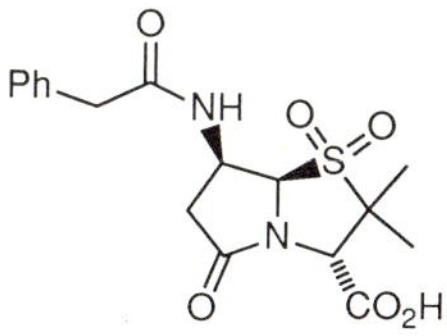 (47) 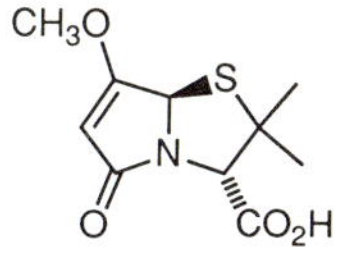 (48)

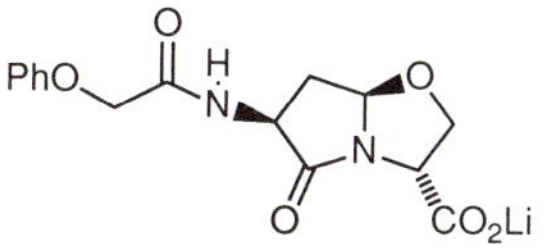

(49)  (50) 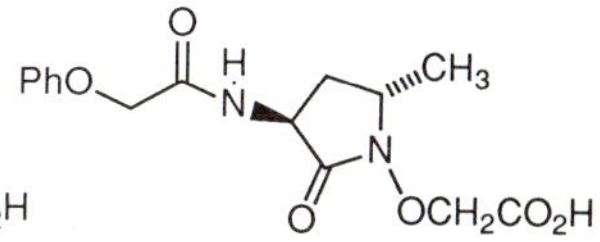 (51)

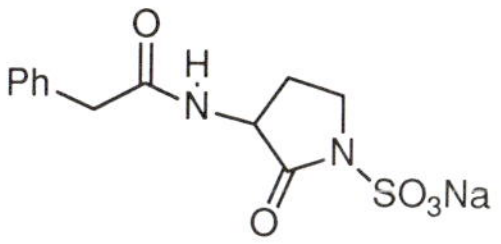

(52) 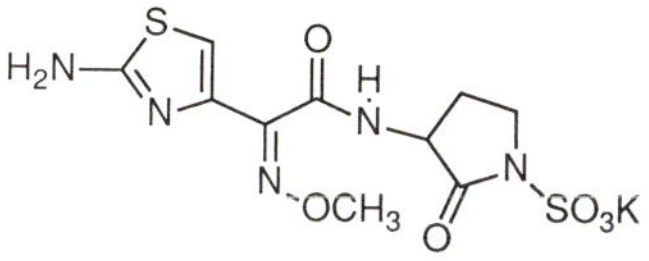 (53)

9.5 Imidazolidinones

Ghosez's group[11,54,55] has described yet another class of potential β-lactam mimics, the bicyclic imidazolidinones (**54**). These compounds have the potential to form two different acyl-enzyme intermediates, (**55**) and (**56**), on ring opening by a PBP or β-lactamase. In both instances a carbamate linkage is formed between the substrate and enzyme, which is much less susceptible to nucleophilic hydrolysis than the simple ester linkage formed with a β-lactam. This design characteristic is desirable, as good inhibitors form acyl-enzyme intermediates that are not readily hydrolyzed to regenerate the free enzyme. One should note, however, that the urea functionality, present in the imidazolidinone, is also much less susceptible to the initial nucleophilic attack by the serine residue in the active site of the target enzymes.

All of the bicyclic imidazolidinones derived from penicillin were devoid of antibacterial activity, e.g. (**57**).[54,55] Three analogues (**58**), (**59**) and (**60**) inhibited selected β-lactamases to the extent of 30–60% at high concentration (1 mM). Compound (**58**) inhibits the enzyme isolated from *Bacillus cereus*, (**59**) inhibits the *Enterobacter cloacae* enzyme, and (**60**) inhibits both of these β-lactamases.[11] The authors suggested that the poor biological results might be due to the fact that the acylamino side chain is incorrectly oriented. Molecular modeling[55] showed that a better mimic of penicillins would be obtained by attaching the side chain to N-7, as in analogue (**61**). Bicyclic imidazolidinone (**61**) was also inactive.

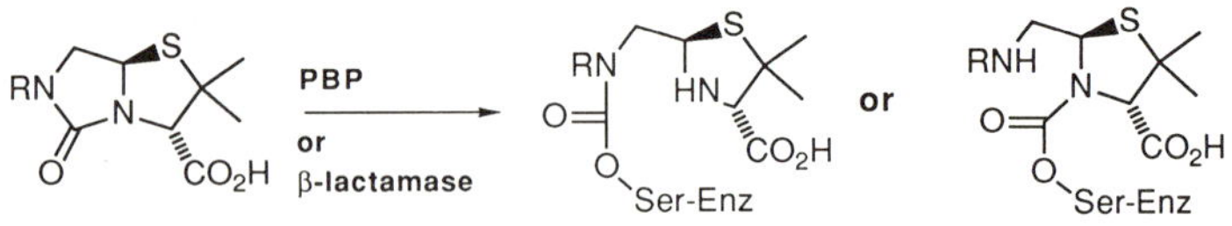

PBP or β-lactamase

(54) (55) or (56)

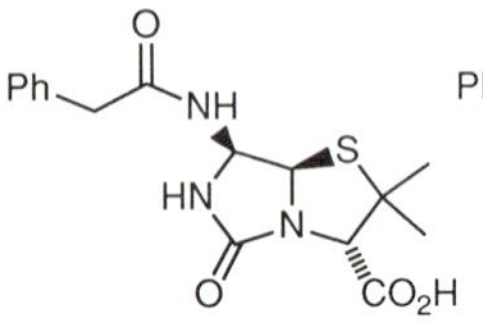

(57) 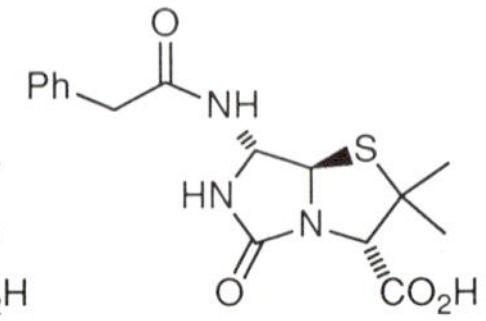 (58) 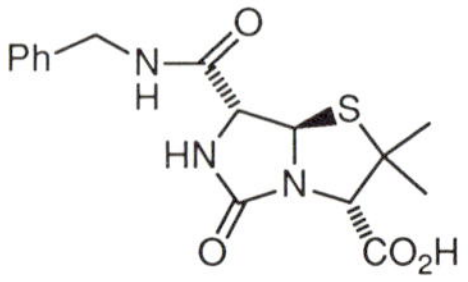 (59)

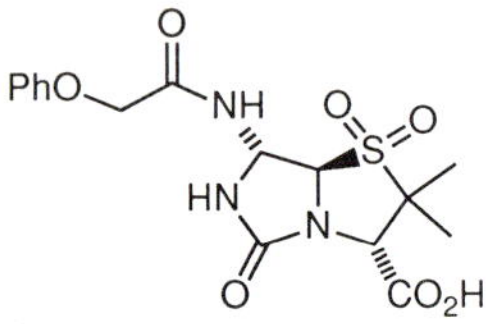

(60)

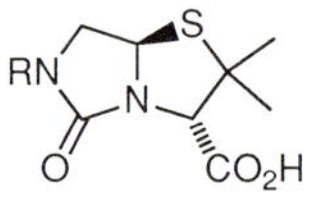

(61) R = $PhCH_2CONH$-
(62) R = $PhCH_2CO$-
(63) R = $PhCH_2SO_2$-

Analogues (**62**) and (**63**) were synthesized[56] in an attempt to enhance the electrophilic character (propensity to form the acyl-enzyme intermediate) of the imidazolidinone carbonyl moiety. Neither compound exhibited antibacterial activity.

9.6 Oxaziridines and epoxides

The β-lactam mimics discussed thus far were all designed to acylate the serine residue present in the active site of the PBPs and many β-lactamases. Ghosez's group[55] has been involved in the design and synthesis of potential alkylating agent inhibitors (**66**), which mimic the natural substrate, D-Ala–D-Ala (**64**), or a β-lactam (**65**). An alkylating agent would in theory be an irreversible inhibitor of the PBPs and β-lactamases. The design was based purely on topological analogy with known β-lactam antibiotics. It neglects important aspects such as the geometrical requirements of potential hydrogen bonds between the enzyme and substrate, and proton transfer, which would assist ring opening of the oxaziridine/epoxide moiety.

The synthesis of oxaziridine (**67a**) has been reported;[57] however, attempts to generate the free acid (**67b**) resulted in decomposition. Thus, the biodegradable pivaloyloxymethyl ester (**67c**) was prepared. This compound showed weak bacteriostatic activity (MIC = 100 mM) vs. *Klebsiella aerogenes* and *Klebsiella pneumoniae* when incubated in the presence of serum.

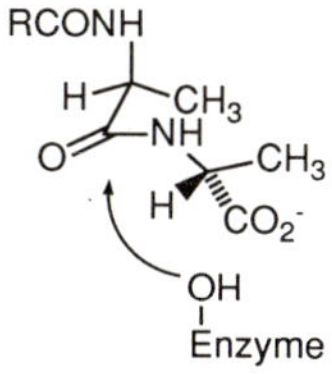

(64) 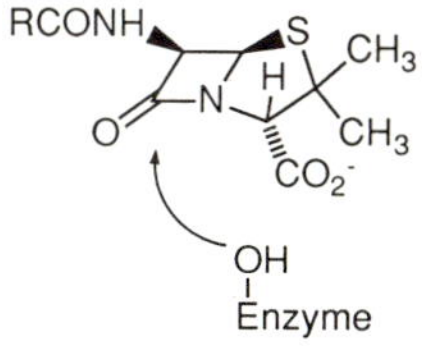 **(65)** 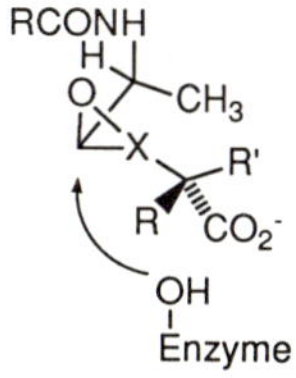 **(66)** X = N, CH

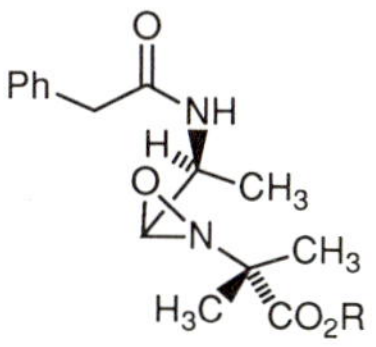

(67a) R = allyl
(67b) R = H
(67c) R = CH_2OCOt-Bu

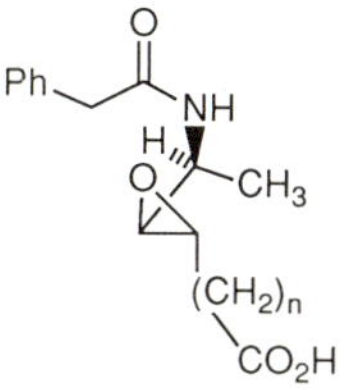

(68a) n = 1
(68b) n = 2

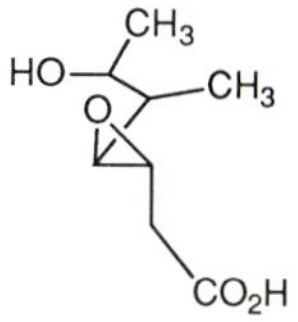

(69)

Epoxides (**68a, b**) bearing acylamino side chains, which might assist epoxide ring opening, were found to be too unstable for biological evaluation.[58] Thus, epoxide (**69**) bearing the thienamycin hydroxyethyl side chain was prepared and found to be stable. Epoxide (**69**) is devoid of antibacterial activity; however, it does exhibit weak inhibition of the *Bacillus cereus* β-lactamase at 1 mM concentration.[11,55] It remains unclear whether the poor biological activity of these alkylating agents is a result of structural properties that affect their interaction with the target proteins, or of chemical instability.

9.7 Cyclobutanones

Several groups have reported the synthesis of cyclobutanone analogues, (**70**), of β-lactams. They reasoned that the relatively reactive cyclobutanone carbonyl could form a stable hemiketal (**71**) on binding to the active site serine residue of a PBP or β-lactamase.[59,60] As the hemiketal is analogous to the tetrahedral intermediate formed when a PBP or β-lactamase serine residue attacks a β-lactam, additional stabilization of the hemiketal by the enzyme might be expected. This, in turn, could lead to enhanced antibacterial activity or β-lactamase inhibition. These studies were also designed to determine the role of the β-lactam nitrogen atom in interactions with the target enzymes.

Gordon *et al.*[61], of Squibb, prepared deaza-carbapenam (**72**) and deaza-carbapenem (**73**); these compounds exhibited neither antibacterial, nor β-lactamase inhibitory activity. More recently, Cocuzza and Boswell[62]

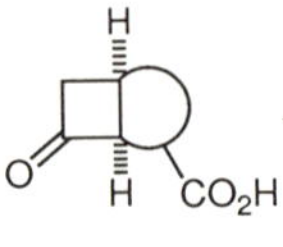

(70) 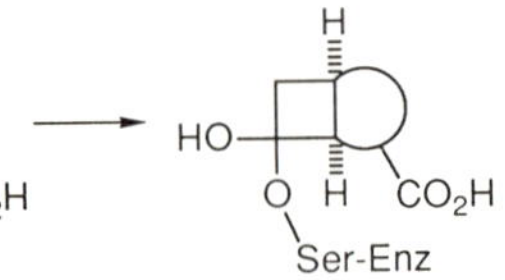 → (71) vs. tetrahedral intermediate

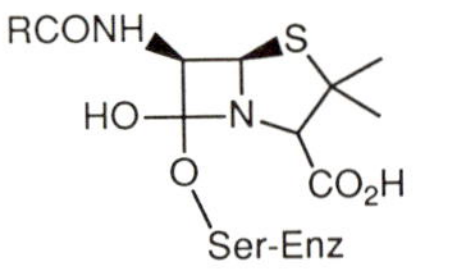

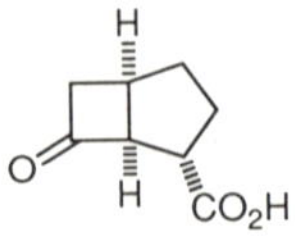

(72) 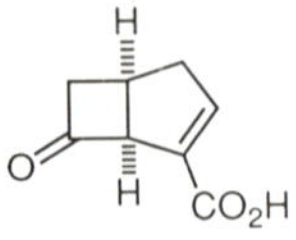 (73) 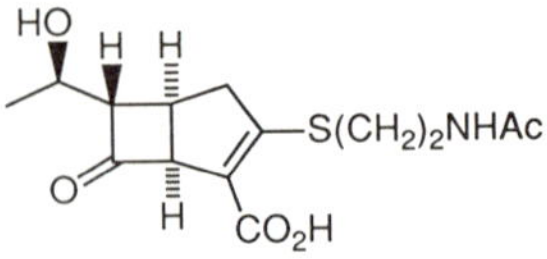 (74)

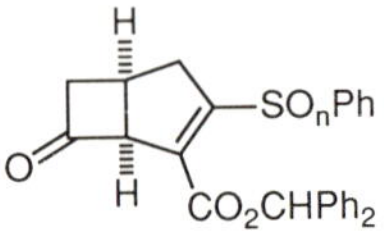

(75) n = 1
(76) n = 2

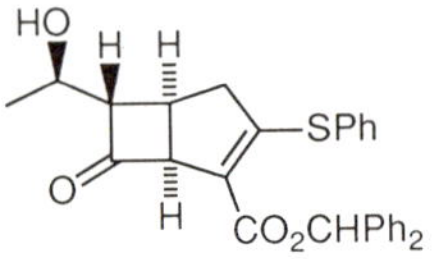

(77) 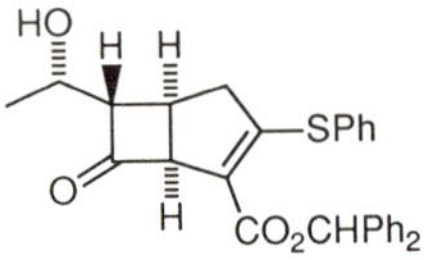 (78)

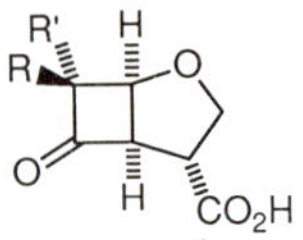

(79) R = $PhCH_2CONH$, R' = H
(80) R = H, R' = $PhCH_2CONH$
(81) R = Cl, R' = H

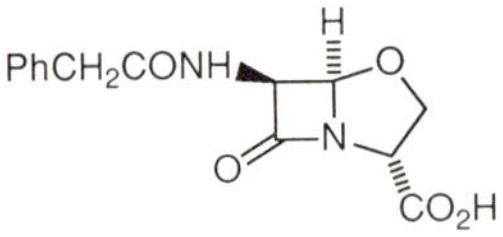

(82) 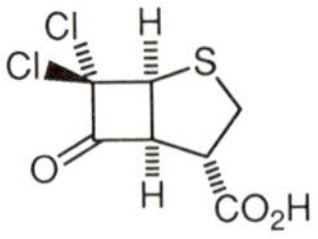 (83)

synthesized a series of deaza-thienamycin analogues. None of the cyclobutanone carboxylic acids, e.g. **(74)**, exhibited antibacterial activity. It is interesting to note that several of the corresponding benzhydryl esters **(75–78)** were active against *Staphylococcus aureus* (MICs 25–50 $\mu g\,ml^{-1}$). Cocuzza and Boswell also noted that several of these esters (compounds not specified) demonstrated synergistic activity with penicillin G against β-lactamase-producing strains of *Staphylococcus aureus*.

Lowe and Swain[63] synthesized deaza-oxapenam analogues **(79)** and **(80)**. An acylamino side chain was incorporated for direct comparison with the known active antibiotic 1-oxabisnorpenicillin G **(82)**. The deaza-analogues **(79)** and **(80)** were devoid of antibacterial activity; however, a mixture (approximately 2 : 1, **(79)** : **(80)**) of these compounds did inhibit the *Streptomyces* R61 D,D-carboxypeptidase enzyme ($IC_{50} = 260\,mg\,l^{-1}$). Lowe and Swain also prepared chlorocyclobutanone **(81)**.[63] Cyclobutanones **(79)**, **(80)** and **(81)** show slow, time-dependent inhibition of the *Escherichia coli* R-TEM-2 β-lactamase, as well as the Type I enzyme isolated from *Bacillus cereus*. Dmitrienko and coworkers[64] have prepared the 1-thia-cyclobutanone analogue **(83)**. The binding of this compound to the active site of the *Streptomyces* R61 D,D-carboxypeptidase enzyme has been observed by X-ray crystallography.[65]

9.8 β-Lactones

In the course of screening for new β-lactam antibiotics, the Squibb group has discovered β-lactone antibiotics Obafluorin (**84**)[66] and SQ 26,517 (**85**).[67] Obafluorin is a metabolite isolated from cultures of *Pseudomonas fluorescens* (ATC39502). It showed weak antibacterial activity against a wide variety of organisms when tested by the disc diffusion method.[68] Its instability in solution precluded observation of antibacterial activity in a standard agar dilution assay (all MICs $> 100\ \mu g\, ml^{-1}$). In spite of this instability, obafluorin did provide some protection to mice infected with *Streptococcus pyogenes* ($ED_{50} = 50\, mg\, kg^{-1}$ IV).[68] The β-lactone ring was also found to be susceptible to hydrolysis by several β-lactamases. SQ 26,517 (**85**) is produced by *Bacillus* sp. SC11,480 and also demonstrates very weak antibacterial activity. MICs ($\mu g\, ml^{-1}$, by agar dilution) vs. *Streptococcus agalactiae*, 50; *Micrococcus luteus*, 100, *Proteus vulgaris*, 100.[69] The generally poor activity of β-lactones, and their susceptibility to β-lactamases limits their potential for use as antibacterial agents.

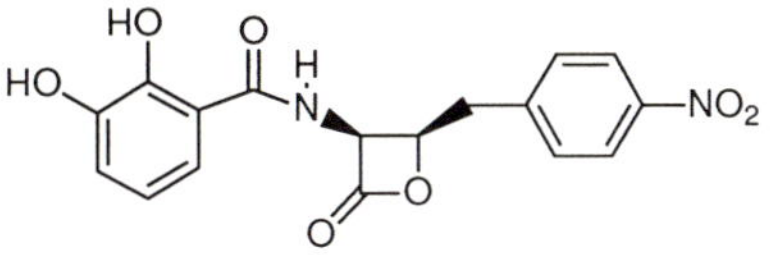

(**84**) 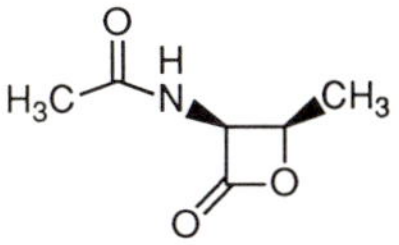 (**85**)

9.9 Summary

It is evident that considerable effort has been dedicated to the discovery of non-β-lactam structures that might mimic β-lactam antibiotics. Two classes of compounds have emerged; these clearly act by inhibiting PBPs, and exhibit clinically relevant levels of antibacterial activity. The first, bicyclic pyrazolidinones, were the result of a rational design exercise; the second, lactivicins, are semisynthetic analogues of a natural product. It is interesting to note that both classes of compounds possess a heteroatom (N, O respectively) covalently bound to the γ-lactam nitrogen atom. Perhaps it is this additional inductive activation of the lactam moiety which allows the pyrazolidinones and lactivicins to express their potent levels of antibacterial activity. The findings reported here suggest that the bicyclic pyrazolidinones and lactivicins represent two exciting new classes of synthetic antibacterial agents that could yield significant candidates for use in human medicine.

References

1. J.-M. Frère, B. Joris, L. Varetto and M. Crine, *Biochemical Pharmacology* (1988) **37** 125–132.
2. D.B. Boyd, in *The Chemistry and Biology of β-Lactam Antibiotics*, Vol. 1 (Eds R.B. Morin and M. Gorman), Academic Press, New York (1982), pp. 437–545.
3. J.M. Indelicato, A. Dinner, L.R. Peters and W.L. Wilham, *J. Med. Chem.* (1977) **20** 961.
4. D.B. Boyd, *J. Med. Chem.* (1984) **27** 63.
5. D.B. Boyd, C. Eigenbrot, J.M. Indelicato, J.M. Miller, C.E. Pasini and S.R. Woulfe, *J. Med. Chem.* (1987) **30** 528.
6. J.M. Indelicato and C.E. Pasini, *J. Med. Chem.* (1988) **31** 1277.
7. V. Du Vineaud and F.H. Carpenter, in *The Chemistry of Penicillin*, (Eds H.T. Clarke, J.R. Johnson and R. Robinson), Princeton University Press, Princeton, NJ (1949), p. 1004.
8. H.H. Wasserman, B. Suryanarayana, K.C. Koch and R.L. Tse, *Chemistry and Industry* (1956) 1022.
9. D. Todd and S. Teich, *J. Am. Chem. Soc.* (1953) **75** 1895–1900.
10. N.E. Allen, D.B. Boyd, J.B. Campbell, J.B. Deeter, T.K. Elzey, B.J. Foster, L.D. Hatfied, J.N. Hobbs Jr., W.J. Hornback, D.C. Hunden, N.D. Jones, M.D. Kinnick, J.M. Morin Jr., J.E. Monroe, J.K. Swartzendruber and D.G. Vogt, *Tetrahedron* (1989) **45** 1905–1927.
11. J. Marchand-Brynaert and L. Ghosez, in *Recent Progress in the Chemical Synthesis of Antibiotics* (Eds M. Ohno and G. Lukais), Springer-Verlag, Berlin (1990) pp. 729–794.
12. J.E. Baldwin, G.P. Lynch and J. Pitlik, *J. Antibiot.* (1991) **44** 1–24.
13. L.N. Jungheim, S.K. Sigmund and J.W. Fisher, *Tetrahedron Lett.* (1987) **28** 285–288; L.N. Jungheim, S.K. Sigmund and N.D. Jones, *Tetrahedron Lett.* (1987) **28** 289–292; L.N. Jungheim and S.K. Sigmund, *J. Org. Chem.* (1987) **52** 4007–4013; L.N. Jungheim, C.J. Barnett, J.E. Gray, L.H. Horcher, T.A. Shepherd and S.K. Sigmund, *Tetrahedron* (1988) **44** 3119–3126.
14. L.N. Jungheim, R.E. Holmes, J.L. Ott, R.J. Ternansky, S.E. Draheim, D.A. Neel, S.K. Sigmund and T.A. Shepherd, Abstract 601, *XXVI Interscience Conference on Antimicrobial Agents and Chemotherapy* (ICAAC), 1986.
15. D.A. Preston, F.T. Counter, A.M. Felty-Duckworth and J.R. Turner, Abstract 603, *XXVI Interscience Conference on Antimicrobial Agents and Chemotherapy* (ICAAC), 1986.
16. N.E. Allen, J.N. Hobbs, Jr., D.A. Preston, J.R. Turner and C.Y.E. Wu, *J. Antibiot.* (1990) **43** 92–99.
17. F.T. Counter, P.W. Ensminger, J.S. Kasher and J.L. Ott, Abstract 1214, *XXVII Interscience Conference on Antimicrobial Agents and Chemotherapy* (ICAAC), 1987).
18. L.N. Jungheim and R.J. Ternansky, Lilly Research Laboratories, unpublished results.
19. R.J. Ternansky and S.E. Draheim, *Tetrahedron Lett.* (1988) **29** 6569–6572.
20. R.J. Ternansky and S.E. Draheim, in *Recent Advances in the Chemistry of β-Lactam Antibiotics* (Eds P.H. Bently and R. Southgate) Royal Society of Chemistry Special Publication No. 70, London (1989), pp. 139–156.
21. L.N. Jungheim, R.J. Ternansky and R.E. Holmes, *Drugs of the Future* (1990) **15** 149–157.
22. T.A. Shepherd and L.N. Jungheim, *Tetrahedron Lett.* (1988) **29** 5061–5064.
23. L.N. Jungheim, *Tetrahedron Lett.* (1989) **30** 1889–1892.
24. R.E. Holmes and D.A. Neel, *Tetrahedron Lett.* (1990) **31** 5567–5570.
25. R.J. Ternansky and S.E. Draheim, *Tetrahedron Lett.* (1990) **31** 2805–2808.
26. C.Y.E. Wu, W.E. Alborn, Jr., J.E. Flokowitsch and D.A. Preston, Abstract 1215, *XXVII Interscience Conference on Antimicrobial Agents and Chemotherapy* (ICAAC), 1987.
27. F.T. Counter, Lilly Research Laboratories, unpublished observations.
28. Y. Nozaki, N. Katayama, H. Ono, S. Tsubotani, S. Harada, H. Okazaki and Y. Nakao, *Nature* (1987) **325** 179–180.
29. S. Harada, S. Tsubotani, T. Hida, H. Ono and H. Okazaki, *Tetrahedron Lett.* (1986) **27** 6229–6232.
30. S. Harada, S. Tsubotani, T. Hida, K. Koyama, M. Kondo and H. Ono, *Tetrahedron* (1988) **44** 6589–6606.

31. Y. Nakao, in *Recent Advances in the Chemistry of β-Lactam Antibiotics* (Eds P.H. Bently and R. Southgate) Royal Society of Chemistry Special Publication No. 70, London (1989), pp. 119–138.
32. Y. Nozaki, N. Katayama, S. Harada, H. Ono and H. Okasaki, *J. Antibiot.* (1989) **42** 84–93.
33. H. Natsugari, Y. Kawano, A. Mormoto, K. Yoshioka and M. Ochiai, *J. Chem. Soc., Chem. Comm.* (1987) 62–63.
34. N. Tamura, Y. Matsushita, Y. Kawano and K. Yoshioka, *Chem. Pharm. Bull.* (1990) **38** 116–122.
35. N. Tamura, Y. Matsushita, K. Yoshioka and M. Ochiai, *Tetrahedron* (1988) **44** 3231–3240.
36. J.E. Baldwin, C. Lowe and C.J. Schofield, *Tetrahedron Lett.* (1990) **31** 2211–2212.
37. D.B. Boyd, T.K. Elzey, L.D. Hatfield, M.D. Kinnick and J.M. Morin Jr., *Tetrahedron Lett.* (1986) **27** 3453–3456.
38. D.B. Boyd, B.J. Foster, L.D. Hatfield, W.J. Hornback, N.D. Jones, J.M. Munroe and J.K. Swartzendruber, *Tetrahedron Lett.* (1986) **27** 3457–3460.
39. J.E. Baldwin, C. Lowe, C.J. Schofield and E. Lee, *Tetrahedron Lett.* (1986) **27** 3461–3464.
40. M. Lang, K. Prasad, W. Holick, J. Gosteli, I. Ernest and R.B. Woodward, *J. Am. Chem. Soc.* (1979) **101** 6296.
41. H.R. Pfaendler, J. Gosteli and R.B. Woodward, *J. Am. Chem. Soc.* (1980) **102** 2039.
42. See for example: T. Kametaini, A. Nakayama, H. Matsumoto and T. Honda, *Chem. Pharm. Bull.* (1983) **31** 2578.
43. J.E. Baldwin, M.F. Chan, G. Gallacher, M. Otsuka, P. Monk and K. Prout, *Tetrahedron* (1984) **40** 4513–4525.
44. M.I. Page, *Acc. Chem. Res.* (1984) **17** 144.
45. J.E. Baldwin, R.M. Adlington, R.H. Jones, C.J. Schofield, C. Zaracostas and C.W. Greengrass, *Tetrahedron* (1986) **17** 4879–4888.
46. J.V. Heck, US Patent 4,428,960, January 31, 1984. See also *Chem. Abstr.* (1984) **100** 191655.
47. S. Hashiguchi, H. Natsugari and M. Ochiai, *J. Chem. Soc., Perkin Trans. 1* (1988) 2345–2352.
48. E.M. Gordon and J. Pluscec, *Tetrahedron Lett.* (1983) **24** 3419–3422.
49. Y. Maki, M. Sako, N. Kurahashi and K. Hirota, *J. Chem. Soc., Chem. Commun.* (1988) 110–111.
50. J.E. Baldwin, R.T. Freeman and C. Schofield, *Tetrahedron Lett.* (1989) **30** 4019–4020.
51. S. Coulton, I. Francois and R. Southgate, *Tetrahedron Lett.* (1990) **31** 6923–6926.
52. M.J. Crossley, R.L. Crumbie, Y.M. Fung, J.J. Potter and M.A. Pleger, *Tetrahedron Lett.* (1987) **28** 2883–2886.
53. L.N. Jungheim, Lilly Research Laboratories, unpublished observations.
54. J. Marchand-Brynaert and L. Ghosez, *Bull. Soc. Chim. Belg.* (1985) **94** 1021–1031.
55. J. Marchand-Brynaert, Z. Bounkhala-Khrouz, J.C. Carretero, J. Davies, D. Ferroud, B.J. van Keulen, B. Serckx-Poncin and L. Ghosez, in *Recent Advances in the Chemistry of β-Lactam Antibiotics* (Eds P.H. Bentley and R. Southgate), Royal Society of Chemistry Special Publication No. 70, London (1989), pp. 157–170.
56. J. Marchand-Brynaert, H. Vanlierde and L. Ghosez, *Bull. Soc. Chim. Belg.* (1988) **97** 1081–1093.
57. J. Marchand-Brynaert, Z. Bounkhala-Khrouz, B.J. van Keulen, H. Vanlierde and L. Ghosez, *Israel J. Chem.* (1989) **29** 247–255.
58. J. Marchand-Brynaert, D. Ferroud, B. Serckx-Poncin and L. Ghosez, *Bull. Soc. Chim. Belg.* (1990) **99** 1075–1084.
59. D. Agathocleous, S. Buckwell, P. Proctor and M.I. Page, in *Recent Advances in the Chemistry of β-Lactam Antibiotics* (Eds A.G. Brown and S.M. Roberts), Royal Society of Chemistry Special Publication No. 52, London (1985), pp. 18–31.
60. G. Lowe and S. Swain, in *Recent Advances in the Chemistry of β-Lactam Antibiotics* (Eds A.G. Brown and S.M. Roberts), Royal Society of Chemistry Special Publication No. 52, London (1985), pp. 209–221.
61. E.M. Gordon, J. Pluscec and M.A. Ondetti, *Tetrahedron Lett.* (1981) **22** 1871–1874.
62. A.J. Cocuzza and G.A. Boswell, *Tetrahedron Lett.* (1985) **26** 5363–5366.
63. G. Lowe and S. Swain, *J. Chem. Soc., Perkin Trans. 1* (1985) 391–398.

64. G. Lange, M.E. Savard, T. Viswanatha and G.I. Dmitrienko, *Tetrahedron Lett.* (1985) **26** 1791–1794.
65. J.B. Bartolone, G.J. Hite, J.A. Kelly and J.R. Knox, in *Recent Advances in the Chemistry of β-Lactam Antibiotics* (Eds A.G. Brown and S.M. Roberts), Royal Society of Chemistry Special Publication No. 52, London (1985), pp. 318–327.
66. J.S. Wells, W.H. Trejo, P.A. Principe and R.B. Sykes, *J. Antibiot.* (1984) **37** 802–803.
67. W.L. Parker, M.L. Rathnum and W.C. Liu, *J. Antibiot.* (1982) **35** 900–902.
68. A.A. Tymiak, C.A. Culver, M.F. Malley and J.Z. Gougoutas, *J. Org. Chem.* (1985) **50** 5491–5495.
69. J.S. Wells, J.C. Hunter, G.L. Astle, J.C. Sherwood, C.M. Ricca, W.H. Trejo, D.P. Bonner and R.B. Sykes, *J. Antibiot.* (1982) **35** 814–820.

10 Classical β-lactam structures

E.W. COLVIN

10.1 Introduction

This chapter will discuss the synthesis and chemistry of the penicillins (**1**), the cephalosporins (**2**) and the cephamycins (**3**). Other biologically active β-lactams are discussed in chapter 8. For general reading on classical β-lactams and specific routes to monocyclic β-lactams, the further reading section should be consulted.

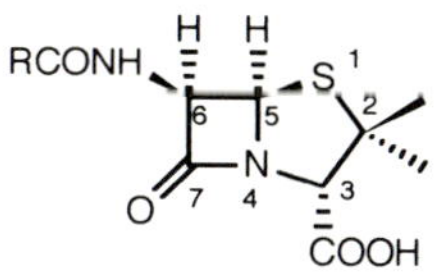

(**1**)

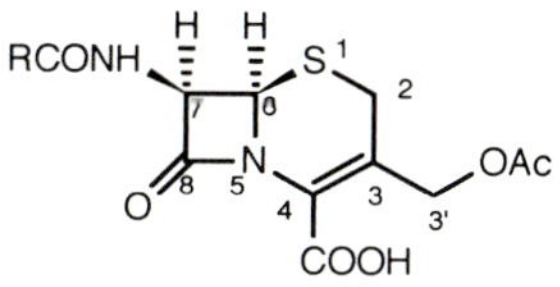

(**2**)

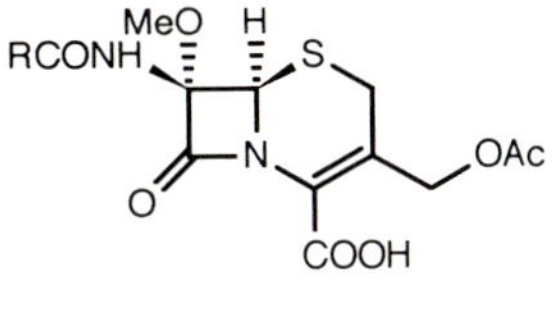

(**3**)

10.2 Sheehan's synthesis of penicillin V (4)

Sheehan and Henery-Logan[1] published the first rational synthesis (Scheme 10.1) of a natural penicillin, penicillin V (**4**), in 1957. The synthesis was conceived at a time when it was generally believed that the instability of penicillin was due to the presence of the strained four-membered lactam ring, therefore the creation of this structural feature was postponed for as long as possible in the synthetic sequence. D-Penicillamine (**5**) was condensed with t-butyl phthalimidoaldehydomalonate (**6**) to give thiazolidine (**7**). This was a mixture of only two of the four possible stereoisomers, one of which corresponded to the configuration found in natural penicillin; the other could be epimerised into the correct isomer by heating in the presence of pyridine. Hydrazinolysis of the phthalimido group, followed by acylation of the free amine with phenoxyacetyl chloride, produced phenoxyacetamide (**8**). The t-butyl ester was then cleaved with dry hydrogen chloride to give diacid (**9**), thus leaving formation of the β-lactam ring as the final step in the synthesis. This was achieved using dicyclohexylcarbodiimide, a reagent introduced by Sheehan and Hess[2] for the formation of amides from amines and carboxylic acids.

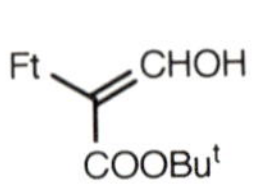

(**5**) 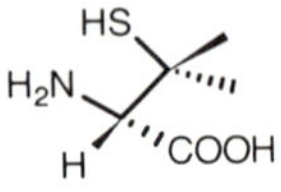 (**6**) 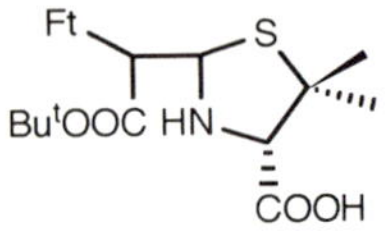 (**7**)

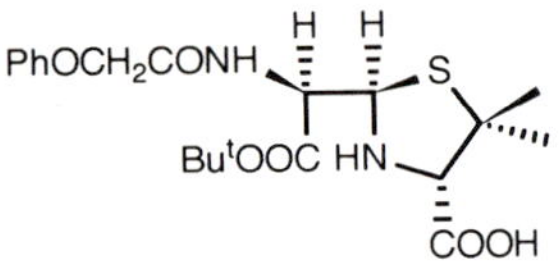

(**8**) 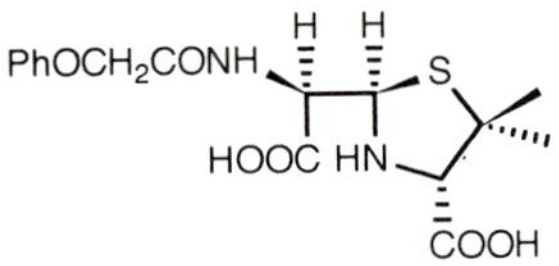 (**9**)

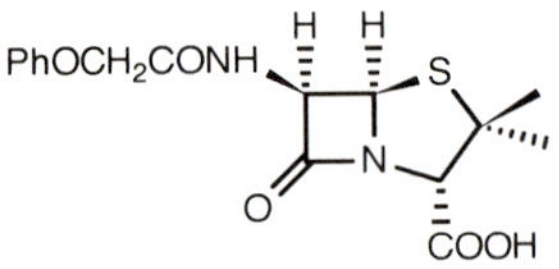

(**4**)

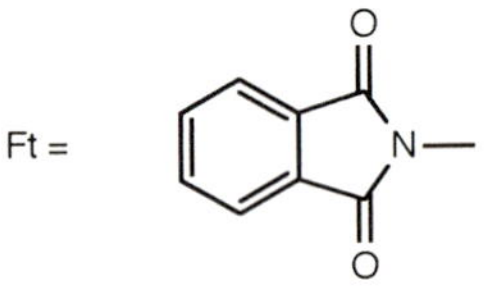

Scheme 10.1

Research groups working independently in two pharmaceutical companies, Roussel[3] and Squibb[4], have extended Sheehan's methodology to the synthesis of cephalosporin. However, the desired cephem carboxylic acid (**10**) could be obtained in only low yield by hydrolysis of the penultimate product, lactone (**11**). In addition, the routes suffer a lack of stereospecificity.

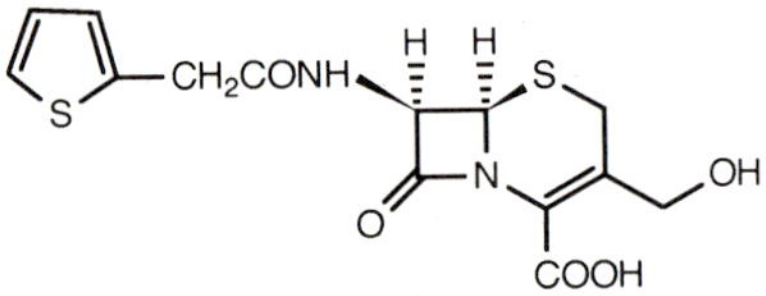

(**10**) 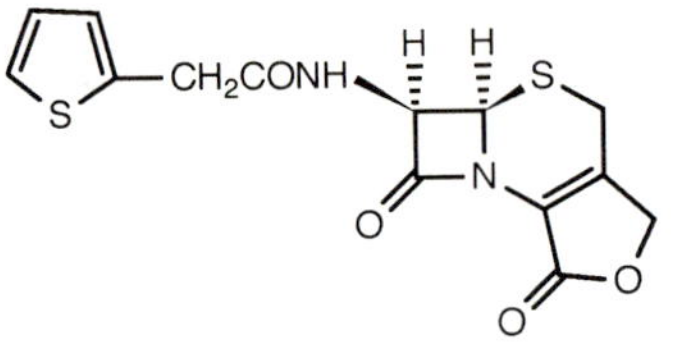 (**11**)

10.3 Woodward's synthesis of cephalosporin C (12)

One of the many outstanding features of this synthesis[5] is the complete stereocontrol achieved by employing as starting material L-cysteine (**13**), masked and activated at its methylene group as the cyclic thiazolidine (**14**) (Scheme 10.2). This chiral building block reacted stereoselectively with dimethyl azodicarboxylate to give (**15**); oxidative cleavage using lead tetra-acetate gave acetate (**16**) accompanied by a small amount of its *cis* epimer. Transesterification liberated alcohol (**17**). This was transformed into amino-ester (**20**) by conversion into mesylate (**18**), displacement with inversion to give azide (**19**) and, finally, reduction with aluminium amalgam. After much experimentation, it was found that triisobutylaluminium effected smooth cyclisation of amino-ester (**20**) to β-lactam (**21**). This is a key intermediate, a corner stone in penam and cephem synthesis, containing as it does the basic structural features common to both penicillins and cephalosporins; it has also been synthesised[6] from 6-t-butoxycarbonylaminopenicillanic acid (**22**). One final point to note here is that the β-lactam ring has been created at a relatively early stage — indeed, this is a strategy common to most subsequent syntheses.

To return to the synthesis under discussion, conjugate addition of the lactam nitrogen function to dialdehyde (**23**) (the condensation product of malondialdehyde and trichloroethyl glyoxylate) gave the fully enolised dialdehyde (**24**). This species now contains all the elements necessary for formation of the cephem skeleton. Treatment of (**24**) with trifluoroacetic acid (TFA) liberated both the amino and mercapto groups, and also induced cyclisation to the Δ^2-cephem (**25**). Acylation with a protected D-α-amino-adipic acid, followed by aldehyde reduction with diborane and acetylation, gave cephem (**26**). The synthesis of cephalosporin C (**12**) was completed by base catalysed isomerisation of the double bond, followed by reductive removal of the ester protective groups using zinc in acetic acid. This last step was the first demonstration of the utility of trichloroethyl ester protection, and its reductive cleavage under mild conditions.

10.4 Biomimetic syntheses

In 1976, Baldwin *et al.*[7] reported the first stereocontrolled total synthesis of a penicillin system. Although this employs a similar activation of the methylene group of L-cysteine, it differs from Woodward's synthesis in that it involves a peptide to β-lactam conversion, and is therefore more bio-genetically patterned. The peptide (**29**) was created by condensation of the protected L-cysteine (**27**) with D-isodehydrovaline methyl ester (**28**) (Scheme 10.3). Functionalisation of the methylene group of the thiazolidine was achieved by treatment with benzoyl peroxide to provide benzoate (**30**).

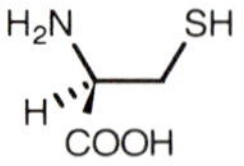

(13)

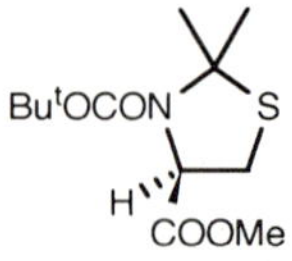

(14)

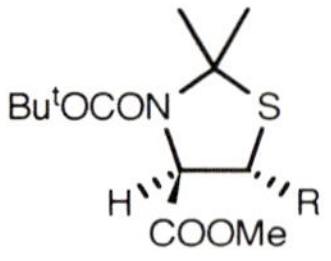

(15) R = NCOOMe (NHCOOMe)

(16) R = OAc

(17) R = OH

(18) R = OSO_2Me

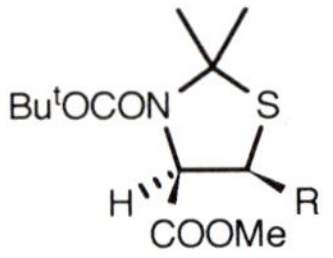

(19) R = N_3

(20) R = NH_2

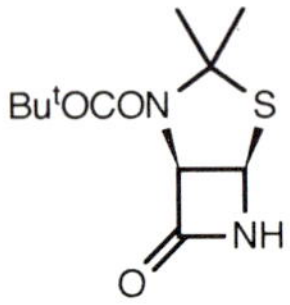

(21)

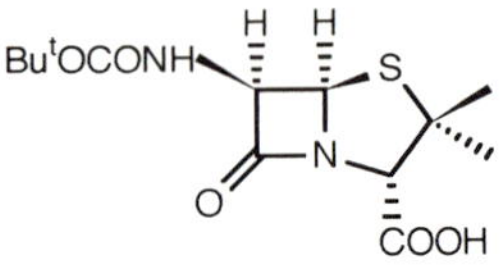

(22)

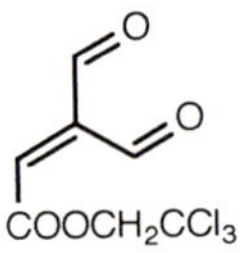

(23)

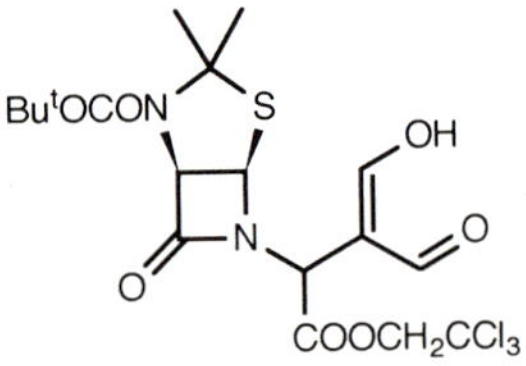

(24)

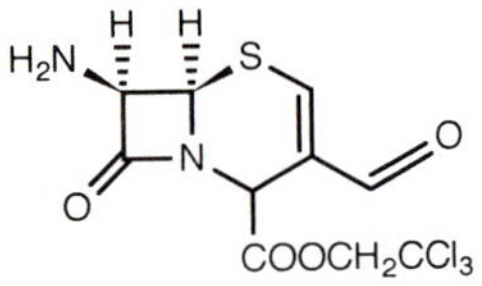

(25)

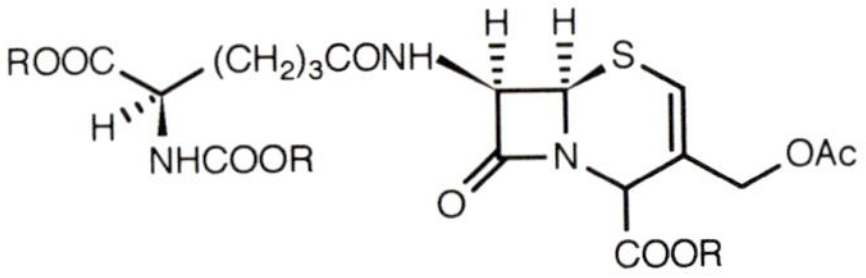

(26) R = CH_2CCl_3

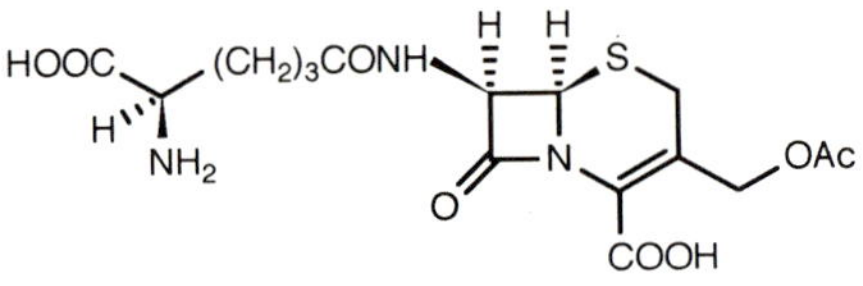

(12)

Scheme 10.2

Exposure to HCl formed chloride (**31**), which underwent cyclisation to β-lactam (**32**) when the amide function was deprotonated with sodium hydride. For penicillin formation, the only necessary further transformations are cleavage of the acetonide moiety and cyclisation. Oxidation with 3-chloroperoxybenzoic acid (*m*-CPBA) gave sulphoxide (**33**), which underwent acid catalysed fragmentation/rearrangement to β-ketosulphide (**34**). Formation of the epimeric oxiranes (**35**) and rearrangement then gave the epimeric aldehydes (**36**). *m*-CPBA oxidation produced a diastereoisomeric mixture of sulphoxides — thermal *syn* elimination of either of these destroys the chirality at sulphur, forming a single sulphenic acid. This acid was not isolated, electrophilic attack on the double bond producing sulphoxide (**37**) directly. Sulphoxide reduction then gave the penicillin (**38**).

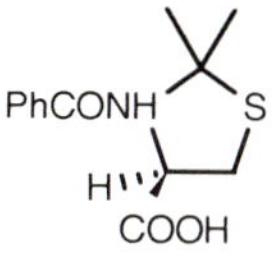

PhCONH S H COOH

(**27**)

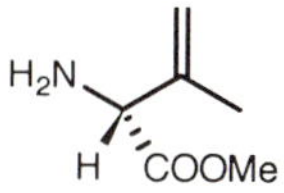

H_2N H COOMe

(**28**)

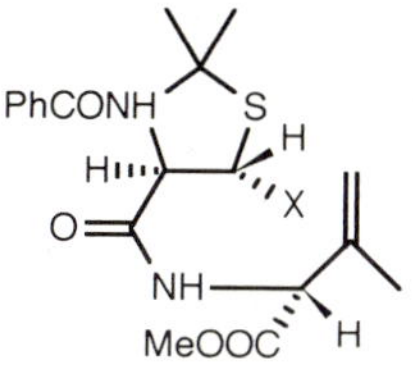

PhCONH S H H X O NH MeOOC H

(**29**) X = H

(**30**) X = OCOPh

(**31**) X = Cl

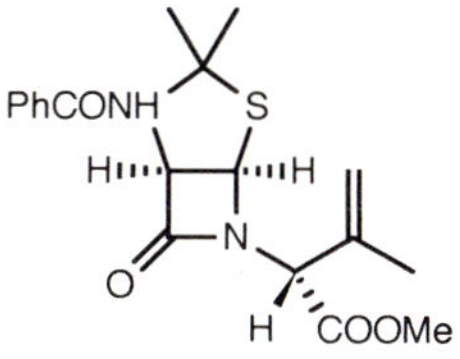

PhCONH S H H N O H COOMe

(**32**)

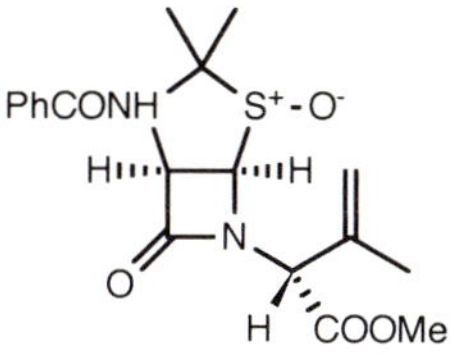

PhCONH S^+-O^- H H N O H COOMe

(**33**)

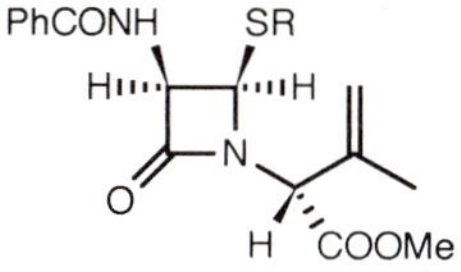

PhCONH SR H H N O H COOMe

(**34**) R = CH_2COCH_3

(**35**) R = CH_2CCH_3 (O epoxide)

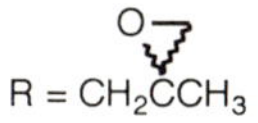

(**36**) R = CH_2CCH_3 (CHO, H)

PhCONH H H O^- S^+ N O COOMe

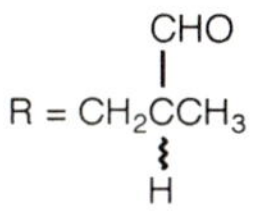

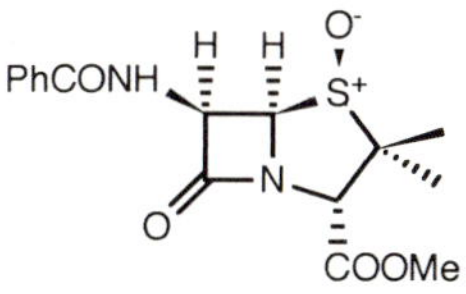

(**37**)

PhCONH H H S N O COOMe

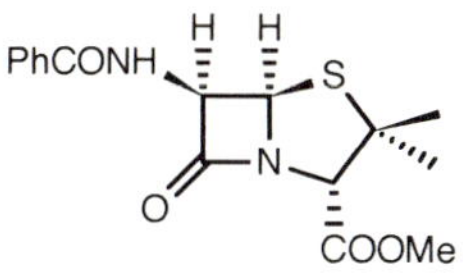

(**38**)

Scheme 10.3

Independently, Kishi and coworkers[8] described the double cyclisation of thioamide (**39**) under sodium hydride conditions (Scheme 10.4). The product thiazoline (**40**) was transformed further to the 3-deacetoxy-7-methoxycephalosporin derivative (**41**) and the 6-methoxypenicillin derivative (**42**). Kishi and coworkers[9] also reported bromination of amide (**43**) to (**44**) using N-bromosuccinimide, followed by potassium hydride mediated cyclisation to the thiazoline β-lactam (**45**).

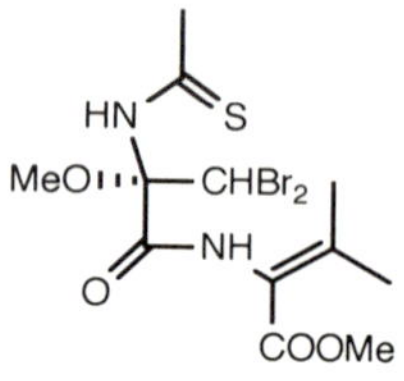

HN S MeO CHBr$_2$ NH O COOMe

(39)

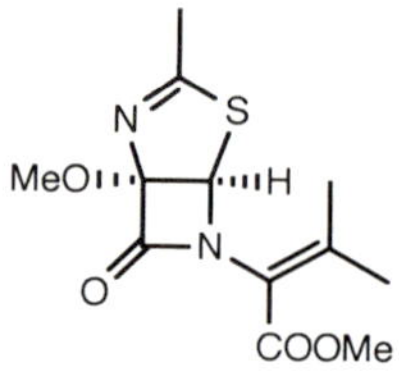

N S MeO H N O COOMe

(40)

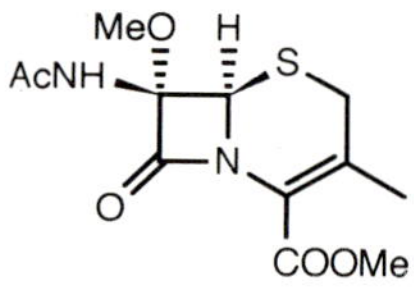

MeO H AcNH S N O COOMe

(41)

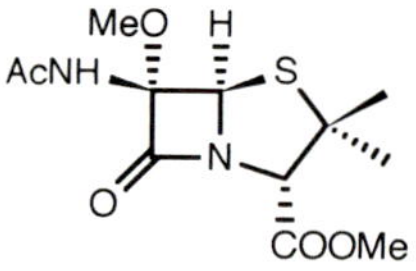

MeO H AcNH S N O COOMe

(42)

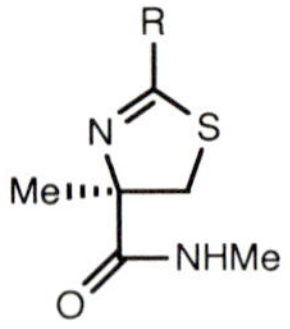

R N S Me NHMe O

(43)

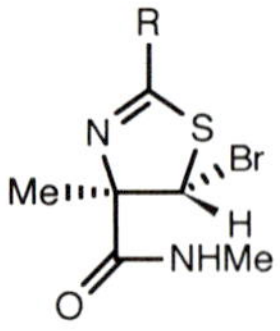

R N S Br Me H NHMe O

(44)

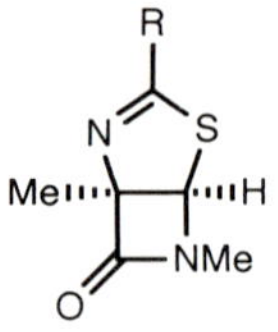

R N S Me H NMe O

(45)

R = C_6H_4 *p* OMe

Scheme 10.4

10.5 Merck synthesis of (±)-cephalothin (46)

The synthetic approach adopted by Merck Sharp & Dohme chemists[10] hinges on β-lactam formation by cycloaddition of imines with ketenes or ketene equivalents. This reaction allowed the first synthesis of a β-lactam by Staudinger[11] in 1907. It has been intensively studied and extended by Bose and coworkers; in particular, the annelation of imines with azidoacetyl chloride[12] or an equivalent allows direct formation of β-lactams possessing nitrogen substitution at the 3-position (monocyclic β-lactam numbering).

Heating triazine (**47**) with diethyl phosphite gave phosphonate (**48**), isolated as its hydrochloride salt (Scheme 10.5). Hydrogenolysis and neutralisation furnished (aminomethyl)phosphonate (**49**), which was converted into the benzaldehyde Schiff base. Acylation by sequential treatment with phenyl lithium and benzyl chloroformate, followed by deprotection, gave the key synthon (**50**),[13] which was converted into thioformamide (**51**) by treatment with ethyl thionoformate.

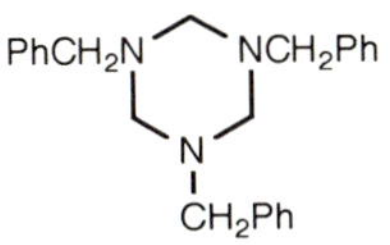

(**47**) 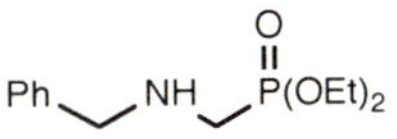 (**48**) 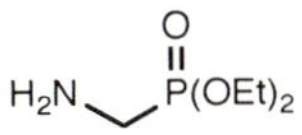 (**49**)

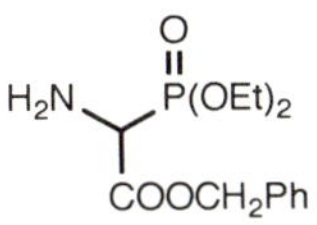

(**50**) 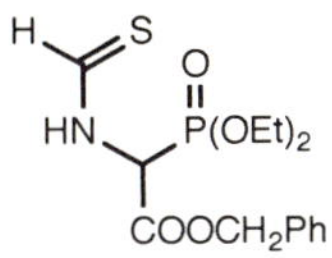 (**51**) 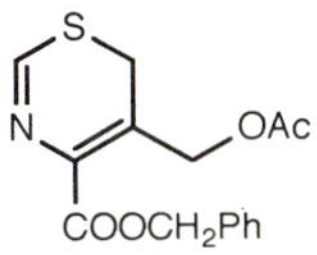 (**52**)

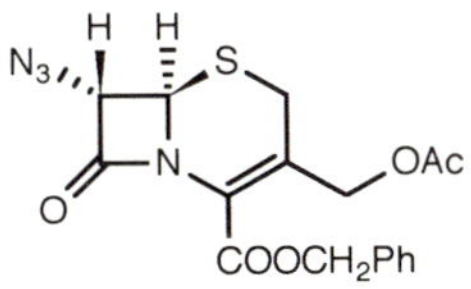

(**53**) 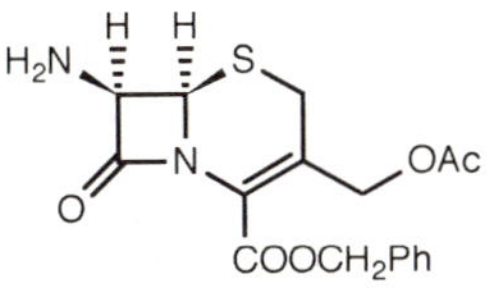 (**54**)

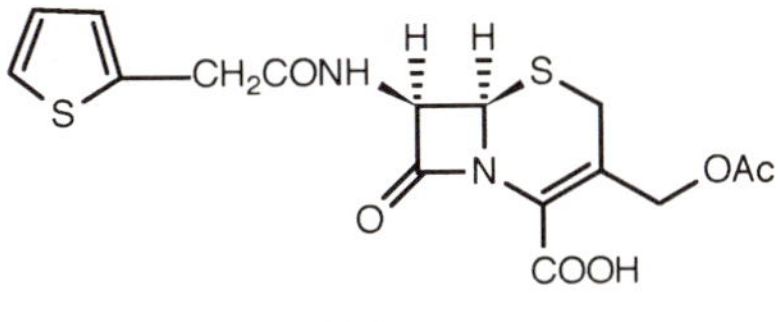

(**46**)

Scheme 10.5

Condensation with 1-chloro-3-acetoxypropanone produced thiazine (**52**). Annelation using azidoacetyl chloride in the presence of triethylamine yielded the 7-α-azidocephem (**53**). Epimerisation at C-7 was achieved[14] by reduction to the amine, formation of the Schiff base with *p*-nitrobenzaldehyde, deprotonation with phenyl lithium, and kinetic reprotonation;

this gave a mixture of the 7-β and 7-α isomers in a 55 : 45 ratio. Liberation of the amine and chromatographic separation gave the 7-β-aminocephem (**54**). Acylation with thienylacetyl chloride followed by hydrogenolysis produced (±)-cephalothin (**46**), which, as its sodium salt, possessed one-half of the biological activity of the homochiral form.

This synthesis is convergent and, as such, is ideally suited to the preparation of nuclearly modified analogues of cephalosporin. Prior to 1973, it was widely believed that the sulphur atom in cephalosporin was necessary for antibiotic activity. To test this hypothesis, the Merck group synthesised analogues in which the sulphur atom had been replaced by simple isosteric groups such as an oxygen atom or a methylene group.

(±)-1-Oxacephalothin (**55**) was prepared[15] as follows (Scheme 10.6). The thioformimidate (**57**), obtained by treatment of thioformamide (**56**) with

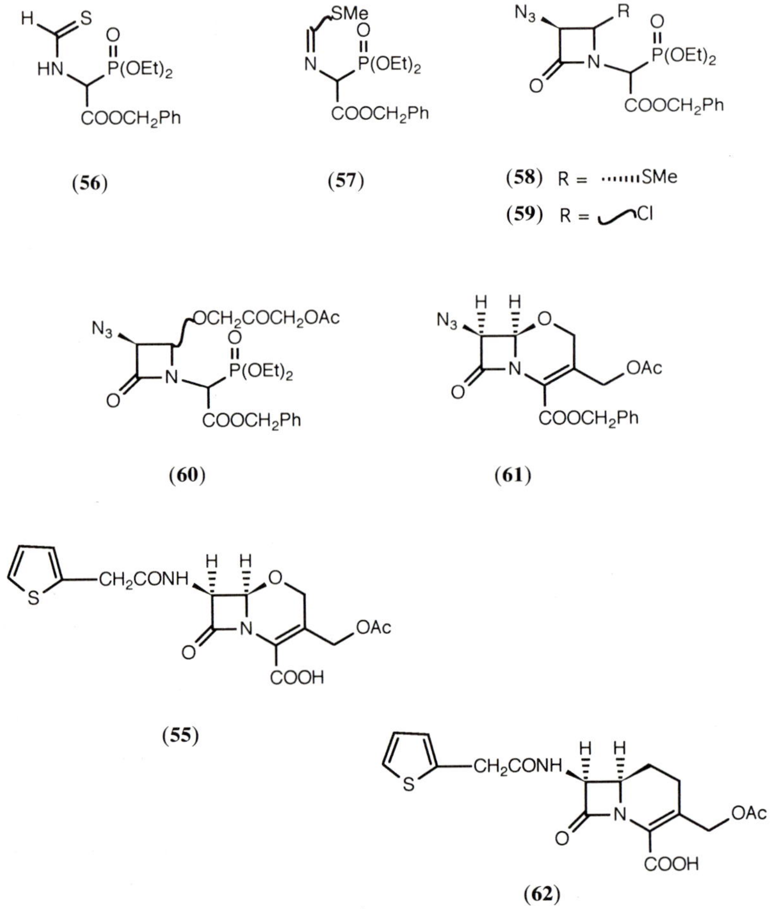

H S O HN P(OEt)$_2$ COOCH$_2$Ph

(56)

SMe O N P(OEt)$_2$ COOCH$_2$Ph

(57)

N_3 R O N P(OEt)$_2$ O COOCH$_2$Ph

(58) R = ⋯SMe

(59) R = ∽Cl

N_3 OCH$_2$COCH$_2$OAc O P(OEt)$_2$ N O COOCH$_2$Ph

(60)

N_3 H H O N OAc O COOCH$_2$Ph

(61)

S CH$_2$CONH H H O N OAc O COOH

(55)

S CH$_2$CONH H H N OAc O COOH

(62)

Scheme 10.6

methyl iodide and potassium carbonate, reacted with azidoacetyl chloride and triethylamine to give the *trans*-β-lactam (**58**). Non-stereospecific conversion into the chlorides (**59**) was achieved using chlorine gas. Silver (I) catalysed displacement with 1-acetoxy-3-hydroxypropanone gave lactam (**60**) as a 1 : 1 *cis*/*trans* mixture. Horner–Emmons cyclisation produced the *cis*-oxacephem (**61**), after chromatographic removal of the accompanying *trans* isomer. Hydrogenolytic ester deprotection was accompanied by concomitant reduction of the azide. Acylation of the resulting amino acid with thienylacetyl chloride gave (±)-1-oxacephalothin (**55**). This oxacephem possessed antibacterial activity comparable with (±)-cephalothin; against a strain of *E. coli*, the oxygen analogue was actually more potent.

This finding indicates that sulphur is not essential for activity, but does not exclude the possibility that oxygen is capable of replacing sulphur in binding to an electrophilic site on the enzyme with which these molecules interact. However, the synthetic route was adapted[16] to allow the preparation of (±)-1-carbacephalothin (**62**). This too was found to retain most of the activity of the parent compound, thereby proving that the presence of sulphur is not necessary for biological activity.

10.6 The Hoechst synthesis

The discovery by Graf[17] that chlorosulphonyl isocyanate (CSI) reacted with alkenes to give β-lactams, from which the N-chlorosulphonyl group can be readily removed,[18] stimulated chemists at Hoechst in what was then West Germany to exploit this cycloaddition reaction for the preparation of functionalised β-lactams.

After much experimentation, the following synthesis[19] of a cephem was achieved. Reaction of vinyl acetate with CSI produced 4-acetoxyazetidin-2-one (**63**), after cleavage of the chlorosulphonyl group (Scheme 10.7). Reaction of this with the 1,4-dithiane derivative (**64**) in the presence of base gave (**65**). Lactam (**65**) was then converted into phosphorane (**66**) by Woodward's method (see section 10.7). Pyrolysis of (**66**) gave a mixture of the Δ^2- and Δ^3-cephems (**67**) and (**68**), which could be converted into pure (**68**) by treatment with base. Attempts to functionalise the C-7 position of (**68**) via deprotonation failed, due to preferential deprotonation of the six-membered ring. This latter process could be blocked by acylation of (**68**) to give diester (**69**). Deprotonation at C-7 followed by reaction with *p*-toluenesulphonyl azide gave the *trans* azide (**70**). Reduction, hydrolysis and decarboxylation, and finally re-esterification gave amine (**71**). Further straightforward transformations led to an active cephalosporin. The many steps of this synthesis render it commercially impractical. However, the utility[20] of 4-acetoxyazetidin-2-one as a readily accessible building block has been amply demonstrated in other syntheses.

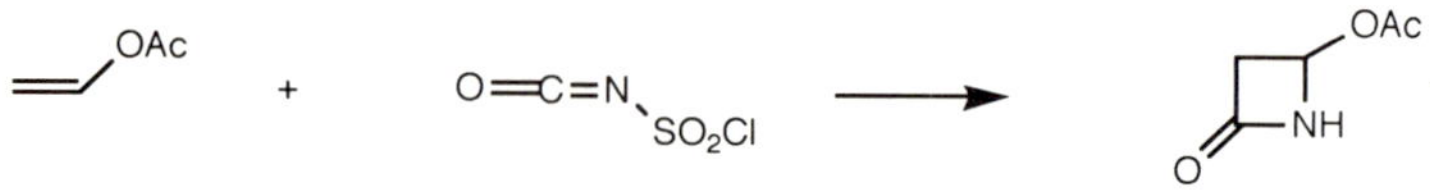

OAc + O=C=N SO$_2$Cl → OAc NH O

(63)

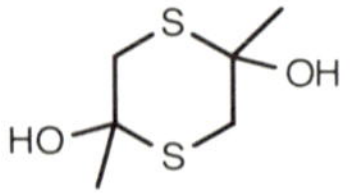

S OH HO S

(64)

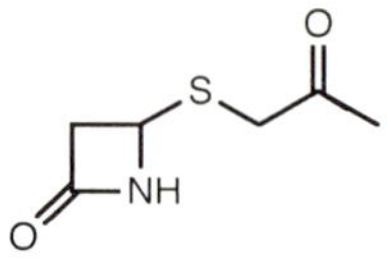

O S NH O

(65)

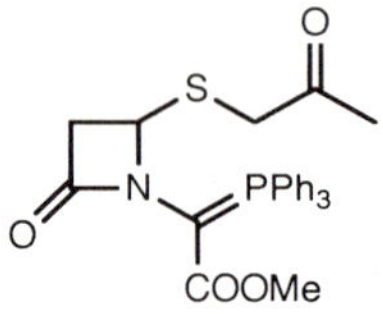

O S N PPh$_3$ O COOMe

(66)

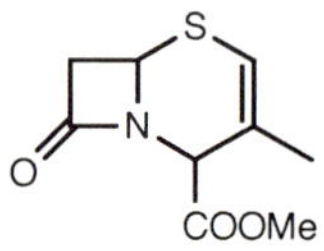

S N O COOMe

(67)

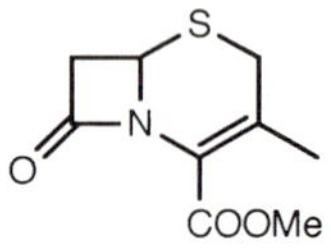

S N O COOMe

(68)

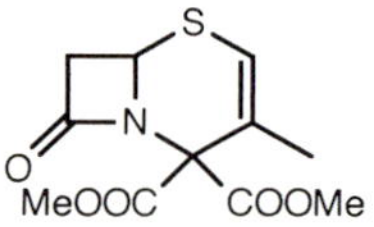

S N O MeOOC COOMe

(69)

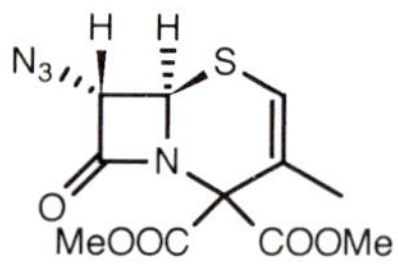

H H N$_3$ S N O MeOOC COOMe

(70)

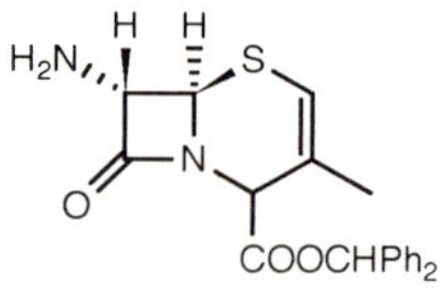

H H H$_2$N S N O COOCHPh$_2$

(71)

Scheme 10.7

10.7 Woodward's penem synthesis

The β-lactam ring in the penicillins is more reactive towards nucleophilic opening than are simple monocyclic azetidin-2-ones. This has been explained in terms of the fused five-membered ring preventing the nitrogen substituents from adopting a planar conformation, thereby reducing the delocalisation of the nitrogen lone pair into the adjacent carbonyl π-orbital.[21] In the cephalosporins, the nitrogen substituents *can* adopt a planar conformation; however, the β-lactam ring is again more reactive than that of a monocyclic β-lactam. An explanation for the lability of the amide bond in the cephalosporins is that the nitrogen lone pair is delocalised into the π-system of the adjacent Δ^3-double bond. This reduces delocalisation into the carbonyl π-system, rendering the carbonyl group more susceptible to nucleophilic attack. Woodward reasoned that a compound incorporating both of these activating features would be very reactive towards nucleophiles, and might therefore be an extremely potent antibiotic. The target was therefore the penem **(72)**, a β-lactam possessing a fused five-membered ring containing an enamine group. After extensive investigation, a high-yielding synthesis of **(72)** and some homologues was developed.[22] The synthetic strategy employed (Scheme 10.8) is an excellent illustration of the

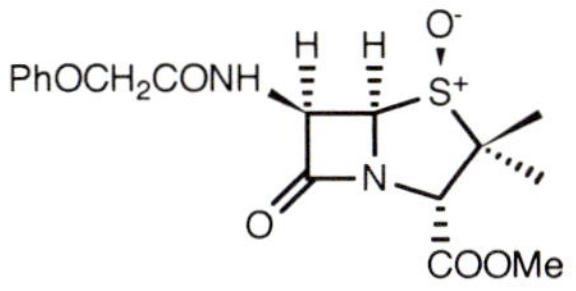

(73)

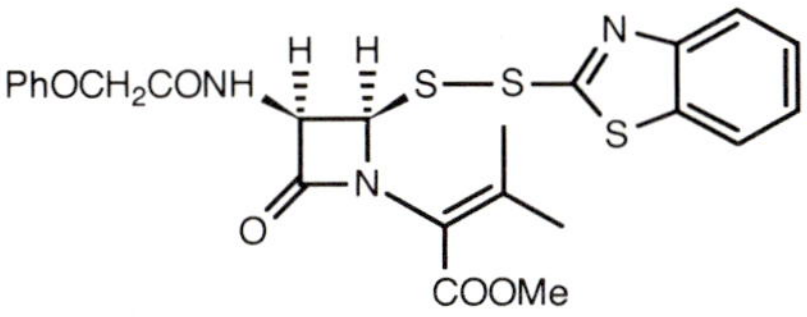

(74)

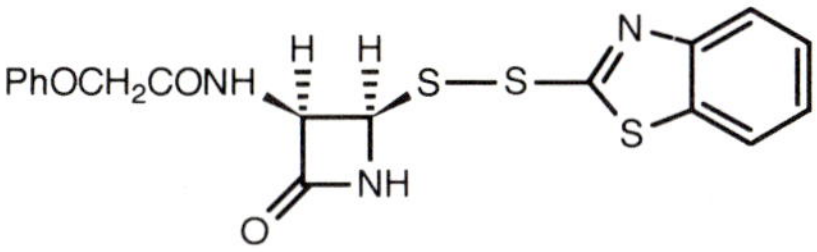

(75)

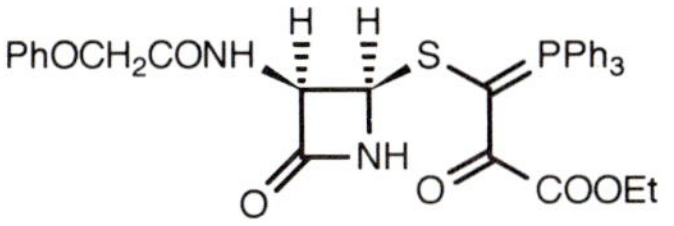

(76)

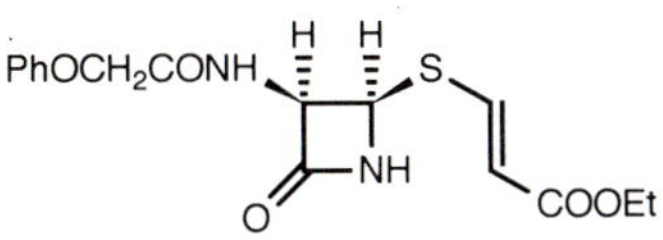

(77)

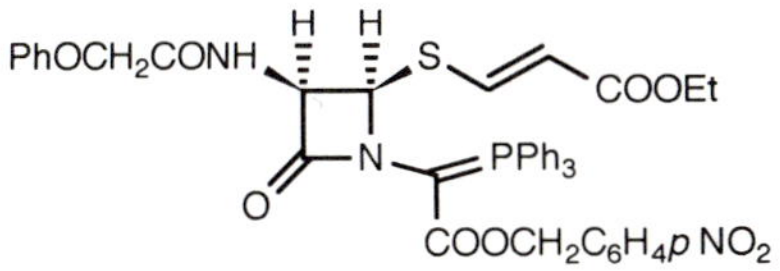

(78)

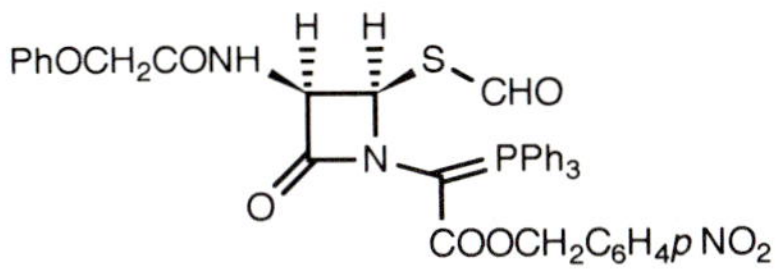

(79)

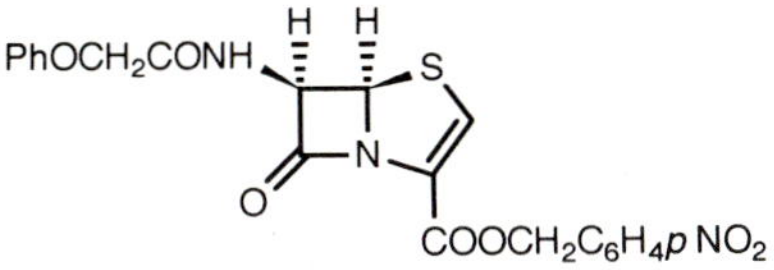

(80)

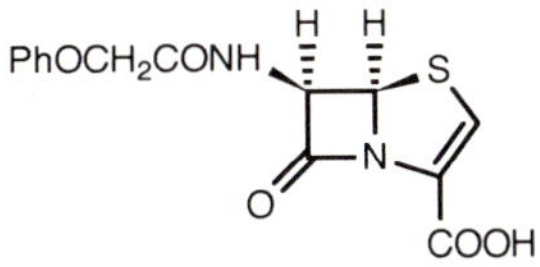

(72)

Scheme 10.8

novel methodology introduced by Woodward for the transformation of natural penicillins into other biologically active species.

Thermal [2,3]-sigmatropic rearrangement of penicillin V *S*-oxide (**73**) in the presence of 2-mercaptobenzothiazole,[23] followed by double bond conjugation using triethylamine gave the benzothiazolyl disulphide (**74**). Ozonolysis followed by methanolysis removed the amide substituent to give (**75**). Reaction of (**75**) with ethyl triphenylphosphoranylidene pyruvate resulted in cleavage of the S–S bond with formation of the new phosphorane (**76**). Borohydride reduction of (**76**) in an acidic medium then gave thioacrylate (**77**). To provide the missing two-carbon unit in a form suitable for final ring closure, a three-step procedure developed earlier[24] and much used since was adopted.

N-Protio β-lactams react readily with glyoxylate esters to form epimeric carbinolamine adducts. Treatment of these with thionyl chloride produces the epimeric chlorides, reaction of which with triphenylphosphine in the presence of base gives a single, normally crystalline and stable, phosphorane. Application of such methodology to thioacrylate (**77**) using *p*-nitrobenzyl glyoxylate gave the phosphorane (**78**). Ozonolysis of the acrylate side chain was performed in the presence of TFA, which protected the ozone-sensitive phosphorane as its phosphonium salt. After ozonolysis, mild base treatment regenerated the phosphorane and gave (**79**). A facile intramolecular Wittig reaction then afforded penem ester (**80**). Choice of the *p*-nitrobenzyl ester proved critical in accessing the free acid — catalytic hydrogenolysis proceeding smoothly to provide penem acid (**72**). Antibacterial *in vitro* tests showed activity against gram-positive strains, demonstrating that biological activity *was* inherent in the new β-lactam system. However, its potential utility was severely reduced by its extreme lability. The synthesis of homochiral (**72**), by a very short sequence of reactions, represents an outstanding achievement in β-lactam synthesis.

10.8 The conversion of penicillins into cephalosporins

6-Aminopenicillanic acid (6-APA) (**81**) is produced in very large quantities by fermentation, and is therefore quite inexpensive. This is not the case with 7-aminocephalosporanic acid (7-ACA) (**82**) and, accordingly, considerable effort[25] has been expended in the search for methods of conversion of penicillins into cephalosporins.

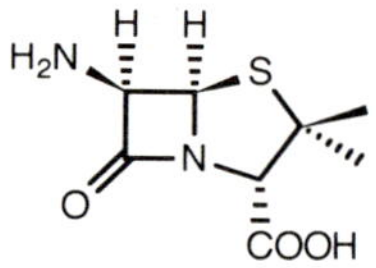

H_2N H H S N O COOH

(**81**)

H_2N H H S N O OAc COOH

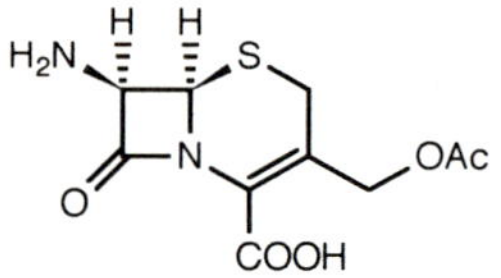

(**82**)

Success was first achieved by Morin and coworkers[26] at the Lilly Research Laboratories in 1963. Oxidation of penicillin ester (**83**) with sodium metaperiodate gave the *S*-sulphoxide (**84**).[27] Treatment of (**84**) with acetic anhydride under normal Pummerer rearrangement conditions did not result in any reaction. However, when heated in acetic anhydride, sulphoxide (**84**) was converted in 60% yield into a 2 : 1 mixture of the isomeric penam (**85**) and cepham (**86**) (Scheme 10.9). Cepham (**86**), on treatment with triethylamine, eliminated acetic acid to give the 3′-desacetoxycephalosporin (**87**). Alternatively, penicillin sulphoxide (**84**) could be converted directly into (**87**) by heating (**84**) in xylene at reflux in the presence of a catalytic amount

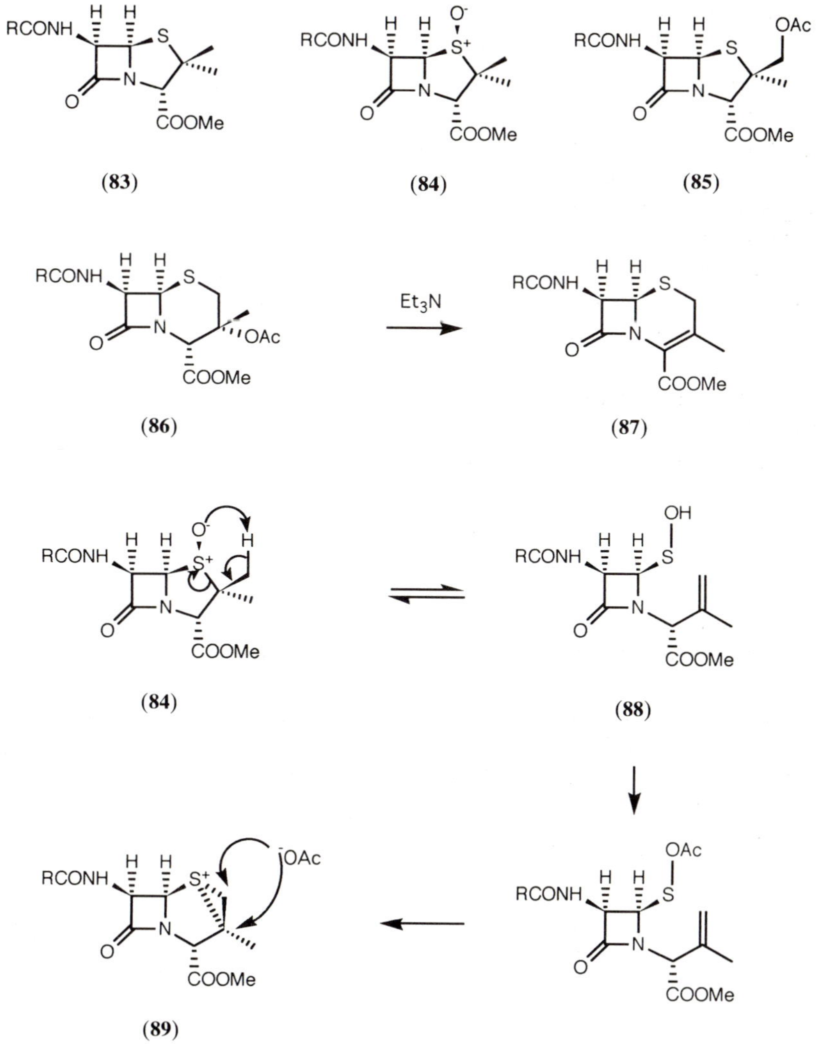

RCONH H H S O N COOMe

(**83**)

RCONH H H O⁻ S+ O N COOMe

(**84**)

RCONH H H S OAc O N COOMe

(**85**)

RCONH H H S OAc O N COOMe

(**86**)

Et_3N

RCONH H H S O N COOMe

(**87**)

RCONH H H O⁻ S+ H O N COOMe

(**84**)

RCONH H H OH S O N COOMe

(**88**)

RCONH H H S+ ⁻OAc O N COOMe

(**89**)

RCONH H H OAc S O N COOMe

Scheme 10.9

of *p*-toluenesulphonic acid. The mechanism of this remarkable reaction involves establishing an equilibrium between sulphoxide (**84**) and sulphenic acid (**88**) by (2,3)-sigmatropic rearrangement, followed by formation of episulphonium ion (**89**). The products observed can be accounted for by alternative fates of this ion.

Further evidence[28] for this pathway was provided by the observation that penicillin *R*-sulphoxides were very unstable to heat, rapidly isomerising to the thermodynamically stable *S*-isomers. When this thermolytic rearrangement was carried out in *o*-deuterated t-butanol, the product *S*-sulphoxide (**90**) possessed one deuterium atom in the 2-β-methyl group (Scheme 10.10).

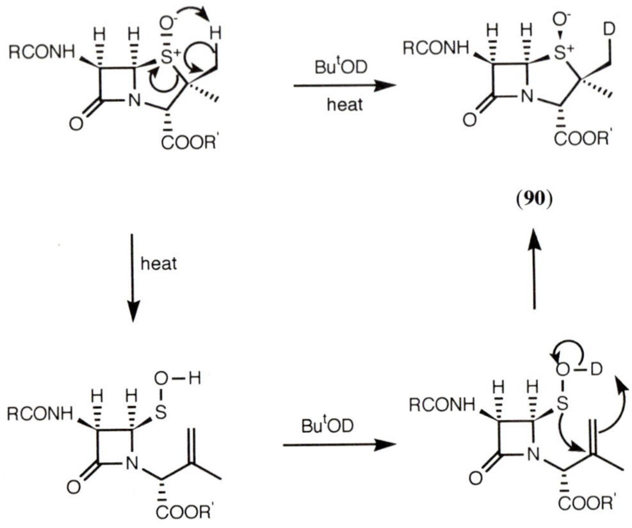

RCONH H H O⁻ S+ H N O COOR'

Bu^tOD heat

RCONH H H O⁻ S+ D N O COOR'

(**90**)

heat

RCONH H H O–H S N O COOR'

Bu^tOD

RCONH H H O–D S N O COOR'

Scheme 10.10

Δ^3-Desacetoxycephalosporins such as (**87**) do not undergo free radical bromination at the 3′-position. However, if base induced isomerisation to the Δ^2-isomer (**91**) is performed, this now undergoes normal allylic bromination. Simple acetolysis then gives the correct substitution pattern, as a mixture of Δ^2- and Δ^3-isomers (**92**). Oxidation[29] gives the pure Δ^3-isomer (**93**). Sulphoxide reduction and deprotection completed this first synthesis[30] of the cephalosporin (**94**) from penicillin (Scheme 10.11).

This ring expansion reaction also provides the basis of the commercial production[31] of a desacetoxycephalosporin, the orally active cephalexin (**95**), from a penicillin.

Further major advances have been made by Kukolja, Koppel and co-workers, also at the Lilly Research Laboratories. Treatment of penicillin sulphoxide (**96**) with N-chlorosuccinimide, or another source of electrophilic halogen, in refluxing CCl_4 gave a mixture of the sulphinyl chlorides (**97**), epimeric at sulphur.[32] Treatment of (**97**) with a variety of Lewis acids

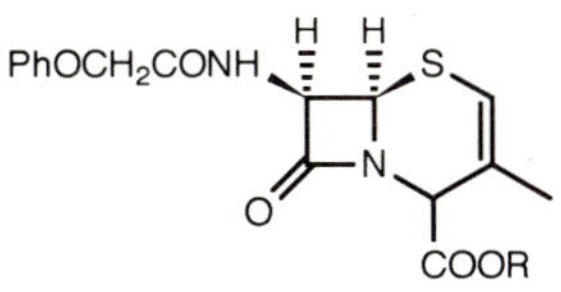

(91) 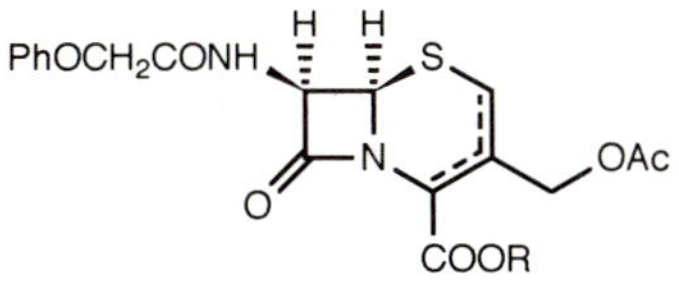 (92)

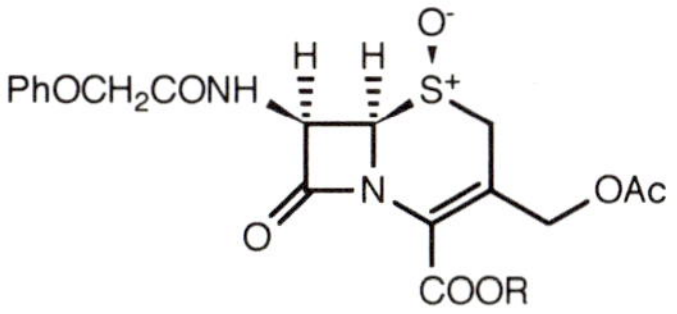

(93) 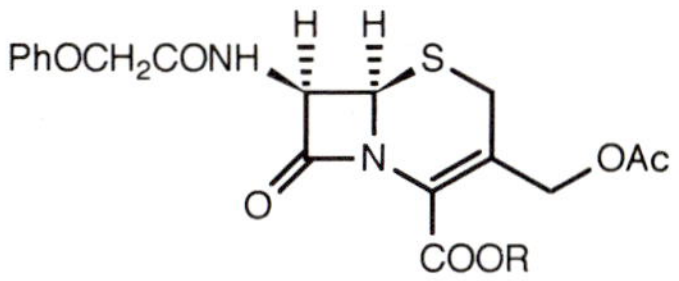 (94)

Scheme 10.11

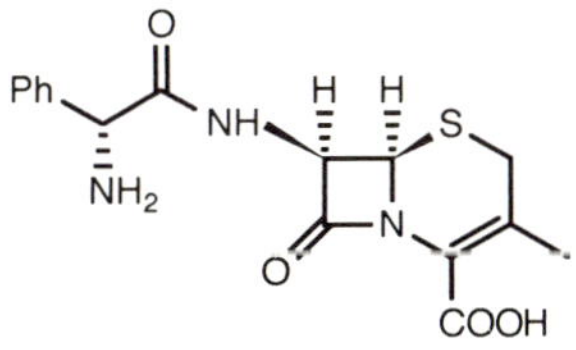

(95)

resulted in cyclisation to the 3-methylenecepham sulphoxides (**98**). Sulphoxide reduction using PBr_3 gave a single 3-methylene cepham (**99**) (Scheme 10.12).

Surprisingly, the 3-methylene group did not add halogens under the normal conditions. However, treatment of (**99**) with 1,8-diazabicyclo-(5.4.0)undec-7-ene and bromine at low temperature gave (**100**), by electrophilic quenching of the intermediate allylic anion. The allyl bromide underwent facile displacement with a variety of nucleophiles. For example, treatment[33] of (**100**) with silver acetate in acetic acid gave cephem (**101**).

On the other hand, treatment of (**98**) with a 2 : 1 mixture of acetic anhydride and acetic acid at reflux gave, in high yield, a mixture of (**101**) and (**103**).[34] This reaction probably proceeds via the Pummerer-type sulphinium cation (**102**), which is quenched by conjugate addition of acetate ion. At the reaction temperature, isomerisation of the initially produced Δ^2-double bond occurs, to give the Δ^2- and Δ^3-isomers in a 3 : 1 ratio. This overall sequence represents a simple and direct two-step conversion of penicillins into 3-acetoxymethylcephems, a substitution pattern not directly accessible by the Morin reaction. The mixture of isomers can be converted by oxidation using *m*-CPBA into the single sulphoxide (**104**).

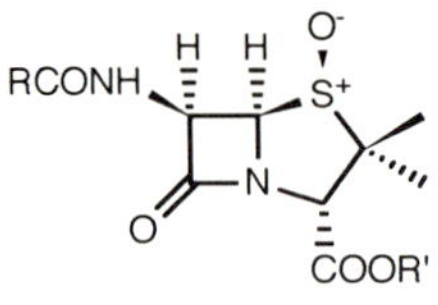

(96) 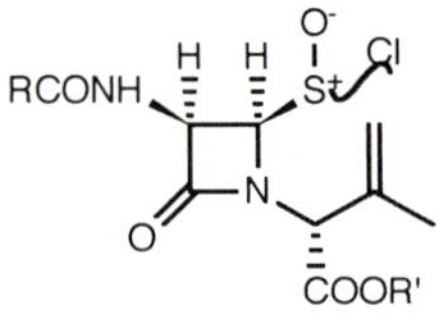 (97) 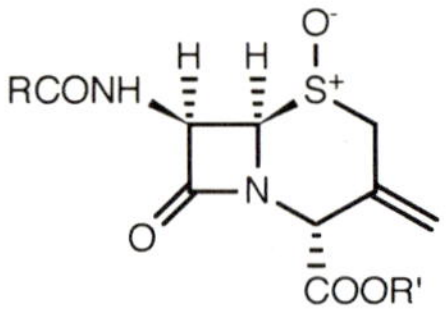 (98)

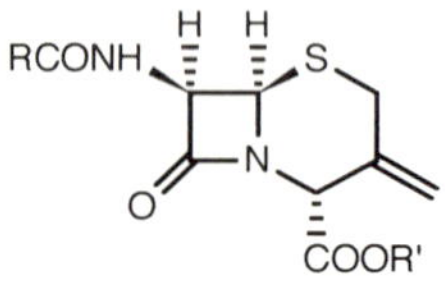

(99) 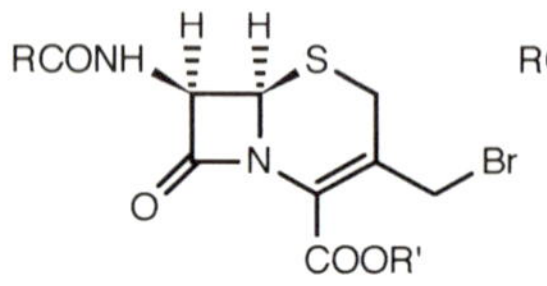 (100) 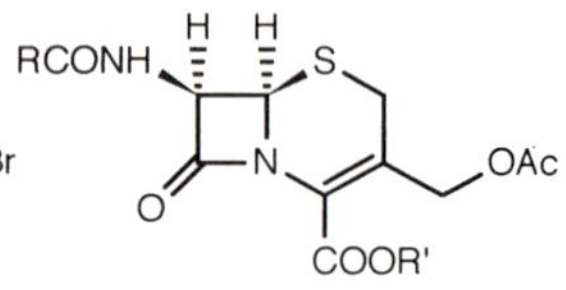 (101)

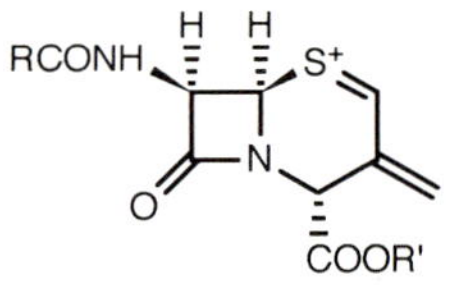

(102) 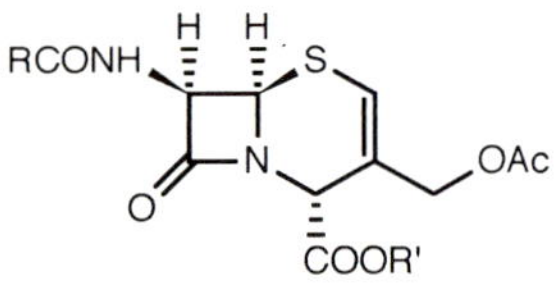 (103)

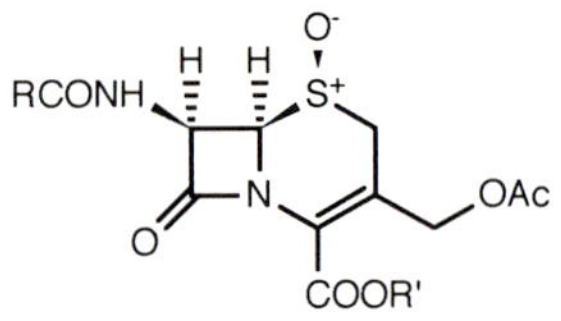

(104)

Scheme 10.12

10.9 Cephamycin antibiotics

Both penicillins and cephalosporins act upon bacteria by interfering with cell-wall production, by inhibiting the transpeptidase enzymes involved. The natural substrate of these enzymes in cell-wall construction is an N-acyl–R-alanyl–R-alanine, and it was proposed[35] by Tipper and Strominger in 1965 that the β-lactam antibiotics mimic this substrate in their role as enzyme inhibitors. Extending this concept, they further proposed that 6-α-methylpenicillin and 7-α-methylcephalosporin derivatives should be more active antibacterial agents than the parent species.

Intensive efforts ultimately led to methodology to allow such methyl group introduction. However, the methylated products proved to have significantly lower activity than their parents. Simultaneously with this methodological success, 7-α-methoxycephalosporins (cephamycins) were isolated from *Streptomyces* species both at the Lilly Research Laboratories, and at Merck, Sharp & Dohme. These new cephalosporins proved to be active against some formerly resistant strains of gram-negative bacteria. Accordingly, further research on 6(7)-alkyl derivatives ceased, but the methodology devised for their synthesis was put to good use in cephamycin production. There are several efficient ways of achieving 6(7)-α-methoxylation, and these have been well reviewed.[36]

10.10 Merck synthesis of (±)-cefoxitin (105)

Deprotonation of either C-7 epimer of racemic Schiff base (**106**)[10,37] followed by treatment with methanesulphenyl chloride gave methylthio derivative (**107**) (Scheme 10.13). Standard transformations then gave amide (**108**), methanolysis of which in the presence of one equivalent of thallium-

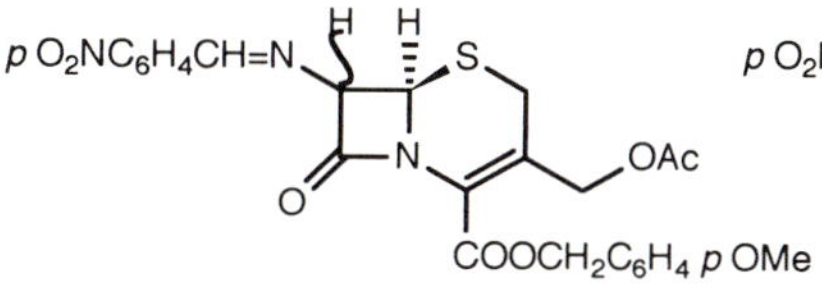

(**106**) 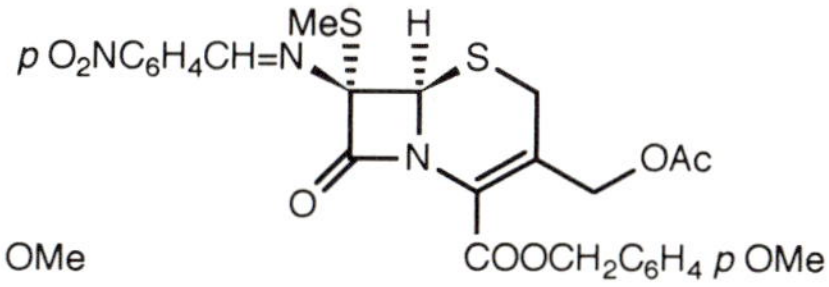 (**107**)

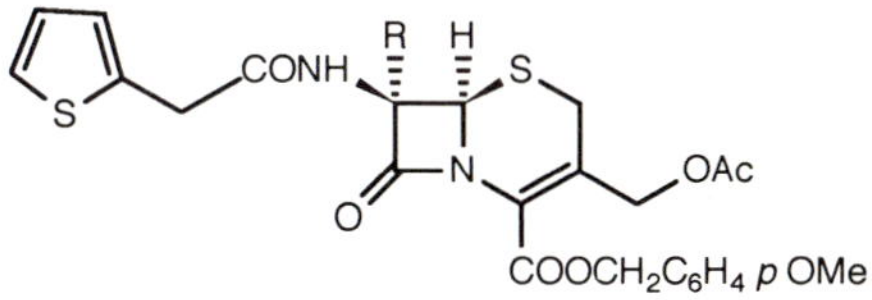

(**108**) R = SMe

(**109**) R = OMe

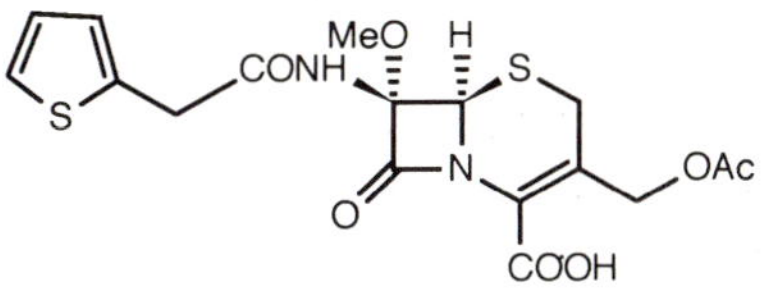

(**105**)

Scheme 10.13

(III) nitrate gave methoxy amide (**109**). After further manipulation, (±)-cefoxitin (**105**) was obtained; as its sodium salt, it displayed approximately one-half of the activity of natural sodium cefoxitin.

10.11 Shionogi synthesis of a 7-α-methoxy-1-oxacephem (110)

In 1979, Shionogi chemists in Japan reported[38] that the non-natural 7-α-methoxy-1-oxacephem (**110**) possessed marked antibacterial activity, particularly against gram-negative microorganisms including β-lactamase-producing, resistant strains. Shortly thereafter, the same group described a stereocontrolled and industrially feasible synthesis of this new antibiotic, starting from a penicillin.[39]

Penicillin G diphenylmethyl ester (**111**) was converted into the 6-*epi*-penicillin sulphoxide (**112**) by base catalysed isomerisation followed by oxidation at sulphur (Scheme 10.14). The *epi*-oxazoline (**113**), obtained by heating sulphoxide (**112**) in the presence of triphenylphosphine,[40] was functionalised at the allylic methyl group by reaction with chlorine gas to give dichloride (**114**). This smooth allylic chlorination probably arises by an ene-type process. Treatment of dichloride (**114**) with aqueous sodium hydrogen carbonate furnished the new *epi*-oxazoline (**115**). Direct conversion into the required alcohol (**118**) failed, but it was obtained by conversion of chloride (**115**) into nitrate (**117**) via iodide (**116**). Reduction of nitrate (**117**) with zinc in acetic acid then gave (**118**), which underwent stereospecific boron trifluoride etherate catalysed cyclisation to the exo-methylene oxacepham (**119**). Chlorination to dichloride (**120**), followed by base induced elimination of HCl, gave the 7-*epi*-1-oxacephem (**121**). This was converted into the 7-α-methoxy compound (**122**) by sequential treatment with lithium methoxide and t-butyl hypochlorite.[41] The synthesis of (**110**) was completed by chloride displacement with sodium 1-methyl-1H-tetrazole-5-thiolate, phosphorus pentachloride mediated deacylation to the 7-amino compound, reacylation, and, finally, cleavage of the ester protective group.

10.12 1,1-Dioxo-*trans*-7-methoxycephalosporanic acid t-butyl ester (123)

Modified cephalosporins, in particular 1,1-dioxo-*trans*-7-methoxycephalosporanic acid t-butyl ester (**123**), have been shown[42] to possess strong *in vitro* inhibition of human leukocyte elastase; this enzyme has been implicated in the pathogenesis of diseases such as emphysema and rheumatoid arthritis. An efficient four-step synthesis (Scheme 10.15) of this compound from 7-ACA has been reported.[43]

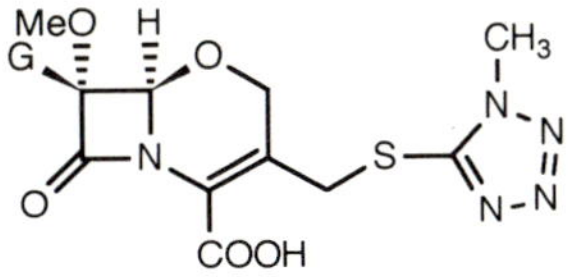

(110) 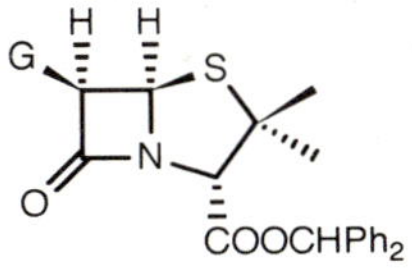 (111)

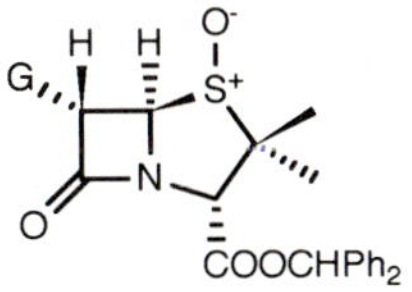

(112) 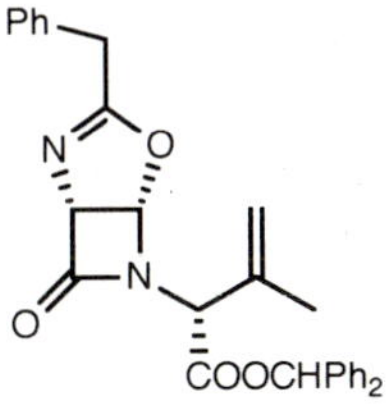 (113) 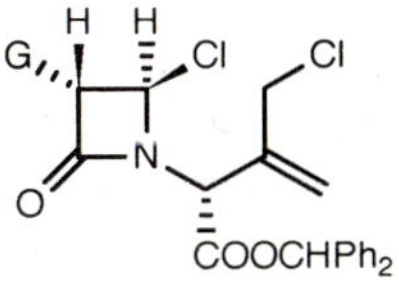 (114)

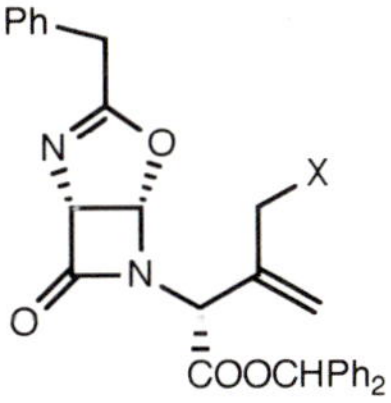

(115) X = Cl
(116) X = I
(117) X = ONO_2
(118) X = OH

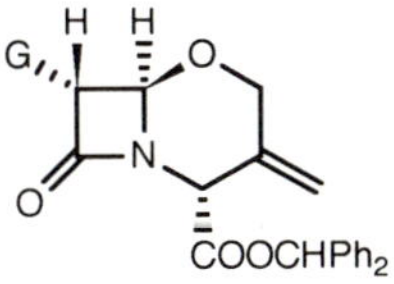

(119) 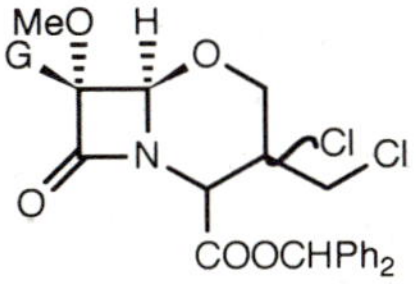 (120)

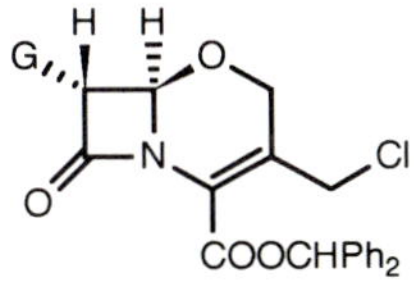

(121) 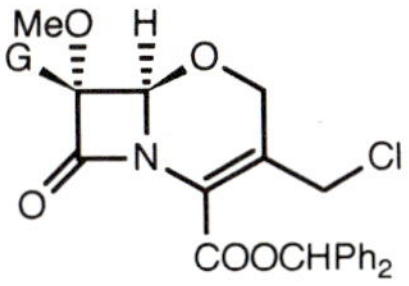 (122)

G = $PhCH_2CONH$

Scheme 10.14

Treatment of 7-ACA with isobutylene and sulphuric acid yielded crystalline 7-ACA t-butyl ester (**124**). A remarkably selective catalytic oxidation to the amino sulphone (**125**) was achieved using sodium tungstate and hydrogen peroxide. Diazotisation, using isopropyl nitrite, followed by a rhodium-catalysed carbene/carbenoid insertion into methanol gave the desired ester (**123**) as a >17 : 1 *trans*/*cis* isomeric mixture.

A related preparation of *trans*-7-methoxy-3′-desacetoxycephalosporanic

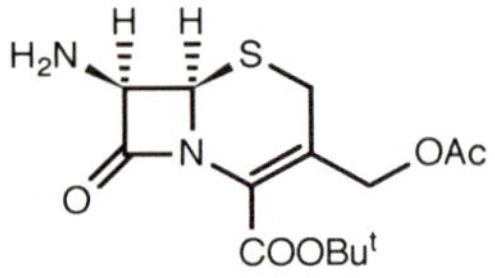

(124) 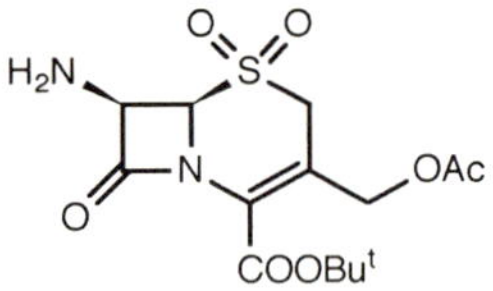 (125)

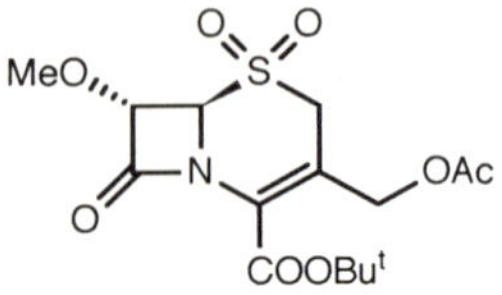

(123) 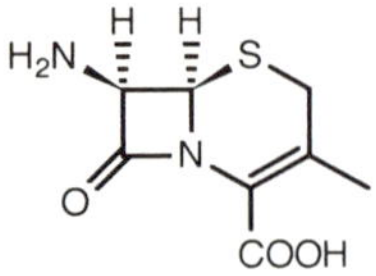 (126)

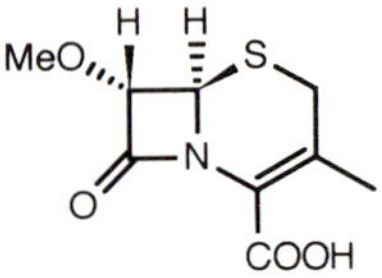

(127) 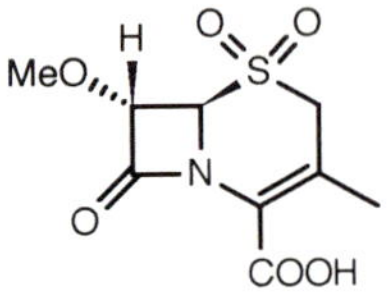 (128)

Scheme 10.15

acid (**127**), by direct diazotisation of 7-amino-3′-desacetoxycephalosporanic acid (7-ADCA) (**126**) using sodium nitrile in methanol/perchloric acid,[44] has recently been communicated.[45] Oxidation then gave the sulphone (**128**) in 25% isolated yield from 7-ADCA.

Introduction of the 3′-acetoxy group is not a problem. Although 7-ADCA does not undergo radical bromination at C-3′, at the sulphone oxidation level allylic bromination proceeds smoothly. Subsequent nucleophilic displacement with triethylammonium acetate in acetic acid provides this functionality in good yield.[46]

References

1. J.C. Sheehan and K.R. Henery-Logan, *J. Am. Chem. Soc.* (1957) **79** 1262–1263; (1959) **81** 3089–3094.
2. J.C. Sheehan and G.P. Hess, *J. Am. Chem. Soc.* (1955) **77** 1067–1068.
3. R. Heymes, G. Amiard and G. Nomine, *Compt. Rend., Ser. C* (1966) **263** 170–172.
4. J.E. Dolphini, J. Schwartz and F. Weisenborn, *J. Org. Chem.* (1969) **34** 1582–1586.
5. R.B. Woodward, K. Heusler, J. Gosteli, P. Naegeli, W. Oppolzer, R. Ramage, S. Ranganathan and H. Vorbrüggen, *J. Am. Chem. Soc.* (1966) **88** 852–853.
6. K. Heusler, *Helv. Chim. Acta* (1972) **55** 388–408.
7. J.E. Baldwin, M.A. Christie, S.B. Haber and L.I. Kruse, *J. Am. Chem. Soc.* (1976) **98** 3045–3047.
8. S. Nakatsuka, H. Tanino and Y. Kishi, *J. Am. Chem. Soc.* (1975) **97** 5008–5010.

9. S. Nakatsuka, H. Tanino and Y. Kishi, *J. Am. Chem. Soc.* (1975) **97** 5010–5012.
10. R.W. Ratcliffe and B.G. Christensen, *Tetrahedron Lett.* (1973) 4649–4652.
11. H. Staudinger, *Liebigs Ann. Chem.* (1907) **356** 51–123.
12. A.K. Bose and B. Anjaneyulu, *Chem. Ind. (London)* (1966) 903; A.K. Bose, B. Anjaneyulu, S.K. Bhattacharya and M.S. Manhas, *Tetrahedron* (1967) **23** 4769–4776; M.S. Manhas, H.P.S. Chawia, S.G. Amin and A.K. Bose, *Synthesis* (1977) 407–409.
13. R.W. Ratcliffe and B.G. Christensen, *Tetrahedron Lett.* (1973) 4645–4648; for a more direct synthesis of this compound see E.W. Colvin, G.W. Kirby and A.C. Wilson, *Tetrahedron Lett.* (1982) **23** 3835–3836.
14. R.A. Firestone, N.S. Maciejewicz, R.A. Ratcliffe and B.G. Christensen, *J. Org. Chem.* (1974) **39** 437–440.
15. L.D. Cama and B.G. Christensen, *J. Am. Chem. Soc.* (1974) **96** 7582–7584.
16. R.N. Guthikonda, L.D. Cama and B.G. Christensen, *J. Am. Chem. Soc.* (1974) **96** 7584–7585.
17. R. Graf, *Liebigs Ann. Chem.* (1963) **661** 111 157; *Angew. Chem., Int. Ed. Engl.* (1968) **7** 172–182; see also W.A. Szabo, *Aldrichimica Acta* (1977) **10** 23–29.
18. For the best cleavage method see T. Durst and M.J. O'Sullivan, *J. Org. Chem.* (1970) **35** 2043–2044.
19. D. Bormann, *Liebigs Ann. Chem.* (1974) 1391–1398; D. Bormann, B. Knabe, M. Schorr, E. Schrinner and M. Worm, in *Recent Advances in the Chemistry of β-Lactam Antibiotics* (Ed. J. Elks), Royal Society of Chemistry Special Publication No. 28, London (1977), pp. 46–67.
20. S.J. Mickel, *Aldrichimica Acta* (1985) **18** 95–99.
21. See, however, N.C. Cohen, *J. Med. Chem.* (1983) **26** 259–264; M.I. Page, P. Webster and L. Ghosez, *J. Chem. Soc., Perkin Trans 2* (1990) 805–811, 813–823 and references therein.
22. I. Ernest, J. Gosteli, C.W. Greengrass, W. Holick, D.E. Jackman, H.R. Pfaendler and R.B. Woodward, *J. Am. Chem. Soc.* (1978) **100** 8214–8222; for a more expansive and characteristically very readable account see R.B. Woodward, in *Recent Advances in the Chemistry of β-Lactam Antibiotics* (Ed. J. Elks), Royal Society of Chemistry Special Publication No. 28, London (1977), pp. 167–180. See also M.W. Foxton, C.E. Newall and P. Ward, in *Recent Advances in the Chemistry of β-Lactam Antibiotics* (Ed. G.I. Gregory), Royal Society of Chemistry Special Publication No. 38, London (1981), pp. 281–290.
23. T. Kamiya, T. Teraji, Y. Saito, M. Hashimoto, O. Nakaguchi and T. Oku, *Tetrahedron Lett.* (1973) **14** 3001–3004; D.H.R. Barton, P.G. Sammes, M.V. Taylor, C.M. Cooper, G. Hewitt, B.E. Looker and W.G.E. Underwood, *J. Chem. Soc., Chem. Commun.* (1971) 1137–1139.
24. R.B. Woodward, K. Heusler, I. Ernest, K. Burri, R.J. Friary, F. Haviv, W. Oppoizer, R. Paoni, K. Syhora, R. Wenger and J.K. Whitesell, *Nouveau J. Chim.* (1977) **1** 85–88; K. Heusler and R.B. Woodward, German Offenlegungsschrift 1935970 (1970).
25. R.D.G. Cooper and D.O. Spry, in *Cephalosporins and Penicillins, Chemistry and Biology* (Ed. E.H. Flynn), Academic Press, New York (1972), pp. 183–254; R.D.G. Cooper and G.A. Koppel, in *The Chemistry and Biology of β-Lactam Antibiotics*, Vol. 1, *Penicillins and Cephalosporins* (Eds R.B. Morin and M. Gorman), Academic Press, New York (1982), pp. 2–92; S. Kukolja and R.R. Chauvette, in *The Chemistry and Biology of β-Lactam Antibiotics*, Vol. 1, *Penicillins and Cephalosporins* (Eds R.B. Morin and M. Gorman), Academic Press, New York (1982), pp. 93–198.
26. R.B. Morin, B.G. Jackson, R.A. Mueller, E.R. Lavagnino, W.B. Scanlon and S.L. Andrews, *J. Am. Chem. Soc.* (1963) **85** 1896–1897; (1969) **91** 1401–1407.
27. R.D.G. Cooper, P.V. DeMarco, J.C. Cheng and N.D. Jones, *J. Am. Chem. Soc.* (1969) **91** 1408–1415.
28. D.H.R. Barton, F. Comer, D.G.T. Greig, G. Lucente, P.G. Sammes and W.G.E. Underwood, *J. Chem. Soc., Chem. Commun.* (1970) 1059–1060 and references therein; D.H.R. Barton, D.G.T. Greig, G. Lucente, P.G. Sammes, M.V. Taylor, C.M. Cooper, G. Hewitt and W.G.E. Underwood, *J. Chem. Soc., Chem. Commun.* (1970) 1683–1684.
29. G.V. Kaiser, R.D.G. Cooper, R.E. Koehler, C.F. Murphy, J.A. Webber, I.G. Wright and E.M. Van Heyningen, *J. Org. Chem.* (1970) **35** 2430–2433; C.F. Murphy and R.E. Koehler, *J. Org. Chem.* (1970) **35** 2429–2430.
30. J.A. Webber, E.M. Van Heyningen and R.T. Vasileff, *J. Am. Chem. Soc.* (1969) **91** 5674–5675.

31. R.R. Chauvette, P.A. Pennington, C.W. Ryan, R.D.G. Cooper, F.L. José, I.G. Wright, E.M. Van Heyningen and G.W. Huffman, *J. Org. Chem.* (1971) **36** 1259–1267.
32. S. Kukolja, S.R. Lammert, M.R.B. Gleissner and A.I. Ellis, *J. Am. Chem. Soc.* (1976) **98** 5040–5041; S. Kukolja, in *Recent Advances in the Chemistry of β-Lactam Antibiotics* (Ed. J. Elks), Royal Society of Chemistry Special Publication No. 28, London (1977), pp. 181–188.
33. G.A. Koppel, M.D. Kinnick and L.J. Nummy, *J. Am. Chem. Soc.* (1977) **99** 2822–2823; idem, in *Recent Advances in the Chemistry of β-Lactam Antibiotics* (Ed. J. Elks), Royal Society of Chemistry Special Publication No. 28, London (1977) 101–110.
34. G.A. Koppel and L.J. McShane, *J. Am. Chem. Soc.* (1978) **100** 288–289.
35. D.J. Tipper and J.L. Strominger, *Proc. Natl. Acad. Sci. USA* (1965) **54** 1133–1141.
36. E.M. Gordon and R.B. Sykes, in *Chemistry and Biology of β-Lactam Antibiotics*, Vol. 1, *Penicillins and Cephalosporins* (Eds R.B. Morin and M. Gorman), Academic Press, New York (1982), pp. 199–370; T. Hiraoka, Y. Sugimura, T. Saito and T. Kobayashi, *Heterocycles* (1977) **8** 719–742; T. Hiraoka, K. Iino, Y. Iwano, T. Saito and Y. Sugimura, in *Recent Advances in the Chemistry of β-Lactam Antibiotics* (Ed. J. Elks), Royal Society of Chemistry Special Publication No. 28, London (1977), pp. 123–128.
37. R.W. Ratcliffe and B.G. Christensen, *Tetrahedron Lett.* (1973) 4653–4656.
38. M. Narisada, T. Yoshida, H. Onoue, M. Ohtani, T. Okada, T. Tsuji, I. Kikkawa, N. Haga, H. Satoh, H. Itani and W. Nagata, *J. Med. Chem.* (1979) **22** 757.
39. M. Yoshioka, T. Tsuji, S. Uyeo, S. Yamamoto, T. Aoki, Y. Nishitani, S. Mori, H. Satoh, Y. Hamada, H. Ishitobi and W. Nagata, *Tetrahedron Lett.* (1980) **21** 351–354.
40. Y. Hamashima, S. Yamamoto, S. Uyeo, M. Yoshioka, M. Murakami, H. Ona, Y. Nishitani and W. Nagata, *Tetrahedron Lett.* (1979) **20** 2595–2598.
41. G.A. Koppel and R.E. Koehler, *J. Am. Chem. Soc.* (1973) **95** 2403–2404; R.A. Firestone and B.G. Christensen, *J. Org. Chem.* (1973) **38** 1436–1437; J.E. Baldwin, F.J. Urban, R.D.G. Cooper and F.L. José, *J. Am. Chem. Soc.* (1973) **95** 2401–2403.
42. J.B. Doherty, B.M. Ashe, L.W. Argenbright, P.L. Barker, R.J. Bonney, G.O. Chandler, M.E. Dahlgren, C.P. Dorn, Jr., P.E. Finke, R.A. Firestone, D. Fletcher, W.K. Hagmann, R. Mumford, L. O'Grady, A.L. Maycock, J.M. Pisano, S.K. Shah, K.R. Thompson and M. Zimmerman, *Nature* (1986) **322** 192–194.
43. T.J. Blacklock, J.W. Butcher, P. Sohar, T.R. Lamanec and E.J.J. Grabowski, *J. Org. Chem.* (1989) **54** 3907–3913.
44. I. McMillan and R.J. Stoodley, *J. Chem. Soc. (C)* (1968) 2533–2537; G. Cignarella, G. Pifferi and E. Testa, *J. Org. Chem.* (1962) **27** 2668–2669.
45. M. Alpegiani, P. Bissolini, M. D'Anello, G. Rivola, D. Borghi and E. Perrone, *Tetrahedron Lett.* (1991) **32** 3883–3886.
46. M. Botta, F. De Angelis, A. Gambacorta, F. Gianessi and R. Nicoletti, *Gazz. Chim. Ital.* (1985) **115** 169–172.

Further reading

For general reading on classical β-lactams, see *Cephalosporins and Penicillins, Chemistry and Biology* (Ed. E.H. Flynn), Academic Press, New York (1972) (excellent coverage, especially of patented work); P.G. Sammes, *Chem. Rev.* (1976) **76** 113–155; A.K. Mukerjee and A.K. Singh, *Tetrahedron* (1978) **34** 1731–1767; F.A. Jung, W.R. Pilgrim, J.P. Poyser and P.J. Siret, in *Topics in Antibiotic Chemistry*, Vol. 4 (Ed. P.G. Sammes), Ellis Horwood, Chichester (1980); *Chemistry and Biology of β-Lactam Antibiotics*, Vols 1 and 2 (Eds R.B. Morin and M. Gorman), Academic Press, New York (1982); W. Dürkheimer, J. Blumbach, R. Lattrell and K.H. Scheunemann, *Angew. Chem. Int. Ed. Engl.* (1985) **24** 180–202; R. Southgate and S. Elson, *Prog. Chem. Org. Nat. Prods.* (1985) **47** 1–106; *Recent Advances in the Chemistry of β-Lactam Antibiotics* (Ed. J. Elks), Royal Society of Chemistry Special Publication No. 28, London (1977); *Recent Advances in the Chemistry of β-Lactam Antibiotics* (Ed. G.I. Gregory), Royal Society of Chemistry Special Publication No. 38, London (1981); *Recent Advances in the Chemistry of β-Lactam Antibiotics* (Eds A.G. Brown and S.M. Roberts), Royal Society of Chemistry Special Publication No. 52, London (1985); *Recent Advances in the Chemistry of*

β-Lactam Antibiotics (Eds. P.H. Bentley and R. Southgate), Royal Society of Chemistry Special Publication No. 70, London (1989).
For specific routes to monocyclic β-lactams, see K.G. Holden, in *Chemistry and Biology of β-Lactam Antibiotics* Vol. 2 (Eds R.B. Morin and M. Gorman), Academic Press, New York (1982), pp. 114–131; G.A. Koppel, in *Heterocyclic Compounds*, Vol. 42, Part 2 (Ed. A. Hassner), John Wiley and Sons, New York (1983), pp. 219–441; A.G.M. Barrett and M.A. Sturgess, *Tetrahedron* (1988) **44** 5615–5649; D.J. Hart and D.-C. Ha, *Chem. Rev.* (1989) **89** 1447–1465; M.J. Brown, *Heterocycles* (1989) **29** 2225–2244; F.H. van der Steen and G. van Koten, *Tetrahedron* (1991) **47** 7503–7524.

Index